STUDENT STUDY GUIDE

Julie Klare
Gwinnett Technical College

DonnaJean Fredeen
Southern Connecticut State University

CHEMISTRY

McMURRY • FAY

SIXTH EDITION

Prentice Hall

Boston Columbus Indianapolis New York San Francisco Upper Saddle River
Amsterdam Cape Town Dubai London Madrid Milan Munich Paris Montréal Toronto
Delhi Mexico City São Paulo Sydney Hong Kong Seoul Singapore Taipei Tokyo

Editor in Chief: Adam Jaworski
Acquisitions Editor: Terry Haugen
Marketing Manager: Erin Gardner
Senior Project Editor: Jennifer Hart
Assistant Editor: Lisa Pierce
Editorial Assistant: Catherine Martinez
Marketing Assistant: Nicola Houston
Managing Editor, Chemistry and Geosciences: Gina M. Cheselka
Project Manager, Science: Wendy A. Perez
Operations Specialist: Maura Zaldivar
Cover Image: Sandralize/iStockphoto.com

Pearson Prentice Hall
Upper Saddle River, NJ 07458

Printed in the United States of America

10 9 8 7 6 5 4 3

ISBN-13: 978-0-321-72724-4
ISBN-10: 0-321-72724-X

Prentice Hall
is an imprint of

www.pearsonhighered.com

Table of Contents

PREFACE

This book is intended to provide a pared-down and condensed version of the material in the textbook, ***CHEMISTRY*** by McMurry and Fay. It is often helpful for students to have a somewhat less intimidating introduction to what can be an overwhelming subject, but one that, despite its fearsome reputation, can be thoroughly enjoyed. After studying chemistry, it is impossible to look at things the same way again: the entire day contains chemistry from start to finish from the time we step into the shower, stir sugar into our coffee, burn gasoline to get to work or school and home again, cook our evening meal, and take one last glance at the LEDs of the clock on the nightstand before sleep. Giving this subject the time and attention that it requires will also mean success in the course. Chemistry is not a subject that can be "crammed" at the last minute (trust the voice of experience): it must be studied routinely to make sure that no concepts are missed. Chemical concepts have a way of reappearing that can be very trying if they were not learned the first time around.

Each chapter of this study guide was written to correspond directly, section-by-section, to the textbook. Each study guide chapter contains the following sections:

- Chapter Learning Goals – this section lists the major concepts in the chapter, arranged in broad categories.
- Chapter in Brief – an introduction to the material that can be enhanced and expanded as needed with lecture notes and more in-depth material from the parent text. Example problems are worked step-by-step within the chapter and paired with similar guided workbook problems. The solutions to the workbook problems are found in Appendix A. These problems are embedded in the relevant portions of the chapter, to provide step-by-step instruction in problem solving and increase confidence.
- Putting it Together – these problems are found at the end of some chapters, and tie together the concepts from the current chapter and previous chapters. Worked-out solutions to these problems are found in Appendix B.
- Self-Test – this section provides an opportunity for the student to check comprehension of both concepts and problem solving. By using this section as a practice exam, it is possible to discover what portions of the chapter may require further review, and which have been mastered. "Challenge" problems located at the end of the self-test provide an opportunity to "think like a scientist." These may require mathematics or logic beyond what is covered in the chapter. Complete solutions to the self-tests and challenge problems, with problems worked out step-by-step are found in Appendix C.
- Inquiry-Based Problems – Occasional problems that demonstrate the application of concepts in a laboratory setting. Worked explanations for these problems are found in Appendix D.

The production of such a manual is never a one-person project, and I am indebted to the author of previous editions, DonnaJean Fredeen, who developed the format to which I adhered. It took two editors to see this project through: Carol DuPont, who entrusted me with the project, and Lisa Pierce who cracked the whip to get it done. These two tolerant ladies requested, cajoled, insisted, and shoved me through the project, and still had the good grace to be grateful when something – anything! – got done.

My dear daughters learned to cook while this project was under way, largely as a matter of survival. They did not see the top of the dining-room table for months, and got used to me sweeping the assorted books and papers into a chair, closing my laptop and putting my dinner plate on top of it when we were pretending to have a family meal. They have no choice but to put up with me, but this book is dedicated to someone who puts up with me even though he does not need to: James Dunstan, who gave my life back.

Julie Klare
Gwinnett Technical College
Medix College

CHAPTER 1

CHEMISTRY: MATTER AND MEASUREMENT

Chapter Learning Goals

A. Experimentation Involving Observations

1. Describe how the scientific method can be used to understand the natural world.

B. Matter: Understanding Composition

1. Explain what an element is.
2. Understand the structure and organization of the periodic table.
3. Identify the group and period to which an element belongs.
4. Identify the regions of the periodic table.
5. Identify elements as metals, semimetals, or nonmetals based on their location.

C. Matter: Understanding Properties

1. Identify properties as extensive or intensive, chemical or physical.
2. Predict basic physical and chemical properties from location on the periodic table.

D. Scientific Measurements

1. List the basic SI units of measure, and give the numerical equivalent of the common prefixes used with these units.
2. Express numbers in scientific notation. (See Appendix A in your textbook.)
3. Interconvert between Fahrenheit, Celsius, and Kelvin temperatures.
4. Understand and use derived units.
5. Determine the number of significant digits in a measured quantity.
6. State the result of a calculation to the correct number of significant digits.
7. Use dimensional analysis to solve conversion problems.

Chapter in Brief

Chemistry is the study of the composition, properties, and transformations of matter. After a brief introduction to the scientific method, the chapter continues with an explanation of the way matter is categorized and how various properties, both chemical and physical, can be used in identification. An introduction to the great variety of elements in the periodic table will show how powerful a tool the periodic table is in organizing the elements into families called groups, and the larger categories of metal, nonmetals, and semimetals.

Chemistry knowledge proceeds through experimentation, and experimentation requires both measurement and data analysis. The seven base units of the International System of Units (abbreviated SI) will be introduced and the most immediately useful discussed in some detail. Units derived from these basic units will also be introduced, as will prefixes used to enhance the usefulness of these basic units for very small and very large measurements. Scientific notation, and the rules for significant figures will be explained, and the process of dimensional analysis demonstrated for converting from one unit to another.

1.1 Approaching Chemistry: Experimentation

The natural world is so complex that it cannot be grasped purely through observation. As a result, scientists have come up with the **scientific method**, a way of systematically testing ideas about the natural world.

To use the scientific method, a *hypothesis* must be developed: this is a testable interpretation of previous observations. If the hypothesis has been tested and found to hold up, it can be referred to as a **theory**.

In science, theories can be shown to be sound, but can never be proven. If new experiments discount the theory, it must be changed or set aside.

1.2 Chemistry and the Elements

An **element** is a fundamental substance that can't be chemically changed or broken down. There are 114 known elements, 90 of which occur naturally, the others having been produced by nuclear chemists.

Elements are given one- or two-letter symbols, the first letter capitalized, the second, if there is one, lowercase. These symbols usually come from the English name of the element, but sometimes come from other languages. For example, the symbol for sodium is "Na" from the Latin "natrium."

EXAMPLE:
Name the following elements: F, Mg, I, Te, P

SOLUTION:
Using the table inside the front cover of the textbook, we find that the names for the symbols are:

F — fluorine	Mg — magnesium	I — iodine
Te — tellurium	P — phosphorus	

EXAMPLE:
Find the symbols for the following elements: iron, oxygen, potassium, chlorine, sodium

SOLUTION:
Using the table inside the front cover of the textbook, we find that the symbols for these elements are:

iron — Fe	oxygen — O	potassium — K
chlorine — Cl	sodium — Na	

You may find it helpful to make flashcards with the name of the element on one side and the symbol on the other. You will generally have access to a periodic table, but it may have only the symbols of the elements, and not their names. Focus first on the main group elements (groups 1A through 8A), as these are the ones you will use the most, particularly at the beginning of the course.

1.3 Elements and the Periodic Table

Ten elements have been known since antiquity, but as the pace of chemical discovery increased in the 1700s and 1800s, efforts were made to categorize the elements by their properties. The most successful of these efforts was the **periodic table** published in 1869 by Dmitri Mendeleev.

The periodic table divided the elements into 18 vertical columns called **groups**, and seven horizontal rows called **periods**. The elements in each group have similar chemical properties. The number of elements in each period – two in period one, eight in period two, etc., - is an indication of increasing atomic size, and the structure of atoms.

Groups 1A and 2A on the left, and 3A through 8A on the right of the periodic table are the **main group** elements. Their chemical properties are easy to predict based on their location on the periodic table.

Groups 1B through 8B are the **transition metals**.

The 14 unnumbered groups shown separately on the bottom of the periodic table are the **inner transition metals.**

The periodic table of the elements is the most important organizing principle of chemistry. It is worth printing out a black-and-white periodic table and coloring in these groups to get a solid understanding of where they lie: use different colors for the main group, transition metals, and inner transition metals.

1.4 Some Chemical Properties of the Elements

Elements – and anything else – can be described in terms of their **properties**. Properties are themselves classified in a number of ways:

- Intensive properties do not depend on the size of the sample. An example is melting point: ice melts at 0 °C, whether it is a snowflake or an iceberg.
- Extensive properties do depend on the amount of the sample. To continue the previous example, the amount of energy needed to melt a snowflake or an iceberg is very different, and depends on the sample size.
- Physical properties can be observed without changing the chemical makeup of the sample: melting is a physical property, as is boiling. Whether ice, liquid, or steam, the sample is still water. Physical changes tend to be readily reversible.
- Chemical properties cannot be observed without a change in the composition of the sample. If hydrogen and oxygen are brought together and a spark is applied, water will result. Something has been learned about hydrogen, but you no longer have hydrogen. These tests are typically also difficult to reverse.

EXAMPLE:

There is overlap in the kinds of properties that are observed. Classify these as intensive or extensive, and also as physical or chemical:

Color Combustibility Smell Length Rusting

SOLUTION:

Using the descriptions of these properties above:

- Color is an *intensive physical* property.
- Combustibility is an *intensive chemical* property.
- Smell is an *intensive physical* property.
- Length is an *extensive physical* property.
- Rusting is an *intensive chemical* property.

Four groups in the periodic table are described here using both physical and chemical properties:

- Group 1A: Alkali metals. These are silvery, soft metals. All react strongly with water to form very alkaline solutions, which gives the group its name, and also makes it impossible to find these metals in pure form naturally.
- Group 2A: Alkaline earth metals. These are also lustrous, silvery metals that are not found free in nature. They also react with water, though not as strongly as the alkali metals.
- Group 7A: Halogens. These are colorful, corrosive nonmetals. They are found in nature only in combination with other elements, including the alkali metals and the alkaline earth metals. These compounds are known as salts: the word "*halogen*" is from the Greek word for salt.
- Group 8A: Noble gases. As the name implies, these elements are almost completely unreactive.

As above, it is a good idea to print out a black-and-white periodic table and color in these categories to help you remember them.

The elements are also divided into three major categories:

- Metals. These are found on the left side of the periodic table, bounded by a zigzag line running from boron at the top to astatine at the bottom. They are generally shiny, malleable, ductile (able to be drawn into wires), and conduct electricity and heat. All except mercury are solid.
- Nonmetals. These are found to the right of the zigzag line. They are typically dull in appearance (when solid), brittle, and poor conductors of heat and electricity. They have a variety of colors, and are divided between solids, liquid (bromine), and gases.
- Semimetals. These nine elements are found sitting above the zigzag line, and also include germanium and antimony underneath the line.

Here again, making yourself a periodic table with the metals, nonmetals, and semimetals colored in with different colors will go a long way to helping you remember exactly where these groups lie.

EXAMPLE:

Determine the type of the following elements (metal, nonmetal, or semimetal):

- iridium
- antimony
- iodine

SOLUTION:

By determining the position of the element in the periodic table, you can determine the class of the element. (It also is helpful to know the symbol for each of the elements.)

Iridium – Ir; iridium is to the left of the zigzag line in the periodic table and is a metal.
Antimony – Sb; antimony is adjacent to the zigzag line in the periodic table and is a semimetal.
Iodine – I; iodine is found to the right of the zigzag line in the periodic table and is a nonmetal.

EXAMPLE:

Determine which class of elements each of these descriptions matches:

- Corrosive yellow gas;
- Shiny, malleable, conducts electricity.

SOLUTION:

From the outline above and the discussion in your book, gases are all nonmetals. If something is malleable, that is, easily shaped without breaking, it is solid, and a metal.

1.5 Experimentation and Measurement

To make results easier to use worldwide, in 1960 a system of measurements was adopted known as the International System or SI system. It is based on the metric system, and has seven fundamental units – kilogram, which measures mass; meter, which measures length; kelvin, which measures temperature; mole, which measures the amount of a substance; seconds, used to measure time; the ampere, which is a unit of electrical current; and the candela, which is a unit of luminous intensity. Not all of these are used in the introductory study of chemistry, and there are units, such as those for volume or energy, that appear to be missing, but all units used in science can be derived from these seven.

To make the units more broadly applicable to many different fields from atomic chemistry to astrophysics, a system of prefixes is available. Memorize them!

These prefixes are used for large units:

- giga (G) $= 10^9$
- mega (M) $= 10^6$
- kilo (k) $= 10^3$

These prefixes are used for small units:

- deci (d) $= 10^{-1}$
- centi (c) $= 10^{-2}$
- milli (m) $= 10^{-3}$
- micro (μ) $= 10^{-6}$
- nano (n) $= 10^{-9}$

To make numbers easier to read, scientific notation can also be used. Review Appendix A for a more in-depth review, but in brief, scientific notation takes a number and puts it in an exponential format in which there is one digit to the left of the decimal place, and an exponent to indicate where the decimal belongs. So 306 becomes $3.06 \times$ x 10^2, and 0.00306 becomes 3.06×10^{-3}.

If a number is a measurement, it must have a unit. The difference between 20 feet, 20 meters, and 20 miles is important! Without the unit, a measured number is meaningless.

EXAMPLE:

Express the following length measurements in scientific notation: 3652 m and 0.000 054 95 m. What unit prefixes could reasonably be used with these numbers?

SOLUTION:

We will use the rules outlined in Appendix A in your textbook.

For the number 3652 we need to move the decimal point three places to the left so that we obtain the number 3.652, which is between 1 and 10. We now must multiply 3.652 by 10^3. Therefore, the answer is 3.652×10^3 m. Looking at the list of possible prefixes, 10^3 corresponds to *kilo*. Therefore this number could also be expressed as 3.652 kilometers.

For the number 0.000 054 95, we need to move the decimal point 5 places to the right to obtain the number 5.495, which is between 1 and 10. Because we moved the decimal point to the right, we must multiply by 10^{-5}. Therefore, the answer is 5.495×10^{-5}. This does not correspond neatly to any of the standard prefixes, but we can adjust the number so that it will. If we move the decimal point, we disrupt our scientific notation, but can make our number 54.95×10^{-6} m. This corresponds to *micro* (μ), so this number could also be expressed as 54.95 μm.

EXAMPLE:

State the SI unit and abbreviation used for a) length, b) temperature, c) amount of substance, and d) mass.

SOLUTION:

From Table 1.3 on page 11 in your textbook, we find that the SI unit for a) length is the meter; b) temperature is the kelvin (K); c) amount of substance is the mole (mol); and d) mass is the kilogram (kg).

1.6 Measuring Mass

Mass is the amount of matter in an object. Matter, in its turn, is everything that has a physical presence. The SI unit of mass is the kilogram (2.205 U.S. lbs). This is much too big for most uses in chemistry, so grams (g), milligrams (mg), and micrograms (μg) are also commonly used.

Mass is commonly confused with weight. Weight has to do with gravitational pull, and will vary depending on the strength of gravity. An object will have the same *mass* on Earth or on the moon, whereas it will have much less *weight* on the moon. The confusion is rarely a problem, as it is unusual to be in a gravitational field other than that of the Earth.

1.7 Measuring Length

The meter is the SI unit for length, and at approximately 40 inches, also over-large for most purposes in chemistry. Again, prefixes help to pare this unit down to something more immediately useful: centimeters and millimeters are more commonly used in chemistry.

1.8 Measuring Temperature

Although the official SI unit for the measurement of temperature is the kelvin, degrees Celsius are more commonly used. The two scales are easily interconverted, as both have 100 degrees between the freezing point of water and the boiling point of water. The difference in the scales is the temperature assigned the value zero.

- Celsius: 0 °C is the freezing point of water
- Kelvin: 0 K is what is known as *absolute zero,* the coldest possible temperature. It is equivalent to -273.15 °C. *Note: the degree symbol is not used with the Kelvin scale. This temperature would be referred to as "zero Kelvin."*

Conversions between Kelvin and Celsius are very straightforward: Celsius temperatures are 273.15 °C higher than Kelvin temperatures. Thus:

$$\text{Temperature in K} = \text{temperature in °C} + 273.15$$

$$\text{Temperature in °C} = \text{temperature in K} - 273.15$$

The older Fahrenheit scale sets 32 °F as the freezing point of water, and 212 °F as the boiling point of water. So where the Celsius and Kelvin scales have 100 degrees between the freezing and boiling points of water, the Fahrenheit scale has 180°. An additional complication is the shifting up of the freezing point of water from 0 °C to 32 °F.

Conversions between Fahrenheit and Celsius require two adjustments: one to adjust for the difference in degree size, and the other to adjust for the difference in zero point:

Adjust for change in degree size:

180 °F encompasses the same range as 100 °C. Two conversions, from °F to °C and from °C to °F are shown below:

$$1°\,C \times \frac{180°\,F}{100°\,C} = \frac{9}{5}\,°F$$

$$1°\,F \times \frac{100°\,C}{180°\,F} = \frac{5}{9}\,°C$$

Adjust for change in zero point (the numerical difference in the freezing point of water).

Add 32 when converting from °C to °F.

Subtract 32 when converting from °F to °C.

To convert from Celsius to Fahrenheit, do a size adjustment, followed by a zero-point adjustment.

$$°F = \left(\frac{9}{5} \times °C\right) + 32$$

To convert from Fahrenheit to Celsius, — do a zero-point adjustment, followed by a size adjustment.

$$°C = \frac{5}{9} \times (°F - 32)$$

EXAMPLE:

Gallium, a metal whose salts are used to produce Blu-ray lasers, melts at 302.9 K. What is this temperature in °F?

SOLUTION:

This is a good time to use the "thinking approach." Not having been given a conversion from Kelvin to Fahrenheit, the easiest approach is to do the quick and easy conversion from Kelvin to Celsius, then do the more complex conversion from Celsius to Fahrenheit.

Zero on the Celsius scale is much higher than zero in the Kelvin scale, so to convert from Kelvin to Celsius, subtract 273.15:

302.9 K – 273.15 = 29.75 °C

Now convert to Fahrenheit.

A Fahrenheit degree is smaller than a Celsius degree. Just think about the freezing point of water. Water freezes at 0° on the Celsius scale and at 32° on the Fahrenheit scale. Therefore, when converting from Celsius to Fahrenheit, the temperature should be higher.

Because we are converting from °C to °F, we should first do a size correction (9/5 × °C), followed by a zero-point correction (+ 32). Now, apply the formula given above and see if your answer compares well with your thought process.

$$°F = \left(\frac{9}{5} \times 29.75\right) + 32 = 85.55\ °F$$

Your calculated value for the temperature on the Fahrenheit scale agrees well with the thinking approach we used. Gallium is an interesting metal: as you can see, its melting point is a little bit lower than body temperature. As a result, a piece of gallium metal will melt in your hand.

Workbook Problem 1.1

You wish to impress upon a friend in Russia the misery of the heat wave you are enduring in Chicago. What is the Celsius equivalent of a grim 102 °F? If you really want to feel hot, what is this temperature in Kelvin?

Strategy: Consider the differences in the degree size and zero point adjustments for the Fahrenheit and Celsius scales.

Step 1: Based on the discussion of the Fahrenheit scale and the Celsius scale, determine if °C should be higher or lower than °F.

Step 2: Apply the formula for the conversion of °F to °C. (Remember, for this conversion, you do a zero-point correction followed by a size correction.)

Step 3: Does your answer in step 2 agree with your answer in step 1?

Step 4: Apply the formula for the conversion of °C to K.

1.9 Derived Units: Measuring Volume

Derived units are those made by combining the seven basic units. A simple example is speed, which would be expressed in SI units as meters per second or m/s.

Volume is also a derived unit that comes from the unit of length: remember from geometry that volume is equal to length, times width, times height. If this is done with centimeters as the lengths, the result is cm^3. This is exactly equivalent to a milliliter (1×10^{-3} L). One liter is equivalent to a dm^3 – a cubic decimeter.

In the chemistry lab, volume is measured using graduated cylinders, syringes, volumetric flasks, and burets.

1.10 Derived Units: Measuring Density

Density is measured by dividing the mass of a substance by its volume. For liquids and solids, density is expressed in g/mL or g/cm^3. (Recall that these units will be identical, as 1 mL = 1

cm^3.) It is an intrinsic physical property that can be used to identify a substance. Density can also be used to convert back and forth between mass and volume.

Density is temperature-dependent. Most substances increase in volume as they are heated, and decrease in volume as they are cooled, so when reporting a density, the temperature must also be noted.

The properties of water make it unique: water shrinks down to 4 °C, then begins to expand. Solid water (ice) is about 10% larger than the same mass of liquid water. This is why ice floats. More generally, substances with a lower density will float on top of substances with higher density.

EXAMPLE:

A student determined that a metal cylinder having a mass of 25.478 g has a volume of 9.597 mL. What is the density of this metal cylinder?

SOLUTION:

Knowing both the mass and volume of the metal cylinder, we can determine the density by simply dividing the mass by the volume. Make sure to keep the units with the numbers, as this will prevent many careless errors.

$$\text{Density} = \frac{25.478\text{ g}}{9.597\text{ mL}} = 2.655\frac{\text{g}}{\text{mL}}$$

Given a chart of metal densities, it would be possible to identify the metal directly from its density, as density is an intrinsic property. The density of aluminum (Al) is about 2.7 g/cm^3 at room temperature, making it likely that the cylinder is aluminum.

EXAMPLE:

A student is given containers of isopropyl alcohol (approximate density 0.79 g/mL) and vegetable oil (approximate density 0.90 g/mL). Vegetable oil will not mix with isopropyl alcohol (rubbing alcohol). If both liquids are poured into a graduated cylinder, which will sink? Which will float? If an ice cube (approximate density 0.91 g/mL) is dropped into the cylinder, where will it settle?

SOLUTION:

It is helpful to understand how density works: a substance that is more dense (has a higher number of grams per milliliter) will sink in something that is less dense. So when the liquids are put into the same container, the more dense liquid (vegetable oil) will sink, and the isopropyl alcohol will float, because the substances will not mix. The ice cube has an intermediate density: it will sink in the alcohol, but float on the oil.

Workbook Problem 1.2

The density of lead at 25 °C is 11.34 g/mL. How many grams of lead are found in a volume of 53.43 mL?

Strategy: Using the information provided about density, set up a mathematical equation that allows you to solve for the mass of lead.

Step 1: Substitute the information given in the problem into the mathematical equation you set up.

Step 2: Solve for the mass of lead.

Does your answer make physical sense?

1.11 Accuracy, Precision, and Significant Figures in Measurement

Accuracy and **precision** are concepts related to the reliability of a measurement. Accuracy refers to the closeness of a number to its accepted value, while precision refers to the closeness of a set of measurements to one another. If a piece of metal is agreed to weigh 4.5 grams, and you weigh it three times getting answers of 4.787 g, 4.763 g, and 4.775 g, you have measurements that are close to one another - they are precise – but they do not agree with the accepted value – they are not accurate.

Significant figures are the number of digits recorded for a measurement. For the example masses above, there are four significant figures. When reading a digital instrument, write down everything: if the instrument reports it, it is significant. When reading an analog scale, such as a ruler, record every number of which you are certain, and one that is a good estimate.

When trying to determine the number of significant figures in a measurement, refer to the four rules in your textbook on page 18. When writing in scientific notation, only significant figures are recorded. A special case is *exact numbers*. These are counting numbers that can have no fractional part: the number of students in a class, for example, will be an exact number, as there will never be a fractional student present. These numbers are said to have an infinite number of significant figures.

EXAMPLE:

How many significant figures do the following measurements have?

10.002 g, 0.00672 mL, 5.3000 m, 83,500 s

SOLUTION:

10.002 g — five significant figures (rule 1)

0.006 72 mL — three significant figures (rule 2); If you write the number as 6.72×10^{-3} mL, it is easier to see the number of significant figures. This could also be written as 6.72 μL.

5.3000 m — three significant figures (rule 3)

83,500 s — uncertain (rule 4); If the number is 8.3500×10^4, there are five significant figures. If the number is 8.350×10^4, there are four significant figures. If the number is 8.35×10^4, there are three significant figures.

1.12 Rounding Numbers

When working with numbers, often you will get an answer from a calculator that has far more digits in it than are reasonable. For example, if you are calculating a density from measurements that are 1 gram and 3 cm^3, a calculator will give you an answer of 0.333 333 333 333 g/cm^3. A few simple rules help deal with these situations:

- When multiplying or dividing, the answer must have the same number of significant figures as the smallest number used. In the example above, the answer can have no more than one significant figure, so the density must be reported as 0.3 g/cm^3.

- When adding or subtracting, the answer must not have more digits to the right of the decimal point than either of the original numbers. So if you added the following: 0.3, 5.75, and 7.267, the answer could have no more than one digit on the right of the decimal point: 13.317 would become 13.3.

- When truncating a number, round up if the number past the last possible digit is a five with additional numbers or higher (e.g., 13.356 becomes 13.4), round down if the number is 5 with nothing following or lower (e.g., 13.3500 becomes 13.3).

- Most importantly, **do not round or truncate any numbers until the calculation is complete.**

EXAMPLE:

If a high-speed train in Japan covers 532 miles in 3.7 hours, what is the average speed of the train?

SOLUTION:

three significant figures *answer needs two significant figures*

$$\frac{532 \text{ miles}}{3.7 \text{ hours}} = 140 \frac{\text{miles}}{\text{hour}}$$

two significant figures

If this problem is plugged in to a calculator, the answer provided will have far more digits than can possibly be justified: 143.783 783 8 miles/hour suggests far more precision than the numbers permit.

EXAMPLE:

What is the final answer to the following problem using the correct number of significant figures?

$$\begin{array}{r} 103.7835 \text{ g} \\ -\underline{1.52 \quad \text{g}} \end{array}$$

SOLUTION:

103.7835 g	*ends 4 places past decimal point*
– 1.52 g	*ends 2 places past decimal point*
102.26 g	*ends 2 places past decimal point*

The final answer cannot have more digits to the right of the decimal that the smallest original number. However, in doing calculations, use all figures until the end of the problem. Only round off the final answer.

EXAMPLE:

The mass of a sample of granite is found to be 15.896 g. If the volume is determined to be 5.8 mL, what can you report for the density of the granite?

SOLUTION:

$$\text{Density} = \frac{15.896\text{ g}}{5.8\text{ mL}} = 2.740\ 689\ 65\frac{\text{g}}{\text{mL}}$$

Decide how many significant figures should be in your answer. The denominator has only two significant figures; therefore, the answer must also have only two significant figures.
Round off your answer. The first digit to be dropped is less than 5. 2.740 689 65 g/mL becomes 2.7 g/mL. This situation is not unusual: the determination of the mass of a granite sample can be done far more precisely than the determination of the volume of an irregular sample.

EXAMPLE:

Round off the numbers to three significant figures: (a) 83.567 g (b) 0.000 358 500 g (c) 28,572 g

SOLUTION:

(a) Because the last digit is a 5 with following digits, the mass rounds to 83.6 g.

(b) Because the last digit is a 5 with nothing following, the number is truncated to 0.000 358 g.

(c) The rounding of this number to 28,600 g introduces ambiguity about the number of significant figures. To prevent this, the number can be reported as 2.86×10^4 g, or as 28.6 kg.

Workbook Problem 1.3

A quantity of carbon dioxide gas, which has a density of 1.977 g/L, is found to have a mass of 4.843 65 g. Determine the volume of the gas, and report this volume with the correct number of significant figures.

Strategy: Set up a mathematical equation to determine the volume of the gas. Apply the rules found above to determine the correct number of significant figures.

Step 1: Solve for the volume of the gas.

Step 2: Apply the rules for multiplication and division to determine the number of significant figures in your answer.

Step 3: If necessary, use the rules for rounding off numbers

1.13 Calculations: Converting from One Unit to Another

To convert from one unit to another requires the use of a conversion factor. The **dimensional-analysis method** clarifies the use of these factors by setting up multiplications such that units cancel out.

Conversion factors can be written as fractions that can be interchanged:

$$1 \text{ in.} = 2.54 \text{ cm}$$

This tells you that there is one inch per 2.54 cm ("per" always implies division) OR 2.54 centimeters per inch.

$$\text{Conversion factos: } \frac{1 \text{ in}}{2.54 \text{ cm}} \text{ or } \frac{2.54 \text{ cm}}{1 \text{ in}}$$

The quantity in the numerator is equal to the quantity in the denominator, making this equivalent to multiplying by 1.

Choose the version of the conversion factor that will eliminate the unwanted units and leave the desired units. If the units are correct, the numbers will necessarily follow. Make sure that the units are multiplied and divided as the problem is worked: bizarre final units (g^2/cm^3, for example) are an indication that an error was made.

EXAMPLE:

On her first trip to the United States, a student rents the least expensive car available. When she gets on the highway, she discovers that the car's top speed is 43 miles per hour. When she calls home, what speed in kilometers per hour does she report to her family?

SOLUTION:

Using the conversion table on the inside back cover, we learn that 1 mi = 1.6093 km. Set up the equation beginning with the information given (43 mi/h), and write the conversion factor so that the unit mi cancels.

$$43 \frac{\text{mi}}{\text{h}} \text{ x } \frac{1.6093 \text{ km}}{1 \text{ mi}} = 69 \frac{\text{km}}{\text{h}}$$

Ballpark Check: There are more kilometers than miles in the same length, but the number will be less than twice as much. So the answer in kilograms will be greater than the number of miles, but less than twice as much.

EXAMPLE:

In the United Kingdom and Ireland, the unit of "stone," though forbidden for use in trade, is still commonly used to report the weights of people. It is equal to 14 pounds. If a friend in Ireland, complaining about having put on the Freshman Stone tells you that he now weighs 11 stone, how much does he weigh in kilograms?

SOLUTION:

Given that 14 lbs = 1 stone and 2.205 lb = 1 kg, set up the equation knowing that you want to convert 11 stone to pounds, then pounds to kg and that the units stone, and lb should cancel.

$$11 \text{ stone x } \frac{14 \text{ lbs}}{\text{stone}} \text{ x } \frac{1 \text{ kg}}{2.205 \text{ lb}} = 70 \text{ kg}$$

Ballpark Check: A stone is 14 pounds, and there are about 2 pounds per kilogram, so the original number of stone will be multiplied by about 7 to get the equivalent quantity in kilograms. So the answer will be close to the original number of stone, multiplied by about 7. In this case, the number of significant figures in the original number determines the number of significant figures.

Warning: It's easy to get the "right" answer using dimensional analysis without really understanding what you're doing.

Workbook Problem 1.4

If a runner completes a 5 mile race in 27 minutes 46 seconds, how long can she expect that it will take her to run a 10 kilometer race?

Strategy: Using the conversion table on the inside back cover of your textbook, determine the conversion factor for meters to miles, and convert time to one unit. Use the time and distance for the 5 mile race to determine the runner's speed, convert that speed to km/min, and determine the time expected for the distance in km.

Step 1: Calculate the amount of time it will take for the athlete to run that distance.

Self–Test

This section is intended to test your knowledge of the material covered in this chapter. Think through these problems, and make certain you understand what is going on. Ask yourself if your answer makes sense. Many of these questions are linked to the chapter learning goals. Therefore, successful completion of these problems indicates you have mastered the learning goals for this chapter. You will receive the greatest benefit from this section if you use it as a mock exam. You will then discover which topics you have learned thoroughly and which topics you need to study in more detail.

True-False

1. The symbol for ruthenium is RE.

2. Sodium is an alkali metal.

3. The number 0.002 05 contains three significant figures.

4. Aluminum is a nonmetal.

5. Semimetals have properties somewhere between those of metals and nonmetals, and therefore, do not conduct electricity.

6. *Weight* and *mass* are the same thing.

7. The unit most commonly used for laboratory volume measurements is the m^3.

8. 1×10^6 g is equal to 1 Mg.

9. The symbol for sodium, a great deal of which can be found in the sea, comes from its Latin name, *natrium*.

10. The number 4000 can be expressed in scientific notation as 4×10^3.

Multiple Choice

1. The symbol for gold is
   a. G
   b. Go
   c. Gl
   d. Au

2. The symbol S represents the element
   a. silicon
   b. sulfur
   c. tin
   d. antimony

3. The symbol for neon is
   a. Ne
   b. N
   c. No
   d. Nn

4. Volume is
   a. anything physically real
   b. the pull of gravity on an object by the earth or other celestial body
   c. how much space an object takes up
   d. the amount of matter in an object

5. The following measurements were made by a student during a laboratory exercise:

   23.05 cm, 22.93 cm, 35.02 cm

   The reference value in this experiment is 30.00 cm. These numbers represent data that are
   a. precise and accurate
   b. precise but not accurate
   c. accurate but not precise
   d. neither precise nor accurate

6. If 3.73 were multiplied by 153.2, the answer would contain
   a. three significant figures
   b. four significant figures
   c. five significant figures
   d. six significant figures

7. The element rhenium is a
   a. nonmetal
   b. semimetal
   c. halogen
   d. metal

8. The alkaline earth metals
   a. are the six larger groups on the right of the periodic table
   b. are the elements found in group 2A
   c. are the 14 groups shown separately at the bottom of the table
   d. are the elements in group 6A

9. An intrinsic property is one that
   a. changes the chemical makeup of the sample
   b. depends on the size of the sample
   c. does not depend on the size of the sample
   d. can be determined without changing the chemical makeup of the sample

10. The SI unit for mass is
    a. oz
    b. lb
    c. g
    d. kg

11. Calcium is a(n)
    a. alkali metal
    b. halogen
    c. chalcogen
    d. alkaline earth metal

12. Iodine, I, is a(n)
    a. alkali metal
    b. halogen
    c. transition metal
    d. actinide element

13. The symbol for tin is
    a. Sn
    b. T
    c. Ti
    d. Tn

14. The SI prefix *nano* corresponds to the multiplier
    a. 10^{-3}
    b. 10^{-6}
    c. 10^{-9}
    d. 10^{-12}

15. A Celsius degree is
    a. larger than a Fahrenheit degree, but the same as a Kelvin degree.
    b. smaller than a Kelvin degree and the same as a Fahrenheit degree.
    c. larger than a Kelvin degree.
    d. smaller than a Fahrenheit degree.

Matching

Scientific Method	a. the amount of matter contained in an object
Theory	b. property that describes the reactivity of a sample
Noble Gases	c. how close a measurement is to an accepted value
Transition Metals	d. how well a number of independent measurements agree with one another
Matter	e. an experiment-based method for increasing understanding of the natural world
Mass	f. unreactive elements found in group 8A
Precision	g. a way to carry out calculations involving different units
Accuracy	h. a consistent explanation of known observations based on many independent experiments. Useful, but cannot be proven
Dimensional analysis	i. term used to describe anything that takes up space
Chemical property	j. central ten groups on the periodic table

Fill-in-the-Blank

1. A ________ is always open to modification by new experimental data.
2. There are 90 naturally-occurring _____________.
3. The vertical columns of the periodic table are known as _________, while the horizontal rows are referred to as the ____________.
4. The third-period element in Group 7A is ______________, the symbol for which is_____.
5. The melting point of a substance is a(n) ___________, ___________ property.
6. Flammability is a ___________ property.
7. Shiny, malleable, ductile, and good conductors of electricity all describe ___________.

8. The SI units for mass, temperature, and length are the __________, the ___________, and the ____________.

9. There are three commonly used temperature scales. _____________ and ____________ have degrees of the same size, while ______________ has degrees that are smaller.

10. The intrinsic property __________ can be determined by dividing a substance's mass by its volume.

11. If an instrument gives you measurements that average to the accepted reference value, but are not close to each other, you would say that your data are ____________, but not ______________.

12. To write numbers that are very small or very large, it is helpful to use _______________ _______________.

13. When finished working a problem, it is often necessary to ___________ the answer to the correct number of ___________ ____________.

14. 1 inch = 2.54 cm is an example of a ____________ ____________.

15. When adding numbers together, the answer cannot have more significant figures on the right of the decimal point than ____.

16. The periodic table arranges the elements into _________ based on their chemical properties.

17. The elements above the zigzag line that divides the periodic table are called _____________.

18. The most appropriate prefix to use for this quantity – 360,000,000 watts – is ______, abbreviated ___.

19. If you know the density of a liquid and its volume, you can calculate its _________.

20. A graduated cylinder and a syringe can both be used to measure ___________.

Problems

1. 246 cm = __________mm = __________m = __________km

2. Perform the following calculations and report your answer with the correct number of significant figures:

   a. $\frac{8.3}{2.793} =$ b. $23.945\,62 \times 8.3421 =$ c. $3.2 + 5.3875 =$ d. $9434.21 - 4.199 =$

3. The freezing point of paraffin wax, commonly used in making preserves, is 56 °C. What is this in Fahrenheit?

4. The land area of Zambia is 290,587 mi^2. Convert this to km^2.

5. How many nanometers are there in 6.73×10^{-7} m?

6. The boiling point of ethanol is 78.5 °C. Convert this temperature to °F and K.

7. In 2009, the world's record for the 100 meter dash was 9.58 seconds, and the world's record for the 100 meter freestyle swim was 47.85 seconds. How fast were the record holders moving in miles per hour?

8. The mass of an amorphous chunk of a yellow element is found to be 9.634 g. When it is placed in a graduated cylinder containing 11.5 mL of water, the water level rises to 16.2 mL. What is the density of the substance in g/cm^3?

9. On June 28, 2010, the Voyager 2 Spacecraft completed 12,000 days of continuous operation and reached a distance of 13.6 billion kilometers from the Earth. If communications travel at the speed of light (3.00×10^8 m/s), how long does it take for data to get from the spacecraft to ground control?

10. A pycnometer is a device used to make precise density determinations. A pycnometer weighs 23.75 g when empty and 37.82 g when filled with water at 20 °C. The density of water at 20 °C is 0.9982 g/mL. When 6.345 g of beryllium are placed in the pycnometer, a total mass of 40.72 g is obtained when the pycnometer is again filled with water. What is the density of beryllium?

11. If the density of titanium is 4.55 g/mL, and a metal plate used to repair a broken skull weighs 10.67 g, what is the volume of the plate?

12. Olive oil has a density of 0.918 g/mL. It will mix with trichloroethane, a solvent with a density of 1.3492 g/mL. In what proportions would 100 total grams of these substances have to be mixed to obtain the same density as water – 0.997 g/mL?

13. Give examples of three physical properties that can be used to help identify a substance.

14. What is the physical basis of 0 °C? How is this different from 0 K?

15. When working through calculations, at what step are the numbers rounded or truncated to the correct number of significant figures? Why?

Challenge Problem

Gallium has a melting point of 29.78 °C and a boiling point of 2403 °C. This very large liquid range coupled with the fact that gallium is a liquid near room temperature makes this metal ideal for the construction of a high temperature thermometer.

You need to measure the melting points of some metals and have decided to construct a thermometer using gallium as the liquid. However, you don't want the thermometer to be too awkward to handle, so you have decided to define the melting point of gallium as 0 °Ga and the boiling point as 1000 °Ga. To ensure the accuracy of your thermometer, you measure the melting point of copper. The melting point of copper on your thermometer is 447 °Ga. The reported melting point for copper is 1085 °C. Is your thermometer accurate?

HINT: You need to determine both a size adjustment and a zero-point adjustment in the same manner outlined for the conversion between °F and °C in your textbook. You can then write the equation for conversion between °Ga and °C.

Inquiry Based Problem

On your first day of General Chemistry laboratory, your instructor informs you that you will not be required to buy a textbook for the lab. Instead, you will be assigned problems throughout the semester related to the material you have learned in the lecture. Using that material, you will design your laboratory experiments to solve the problem at hand. Your first problem is stated below.

You are given a metal cylinder and a table of densities for metal cylinders from the *CRC Handbook of Chemistry and Physics*. You are also provided with a caliper (an instrument for measuring thicknesses and external diameters calibrated to 1 μm), a balance, and a graduated cylinder calibrated to 1 mL. You are required to determine the identity of the metal cylinder using two different procedures. Which procedure is more accurate?

The following questions/information may help you determine two procedures to use:

Knowing the definition of density, what two measurements do you need to make?

Are there any mathematical formulas which may be helpful in this experiment? (Please refer to problem 13 above.)

Is there more than one way to determine these measurements given the equipment provided?

How many measurements should you make?

Which method is more accurate?

Once you have written the two procedures, you collected the following data. From this data and the two procedures, determine the identity of the metal cylinders.

Mass	**Volume – H_2O displaced**	**Diameter**	**Length**	***CRC Handbook Data***	
				Element/Alloy	**Density**
47.83 g	5.5 mL	1.0633 cm	6.3945 cm		
47.52 g	5.3 mL	1.0052 cm	6.3855 cm	Stainless Steel Type 304	7.9 g/cm^3
47.99 g	5.4 mL	1.0608 cm	6.3891 cm	Yellow Brass (high brass)	8.47 g/cm^3
				Aluminum bronze	7.8 g/cm^3
				Beryllium copper 25	8.23 g/cm^3
				Red brass, 85%	8.75 g/cm^3

CHAPTER 2

ATOMS, MOLECULES, AND IONS

Chapter Learning Goals

A. Chemical Laws

1. Explain the law of a) definite proportions, b) multiple proportions, c) conservation of mass.
2. For two different compounds comprised of the same two elements, show that the law of multiple proportions is obeyed.
3. Use the law of conservation of mass to determine amounts of masses created or consumed in chemical reactions.

B. Atomic Structure

1. Explain what information about the atom was revealed by the experiments of a) Thomson, b) Millikan, c) Rutherford.
2. Describe the atom in terms of the mass concentration of the nucleus and the volume occupied by the electrons.
3. Describe the charge, relative mass, and location of the three subatomic particles.

C. Elements

1. Understand the difference between atoms and elements.
2. Given the symbol for a neutral isotope of an atom or ion, determine the number of protons, neutrons, and electrons.
3. Given the mass and natural abundance of all isotopes of a given element, calculate the average atomic mass of that element.
4. For any element, calculate a) the mass in grams of a single atom and b) the number of atoms in a given number of grams.
5. Use the mole concept to compare elements by mass and number of atoms.

D. Atoms, Compounds, and Molecules

1. State the difference between a) compounds and mixtures, b) heterogeneous and homogeneous mixtures, and c) atoms and molecules.
2. Have a working understanding of the differences between ionic and molecular substances: be able to identify which substances are ionic and which are molecular.
3. Be able to define acids and bases: identify which substances are acids and which are bases.
4. Give the formulas and names of a) common polyatomic ions, b) ionic compounds, c) binary molecular compounds, d) common acids.

Chapter in Brief

Gaining a basic understanding of the way that the science of chemistry developed helps make the behavior of chemical reactions understandable. From these behaviors, atomic theory was developed and expanded with modern techniques. Atoms will be described, and the three basic subatomic particles – protons, neutrons, and electrons – introduced, along with the ways in which these vary in ions and isotopes. Once atoms are understood, compounds, mixtures, chemical bond types, acids and bases, and chemical nomenclature follow.

2.1 Conservation of Mass and the Law of Definite Proportions

By carefully examining the behavior of gases, Robert Boyle was able to propose an atomic theory, as well as the existence of a number of elements. From his work came two fundamental laws of chemistry:

- **Law of Mass Conservation** — Mass is neither created nor destroyed in chemical reactions – this is a cornerstone of chemical science, and makes it possible to perform calculations related to reactions.

- **Law of Definite Proportions** — Different samples of a pure chemical substance always contain the same proportion of elements by mass. What this means in layman's terms is that specific mass ratios of elements will form specific compounds. If hydrogen and water are combined to make water, 8 grams of oxygen will always combine with 1 gram of hydrogen. Where this can get tricky is when compounds will form more than one compound with each other. Carbon monoxide and carbon dioxide, for example, both only contain carbon and oxygen, but they will have different mass ratios.

EXAMPLE

5.63 grams of methane are burned in oxygen to form carbon dioxide and water. If the reaction produces 15.5 grams of carbon dioxide, and 12.7 grams of water, how much oxygen was required for the reaction to be completed?

SOLUTION:

According to the **Law of Mass Conservation** mass is neither created nor destroyed in chemical reactions. Therefore, the total mass of the reactants (the starting material) must equal the total mass of the products (the substances produced after the reaction has occurred). In this case

g methane + g oxygen = g carbon dioxide + g water

(5.63 g methane) + (x g oxygen) =
(15.5 g carbon dioxide) + (12.7 g water)

x g oxygen = (15.5 g + 12.7 g) – 5.6 g

x g oxygen = 22.6 g

EXAMPLE:

If 23 grams of sodium combine with 37.5 grams of chlorine to form sodium chloride, what is the mass ratio of the two elements in the compound? How much chlorine will be required to react with 8.5 grams of sodium?

SOLUTION:

The mass ratio will simply be the mass of either element divided by the mass of the other. So the mass ratio can either be expressed as 1.63 grams of chlorine per gram of sodium (37.5 ÷ 23), or as 0.613 grams of sodium per gram of chlorine (23 ÷ 37.5). For a problem like this, it is a good idea to keep "artificial units" with the number, so you can figure out how best to use it. Using this idea, these ratios are best expressed as 1.63 g Cl/g Na (1.63 grams of chlorine per gram of sodium), or 0.613 g Na/g Cl.

To figure out how much chlorine will be needed, either of the ratios can be used, but care must be taken to make sure the units cancel:

$$8.5 \text{ g Na x } 1.63 \frac{\text{g Cl}}{\text{g Na}} = 13.85 \text{ g Cl}$$

This answer has more digits than it should: significant figure rules reduce the answer to 14 grams of chlorine.

2.2 Dalton's Atomic Theory and the Law of Multiple Proportions

Dalton's Atomic Theory was proposed in 1808, and contains a number of precepts that have survived essentially unchanged to this day:

- Elements are made of tiny particles called atoms.
- Each element is characterized by the mass of its atoms: atoms of the same element have the same mass, while atoms of different elements have different masses.
- When atoms join together to form compounds, they do so in small, whole-number ratios. It is never possible to have a fraction of an atom participating in a reaction.
- Chemical reactions rearrange the combinations of atoms, but the atoms themselves are not changed.
- Law of Multiple Proportions — If two elements combine in different ways to form different substances, the mass ratios are small, whole-number multiples of each other.

EXAMPLE

Nitrogen and oxygen form a large number of different compounds with one another. In one such compound, 2.8 grams of nitrogen combines with 3.2 grams of oxygen. In a different compound, 2.3 grams of nitrogen combines with 5.3 grams of oxygen. Do these compounds obey the law of multiple proportions?

SOLUTION:

Find the O:N mass ratio in each compound. First compound: O:N mass ratio $= \frac{3.2 \text{ g O}}{2.8 \text{ g N}} = 1.14$

Second compound: O:N mass ratio $= \frac{5.3 \text{ g O}}{2.3 \text{ g N}} = 2.3$

Determine if the two C:O mass ratios are small, whole-number multiples of each other.

$$\frac{\text{O : N mass ratio in second compound}}{\text{O : N mass ratio in first compound}} = \frac{2.30}{1.14} = 2.00$$

Note: the ratios can be done in any way, and give the same information. The ratio of the oxygen to nitrogen in the second compound is twice as high as in the first compound. If this were done the other way around, the answer obtained would state that the ratio in the first compound is half that of the second compound. This is the same answer. With a problem of this type, the important thing is to show that the ratios are small: 1:2 and 2:1 are both small, whole-number ratios.

Workbook Problem 2.1

Carbon and hydrogen form a large number of compounds. In one such compound, 3.43 g of carbon combines with 0.857 g of hydrogen. In another, 4.80 g of carbon combines with 0.400 g of hydrogen. Show that the law of multiple proportions is being obeyed.

Strategy: Determine the carbon-to-hydrogen mass ratios for both compounds, and compare these ratios to each other.

Step 1: Determine the carbon to hydrogen ratio for the first compound.

Step 2: Determine the carbon-to-hydrogen ratio for the second compound.

Step 3: Divide the ratio found for the first compound by the ratio for the second compound.

Step 4: Can your answer be converted to a small, whole-number ratio?

2.3 The Structure of Atoms: Electrons

Dalton had no way to investigate what atoms are made of, but **cathode ray tubes** were the key by which J. J. Thomson was able to unlock the mystery of the electron. They have the following properties:

- High voltage causes electric current to be emitted from electrodes made of two thin pieces of metal.
- These electrons travel toward the positive electrode, so must have a negative charge, and travel like a stream of infinitesimally small particles.
- Many different metals may be used to make electrodes that will emit electrons: electrons are a fundamental part of matter.
- Cathode rays can be deflected by bringing either a magnet or an electrically charged plate near the tube.

Thomson reasoned that this deflection must depend on the strength of the deflecting magnetic or electric field, the size of the negative charge on the electron, and the mass of the electron. By making a large number of very careful measurements, Thomson determined that the charge-to-mass ratio, *e/m* of the electron = $1.758\ 819 \times 10^{8}$ C/g (C = coulombs, a measure of electrical current that will be discussed further in later chapters).

Robert Millikan built on Thomson's work by building an apparatus that would enable him to coat tiny drops of oil with electrons, then determine the charge on the drop. With the use of repeated measurements, he was finally able to determine that the charges on the drops were always multiples of $1.602\ 177 \times 10^{-19}$ C (actual value is negative, as the charge on an electron is negative).

With the values for the charge of an electron, and the charge-to-mass ratio, it was then possible to calculate the mass of an electron. These infinitesimally tiny particles weigh $9.109\ 390 \times 10^{-28}$ g. This calculation is shown in your book: note that this was determined using nothing more than simple unit-analysis.

2.4 The Structure of Atoms: Protons and Neutrons

If electrons are negatively charged, there must be other particles that provide positive charge to maintain electrical neutrality of matter. Ernest Rutherford successfully investigated this with the use of alpha (α) particles. These have the following characteristics:

- They are 7000 times more massive than electrons.
- They have a positive charge that is twice the magnitude of, but opposite in sign to, the charge on an electron (electrons have a charge of -1, alpha particles have a charge of +2).
- They are naturally emitted by some radioactive elements.

In his experiment, Rutherford directed a beam of alpha particles at a sheet of gold foil only a few atoms thick. Most passed through the foil, but a very few (0.005%) were deflected, and a tiny number bounced back at the source.

Rutherford explained this by proposing that atoms are almost all empty space, with positive charge and most of the mass concentrated in a core that he called the **nucleus**. Most of the alpha particles passed through the empty space, but those that were deflected were affected both by the mass and the charge of the particles.

Further experiments showed that the nucleus contains not only positively charged protons, but also uncharged neutrons.

Summary of the nuclear model:

- Three subatomic particles: electrons, protons, and neutrons
- Electrons and protons have equal but opposite charges, with electrons carrying negative charges, and protons carrying positive charges.
- Protons and neutrons have almost the same mass, which is much larger than the mass of an electron.
- The nucleus of the atom (protons and neutrons) provide most of the mass of the atom, but very little of the volume.
- The electrons provide almost none of the mass of an atom, but almost all of the volume.

2.5 Atomic Number

Elements are defined by the number of protons their nuclei contain. This is also known as their **atomic number**, abbreviated "Z". The elements in the periodic table are arranged in order of atomic number, which resolved a few discrepancies that occurred when they were arranged by mass. For example, tellurium (Z = 52, mass = 127.6) and iodine (Z = 53, mass = 126.9): iodine very much behaves like a halogen, but if the periodic table is arranged by mass, it will not fall in that column. The discovery of atomic numbers straightened all this out.

In a neutral (uncharged) atom, the number of protons will be the same as the number of electrons.

Most atomic nuclei also contain neutrons. The **mass number** of an atom, abbreviated "*A*" is found by adding the atomic number to the number of neutrons, *N*. The mass of electrons is so small compared to these other particles that they can be ignored.

$$A = Z + N$$

Elements are defined by their protons: any variation in the number of protons and you will have a different element. But the number of neutrons in the nucleus can change without changing the identity of the element. Atoms with different numbers of neutrons are called **isotopes**. The number of neutrons has very little effect on the chemical properties of an atom.

Isotope symbols are written using the following conventions, where "X" indicates the elemental symbol:

$$^{A}_{Z}X$$

$^{A}_{Z}X$ From this, it is possible to calculate the number of neutrons ($A - Z$). It is not difficult to remember which of these numbers is which: the smaller will always be the atomic number, the larger will always be the mass number. Sometimes the atomic number is omitted, as it can be taken from the periodic table, and does not vary.

EXAMPLE:

Determine the number of protons, neutrons, and electrons in the following isotopes:

$^{71}_{31}Ga$ $^{191}_{77}Ir$ $^{88}_{39}Y$

SOLUTION:

The subscript is the atomic number (Z), and the superscript is the mass number (A).
Z = the number of protons = the number of electrons, and $A - Z$ = the number of neutrons.

$^{71}_{31}Ga$: $Z = 31$; number of protons = number of electrons = 31

$A - Z = 71 - 31 = 40$; number of neutrons = 40

$^{191}_{77}Ir$: $Z = 77$; number of protons = number of electrons = 77

$A - Z = 191 - 77 = 114$; number of neutrons = 114

$^{88}_{39}Y$: $Z = 39$; number of protons = number of electrons = 39

$A - Z = 88 - 39 = 49$; number of neutrons = 49

Workbook Problem 2.2

Oxygen is the most abundant element by mass in the biosphere. All classes of biomolecules, proteins, sugars, fats, and nucleic acids, contain oxygen, and oxygen makes up almost 90% of the mass of the world's oceans. Oxygen is made up of three stable isotopes, with mass numbers 16, 17, and 18. Write the standard symbols for these isotopes.

Strategy: Use the periodic table to determine the atomic number and the number of neutrons in each isotope. Give the standard symbol for each.

Step 1: First, it's necessary to know the chemical symbol for oxygen.

Step 2: Use the periodic table to determine the atomic number for oxygen.

Step 3: Use the mass numbers given to determine the number of neutrons in each isotope.

Step 4: The standard symbol is written with the mass number as a superscript and the atomic number as a subscript, both to the left of the symbol.

Workbook Problem 2.3

The metalloid X has been used as a cosmetic since at least 3000 B.C. When bound to sulfur atoms in a 2:3 ratio, it is known as "kohl." An atom of the metal contains 51 protons, and its most stable isotope contains 70 neutrons. Identify the metal, and write the symbol in standard format.

Strategy: Determine the atomic number from the information given.

Step 1: Knowing the atomic number, identify the element by using the periodic table.

Step 2: The standard symbol is written with the mass number as a superscript and the atomic number as a subscript, both to the left of the symbol.

2.6 Atomic Mass

Because the masses of individual atoms and their components are so tiny, grams are an inconvenient unit. As a result, **atomic mass units** are used instead. These are abbreviated amu in chemistry, and are known as Daltons (Da) in biochemistry. Atomic mass units are defined as exactly 1/12 the mass of an atom of ${}^{12}_{6}C$. 1 amu = $1.660\,54 \times 10^{-24}$ g

Most elements occur as a mixture of isotopes, so the atomic mass reported on the periodic table is a weighted average of the naturally occurring isotopes.

Atomic mass of an element = Σ(mass of each isotope $\times$ the abundance of the isotope). The Greek letter Σ (sigma) means "the sum of."

EXAMPLE:

Calculate the mass in grams of a single atom of cadmium.

SOLUTION:

From the mass given in the periodic table, we know that 1 atom of Cd has a mass of 112.411 amu. We also know that 1 amu = 1.6605×10^{-24} g. The final unit we want to end up with is g/atom of Cd. Therefore, we should set up our problem in the following manner:

$$\frac{112.4\text{ amu}}{1\text{ atom Cd}} \times \frac{1.6605 \times 10^{-24}\text{ g}}{1\text{ amu}} = 1.867 \times 10^{-22}\text{ g}$$

EXAMPLE:

Boron has two stable isotopes: B-10 and B-11. Boron-10 is 19.9% of a sample of boron, and has an atomic mass of 10.0129 amu. The remaining 80.1% is boron-11, with an atomic mass of 11.0093 amu. Calculate the average atomic mass of boron.

SOLUTION:

To work out the average atomic mass of boron, you multiply the individual masses by their abundance, and add them together:

$$\sum (0.199 \text{ x } 10.0129\text{amu}) + (0.801 \text{ x } 11.0093\text{amu}) = 10.81 \text{ amu}$$

This agrees with the value on the periodic table, which is always a good sign.

Workbook Problem 2.4

A 14-carat gold ring is 14/24 gold. If a 14-carat gold ring weighs 3.65 g, how many gold atoms does it contain?

Strategy: Figure out how many grams of gold are present, then determine the conversion factors needed to convert from grams of sample to number of atoms.

Step 1: Set up a mathematical equation such that all of the units except number of atoms cancel.

Workbook Problem 2.5

Lead has four naturally occurring isotopes. ^{208}Pb has an atomic mass of 207.977 amu, and is 52.4% of an average lead sample. ^{207}Pb has an atomic mass of 206.976 and an abundance of 22.1%. ^{206}Pb has an atomic mass of 205.974, and an abundance of 24.1% . The remaining lead will be ^{204}Pb with an atomic mass of 203.973 amu. Calculate the average atomic mass of Pb.

Strategy: Remember, the atomic mass of an element = Σ(mass of each isotope × the abundance of the isotope), and all abundances must add up to 100%.

Step 1: Use the information given to determine the abundance of ^{204}Pb.

Step 2: Use the information given to determine the average atomic mass.

2.7 Compounds and Mixtures

It is very unusual to find pure elements: there are many more types of matter than there are elements on the periodic table. These substances can all be classified as either **pure substances**, or **mixtures**.

Pure substances are either **elements** or **chemical compounds**. Elements cannot be changed into another substance until Chapter 22. Chemical compounds are made up of atoms that are chemically bonded together. They are formed by chemical reactions, and cannot be separated except by chemical means. They also have properties very different from their constituent elements.

To form carbon dioxide, a compound, a solid, and a gas react to form a gas. Carbon and oxygen gas combine in a *specific ratio* – one carbon to two oxygens – every time.

This specificity is reflected in the **chemical formula** which is written to reflect this ratio: CO_2. This symbolically represents the ratio of one carbon to two oxygens that is found in carbon dioxide wherever it is found.

The formation of a compound is represented using a **chemical formula**. In this format, the starting substances, or *reactants*, are on the left, and the final substances, or products, are found on the right:

$$C + O_2 \rightarrow CO_2$$

Because of the Law of Mass Conservation, the number of atoms of each type must be the same on both sides of the equation: here there is one carbon atom on the left and one on the right, and two oxygen atoms on the left and two on the right.

Mixtures, in contrast, are formed when two or more substances are mixed together without any chemical interaction. The individual substances maintain their properties, and can be separated relatively easily.

Heterogeneous mixtures are not uniformly mixed: dirt in water will not be evenly distributed, and will tend to settle out. If it is mixed in, the liquid will be cloudy.

Homogeneous mixtures are uniform throughout, and can be solid (metal alloys), liquid (solutions), or gases (air). As an example, sugar dissolved in water will have the same amount of sugar at the top of the water as at the bottom. The liquid will also be clear, as the salt is perfectly dispersed in the liquid.

EXAMPLE:

Identify the following household items as elements, compounds, homogenous mixtures, or heterogeneous mixtures:

Salad, tea, metal utensils, water, soda can, air

SOLUTION:

Think about the composition of these things you see every day. Salad is a heterogeneous mixture, as it has areas of variable composition: two separate spoonfuls will have different components. Tea is a clear liquid, and does not settle: it is a homogenous mixture. Metal utensils can be made of any number of things from stainless steel to sterling silver, but all are mixtures of metals in unvarying proportions – these are homogenous mixtures as well. Water contains elements bound together to form something with very different properties: it cannot easily be separated back into its constituent parts, so it is a compound. A soda can, barring the paint on the outside, is made of aluminum. This is an element. Air is also a mixture of elements not bonded to one another. It contains nitrogen, oxygen, hydrogen, and many other components: it is a homogenous mixture.

2.8 Molecules, Ions, and Chemical Bonds

When atoms approach one another, the electrons are able to interact to form **chemical bonds**. Two main types are possible: **covalent bonds** occur between nonmetals, and **ionic bonds** form between metals and nonmetals.

In covalent bonds, the atoms share electrons. As two or more specific atoms are bonded together, the units they form are called **molecules**. These are represented in a number of ways (see Figure 2.9 in the textbook), including *ball-and-stick models*, which show the bonds as "sticks" connecting the atoms; and *space-filling models* that show the atoms stuck together without explicitly showing the bonds.

Covalent bonds are commonly shown using **structural formulas**: these use lines between atoms to show how the atoms are connected. This is particularly important in organic chemistry, which is the study of carbon compounds. In these compounds, structure determines behavior, even if the atoms are the same.

EXAMPLE:

Ethanol and dimethyl ether have the same chemical formula: C_2H_6O. Because the atoms in the molecules are connected in different ways, one is a gas used as a propellant in aerosol cans, and the other is a liquid used in mixed drinks. In dimethyl ether, the carbon atoms are linked with the oxygen in the middle, and in ethanol, the carbons are attached to each other with the oxygen attached to one end. Each carbon has four bonds, each oxygen has two bonds, and each hydrogen has one bond. Draw the structures.

SOLUTION:

O
H_3C CH_3
dimethyl ether

H_2
C
H_3C OH
ethanol

EXAMPLE:

Butyl alcohol has the structural formula shown below. Write the chemical formula.

H_2 H_2
C C
HO C CH_3
H_2
butyl alcohol

SOLUTION:

Chemical formulas are written from structural formulas by putting the constituents in alphabetical order. This structure has four carbon atoms, ten hydrogen atoms, and one oxygen atom. Its chemical structure is $C_4H_{10}O$.

Some elements also exist as molecules. Most of these are diatomic, that is, consisting of two atoms per molecule. H_2, O_2, N_2, F_2, Cl_2, Br_2, and I_2 are all diatomic. Rather than memorize this list, take note of where they lie on the periodic table: with the exception of hydrogen, they are all together, and form a number "7". This may be a good addition to the colored-in periodic table stack you are developing (you are, aren't you?).

Ionic bonds are formed when electrons are not shared: they are the result of electrons being transferred from one atom to another. This gives the atoms opposite charges: these opposite charges then serve to hold the atoms together.

Metals give up electrons easily to form **cations** – positively charged particles. This is easy to remember if you recall that metals conduct electricity. They do this because they are not holding on to their electrons very tightly, so electrons are able to flow through them.

Nonmetals readily gain electrons to form **anions** – negatively charged particles.

When anions and cations interact, they do so generally on the basis of opposite charges. They do not form specific bonds, but do all they can to neutralize their opposing charges. A cation will surround itself with anions, and *vice versa.* As a result, ionic bonds result in crystals (like those of common table salt) called **ionic solids**, not molecules. When a formula is written for an ionic solid, it is written so as to neutralize the charges, but does not indicate that a specific anion is bonded exclusively to a specific cation.

Polyatomic ions also occur: these are charged (usually negatively, as they consist of nonmetals), covalently bonded groups of atoms that carry a charge. There are covalent bonds within polyatomic ions; as units, they can form ionic bonds.

EXAMPLE:

Identify the following as covalent molecules or ionic compounds: a) CH_4, b) MgO, c) KBr, d) NO_2

SOLUTION: The easiest way to differentiate between molecules and ionic compounds is to determine whether the formula contains both metals and nonmetals or only nonmetals. If only nonmetals are present, a covalent molecule has formed. If metals and nonmetals are both present, ionic bonding is occurring. CH_4 and NO_2 contain only nonmetals, so are molecules. MgO and KBr are both combinations of metals and nonmetals. These are ionic compounds.

2.9 Acids and Bases

Two ions have special roles in substances known as **acids** and **bases**.

Acids can be defined as compounds that produce H^+ ions when dissolved in water. Acids can produce one or more than one ion when dissolved.

In this same way, **bases** are defined as substances that release OH^- ions (called hydroxide ions) when dissolved in water, and can produce one or more than one hydroxide when dissolved.

Note the fact that the two ions are complementary: when acids and bases are mixed, the following basic reaction occurs:

$$H^+ \text{(acid)} + OH^- \text{(base)} \rightarrow H_2O \text{ (neutral water)}$$

EXAMPLE:

Identify the following as acid or a base: a) HI, b) H_2S, c) $Mg(OH)_2$, d) LiOH

SOLUTION:

A substance is an acid if it is capable of providing an H^+ ion when dissolved. HI and H_2S both contain hydrogen and will produce H^+ when dissolved in water. Therefore, these compounds are acids. LiOH and $Mg(OH)_2$ both contain OH^- and will produce OH^- when dissolved in water. Therefore, these compounds are bases.

Workbook Problem 2.6

Identify the following as either a molecule or ionic compound. If applicable, also identify as either an acid or base. a) HI, b) NO_2, c) $Ba(OH)_2$, d) PCl_3

Strategy: Use the periodic table to determine the classification of the elements present, and apply this knowledge to determine what type of compound is present.

Step 1: Determine whether the formula contains both metals and nonmetals or only nonmetals.

Step 2: Identify the compounds as either molecular or ionic, based on your conclusions in step 1.

Step 3: Determine if the substance is capable of providing either an H^+ or an OH^- ion in water.

Step 4: Identify the compounds as either acids or bases (if applicable), based on your conclusions in step 3.

2.10 Naming Chemical Compounds

Special rules have been developed to name chemical compounds. The simplest of these are **binary ionic compounds**: as the name suggests, these have only two elements, and are ionic. They are named cation first, then anion with the suffix *–ide.* For all practical purposes, these compounds can be named as "*metal nonmetalide.*". No information about the ratio of one element to the other is given in the name, but must be given in the formula.

The formula of a binary ionic compound must be written to give the smallest ratio of anion to cation that provides for electrical neutrality. What this means in plain English is that if you are balancing charges, make sure you don't do more than you need to: Mg_3N_2, not Mg_6N_4. If all your subscripts have a common factor, you need to fix it.

It is expected that the charges on the ions will be known. For main group metals and nonmetals, this information can be taken straight from the periodic table.

- Main group metals carry a positive charge equal to their group number. Alkali metals, for example, are found in group 1A – they all carry charges of +1.
- Nonmetals carry negative charges equal to their group number minus 8. So oxygen, which falls in group 6A will carry a charge of 6 - 8 = -2.

When transition metals are involved as cations, it is not possible to take the charges from the periodic table. As a result, the charge must be given when the name is written: iron(II) chloride indicates that the iron atoms in the compound carry a charge of +2. When the formula is written, the charges on the iron atom must be balanced by negative charges. As chlorine carries a charge of -1 when it is an ion (group 7A – 8 = -1), there will need to be two chlorines. The formula will be $FeCl_2$.

EXAMPLE:

Name the following compounds:

MgO $CuCl_3$ $RbCl$ SnO_2

SOLUTION:

MgO:	magnesium oxide	No Roman numeral is needed: magnesium is a group 2A metal and forms only Mg^{2+}.
$CuCl_3$:	copper(III) chloride	The charge on Cl is –1, and there are three of them. To have electrical neutrality, the charge on copper, which is a transition metal, must be +3. The Roman numeral (III) is used to indicate this charge.
$RbCl$:	rubidium chloride	Rubidium is a group 1A metal and forms only Rb^+.
SnO_2:	tin(IV) oxide	There are 2 O^{2-} ions with a total charge of –4. To have electrical neutrality, the charge on tin must be +4. Tin can form more than one kind of ion; therefore the charge must be indicated with a Roman numeral.

EXAMPLE:

Write formulas for the following compounds:

cesium oxide magnesium nitride vanadium(V) bromide

SOLUTION:

Cs_2O Cesium is a group 1A metal and forms only Cs^+. Oxygen forms O^{2-} oxide ions, so two cesium ions are needed for each oxygen.

Mg_3N_2 Magnesium is a main group, group 2A metal that has a +2 charge, while nitride ions have a –3 charge (group 5 – 8 = -3). To have electrical neutrality, you need 3 Mg^{2+} and 2 N^{3-}. $[(3 \times +2) + (2 \times -3)] = 0$. There will need to be six positive charges and six negative charges to achieve neutrality.

VBr_5 Vanadium(V) has a +5 charge, while bromine has a –1 charge. To have electrical neutrality, you need 1 V^{5+} and 5 Br^-.

Workbook Problem 2.7

Name or write formulas for the following:

MgF_2 sodium chloride chromium(III) oxide CaO TiI_4

bismuth(II) nitride aluminum sulfide CoF_3 CuSe

Strategy: Follow the rules outlined here and in your textbook.

Step 1: When naming compounds, determine if the metal is a main group metal or a transition metal. If the metal is a transition metal, you must indicate the charge on the metal when naming the compound. The charge on the metal is determined from the number and charge on the anion. (Remember, the charges must all add up to zero.)

Step 2: When writing formulas, use the periodic table or the name of the compound to determine the charge on the metal. Use the periodic table to determine the charge on the anion. (Remember, the charges must all add up to zero.)

Step 3: Make sure the formula contains the smallest whole-number ratio of cation to anion.

Binary molecular compounds contain only two nonmetals. They are named by first naming the element closest to the metals on the periodic table, followed by the element farthest from the metals with an *–ide*.

Because these are electron-sharing interactions, they can occur in many different ways (remember the Law of Multiple Proportions?). Numerical prefixes must be used to identify the number of each element in the molecule. *Mono-* is not generally used for the first element. These prefixes should already be familiar, and must be memorized.

Careful! If a binary molecular compound contains hydrogen, it will be named using these conventions if it is a gas, but named as an acid if it is dissolved in water (aqueous). This will be indicated in parentheses after the formula: HCl (*g*) is hydrogen chloride, but HCl (*aq*) is hydrochloric acid!

EXAMPLE:

Name the following compounds:

SF_6 N_2O_5 P_4S_{10} NI_3

SOLUTION:

SF_6: sulfur hexafluoride

N_2O_5: Dinitrogen pentoxide

P_4S_{10}: tetraphosphorus decasulfide

NI_3: nitrogen triiodide

EXAMPLE:

Write formulas for the following compounds:

hydrogen chloride (gas) | xenon hexafluoride | diphosphorus trioxide

SOLUTION:

hydrogen chloride (gas): $HCl(g)$ gaseous, so it is named as a binary molecule

xenon hexafluoride: XeF_6

diphosphorus trioxide: P_2O_3

Workbook Problem 2.8

Name or write formulas for the following:

PF_6 dinitrogen tetroxide selenium dioxide NBr_3 $HF(g)$

Strategy: Follow the rules given here and in your textbook.

Step 1: When naming the molecules, remember that the element that is more anionlike uses the *ide* suffix.

Step 2: Add numerical prefixes.

Ionic compounds containing polyatomic ions are named by following the rules for naming binary ionic compounds: cation, then anion. A number of polyatomic ions are listed in a table in your book: these should be memorized. Flashcards are a great way to do this!

Oxoanions have special naming conventions that may make their names easier to learn: first learn the name and formula of the ion whose name ends with ate (including the charge on the anion). Other similar ions will be named using the following conventions:

- Add one O; add prefix *per*.
- Remove one O; change ending to *ite*.
- Remove two O's; add prefix *hypo* and change *ate* to *ite*.

EXAMPLE:

Name the following compounds:

$NaClO_4$ MgS_2O_3 $Fe_2(CO_3)_3$ $Mn(NO_3)_2$

SOLUTION:

$NaClO_4$: sodium perchlorate.

MgS_2O_3: magnesium thiosulfate

$Fe_2(CO_3)_3$: iron(III) carbonate. The carbonate anion has a –2 charge, and there are three of them for a total charge of -6. Therefore, there must be a +6 charge spread over the two iron cations. Therefore, the charge on the iron cation must be +3. Each iron cation has a charge of +3 (+6/2). The charge on the iron is indicated with Roman numerals because iron is a transition metal.

$Mn(NO_3)_2$: manganese nitrate.

EXAMPLE:

Write formulas for the following compounds:

iron(III) sulfate cesium nitrite sodium acetate

SOLUTION:

iron(III) sulfate: $Fe_2(SO_4)_3$ The sulfate anion has a –2 charge. Iron(III) by definition has a charge of +3, so the charges must be balanced. Three sulfate ions are needed to give a charge of -6, which will balance two Fe^{3+} ions, with a charge of +6.

cesium nitrite: $CsNO_2$ Cesium is a group 1A metal, so carries a charge of +1, as does the nitrite ion.

sodium acetate: $NaCH_3CO_2$ The acetate anion is one of the most complex of the polyatomic ions. It has a –1 charge, and is thus perfectly balanced by the +1 charge of the group 1A sodium ion.

Workbook Problem 2.9

Name or write formulas for the following:

$Fe(NO_3)_3$ potassium chromate magnesium phosphate $Al_2(SO_4)_3$

ammonium nitrate

Strategy: Follow the rules listed above and in your textbook.

Step 1: Identify the cation and the anion. Write the cation first.

Step 2: Determine the charge on the cation, and write the formula to make sure all charges are neutralized.

Oxoacids contain oxygen as well as hydrogen and other elements. When they are dissolved in water, they will release H^+ and a polyatomic oxoanion. There is a table of these in the textbook: their names are related to the names of the corresponding oxoanions thus:

- The suffix *-ite* becomes *-ous acid.*
- The suffix *-ate* becomes *-ic acid.*

Binary acids contain no oxygen. They are named as *hydro anionic acid.* When writing formulas for these acids, they must be followed by (*aq*) to indicate that they are to be named as acids.

EXAMPLE:

Name the following acids:

H_3PO_4 $HBr(aq)$ H_2CrO_4

SOLUTION:

H_3PO_4: This compound is an acid that contains the phosph*ate* anion. It is named phosphor*ic* acid.

HBr(*aq*) When hydrogen bromide is dissolved in water, it becomes hydrobromic acid.

H_2CrO_4 This compound dissociates to form hydrogen ions and chromate ions. It is named chromic acid.

Workbook Problem 2.10

Name or write formulas for the following:

$MnSO_4$	hydrochloric acid	H_2SO_3	$H_2S(g)$
dinitrogen trioxide	cobalt(II) sulfide	MgF_2	sodium sulfite

Strategy: First, determine the type of compound. Is it a simple binary ionic compound, a molecule, an ionic compound containing a polyatomic ion, or an acid (either binary acid or oxoacid)? Once you have determined the type of compound, apply the appropriate nomenclature rules.

Self-Test

This section is intended to test your knowledge of the material covered in this chapter. Think through these problems, and make certain you understand what is going on. Ask yourself if your answer makes sense. Many of these questions are linked to the chapter learning goals. Therefore, successful completion of these problems indicates you have mastered the learning goals for this chapter. You will receive the greatest benefit from this section if you use it as a mock exam. You will then discover which topics you have mastered and which topics you need to study in more detail.

True-False

1. Isotopes are atoms of the same element that carry a different charge.
2. The mass of the products of a chemical reaction will always be the same as the mass of the reactants.
3. The smallest unit of matter that retains its identity as an element is an atom.
4. Two nonmetals will only ever form one type of compound.
5. Electrons and neutrons are found in the nucleus of an atom, protons in a diffuse cloud outside the nucleus.
6. The atomic number – the number of protons in an element – is what gives an element its identity.
7. All atoms of the same element have the same infinitesimally tiny mass.
8. Acids dissociate in water to give hydrogen ions – H^+ - and a balancing anion.
9. Ionic compounds are named using prefixes to denote the number of each element.
10. Main-group metals such as sodium can carry a variety of different charges.

Multiple Choice

1. The isotope $^{71}_{31}Ga$ contains:
   a. 71 protons, 31 neutrons, and 71 electrons
   b. 31 protons, 71 neutrons, and 31 electrons
   c. 31 protons, 40 neutrons, and 31 electrons
   d. 31 protons, 40 neutrons, and 40 electrons

2. The compound copper(I) selenide is:
   a. covalent, with formula CuS
   b. covalent, with formula CuSe
   c. ionic, with formula CuSe
   d. ionic with formula Cu_2Se

3. The proper configuration of an atom is:
   a. neutrons and protons in the nucleus, electrons outside the nucleus
   b. neutrons and electrons in the nucleus, protons outside the nucleus
   c. protons and electrons in the nucleus, neutrons outside the nucleus
   d. none of the above

4. The compound Na_2SO_4 has the following characteristics:
   a. covalent, acid, called sodiosulfuric acid
   b. ionic, neither acid nor base, called sodium sulfate
   c. covalent, neither acid nor base, called sodium sulfide
   d. ionic, basic, called sodium sulfoxide

5. Covalent bonding involves:
   a. interactions between electrons
   b. formation of molecules
   c. nonmetals bonding to nonmetals
   d. all of the above

6. Ionic bonding requires:
   a. similarly sized atoms
   b. metal and nonmetal atoms
   c. oppositely charged particles
   d. isotopes

7. In a homogenous mixture:
   a. the properties of the mixture are completely different from the properties of the components
   b. the composition is uniform throughout, and the components retain their separate identities
   c. the composition is variable, with different proportions in different locations
   d. only liquids can be involved

8. Which of the following correctly identifies the subatomic particles?
   a. proton, positive charge, mass of 1 amu
   b. neutron, negative charge, mass of 1 amu
   c. neutron, no charge, negligible mass
   d. electron, negative charge, mass of 1 amu

9. The correct name of N_2O_5 is
   a. nitrogen oxide
   b. oxygen nitrate
   c. dinitrogen pentoxide
   d. nitropentoxygen

10. HCl(*aq*) is
    a. hydrogen dichloride
    b. hydrochloric acid
    c. hypochlorous acid
    d. hydrogen chlorine

Fill-in-the-Blank

1. Elements are made up of __________, which contain three fundamental particles: ____.
2. The identity of an element is defined by the number of ________________it contains, also called the ________________________; the other particles can vary.
3. When the number of _____________changes, the mass of the element changes. The result is called an ____________. When the number of _______________ changes, the charge changes: the result is known as an ___________.
4. Very few pure elements exist in the world around us: mostly we are surrounded by ________________, which are also considered pure substances, and ________________.
5. When nonmetals combine, the result is a _______________ bond. In this type of bond, electrons are ________________.
6. When a ___________________ and a ___________________ combine, the result is an ________________ bond. In this type of bond atoms are held together by opposite _____________.
7. When a covalently bonded group of atoms also carries a charge, it is known as a __________________. Some of the most common of these are the _______________ ion, NO_3^-, the ________________ ion, NH_4^+, and the _______________ ion, SO_4^{2-}.
8. A substance that comes apart in water to form H^+ ions is called an _______________. A substance that comes apart in water to form OH^- ions is called a _______________.
9. In molecules, the number of each element is denoted using a series of prefixes. One of an element is denoted mono, two is _______, three is _______, five is __________, and so on.
10. When chemical reactions occur, the mass of the _______________ will always be the same as the mass of the products. This is known as the __________________________________.

Matching

Ionic compound	a. compounds containing two elements
Base	b. number of protons in an element
Molecules	c. compound that releases H^+ ions
Atomic number	d. negatively charged subatomic particle
Polyatomic ions	e. discoverer of atomic nucleus
Binary compounds	f. bond formed by sharing electrons
Acid	g. compound formed by opposite charges
Heterogeneous mixture	h. 1/12 the mass of a C-12 atom
Structural formula	i. positively charged ion
Electrons	j. covalently bonded units
Proton	k. provide information about molecular connectivity
Anion	l. uncharged subatomic particle
Atoms	m. positively-charged subatomic particle
Neutron	n. covalently bonded units that carry a charge
Ernest Rutherford	o. discoverer of the electron
Isotopes	p. negatively charged subatomic particle
J. J. Thomson	q. compound that releases OH^- ions
Covalent bond	r. same element, varying masses
Cation	s. proportions of components vary within mixture.
Atomic mass unit	t. smallest particle of any element

Problems

1. Hydrogen gas reacts explosively with oxygen to form water vapor. If an unknown quantity of hydrogen gas is reacted with 3.2 L of oxygen (density 1.43 g/L), and 5.15 g of water are formed, what volume of hydrogen gas (density 0.0893 g/L) was in the initial mixture?

2. Phosphorus and oxygen form two compounds, one of which is a white, crystalline solid, the other of which is a waxy substance that smells like garlic. In the first compound, 4.3 g phosphorus combined with 11.4 g oxygen. In the second, 3.34 g phosphorus combined with 5.33 g oxygen. Demonstrate that these compounds obey the law of multiple proportions.

3. How many protons, neutrons, and electrons are present in the following elements?

   a. ${}^{97}_{42}Mo$ b. ${}^{235}_{92}U^{6+}$ c. ${}^{235}_{93}Np$ d. ${}^{202}_{80}Hg^{4+}$

4. Neon has three stable isotopes, and an average atomic mass of 20.18 amu. If Neon-20 has an atomic mass of 19.992 and an abundance of 90.48%, and Neon-21 has an atomic mass of 20.993 amu and an abundance of 0.27%, what is the abundance and atomic mass of Neon-22?

5. Magnesium, Mg, has three naturally occurring isotopes: ${}^{24}Mg$ has an abundance of 78.99% and an atomic mass of 23.985 amu, ${}^{25}Mg$ with an abundance of 10.00% and an atomic mass of 24.986 amu, and ${}^{26}Mg$ with an abundance of 11.01% and an atomic mass of 25.983 amu. Calculate the average atomic mass of magnesium.

6. Naphthalene is the compound that gives mothballs their distinctive smell. Given its structural formula, below, what is its chemical formula?

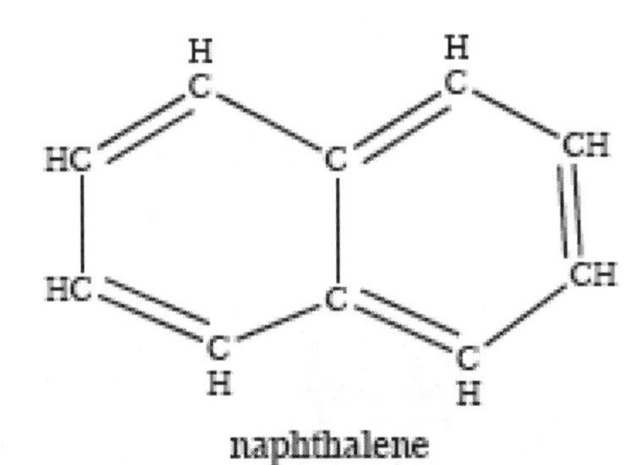

7. Write the formula and name for the compound that forms between each of the following pairs of elements:

   a. Li, Cl b. Na, S c. Ca, O d. Mg, N

8. Name the following compounds, and identify them as ionic or covalent:

   a. CO b. $SrCl_2$ c. CCl_4 d. PtO_2

9. What is the mass (in grams) of 9500 atoms of gold?

10. Formulas for chemical compounds can be worked out using the percent masses of each component. For the organic compound dichloromethane (CCl_2H_2), what percentage of the total mass is provided by each component (use the atomic masses in amu)?

11. Write a formula for the compound that would form between calcium and the following ions:

a. fluoride   b. phosphate   c. nitrate   d. carbonate

12. Name the following compounds:

a. $SnCl_4$ b. $H_3PO_4(aq)$ c. $Mg(C_2H_3O_2)_2$ d. HCl

e. Rb_2S f. CO_2 g. $CrPO_4$ h. F_2O

i. PO_5 j. $NiCl_2$ k. $Fe(OH)_3$ l. $HCl(aq)$

m. PbI_2 n. $HNO_3(aq)$ o. $BaBr_2$ p. $Al(NO_3)_3$

13. Write formulas for the following compounds:

a. calcium nitrite   b. sodium chloride   c. dichlorine monoxide

d. dihydrogen monoxide   e. nickel(II) hydride   f. boron tribromide

g. copper(II) chloride   h. xenon tetrafluoride   i. nitrogen dioxide

j. barium nitride   k. phosphoric acid   l. hydrofluoric acid

m. lithium permanganate   n. carbonic acid   o. phosphorous trichloride

p. carbon tetrabromide

Challenge Problem

The smallest thing that can be resolved by the human eye is about 0.2 mm across. A sphere of osmium metal (density = 22.61g/cm^3) with a diameter of 0.2 mm would contain how many atoms? How many atoms would a sphere of aluminum (density 2.7 g/cm^3) contain? If these atoms were cubes with edge length 0.1 mm, what would the volume of these atoms be?

CHAPTER 3

MASS RELATIONSHIPS IN CHEMICAL REACTIONS

Chapter Learning Goals

A. ***Balancing Chemical Equations***

1. For simple chemical reactions, write and balance chemical equations.

B. ***Formula Units and Moles***

1. Calculate molar mass.
2. Interconvert grams, moles, and numbers of formula units.

C. ***Stoichiometry***

1. Determine the number of moles and grams of one reactant needed to react with a given number of moles and grams of another reactant and the number of moles and grams of product(s) that result from the reaction.
2. Calculate percent yield.
3. Calculate the number of grams of products produced from a given number of grams of reactants when the theoretical yield is less than 100%.
4. Identify the limiting and excess reactants in a reaction mixture.
5. Determine the number of grams of excess reactant remaining at the end of a reaction and the number of grams of product(s) produced.

D. ***Chemical Reactions Performed in Solution***

1. Describe how to prepare a solution of known molarity by a) dissolving a solid in a solvent and b) diluting a more concentrated solution.
2. Interconvert solution molarity, solution volume, solute moles, and solute grams.
3. Determine the volume of one reactant needed to react with a given volume of a second reactant.

E. ***Elemental Analysis***

1. Determine the percent composition and empirical formula of a compound.
2. Understand how to use combustion analysis to obtain the empirical formula of a compound containing carbon, hydrogen, and one other element.
3. From empirical formula and molar mass, determine the molecular formula of a compound.

Chapter in Brief

The central concern of chemistry is the change of one substance into another. As a consequence, chemical reactions are at the heart of the science. In this chapter, you begin your study of chemical reactions by learning how to balance chemical equations. Using balanced chemical equations along with the concepts of the mole and stoichiometry you will learn how to calculate the amounts of reactants needed and the theoretical amount of products produced in a reaction. These calculations are carried out using molar masses for gram ↔ mole conversions or molarities for mole ↔ volume ↔ molarity conversions. You will also learn how to determine an empirical formula from the percent composition of a compound and vice versa. Finally, you will see how combustion analysis, a type of elemental analysis, along with mass spectrometry can be used to determine both the empirical and molecular formula of a compound.

3.1 Balancing Chemical Equations

The Law of Mass Conservation states that the total mass of products and reactants in a chemical reaction must be equal: matter is neither created nor destroyed. Dalton's atomic theory states that in reactions, the atoms are simply rearranged, not altered. As a result, the number and kind of atoms must be the same on both sides of a chemical reaction.

Equations must also be balanced by **formula unit**: not simply the individual atoms, but also the units – whether covalent molecules or ionic units – must be balanced as well.

To balance an equation:

- Write the unbalanced equation using the correct formula units on each side of the reaction
- Find suitable **coefficients** for the formula units: the coefficients are placed in front of the equations: you cannot balance equations by changing the subscripts within the molecules, or you will be changing the compounds.
- Reduce coefficients to their smallest whole-number value, dividing by a common factor if necessary
- Check that the same number of each kind of atom appears on each side of the reaction.

EXAMPLE:

Methane (CH_4) burns in oxygen to produce carbon dioxide and water. Write and balance the equation.

SOLUTION:

Step 1: Write the unbalanced equation.

$CH_4 + O_2 \rightarrow CO_2 + H_2O$

Step 2: Use coefficients to balance the equation. Think about one element at a time. (HINT: <u>It usually helps to save oxygen for last.</u>) Notice on the reactant side that there are four hydrogen atoms, but only two hydrogens on the product side. Begin by placing a 2 in front of H_2O.

$CH_4 + O_2 \rightarrow CO_2 + 2\ H_2O$

You now have four hydrogens on each side, and one carbon on each side, leaving oxygen still unbalanced: there are two oxygens on the reactant side, and four on the product side.

$CH_4 + 2\ O_2 \rightarrow CO_2 + 2\ H_2O$

Step 3: The coefficients are already reduced to their smallest whole-number ratio.

Step 4: Check your answer.

Reactant side	Product side
1 C	1 C
4 O	4 O
4 H	4 H

EXAMPLE:

Write a balanced equation for the combustion of 1-pentanol ($C_5H_{11}OH$).

SOLUTION:

The term *combustion* simply means reaction with oxygen. When hydrocarbons (compounds containing primarily C and H) such as methane and 1-pentanol undergo combustion, carbon dioxide and water are produced.

Step 1: Write the unbalanced equation.

$$C_5H_{11}OH + O_2 \rightarrow CO_2 + H_2O$$

Step 2: Use coefficients to balance the equation. Begin with carbon. There are five carbon atoms on the reactant side, but only one carbon atom on the product side. Begin by placing a 5 in front of CO_2.

$$C_5H_{11}OH + O_2 \rightarrow 5\ CO_2 + H_2O$$

There are 12 hydrogens one the reactant side – do not be confused by the fact that they are broken up - but only two hydrogens on the product side. Place a 6 in front of H_2O.

$$C_5H_{11}OH + O_2 \rightarrow 5\ CO_2 + 6\ H_2O$$

Now balance the oxygens. There are three oxygens on the reactant side - don't forget the methanol oxygen – and 16 oxygens on the product side. To get an even number of oxygens will require doubling the number of pentanol molecules, and all of the coefficients that were taken from it:

$$2\ C_5H_{11}OH + O_2 \rightarrow 10\ CO_2 + 12\ H_2O$$

There are now four oxygens on the reactant side and a whopping 32 oxygens on the product side. To balance this number of oxygens on the product side, subtract out the two oxygens that are associated with the methanol molecules leaving 30 oxygens that need balancing, and put a 15 in front of the molecular oxygen:

$$2\ C_5H_{11}OH + 15\ O_2 \rightarrow 10\ CO_2 + 12\ H_2O$$

Step 3: Reduce the coefficients to their smallest whole-number ratio.

The ratio is 2:15:10:12, which is the smallest whole-number ratio.

Step 4: Check your answer.

Reactant side	Product side
10 C	10 C
24 H	24 H
32 O	32 O

The molecular nature of oxygen, which requires that it be added two at a time to reactions, can make balancing equations a little tricky.

Workbook Problem 3.1
Write a balanced chemical equation for the combustion of hexane (C_6H_{14}).

Strategy: Remember, the term *combustion* is used to indicate reaction with oxygen. When hydrocarbons (compounds containing primarily C and H) undergo a combustion reaction, carbon dioxide and water are the products.

Step 1: Write the unbalanced chemical equation.

Step 2: Use coefficients to balance the equation. (Remember, it helps to save oxygen for last.)

Step 3: Reduce the coefficients to their smallest whole-number ratio.

Step 4: Check your answer.

3.2 Chemical Symbols on a Different Level

The chemical symbols in a chemical equation have meanings at both the microscopic and macroscopic levels. At the microscopic level, these equations represent the actions of individual atoms and molecules. However, these also explain the actions of macroscopic quantities of the same elements and compounds. In the laboratory, it is not possible to observe individual atoms, so the macroscopic properties of chemical reactions, which are the result of the microscopic interactions, are what are observed and measured.

3.3 Avogadro's Number and the Mole

Individual molecules can't be weighed, so a conversion must be made between the ratios of molecules in a balanced reaction and masses of reactants.

The conversion factor for this is a unit called the **mole**.

A mole is equal to 6.022×10^{23} formula units. This quantity is also known as **Avogadro's Number**. 6.022×10^{23} of any substance is equal to one mole. This is also the molecular mass of the substance (from the periodic table) in grams.

EXAMPLE:
What is the mass of one mole of hydrogen?

SOLUTION:
The definition of the mass of one mole of a substance is the amount of that substance contained in the molecular mass (from the periodic table) in grams. But does it really work?

Hydrogen exists as a diatomic molecule: H_2. The molecular mass of hydrogen from the periodic table is 1.008 amu., so the mass of a molecule of hydrogen would be 1.008×2 or 2.016 amu. By definition, this means that one mole of hydrogen will have a mass of 2.016 grams.

Just this once, let's do the conversion anyway.

$$2.016\frac{\text{amu}}{\text{molecule}} \text{x} 6.022\text{x}10^{23}\frac{\text{molecules}}{\text{mole}} \text{x} 1.660539\text{x}10^{-24}\frac{\text{grams}}{\text{amu}} = 2.016\frac{\text{g}}{\text{mol}}$$

Good news: the lengthy conversion from amu to moles, with a conversion factor in it that no one wants to memorize, is completely unnecessary. When you need to know the mass of one mole of a substance, you can pull it straight off the periodic table, no conversions necessary.

Moles make it possible to use macroscopic quantities of substances for microscopic purposes: one mole of hydrogen molecules will have exactly the same number of particles as a mole of water, or oxygen, or table salt. If you need to react substances together, you will have the same number of particles if you have the same number of moles, even if the masses are wildly different.

EXAMPLE:

How many grams of each substance will be required/produced if one mole of hydrogen reacts with one mole of hydrogen to produce hydrogen chloride gas by the following unbalanced equation?

$H_2(g) + Cl_2(g) \rightarrow HCl\ (g)$

SOLUTION:

First, the equation must be balanced. This is simply done by putting a 2 in front of the hydrogen chloride gas produced.

$H_2(g) + Cl_2(g) \rightarrow 2\ HCl\ (g)$

One mole of any substance has a mass equal to its molecular mass in grams, so the one mole of hydrogen will have a mass of 2.016 g. One mole of chlorine gas will be equal to 2 × 35.45 g/mol, or 70.9 g. Despite this huge mass difference, the number of particles of each substance – hydrogen and chlorine – is the same. The chlorine molecules are just *much* heavier.

To determine how much hydrogen chloride gas is produced, there is a coefficient to be considered. The coefficient tells you that for every mole of hydrogen gas or chlorine gas, two moles of hydrogen chloride gas are produced. This makes sense, as when you break apart one molecule of hydrogen and one molecule of chlorine, you have four atoms. Redistribute them, and you have made two hydrogen chloride molecules.

So when we added together one mole of hydrogen and one mole of chlorine, we made two moles of hydrogen chloride. This has a molecular mass of 1.008 + 35.45, or 36.46 g/mol. If there are two moles, there will be 72.9 grams of hydrogen chloride gas.

Take note that once again, the Law of Mass Conservation is at work. On the reactants side, 2.016 g + 70.9 g = 72.9 grams of reactants. On the product side, 72.9 grams of hydrogen chloride gas is produced.

Workbook Problem 3.2

When hydrogen peroxide (H_2O_2) solutions are left exposed to light, the hydrogen peroxide breaks down to form oxygen and water. If 2 liters of a 3% hydrogen peroxide solution contain 60 grams of hydrogen peroxide, and the solution is left in the light until the reaction is complete, how many moles of oxygen are produced? How many moles of water?

Strategy: To determine mole ratios, you first need a balanced equation. After obtaining the mole ratio from the balanced equation, you can use this knowledge to obtain the conversion factors needed to solve the problem.

Part a

Step 1: Write an unbalanced chemical equation based on the information given in the problem.

Step 2: Use coefficients to balance the equation. (Remember to save oxygen for last.)

Step 3: Reduce the coefficients to their smallest whole number ratio if necessary.

Step 4: Check your answer.

Part b

Step 1: Determine the mole ratio for hydrogen peroxide and oxygen from your balanced chemical equation.

Step 2: Use the mass and molecular weight of the hydrogen peroxide solution to determine the moles of hydrogen peroxide present.

Step 3: Use the mole ratio from above as a conversion factor, and calculate the number of moles of oxygen generated by the decomposition reaction.

Step 4: Use the mole ratio from above as a conversion factor, and calculate the number of moles of water generated by the decomposition reaction.

3.4 Stoichiometry: Chemical Arithmetic

The mole relationships in a chemical equation can also be used with molecular weights, to make mass determinations for a chemical reaction. As moles cannot be measured directly in the lab, moles must be converted back to grams for practical use.

Stoichiometry is the term for the use of these ratios.

EXAMPLE:

When concentrated sulfuric acid is added to table sugar, a dehydration reaction occurs according to the following unbalanced equation:

$$C_{12}H_{22}O_{11} + H_2SO_4 \rightarrow C + H_2O + H_2SO_4$$

If 0.50 moles of sucrose are to be reacted with an excess of sulfuric acid, how many grams of sucrose are needed, and how many grams of carbon will result?

SOLUTION: First, balance the equation:

$$C_{12}H_{22}O_{11} + H_2SO_4 \rightarrow 12\ C + 11\ H_2O + H_2SO_4$$

Molar mass of sucrose:

12 carbon atoms	×	12.01 amu	=	144.12
22 hydrogen atoms	×	1.008 amu	=	22.176
11 oxygen atoms	×	16.00 amu	=	176.00
				342.30 g/mol

This molar mass also represents a conversion factor of $\frac{342.3 \text{ grams}}{\text{mole}}$ or $\frac{1 \text{ mole}}{342.3 \text{ g}}$.

To determine the grams of sucrose in 0.50 moles:

$$0.50 \text{ mol x } \frac{342.3 \text{ g}}{\text{mol}} = 170\text{g}$$

To determine the number of grams of carbon that will be produced will take a couple of steps. First, determine how many moles of carbon will be produced using the ratios from the balanced equation:

$$0.50 \text{ mol sucrose x } \frac{12 \text{ mol C}}{\text{mol sucrose}} = 6.0 \text{ mol C}$$

Now convert that number of moles to grams:

$$6.0 \text{ mol C x } \frac{12.01 \text{ g}}{\text{mol}} = 72 \text{ g C}$$

72 grams of carbon will be produced from the reaction of 0.5 moles of sucrose with sulfuric acid. With practice, it will be possible to do this without breaking it up into separate steps. Once you have a balanced equation, you can set up a problem like this to go straight from moles of sucrose to grams of carbon:

$$0.50 \text{ moles sucrose x } \frac{12 \text{ moles C}}{\text{mol sucrose}} \text{ x } \frac{12.01 \text{ g C}}{\text{mol C}} = 72 \text{ g C}$$

Start out by breaking things up into as many steps as you need to when you are starting out, but it will get faster!

The basic steps needed to complete stoichiometry problems can be summarized like this:

- Balance the equation – an unbalanced, or incorrectly-balanced equation will make it impossible to solve a problem correctly.
- Convert to moles: if you are given grams of one substance, and need grams of another substance, you will need the formula weights of both.
- Set up a dimensional-analysis problem. Keep both units and identifiers with all numbers. For example, don't just write 12.01 g/mol, write 12.01 g carbon/mol. You will have numerous conversion factors of this type, so you want to make sure that not only the units, but even the substances cancel out. Using "fake units" will continue to be helpful as you move through the study of chemistry!
- Solve, making sure that your units are correct, and that your answer makes sense.

Workbook Problem 3.3

When elemental iron rusts, it combines with oxygen to form iron(III) oxide. How many grams of rust will be produced if 14 grams of metal nails are left out on a construction site all winter? How many grams of oxygen will have been consumed in this reaction?

Strategy: You can calculate the grams of iron oxide produced by using the molar mass of iron oxide and the mole ratio of iron to iron oxide from the balanced chemical equation as conversion factors.

$$\text{g iron} \rightarrow \text{moles iron} \rightarrow \text{moles iron(III) oxide} \rightarrow \text{g iron(III) oxide}$$

Step 1: Write an unbalanced chemical equation from the information given above.

Step 2: Balance the equation.

Step 3: From the balanced chemical equation, determine the mole to mole ratio of iron to iron(III) oxide.

Step 4: Find the atomic mass of iron.

Step 5: Determine the molecular mass for iron(III) oxide.

Step 6: From the information in steps 3, 4, and 5, create conversion factors. Use these conversion factors, and set up a dimensional-analysis problem, to convert from grams of iron to moles of iron to moles of iron oxide to grams of iron oxide.

Step 7: Using the Law of Mass Conservation, how many grams of oxygen were added to the iron?

3.5 Yields of Chemical Reactions

When stoichiometric calculations are done, an ideal result is provided. This is known as the **theoretical yield**.

Running an experiment will provide an **actual yield**. The actual yield will always be less than the theoretical yield. In the workbook problem above, for instance, there will probably be little pockets of unrusted iron within the nails, which would make the final mass lower than you calculated in assuming that every atom of elemental iron became iron oxide.

To report a result, it is often given in terms of reaction efficiency or **percent yield**. This is calculated very simply:

$$\%\ \text{yield} = \frac{\text{actual yield}}{\text{theoretical yield}} \text{x} 100$$

EXAMPLE:

When 15.0 grams of hydrogen react with oxygen, 120 grams of water result. What is the percent yield of this reaction?

SOLUTION:

Write and balance the equation.

$2\ H_2 + O_2 \rightarrow 2\ H_2O$

Figure out the atomic or molecular weights of all the relevant components:

H_2	=	2.02 g/mol
H_2O	=	18.02 g/mol

You don't need the molecular weight for oxygen.

Set up a dimensional analysis equation to determine the theoretical yield of the reaction:

$$15\ \text{g}\ H_2 \text{x} \frac{1\ \text{mol}\ H_2}{2.02\ \text{g}} \text{x} \frac{2\ \text{mol}\ H_2O}{2\ \text{mol}\ H_2} \text{x} \frac{18.02 \text{g}\ H_2O}{\text{mol}\ H_2O} = 134\ \text{g}$$

Now solve for the percent yield:

$$\%\ \text{yield} = \frac{\text{actual yield}}{\text{theoretical yield}} \text{x}\ 100 = \frac{120\ \text{g}}{134\ \text{g}} \text{x}\ 100 = 89.7\%$$

EXAMPLE:

If burning hydrogen in oxygen to form water generally has a 95% yield, how many grams of water can be expected from the combustion of 23 grams of hydrogen?

SOLUTION:

Use the balanced equation from above: $2\ H_2 + O_2 \rightarrow 2\ H_2O$ and use it to determine the theoretical yield of the equation if 23 grams of hydrogen are used:

$$23\text{ g }H_2 \text{x} \frac{1\text{ mol }H_2}{2.02\text{ g}} \text{x} \frac{2\text{ mol }H_2O}{2\text{ mol }H_2} \text{x} \frac{18.02\text{ g }H_2O}{\text{mol }H_2O} = 205\text{ g}$$

Now the actual yield can be estimated: it will 95% of the theoretical yield:

$$205\text{ g} \times 0.95 = 195\text{ g}$$

Workbook Problem 3.4

The last step in ammonia production is a high-pressure reaction that proceeds according to the following unbalanced reaction:

$$H_2 + N_2 \rightarrow NH_3$$

If 18.0 grams of nitrogen gas are reacted, and 14.6 grams of ammonia result, what is the percent yield?

Strategy: To determine the percent yield, first balance the equation, then determine the theoretical yield of ammonia in the reaction.

Step 1: Balance the equation.

Step 2: Calculate the theoretical yield using dimensional analysis.

Step 3: Calculate the percent yield using the actual yield stated in the problem and the theoretical yield just calculated.

3.6 Reactions with Limiting Amounts of Reactants

To maximize yields, reactions are often carried out with much more of one reactant than necessary based on the balanced equation. In this situation, there is a **limiting reactant** and an excess reactant. The limiting reactant determines the theoretical yield of the reaction: once it is used up, no more reaction is possible.

Problems involving a limiting reactant can be complex: proceed with caution, as no information can be obtained from the masses of the reactants: only their mole ratios contain useful information.

To work these problems, convert all reactant quantities to moles, and run the reaction mathematically, once with each reactant. The reactant that gives you the fewest moles of product is your limiting reactant, and that amount is the theoretical yield of the reaction.

EXAMPLE:

15 grams of hydrogen gas and 37 grams of chlorine gas are reacted together to form hydrogen chloride gas. Which is the limiting reactant? How many grams of hydrogen chloride gas can be produced (theoretical yield)? How many grams of the excess reactant will be left over?

SOLUTION: First, write a balanced chemical equation.

$$H_2 + Cl_2 \rightarrow 2\ HCl$$

Determine the number of moles of each reactant present.

$$15\ g\ H_2 \times \frac{1\ mol\ H_2}{2.02\ g} = 7.4\ mol\ H_2$$

$$37\ g\ Cl_2 \times \frac{1\ mol\ Cl_2}{71\ g\ Cl_2} = 0.52\ mol\ Cl_2$$

The reactants are required in a 1:1 mole ratio, so in this example it is possible to see that chlorine will be the limiting reactant and hydrogen the excess reactant, despite the fact that by mass there is a great deal more chlorine than hydrogen.

These ratios will not always be this straightforward: a foolproof way to find the limiting and excess reactants is to run the reaction stoichiometry with each reactant and look at the amount of product produced by each:

$$7.4\ mol\ H_2 \times \frac{2\ mol\ HCl}{mol\ H_2} = 14.8\ mol\ HCl$$

$$0.52\ mol\ Cl_2 \times \frac{2\ mol\ HCl}{mol\ Cl_2} = 1.04\ mol\ HCl$$

This method works equally well for simple and complex reactions: the number of moles of chlorine available can produce far less than the number of moles of hydrogen. Chlorine is your limiting reactant, and the theoretical yield of the reaction is 1.04 mol HCl gas.

$$1.04\ mol\ HCl \times \frac{36.5\ g\ HCl}{mol} = 38.0\ g\ HCl$$

How many grams of hydrogen will be left over? There are several ways of getting at this information. The most straightforward is to do a quick calculation using the Law of Mass Conservation: we have obtained 38 grams of products from 15 g + 37 g of reactants: 14 grams of reactants must be left over. As hydrogen is in the excess, there are 14 grams of hydrogen left over.

Another way to get at this is to use mole quantities: 0.52 moles of chlorine reacted, so you can set a stoichiometric equation to get at the number of moles of hydrogen that were used. This is extreme overkill for the simple ratio here, but is very useful when the ratio is less straightforward:

$$0.52 \text{ mol } Cl_2 \times \frac{1 \text{ mol } H_2}{1 \text{ mol } Cl_2} = 0.52 \text{ mol } H_2$$

Now that the number of moles of hydrogen used up is known, you can simply subtract from the original number of moles of hydrogen and convert to grams:

$$(7.4 \text{ mol } H_2 - 0.52 \text{ mol } H_2) \times \frac{2.02 \text{ g } H_2}{\text{mol } H_2} = 14 \text{ g } H_2$$

Workbook Problem 3.5

Aluminum metal and hydrochloric acid react to make aluminum chloride and hydrogen gas. If 5.3 grams of aluminum are reacted with 9.2 grams of hydrochloric acid in solution, how many grams of aluminum chloride can be obtained? What is the limiting reactant? How much of the excess reactant (in grams) is left over? What is the percent yield of the reaction if 8.3 g of aluminum chloride are produced?

Strategy: Generate a balanced equation and determine the number of moles available for each reactant. Next, run the reaction mathematically with each of the reactants and see which gives the smallest number of moles of product. This will determine the limiting reactant and the theoretical yield and make it possible to calculate the percent yield and the amount of excess reactant left over. Next, set up a compare mole ratio of the actual amounts to the mole ratios required according to the balanced chemical equation. You can then determine which reactant is in excess and which reactant is present in a limiting amount.

Step 1: Using the information given, write an unbalanced chemical equation.

Step 2: Balance the chemical equation.

Step 3: Determine the number of moles of each reactant present.

Step 4: Determine which reactant gives the smallest number of moles of product. This gives the limiting reactant and the theoretical yield.

Step 5: Determine how many moles of the excess reactant were used.

Step 6: Determine how many moles of the excess reactant remain.

Step 7: Determine how many grams of the excess reactant remain.

Step 8: Determine the percent yield of the reaction.

3.7 Concentrations of Reactants in Solution: Molarity

Many chemical reactions are carried out in solution, as this provides far more opportunity for particles of reactants to collide with one another. If reactants are dissolved, it is essential that their concentration be known.

Chemical quantities in reactions are given in moles, so the most convenient unit of concentration for the study of chemistry should do the same: **molarity** is defined as the moles of **solute** per liter of solution. When written in equation form, it can also be rearranged as needed to be used as a conversion factor:

$$\text{Molarity (M)} = \frac{\text{moles solute}}{\text{L solution}}$$

$$\text{moles solute} = \text{Molarity x L solution}$$

$$\text{Volume of solution} = \frac{\text{moles solute}}{\text{Molarity}}$$

To make up a solution to a particular molarity, the moles of solute needed must be converted to grams. This amount of solute is then placed in a properly sized volumetric flask, and partially filled with water. Once the solute is dissolved completely, the solution is made up to volume by adding solvent up to a calibration line. The reasons for this will be further examined in the examples.

EXAMPLE

How many grams of sucrose (M.W. 342.3 g/mol) are needed to make 500 mL of 0.5 M solution? How would this solution be made?

SOLUTION: Molarity uses moles, not grams, so it is necessary to first figure out how many moles of sucrose are needed, then solve for grams:

$$\text{moles} = 0.5\ \text{M x } 0.500\ \text{L} = 0.25\ \text{mol}$$

Notice that the volume must be converted to liters. The reason for this is more obvious if molarity is expanded to moles/liter:

$$\text{moles} = 0.5\ \frac{\text{mol}}{\text{L}}\ \text{x } 0.500\ \text{L} = 0.25\ \text{mol}$$

Always be *very careful* with units. They matter!

Now convert to grams:

$$0.25\ \text{mol sucrose x } \frac{342.3\ \text{grams sucrose}}{\text{mol sucrose}} = 85.6\ \text{grams sucrose}$$

To make up this solution, put 85.6 grams of sucrose in a 500 mL volumetric flask, and dissolve with water, then fill to 500 mL.

Note: 85.6 grams of sucrose is about a half-cup of sugar, and 500 mL is about 2 cups of water. If the solution were to be made by dropping the sugar into 500 mL of water, the volume of the final solution would be far more than 500 mL, and the concentration would be uncertain. This is why volumetric glassware is used to make these solutions.

EXAMPLE:

How many milliliters of 0.4 M salt solution contain 0.112 moles of salt?

SOLUTION: The identity of the chemical compound here generically referred to as "salt" is not given, nor is it needed. Don't let that throw you off!

$$\text{volume (L)} = \frac{\text{moles}}{\text{Molarity}} = \frac{0.112 \text{ mol}}{0.4 \text{ M}} = 0.28 \text{ L}$$

$$0.28 \text{ L x } \frac{1000 \text{ mL}}{\text{L}} = 280 \text{ mL}$$

Make sure you report the answer in the units asked for: here, as above, that required a conversion.

EXAMPLE:

What is the molarity of a solution of barium chloride made by dissolving 15 grams of the salt into 250 mL of solution?

SOLUTION: First, determine the number of moles of barium chloride in 15 g:

$$15 \text{ g } BaCl_2 \text{ x } \frac{1 \text{ mol } BaCl_2}{208.3 \text{ g } BaCl_2} = 0.072 \text{ mol } BaCl_2$$

The molarity of the solution can now be calculated:

$$\text{Molarity} = \frac{0.072 \text{ mol } BaCl_2}{0.250 \text{ L}} = 0.29 \text{ M}$$

Workbook Problem 3.6

How many grams of sodium chloride are needed to make 500 mL of a 0.30 M sodium chloride solution?

Strategy: Using the definition of molarity and the volume of solution, determine moles and grams of solute.

Step 1: First, determine the number of moles found in 500 mL of a 0.30 M solution.

Step 2: Determine the number of grams of sodium chloride from the number of moles.

3.8 Diluting Concentrated Solutions

Some chemicals are sold as concentrated solutions (often called *stock solutions*) that are meant to be diluted before use. Fortunately, it is very simple to figure out how to dilute a solution. Remember, when a solution is diluted, the amount of *solute* does not change, only the amount of *solvent* does. This leads to a simple equation:

$$M_i \times V_i = M_f \times V_f$$

In words, the initial molarity times the initial volume equals the final molarity times the final volume.

A word of caution: when performing dilutions in the lab, you need to be cautious to always add strong acids to water, and not the other way around. When acids are diluted, they generate large quantities of heat: if water is added to acid, the heat can cause the water to boil and splatter the acid. Be careful!

EXAMPLE:

A lab protocol asks for 100 mL of 5 M HCl. If the stock solution is 12M, how much is needed for the dilution, and how would you perform it?

SOLUTION: Use the above equation, rearranging to solve for the initial volume:

$$M_i \times V_i = M_f \times V_f$$

$$V_i = \frac{M_f \times V_f}{M_i}$$

$$V_i = \frac{5 \text{ M} \times 100 \text{ mL}}{12 \text{ M}} = 42 \text{ mL}$$

Note: no conversion was done of the volumes here. When this equation is used, molarity cancels out, leaving behind whatever volume unit is being used. If this makes you uncomfortable, by all means convert everything to liters, but it isn't strictly necessary.

To make this solution, 42 mL of acid would be carefully measured, then poured slowly into a volumetric flask containing about 40 mL of water. Once the acid was diluted by being added to water, the solution could be made up to volume.

Workbook Problem 3.7

What is the final concentration when 50 mL of 10 M magnesium chloride is diluted to 1 L?

Strategy: Remember that the number of moles of solute present do not change when a solution is diluted.

Step 1: Rearrange the equation $M_i \times V_i = M_f \times V_f$ to solve for the final concentration.

3.9 Solution Stoichiometry

Molarity can be used as a conversion factor in stoichiometry calculations to determine volumes of solutions needed in reactions.

EXAMPLE:
Silver nitrate and calcium chloride, both soluble, react to form silver(I) chloride, an insoluble solid. How much 0.10 M calcium chloride is needed to completely react with 100 mL of 0.5 M silver nitrate?

SOLUTION: First, write the balanced equation for the reaction.

$$CaCl_2 + 2\ AgNO_3 \rightarrow 2\ AgCl + Ca(NO_3)_2$$

Next, calculate the moles of silver nitrate present:

$$0.100\ L \times 0.50\ \frac{mol}{L} = 0.050\ mol\ silver\ nitrate$$

Using the balanced equation, calculate the number of moles of $CaCl_2$ needed to react with 0.050 mol silver nitrate:

$$0.050\ mol\ AgNO_3 \times \frac{1\ mol\ CaCl_2}{2\ mol\ AgNO_3} = 0.025\ mol\ CaCl_2$$

The volume of the $CaCl_2$ solution can now be calculated:

$$\frac{0.025\ mol\ CaCl_2}{0.10\ \frac{mol\ CaCl_2}{L}} = 0.250\ L = 250\ mL\ CaCl_2$$

Workbook Problem 3.8
Hydrochloric acid reacts with zinc metal to produce a solution of zinc chloride and hydrogen gas. What volume of 5.0 M hydrochloric acid is needed to react with 4.75 g of zinc? If the density of hydrogen gas is 0.0899 g/L, what volume of hydrogen gas will be produced?

Strategy: To solve any stoichiometry problem, we need to know mole ratios of reactants and products as determined by the balanced chemical equation. Knowing the actual number of moles present of one reactant or product, we can then determine the number of moles needed of all other reactants and products.

Step 1: Write a balanced chemical equation.

Step 2: Determine the number of moles of zinc present.

Step 3: Determine the number of moles of HCl needed.

Step 4: Calculate the volume of 5.0 M HCl needed for this reaction.

Step 5: To calculate the volume of H_2 gas produced, we first need to calculate the moles, then grams of H_2 gas produced.

Step 6: Determine the volume of H_2 gas produced, using the grams of H_2 produced and the density of H_2 gas.

3.10 Titration

To determine the concentration of an unknown solution, a procedure called **titration** can be used. A carefully measured volume of the unknown solution is slowly reacted with a solution of known concentration. An **indicator** – a molecule that changes color when the reaction is complete – is generally used to indicate the endpoint of the reaction.

EXAMPLE:

Apple juice contains malic acid, a weak acid with formula $C_4H_6O_5$ that contains two acid groups. If 27 mL of 0.10 M sodium hydroxide is needed to neutralize 15 mL of a malic acid solution, what is the concentration of the solution?

SOLUTION: First, write a balanced equation for the neutralization reaction.

$$C_4H_6O_5\,(aq) + 2\,NaOH\,(aq) \rightarrow Na_2C_4H_4O_5\,(aq) + 2\,H_2O\,(l)$$

From the information given, we can calculate the number of moles of sodium hydroxide used to reach the equivalence point of the titration.

$$0.027 \text{ L NaOH x } 0.10\frac{\text{mol}}{\text{L}} = 0.0027 \text{ mol NaOH}$$

From the balanced equation, calculate the moles and molarity of malic acid.

$$0.0027\text{mol NaOH x } \frac{1 \text{ mol malic acid}}{2 \text{ mol NaOH}} = 0.0014 \text{ mol malic acid}$$

$$\text{Molarity} = \frac{0.0014 \text{ mol malic acid}}{0.015 \text{ L}} = 0.093 \text{ M malic acid}$$

Workbook Problem 3.9

A student determines that 16.37 mL of sulfuric acid is required to neutralize a 100.0 mL solution containing 0.153 g of NaOH. Determine the molarity of the nitric acid solution.

Strategy: Balance equation to find mole ratios, determine moles of known compounds, then calculate moles of unknown and determine concentration.

Step 1: Write a balanced chemical equation for the reaction.

Step 2: Calculate the number of moles of NaOH present.

Step 3: Determine the number of moles of sulfuric acid needed to react with that quantity of NaOH.

Step 4: Calculate the molarity of the acid solution.

3.11 Percent Composition and Empirical Formulas

Determining the composition of a new compound requires that the elements in the compound and their relative abundances be known.

This is expressed as **percent composition**: how much of each element is present in a compound by mass. This can also be calculated for a known compound if desired.

Percent composition can be converted into mole ratios by figuring out how many moles of each element would be present in 100 grams of the unknown compound. This mole ratio can then be expressed as an **empirical formula**: the simplest ratio of the elements to one another.

If the molecular mass is known, the empirical formula can be converted into a **molecular formula**: the actual numbers of each element in a molecule of the compound. Sometimes, the two are the same, but other times the molecular formula will be a whole-number multiple of the empirical formula.

EXAMPLE:

A simple hydrocarbon is found to be 75% C by mass and 25% H by mass. What is the empirical formula of the compound? If its molecular mass is determined to be 16 amu, what is its molecular formula?

SOLUTION: Assume that 100 grams of the compound are present. 75% of 100 g is 75 grams of carbon; 25% of 100 grams is 25 g hydrogen. Convert to moles:

$$75\text{ g C} \times \frac{1\text{ mol C}}{12\text{ g C}} = 6.25\text{ mol C}$$

$$25\text{ g H} \times \frac{1\text{ mol H}}{1\text{ g H}} = 25\text{ mol H}$$

This makes it possible to find the ratio of one element to the other. To determine the empirical formula, divide to find the smallest ratio. Be sure to keep element names or symbols with the numbers to avoid confusion:

$$\frac{25 \text{ mol H}}{6.25 \text{ mol C}} = 4 \frac{\text{mol H}}{\text{mol C}}$$

This tells us that there are four hydrogens for every one carbon, which gives an empirical formula of CH_4.

If the unknown hydrocarbon has a molecular weight of 16 amu, we can also determine the molecular formula of the compound. The empirical formula has a molecular weight of 16, so in this case, the empirical formula is the same as the molecular formula.

EXAMPLE:

A compound is found to be 36.8% N and 63.2% O. What is the empirical formula of the compound?

SOLUTION: Assume that 100 grams of the compound are present. 36.8% of 100 g is 36.8 grams of nitrogen; 63.2% of 100 grams is 63.2 g oxygen. Convert to moles:

$$63.2 \text{ g O x } \frac{1 \text{ mol O}}{16 \text{ g O}} = 3.95 \text{ mol O}$$

$$36.8 \text{ g N x } \frac{1 \text{ mol N}}{14 \text{ g N}} = 2.63 \text{ mol N}$$

There is no obvious whole number ratio here, but determine the ratio:

$$\frac{3.95 \text{ mol O}}{2.63 \text{ mol N}} = 1.5 \frac{\text{O}}{\text{N}}$$

A formula can be written with this, but it will need a little extra work:

$NO_{1.5}$ is not a proper formula, as partial atoms are not allowed. Multiply through to get rid of the fraction (this sometimes takes a little trial and error) and make sure that the formula still is the smallest allowable:

N_2O_3 is an allowable formula that maintains the ratio of 1.5 oxygens for each nitrogen.

EXAMPLE:

A slightly more complex compound is determined to be 27% Na, 16.5% N, and 56.5% O. What is the empirical formula of the compound?

SOLUTION: Assume that 100 grams of the compound are present. This will be 27 g Na, 16.5 g N, and 56.5 g O. Convert to moles:

$$27 \text{ g Na x } \frac{1 \text{ mol Na}}{23 \text{ g Na}} = 1.174 \text{ mol Na}$$

$$16.5 \text{ g N x} \frac{1 \text{ mol N}}{14 \text{ g N}} = 1.179 \text{ mol N}$$

$$56.5 \text{ g O x } \frac{1 \text{ mol O}}{16 \text{ g O}} = 3.538 \text{ mol O}$$

There are essentially equal numbers of moles of Na and N – there will be the same number of each of these. What remains is to determine the ratio of oxygens to the other elements:

$$\frac{3.538 \text{ mol O}}{1.179 \text{ mol N}} = 3\frac{\text{mol O}}{\text{mol N}}$$

So there are 3 oxygens for every 1 nitrogen, and 1 sodium for each nitrogen: $NaNO_3$. This is sodium nitrate. The most important thing with these problems is to not confuse the elements relative to one another.

Workbook Problem 3.10

Determine the empirical formula of a compound that is 18.9% Li, 16.2% C, and 64.9% O.

Strategy: Assume you have a 100 g sample, and convert the percentages of the elements into grams.

Step 1: Convert the grams of each element into moles of each element.

Step 2: Find the mole ratios.

Step 3: If necessary, convert the ratios to whole numbers, and write the empirical formula.

EXAMPLE:

What is the percent composition of sulfuric acid?

SOLUTION: The formula for sulfuric acid is H_2SO_4. Convert this into a mass percent by determining the mass of each element and the total molecular mass:

H × 2 × 1.008	=	2.0016
S × 1 × 32.06	=	32.06
O × 4 × 16.00	=	64.00
Molecular weight	=	98.06

Mass % of each element:

$$\%H = \frac{2.0016}{98.06} \text{ x } 100 = 2.04\%$$

$$\%S = \frac{32.06}{98.06} \text{ x } 100 = 32.7\%$$

$$\%O = \frac{64.00}{98.06} \text{ x } 100 = 65.3\%$$

Workbook Problem 3.11

Determine the percent composition of sodium hypochlorite.

Strategy: Write the formula for sodium hypochlorite and determine the mole ratio of the elements in the compound.

Step 1: Determine the formula of the compound.

Step 2: Determine the molecular weight and the weights of each component.

Step 3: Determine the percent composition of each component.

3.12 Determining Empirical Formulas: Elemental Analysis

As has been mentioned before, when a hydrocarbon burns, it produces water and carbon dioxide. This useful fact can be used to determine the ratios of the elements to one another: the moles of carbon dioxide and water produced by such a reaction can be used to determine the moles of hydrogen and carbon in the original compound, and thus the empirical formula of the compound.

Determine moles of CO_2 – this will be the same as the number of moles of carbon.

Determine moles of H_2O – this will be one-half the number of moles of hydrogen.

If there is another element present, it can be determined by subtracting the masses of carbon and hydrogen (go from moles to grams, to determine this) from the total mass of the unknown compound.

If the molecular mass is also known, it is possible to determine not only the empirical formula, but the molecular formula as well.

EXAMPLE:

When a 7.03 gram sample of a hydrocarbon is burned, 21.32 g of carbon dioxide and 10.82 g of water are produced. If the compound contains only carbon and hydrogen, and its molecular weight is 58, what is the molecular formula of the compound?

SOLUTION: Convert the grams of carbon dioxide and water into moles of carbon and hydrogen, then use the ratio to generate an empirical formula. From the molecular weight, develop the molecular formula from the empirical formula if necessary.

$$21.32 \text{ g } CO_2 \text{ x } \frac{1 \text{ mol } CO_2}{44.01 \text{ g } CO_2} = 0.484 \text{ mol } CO_2 \text{ x } \frac{1 \text{ mol C}}{\text{mol } CO_2} = 0.484 \text{ mol C}$$

$$10.82 \text{ g } H_2O \text{ x } \frac{1 \text{ mol } H_2O}{18.00 \text{ g } H_2O} = 0.601 \text{ mol } H_2O \text{ x } \frac{2 \text{ mol H}}{\text{mol } H_2O} = 1.202 \text{ mol H}$$

Calculate the mole ratio of hydrogen to carbon:

$$\frac{1.202 \text{ mol H}}{0.484 \text{ mol C}} = 2.5 \frac{\text{H}}{\text{C}}$$

The initial formula is $CH_{2.5}$ which becomes a proper empirical formula C_2H_5.

The molecular weight of the empirical formula is 29 amu, so the molecular formula will be a multiple of this formula. 29×2 is the proper molecular weight of 58, so there will be two empirical formula units in the molecule, and the molecular formula will be C_4H_{10}.

EXAMPLE:

A compound known as methyl butanoate is one of the compounds responsible for the distinctive smell of pineapple. It contains carbon, hydrogen, and oxygen. When combusted, a 3.73 g sample of this compound produced 8.03 g carbon dioxide and 3.27 g water. If the molecular weight of the compound is 102 amu, what is the molecular formula?

SOLUTION: Convert the grams of carbon dioxide and water into moles of carbon and hydrogen. Convert these back into grams to determine the weight of oxygen in the compound, then determine the number of moles of oxygen. Use these three quantities to develop mole ratios and an empirical formula. Compare the empirical formula weight and the molecular formula weight to determine the molecular formula.

$$8.03 \text{ g } CO_2 \text{ x } \frac{1 \text{ mol } CO_2}{44.01 \text{ g } CO_2} = 0.182 \text{ mol } CO_2 \text{ x } \frac{1 \text{ mol C}}{\text{mol } CO_2} = 0.182 \text{ mol C x } \frac{12.01 \text{ g C}}{\text{mol C}} = 2.19 \text{ g C}$$

$$3.27 \text{ g } H_2O \text{ x } \frac{1 \text{ mol } H_2O}{18.00 \text{ g } H_2O} = 0.182 \text{ mol } H_2O \text{ x } \frac{2 \text{ mol H}}{\text{mol } H_2O} = 0.363 \text{ mol H x } \frac{1.008 \text{ g H}}{\text{mol H}} = 0.366 \text{ g H}$$

Now determine the amount of O by subtracting the masses of carbon and hydrogen from the mass of the starting sample:

$$3.73 \text{ g} - 2.19 \text{ g} - 0.37 \text{ g} = 1.17 \text{ g oxygen}$$

Convert the mass of O to moles of O.

$$1.17 \text{ g O} \times \frac{1 \text{ mol O}}{16.00 \text{ g O}} = 0.0731 \text{ mol O}$$

Mole ratios can now be determined: 0.0731 moles of oxygen, 0.363 moles H, and 0.182 moles C.

$$\frac{0.363 \text{ mol H}}{0.0731 \text{ mol O}} = 5\frac{\text{H}}{\text{O}}$$

$$\frac{0.182 \text{ mol C}}{0.0731 \text{ mol O}} = 2.5\frac{\text{C}}{\text{O}}$$

This gives a preliminary formula of $C_{2.5}H_5O$ and a real empirical formula of $C_5H_{10}O_2$. The empirical formula has a molecular weight of 102, so this is also the molecular formula.

Workbook Problem 3.12

2-butene-1-thiol is one of the compounds that gives skunk spray such a memorable odor. It contains only carbon, hydrogen, and sulfur. When 2.76 grams of this compound were burned (over a weekend, to make sure that all the custodial staff didn't quit), 5.70 g of carbon dioxide and 2.31 g of water were produced. If the compound has a molecular mass of 88, determine the empirical and molecular formulas of this compound.

Strategy: Convert the grams of carbon dioxide and water into moles of carbon and hydrogen. Convert these back into grams to determine the weight of oxygen in the compound, then determine the number of moles of oxygen. Use these three quantities to develop mole ratios and an empirical formula. Compare the empirical formula weight and the molecular formula weight to determine the molecular formula.

Step 1: Find the molar amounts of C and H in CO_2 and H_2O.

Step 2: Carry out mole-to-gram conversions to find the number of grams of C and H in the original sample.

Step 3: Subtract the masses of C and H from the mass of the starting sample to determine the mass of S.

Step 4: Convert the mass of S to moles of S.

Step 5: Find the mole ratios.

Step 6: Use the ratios above to write the empirical formula of the compound.

Step 7: Compare the empirical formula weight to the molecular formula weight to see if a conversion is needed.

Step 8: Write the molecular formula.

Self-Test

This section is intended to test your knowledge of the material covered in this chapter. Think through these problems, and make certain you understand what is going on. Ask yourself if your answer makes sense. Many of these questions are linked to the chapter learning goals. Therefore, successful completion of these problems indicates you have mastered the learning goals for this chapter. You will receive the greatest benefit from this section if you use it as a mock exam. You will then discover which topics you have mastered and which topics you need to study in more detail.

True-False

1. When balancing equations, it is essential that there be the same number of molecules on each side.
2. Chemical equations describe behavior on both the atomic (microscopic) level and the macroscopic level.
3. One mole of two different substances will contain the same number of particles.
4. The coefficients in a balanced chemical equation show the mass ratios of the reactants and products.
5. Theoretical yields of chemical reactions are determined in the laboratory.
6. When reactants are not present in stoichiometric ratios, the limiting reactant will determine the maximum yield of the reaction.
7. Molarity is defined as the number of moles of solute per liter of solvent.
8. Combustion analysis can be used to determine the molecular formula of an unknown compound.
9. The empirical formula and the molecular formula for a compound are sometimes the same.
10. When diluting a concentrated solution, the moles of solute do not change, only the amount of solvent.

Matching

Balanced

a. sum of all the masses of the atoms in a molecule

Yield	b. the smallest ratio of elements in a compound
Mole	c. substance dissolved in solvent to make a solution
Coefficient	d. the amount of product formed in a reaction
Stoichiometry	e. the number in front of a reactant or product that shows its mole ratio with the other components
Dilution	f. procedure for determining the concentration of an unknown solution
Molecular mass	g. reaction with oxygen
Empirical formula	h. indication of number of each type of atom in a molecule
Solute	i. having the same number of all atom types in both products and reactants
Titration	j. experimentally determined amount of product in a reaction.
Subscript	k. going from a more concentrated solution to a less concentrated solution.
Combustion	l. the amount of a substance equal to the molecular or atomic mass in grams
Actual yield	m. the use of mole ratios and molecular weights to carry out chemical calculations

Fill-in-the-Blank

1. To be useful in making calculations, a chemical equation must be ________________.

2. The mole ratios of the components in a reaction are indicated by ______________________ in front of the components. These can be changed if necessary, but the _________________ within the components cannot be changed, as this will change the identity of the component.

3. To convert amounts of substances from grams to moles, the conversion factor is ___________ _________.

4. If reactants are not present in stoichiometric ratios, the amount of product will be determined by the ________________ ________________.

5. The reactant that does not determine the yield of the reaction is known as the ___________ _______________.

6. The ______________________ __________ of a reaction can be calculated using a balanced equation, but it is rarely, if ever, achieved. In the lab, the ____________ __________ will be

obtained. The two together can be used to calculate the ______________ __________ of the reaction.

7. It is possible to determine the ________________ _________________ of a compound from the mass percentages of the elements in it, but to determine the ________________ ______________ of the compound, the _________________ ________ must be known.

8. When making a solution, the moles of _____________ per liter of _______________ is known as the _______________ of the solution.

9. When a hydrocarbon is burned in oxygen, the primary products are ____________ ______________ and __________.

10. The number of particles in one mole of a substance is known as __________________ ____________. It is equal to _______________.

Problems

1. Balance the following equations:

   a. $N_2 + O_2 \rightarrow N_2O_5$

   b. $C_3H_8 + O_2 \rightarrow CO_2 + H_2O$

   c. $H_3PO_4 + NaOH \rightarrow Na_3PO_4 + H_2O$

   d. $Ba(NO_3)_2 + NaOH \rightarrow Ba(OH)_2 + NaNO_3$

2. Determine the molecular or formula mass of the following compounds and convert the given amounts to moles:

   a. 45 g calcium chloride

   b. 23 g sucrose ($C_{12}H_{22}O_{11}$)

   c. 67 g phosphorus trifluoride

   d. 5 grams ammonium carbonate

3. Using the balanced equation above for the production of dinitrogen pentoxide, if four moles of nitrogen and 10 moles of oxygen are combined, what is the theoretical yield of the reaction in moles and grams? If 376 grams of product are obtained, what is the percent yield?

4. When sodium hydrogen carbonate is heated, it decomposes according to the following unbalanced equation:

$$NaHCO_3 \rightarrow Na_2CO_3 + CO_2 + H_2O$$

If 20 grams of sodium hydrogen carbonate are heated, what is the theoretical yield of water in grams? If the reaction has a percent yield of 87%, what is the maximum quantity of water that can be obtained?

5. When 6.7 grams of methane (CH_4) are burned in 20 g of oxygen, what is the limiting reactant? What is the theoretical yield of water and carbon dioxide? How many grams of the excess reactant will remain when the reaction is complete?

6. How many grams of potassium iodide are needed to make 250 mL of 0.35 M solution?

   How many mL of that solution will provide 2.60 g of potassium iodide?

   To make 100 mL of 0.10 M solution, how many mL of 0.35 M stock solution are needed?

7. Elemental iron and aqueous copper(II) sulfate will react to generate aqueous iron(II) sulfate and elemental copper. If 1.37 g iron are added to 100 mL of 0.50 M copper(II) sulfate, how many grams of copper are possible?

8. To convert 1.63 g of zinc completely to zinc chloride, how many mL of 0.50 M HCl are needed? Use the following unbalanced reaction as a starting point:

$$Zn + 2\,HCl \rightarrow ZnCl_2 + H_2$$

9. Aqueous barium nitrate and sodium hydroxide react according to the equation balanced in problem 1. If 50 mL of 0.250 M barium nitrate and 50 mL of 0.10 M sodium hydroxide are mixed, what is the theoretical yield of barium hydroxide in grams?

10. A solution of acetic acid is titrated with 0.100 M sodium hydroxide. If 17.3 mL of the sodium hydroxide are required to titrate 50 mL of the citric acid solution, and the acid and base react in a 1:1 ratio, what is the molar concentration of the acetic acid solution?

11. Determine the % composition of the following compounds:

    Carbon dioxide
    Sodium phosphate
    CH_3COOH
    Adenosine triphosphate: $C_{10}H_{16}N_5O_{13}P_3$

12. A compound containing 40.0% carbon, 6.7% hydrogen, and 53.3% oxygen has a molecular mass of 180 amu. What is the molecular formula of this compound?

13. Pentyl acetate is a compound responsible for the distinctive smell of bananas. It contains only carbon, hydrogen, and oxygen. When 1.57 grams of this compound were subjected to combustion analysis, 3.71 g of CO_2 and 1.51 g of H_2O were measured. What is the empirical formula of this compound?

Challenge Problem

When sodium or lithium are added to water, the following rather exciting reaction occurs according to the following unbalanced equation:

$$M\,(s) + H_2O\,(l) \rightarrow MOH\,(aq) + H_2\,(g)$$

Enough heat is typically generated by this reaction that the hydrogen gas burns as it is generated. If a 4 gram mixture of lithium and sodium is added to water, and the resulting solution titrated, 59.7 mL of 7.5 M HCl is required to neutralize the solution. How many grams of lithium and sodium were contained in the original mixture?

Inquiry Based Problem

You have been hired to help out in a laboratory, and your first assignment is to produce an inventory of all the chemicals in a storage area. Behind all the acids, you find a bottle labeled only "5.0 M acid – careful!" The ordering records show that only nitric, hydrochloric, phosphoric, and sulfuric acids have ever been ordered, but none in 5.0 M concentration.

To figure out which of these acids this might be, you titrate 5.0 mL of the acid with 1.0 M NaOH. What result will you expect from each of these acids? If about 50 mL of the base were needed, could you make a tentative identification of the acid?

Write and balance equations for each of the reactions.

CHAPTER 4

REACTIONS IN AQUEOUS SOLUTION

Chapter Learning Goals

A. Electrolytes and Nonelectrolytes
1. Classify substances as electrolytes or nonelectrolytes.

B. Chemical Equations
1. Write molecular, ionic, and net ionic equations for precipitation, acid–base, and redox reactions.

C. Chemical Reactions in Aqueous Solution: Precipitation and Acid–Base Neutralization
1. State solubility rules, and use them to predict whether a precipitate might form when aqueous salt solutions are mixed.
2. Identify the common strong acids and strong bases.

D. Chemical Reactions in Aqueous Solution: Oxidation–Reduction
1. Assign oxidation numbers to each atom in a chemical species.
2. In a redox reaction, identify the species oxidized, the species reduced, the oxidizing agent, and the reducing agent.
3. Using an activity series, predict whether a redox reaction will occur when a metal is placed in contact with a solution containing an ion of a different metal.
4. Balance redox reactions by the oxidation-number method and by the half-reaction method.
5. Determine the concentration of a species, using data from a redox titration.

Chapter in Brief

This chapter discusses three different types of chemical reactions in aqueous solution: precipitation reactions, acid–base reactions, and oxidation–reduction reactions. These reactions are often written as net ionic equations. As a consequence, it is necessary to know the difference between strong and weak electrolytes and nonelectrolytes. You can predict the outcome of each of these reactions if you know the solubility rules, can recognize acids and bases, and know how to assign oxidation numbers to compounds. You are also shown how to balance oxidation–reduction reactions with either the oxidation–number method or the method of half-reactions. Finally, you will apply the concepts you learned in earlier chapters about stoichiometry to the reactions that are discussed in this chapter.

4.1 Some Ways that Chemical Reactions Occur

In **precipitation reactions**, two soluble components interact to form an insoluble product, which then falls out of solution.

In **acid-base neutralization reactions**, an acid and a base react to form water and a salt (salt is simply a generic term that refers to an ionic compound). In these reactions, both the acid and basic components of the mixture (H^+ and OH^-) are removed from the reaction, combining to form neutral water.

In **oxidation-reduction reactions**, also called **redox** reactions, electrons are transferred between components in a reaction mixture. To find these, look particularly for uncharged elements on one side of the reaction becoming part of ionic compounds and *vice versa*.

4.2 Electrolytes in Aqueous Solution

Substances that dissolve in water to form ions are known as **electrolytes**: solutions containing these compounds will conduct electricity. They are soluble ionic compounds such as sodium chloride (table salt). Acids and bases are electrolytes, as they dissolve to form ions, in particular, the ions that distinguish them as being acids (H^+) or bases (OH^-).

Nonelectrolytes may dissolve in water, but they will not conduct electricity when dissolved, as they do not dissociate into ions. Covalent molecules such as sugars do not dissociate when they dissolve.

Electrolytes can be either **strong** or **weak**: this has to do with how much they dissociate, and nothing else: a strong electrolyte dissolves almost completely, and a weak electrolyte very little. By extension, a strong acid dissociates completely, and a weak acid only partially.

Dissociation is a dynamic process: if a weak acid is dissolved in water, there will be both dissociated and undissociated molecules in the water, and it will conduct electricity to a certain extent. Molecules that are whole will dissociate, and dissociated molecules will come back together. This is shown with a double arrow:

$$CH_3COOH \leftrightarrow CH_3COO^- + H^+$$

This equation is for the dissociation of a weak acid. The molecules are constantly reforming and dissociating again. When looking for weak electrolytes, organic acids have formulas that contain $-CO_2H$ or COO^-, and all are weak electrolytes. Alcohols, on the other hand, have formulas that contain –OH groups: these are nonelectrolytes.

EXAMPLE:

$Mg(OH)_2$ is a strong base. From this definition, what is the concentration of OH^- ions in a 0.450 M solution of $Mg(OH)_2$?

SOLUTION:

A strong electrolyte dissociates completely. So the $Mg(OH)_2$ will not be found as units, but as dissociated ions:

$$Mg(OH)_2 \rightarrow Mg^{2+} + 2OH^-$$

This makes it possible to calculate the concentration of OH^- ions directly:

$$0.450\text{M } Mg(OH)_2 \text{ x } \frac{2\ OH^-}{Mg(OH)_2} = 0.900\text{ M } OH^-$$

Workbook Problem 4.1

Determine the total molar concentration of ions in a 0.75 M solution of aluminum chloride.

Strategy: $AlCl_3$ is a strong electrolyte and completely dissociates in water.

Step 1: Determine the total number of moles of ions formed when $AlCl_3$ completely dissociates in water.

Step 2: Create a conversion factor comparing the total number of moles of ions in solution to 1 mol of $AlCl_3$.

Step 3: Use the conversion factor to calculate the molar concentration of ions in solution.

4.3 Aqueous Reactions and Net Ionic Equations

There are a number of ways of writing equations for reactions that occur in aqueous solutions.

Molecular equations are written as though the components are not dissociated: ions are kept together.

Ionic equations are written to indicate the dissociation of components that dissociate: rather than show strong electrolytes as though they are combined, they are shown as free ions.

Net ionic equations are written so that only the ions that participate in the reaction are shown. Generally, ionic reactions contain ions that do not participate, but are unchanged as the reaction proceeds. These are known as **spectator ions**: they are removed when a net ionic equation is written.

EXAMPLE:

When solutions of lead(II) nitrate and sodium chloride are mixed, solid lead chloride and aqueous sodium nitrate are produced. Write molecular, ionic, and net ionic equations for this reaction.

SOLUTION:

Molecular Eqn: $Pb(NO_3)_2\ (aq) + NaCl\ (aq) \rightarrow PbCl_2\ (s) + NaNO_3\ (aq)$

Balanced Molecular Eqn: $Pb(NO_3)_2\ (aq) + 2\ NaCl\ (aq) \rightarrow PbCl_2\ (s) + 2\ NaNO_3\ (aq)$

Ionic Eqn: $Pb^{2+}\ (aq) + 2\ NO_3^-\ (aq) + 2\ Na^+\ (aq) + 2\ Cl^-\ (aq) \rightarrow PbCl_2\ (s) + 2\ Na^+\ (aq) + 2Cl^-\ (aq)$

Spectator Ions: Na^+ and NO_3^-

Net Ionic Equation: $Pb^{2+}\ (aq) + 2\ Cl^-\ (aq) \rightarrow PbCl_2\ (s)$

Workbook Problem 4.2

When solutions of nitric acid and potassium hydroxide are mixed, a neutralization reaction occurs. Write the net ionic equation for this reaction.

Strategy: From the information given, write a molecular equation and determine if any of the reactants or products are strong electrolytes.

Step 1: Write and balance the molecular equation.

Step 2: Write the strong electrolytes as free ions for the ionic equation.

Step 3: Eliminate any spectator ions to write the net ionic equation.

4.4 Precipitation Reactions and Solubility Guidelines

Predicting the results of a precipitation reaction requires an understanding of **solubility**: ionic compounds that are highly soluble will not precipitate, while those with low solubility will fall out of solution.

Solubility guidelines are given that name as they are only partially predictive, but these guidelines will get you through most problems having to do with precipitation reactions:

- Anything with an alkali metal (group I) cation will be soluble.
- Anything with an ammonium cation (NH_4^+) will be soluble.
- Anything with a nitrate anion (NO_3^-), perchlorate ion (ClO_4^-) or an acetate ion ($CH_3CO_2^-$) will be soluble.
- Halides (Cl^-, Br^-, and I^-) and sulfates (SO_4^{2-}) will be soluble with exceptions.
- With other anions, "heavy" metals will fall out of solution, "light" metals will stay in: be careful with lead, silver, and mercury cations. Unless they are attached to a nitrate, they are unlikely to be soluble.

EXAMPLE:

For which of the following solutions would precipitations occur? Na_2CO_3 and $Pb(NO_3)_2$, NH_4Cl and $NaCH_3CO_2$, $AgNO_3$ and $MgCl_2$.

SOLUTION: Exchange the cations and anions and examine the possible products.

Na_2CO_3 and $Pb(NO_3)_2$ can yield $NaNO_3$ and $PbCO_3$.

Anything attached to lead is a suspect for precipitation, and $NaNO_3$ contains two ions identified as always soluble. $PbCO_3$ will precipitate.

NH_4Cl and $NaCH_3CO_2$ can yield $NH_4CH_3CO_2$ and NaCl

Anything attached to an ammonium ion will be soluble, as will anything attached to a sodium ion: there is not a cation here that can generate a precipitate.

$AgNO_3$ and $MgCl_2$ can yield AgCl and $MgNO_3$.

Although chlorides are usually soluble, silver is always suspicious: silver chloride will precipitate out. Although magnesium is not in any of the lists of "always solubles," it is attached to a nitrate, so it will not precipitate.

Workbook Problem 4.3

Describe how to produce 1.3 grams of lead(II) chloride from 1 M solutions of two soluble salts.

Strategy: Using the solubility guidelines and solution stoichiometry, determine suitable reactants to use in the preparation of lead chloride and the amount of reactants needed.

Step 1: Determine reactants that are soluble and will produce the insoluble lead(II) chloride and another soluble product.

Step 2: Write a balanced chemical equation.

Step 3: Use solution stoichiometry to determine the volume of each reactant needed.

4.5 Acids, Bases, and Neutralization Reactions

The Arrhenius definition of acids and bases, named after Svante Arrhenius, who proposed it in 1887, is the definition that has been being used to this point: acids dissociate in water to form hydrogen ions, and bases dissociate in water to form hydroxide ions. It is possible for acids, sulfuric and phosphoric acids, for example, to dissociate to form more than one hydrogen ion in solution.

H^+, as a bare proton, is an exceptionally strong ion. As a result, it does not truly exist free in solution, but associates itself to a water molecule to form a **hydronium ion** with formula H_3O^+.

It is possible for an acid to release more than one proton into solution: hydrochloric acid, HCl, is called a **monoprotic acid** because it has only one proton to release into solution. Similarly, sulfuric acid, H_2SO_4, is a **diprotic acid**, and phosphoric acid, H_3PO_4, is a **triprotic acid.** Not all protons will be strong: it is more difficult for a proton to be removed from an already-charged unit: H_2SO_4 will dissociate completely to $HSO_4^- + H^+$, but HSO_4^- is a *weak* acid.

Acids and bases can be strong and weak, just as other electrolytes, and the definition means the same thing: it has only to do with the amount of dissociation, and not the corrosiveness of a substance. A concentrated solution of a weak acid will cause a more serious burn than a dilute solution of a strong acid.

Strong acids: perchloric ($HClO_4$), sulfuric (H_2SO_4), hydrochloric (HCl), hydrobromic (HBr), and nitric (HNO_3). These fully dissociate. In the case of sulfuric acid, the first proton dissociates completely, the second proton is *weak*, and dissociates incompletely.

Weak acids: hydrofluoric (HF), phosphoric (H_3PO_4), and acetic (CH_3CO_2H).

Strong bases: hydroxide metal salts are all strong bases.

Weak base: ammonia (NH_3), and any other compound that has the ability to remove hydrogen ions from solution.

EXAMPLE:

Determine if the following are acids, bases, or neutral salts: $Ca(OH)_2$, HCl (*aq*), NaCl, NaOH, $CaBr_2$

SOLUTION: An acid will produce H^+ ions, a base will produce OH^- ions, and a neutral salt will produce neither.

$Ca(OH)_2$ – produces OH^- in solution; base
HCl (*aq*) – produces H^+ in solution; acid
NaCl – produces Na^+ and Cl^- in solution; neutral salt
NaOH – produces OH^- in solution; base
$CaBr_2$ – produces Ca^{2+} and Br^- in solution; neutral salt

Workbook Problem 4.4

Determine the reaction type, and write balanced molecular, ionic, and net ionic equations for the following reactions:

a. LiOH and H_2SO_4
b. NH_4Cl and $AgNO_3$

Strategy: Determine the reaction type by examining the reactants – are they acids or bases or neutral?

Step 1: Write the balanced molecular equation for each reaction.

Step 2: Determine the presence of any strong electrolytes. Write the ionic equation, showing the strong electrolytes in terms of their free ions.

Step 3: Write the net ionic equation by removing spectator ions.

4.6 Oxidation–Reduction (Redox) Reactions

Originally, **oxidation** was the gaining of oxygen and **reduction** was the removal of oxygen. Although these definitions have been significantly broadened, this is still a useful first step when looking for these reactions.

Oxidation is now defined as *losing electrons.* When an element combines with oxygen, to use the historical example, the metal loses electrons to become a cation, and the oxygen gains those same electrons to become an oxide ion:

$$2\ Mg + O_2 \rightarrow 2\ MgO$$

When electrons are lost, the result is a change in charge *toward positive*. This awkward definition is made necessary by elements that will hold more than one charge: a change in charge from -2 to -1 is an oxidation, despite the fact that no positively charged species is generated.

Reduction is now defined as gaining electrons. If the reaction above is reversed, the Mg^{2+} ions gain two electrons to become elemental magnesium.

$$2\ MgO \rightarrow 2\ Mg + O_2$$

In reactions of this kind, the number of electrons lost by one species must be the number gained somewhere else: electrons do not just fly away.

So how is this all managed? There is a method of electron bookkeeping that uses oxidation numbers to show where electrons are going in a reaction. Sometimes, the oxidation number indicates a charge, but not always.

Simplified rules for assigning oxidation numbers (make sure to refer also to the full-bore rules in your textbook):

- Elements have oxidation numbers of 0.
- Ions that contain only one atom have an oxidation number identical to their charge.
- In molecules or ionic compounds, atoms generally carry oxidation numbers equal to their usual ionic state
- Oxygen is usually -2
- Hydrogen is usually +1 (when bonded to a metal, it will be -1: metals are prone to losing electrons)
- Halogens are usually -1
- Neutral compounds will have oxidation numbers on their atoms that sum to 0. Polyatomic ions will have oxidation numbers on their atoms that sum to the charge on the ion.

EXAMPLE:

Determine the oxidation number of each of the atoms in the following: HCl, NaOH, NH_3, H_3PO_4

SOLUTION:

For HCl, halogens are -1, hydrogen is +1. When these are added together, they give 0, correct for a neutral molecule.

H = +1
Cl = -1

For NaOH, Na carries its usual charge as its oxidation number, and is +1. Oxygen is -2, leaving hydrogen. Hydrogen's charge depends on whether it is bonded to a metal or a nonmetal, and here we have both as possibilities. However, this is a neutral compound, so the oxidation numbers must add up to 0, and hydrogen must carry an oxidation state of +1.

Na = +1
O = -2
H = +1

In NH_3, hydrogen is bound to a nonmetal and thus each carries an oxidation number of +1. This is a neutral compound, so N must carry an oxidation number of -3 to balance the three +1 numbers on the hydrogens.

N = -3
H = +1

In H_3PO_4, there are a number of things to consider. This compound is all nonmetals, so the hydrogens will carry oxidation numbers of +1. Oxygen carries a charge of -2, and there are four of them. This leaves phosphorus to balance things out. $(4 \times -2) + (3 \times +1) + P = 0$, so phosphorus carries an oxidation number of +5. This is clearly not a real charge: it is results of this kind that remind us that these are useful bookkeeping methods, but do not necessarily reflect ionic reality.

Workbook Problem 4.5

Determine the oxidation number for each element in VO_4^{3-} and $NaVO_3$.

Strategy: Follow the rules for determining oxidation states found above:

Step 1: Identify the elements that have a fixed oxidation state.

Step 2: Determine the oxidation number of the remaining elements, keeping in mind that the sum of all oxidation numbers must be equal to the ionic charge or to zero for a neutral molecule.

4.7 Identifying Redox Reactions

A **reducing agent** causes reduction to occur. It is oxidized itself by losing one or more electrons, and its oxidation number increases.

Oxidizing agents cause oxidation to occur by accepting electrons. The oxidation number of an oxidizing agent decreases.

When one species in a reaction loses electrons, another gains them. Oxidation and reduction must occur together. When a metal and a nonmetal react with one another, the metal will be the reducing agent, and the nonmetal will be the oxidizing agent (this is easy to remember, as oxygen is a nonmetal), but metals can also oxidize and reduce one another, depending on the strength of their attraction for their electrons.

EXAMPLE:

Identify the species oxidized, the species reduced, the oxidizing agent, and reducing agent in the following reaction:

$$Zn\ (s) + 2\ H^+\ (aq) + 2\ Cl^-\ (aq) \rightarrow Zn^{2+}\ (aq) + 2\ Cl^-\ (aq) + H_2\ (g)$$

SOLUTION: To solve this problem, look for atoms that have had a change in oxidation number:

$Zn \rightarrow Zn^{2+}$ Oxidation number goes from 0 to +2. Zinc is oxidized. Donates electrons to hydrogen: reducing agent.

$H^+ \rightarrow H_2$ Oxidation number goes from +1 to 0. Hydrogen is reduced. Takes electrons from zinc: oxidizing agent.

Workbook Problem 4.6

Identify the species oxidized, the species reduced, the oxidizing agent, and reducing agent in the following reaction:

$$Cu\ (s) + 2\ Ag^{+}\ (aq) + 2\ NO_3^{-}\ (aq) \rightarrow 2\ Ag\ (s) + Cu^{2+}\ (aq) + 2\ NO_3^{-}\ (aq)$$

Strategy: Determine the oxidation number of all species present.

Step1: Identify the species that have a change in oxidation number.

Step 2: Based on the change in oxidation number, identify the species oxidized, the species reduced, and the oxidizing and reducing agent.

4.8 The Activity Series of the Elements

Predicting which metals will take electrons and which will donate electrons when they are together in aqueous solution requires more than a periodic table, unfortunately. The good news is that the work has been done for you, and metals arranged in what is known as an **activity series**.

If a metal is above another metal in the activity series, it will give up electrons to that metal. So if you are looking to see what will happen if you put magnesium ribbon into a copper sulfate solution, you will see that magnesium is above copper, so you will wind up with magnesium ions, and copper metal.

Also in the series is H^+, though not a metal. Metals above H^+ will react with acids to form metal ions and hydrogen gas. Those below it will not. The metals at the bottom of the activity series are the least reactive metals (the least likely to give up electrons) and those at the top the most reactive. Those at the very top are so eager to ionize that they will react with water to form hydrogen gas and hydroxide ions.

Partial Activity Series for Oxidation Reactions

Strongly reducing	$Li \rightarrow Li^{+} + e^{-}$ $K \rightarrow K^{+} + e^{-}$ $Ba \rightarrow Ba^{2+} + 2\ e^{-}$ $Ca \rightarrow Ca^{2+} + 2\ e^{-}$ $Na \rightarrow Na^{+} + e^{-}$	These elements react rapidly with aqueous acid (H^+) or H_2O to release H_2 gas.
	$Mg \rightarrow Mg^{2+} + 2\ e^{-}$ $Al \rightarrow Al^{3+} + 3\ e^{-}$ $Mn \rightarrow Mn^{2+} + 2\ e^{-}$ $Zn \rightarrow Zn^{2+} + 2\ e^{-}$ $Cr \rightarrow Cr^{3+} + 3\ e^{-}$ $Fe \rightarrow Fe^{2+} + 2\ e^{-}$	These elements react with aqueous H^+ ions or with steam to release H_2 gas.
	$Co \rightarrow Co^{2+} + 2\ e^{-}$ $Ni \rightarrow Ni^{2+} + 2\ e^{-}$ $Sn \rightarrow Sn^{2+} + 2\ e^{-}$	These elements react with aqueous H^+ ions to release H_2 gas.
	$\mathbf{H_2 \rightarrow 2\ H^{+} + 2\ e^{-}}$	
	$Cu \rightarrow Cu^{2+} + 2\ e^{-}$ $Ag \rightarrow Ag^{+} + e^{-}$ $Hg \rightarrow Hg^{2+} + 2\ e^{-}$	These elements do not react with aqueous H^+ ions to release H_2 gas.

Weakly reducing	$Pt \rightarrow Pt^{2+} + 2\ e^-$ $Au \rightarrow Au^{3+} + 3\ e^-$

EXAMPLE:

Using the activity series, write balanced chemical equations for the following reactions:

$Cu\ (s) + Co^{2+}\ (aq) \rightarrow$
$Mn\ (s) + Zn^{2+}\ (aq) \rightarrow$
$Sn\ (s) + HCl\ (aq) \rightarrow$
$Pt\ (s) + Hg^{2+}\ (aq) \rightarrow$

$Al\ (s) + Zn^{2+}\ (aq) \rightarrow$
$Ni\ (s) + Mn^{2+}\ (aq) \rightarrow$
$Ni\ (s) + HCl\ (aq) \rightarrow$
$Pt\ (s) + HBr\ (aq) \rightarrow$

SOLUTION: To predict the outcome of these reactions, remember that any element higher in the activity series will react with the ion of any element lower in the activity series. Also remember that metals above the H^+ ion in the activity series will displace the hydrogen ion from an acid to form H_2 gas.

$Cu\ (s) + Co^{2+}\ (aq) \rightarrow$ no reaction; Co^{2+} lies above Cu in the activity series
$Mn\ (s) + Zn^{2+}\ (aq) \rightarrow Mn^{2+}\ (aq) + Zn\ (s)$
$Sn\ (s) + 2\ HCl\ (aq) \rightarrow SnCl_2\ (aq) + H_2\ (g)$
$Pt\ (s) + Hg^{2+}\ (aq) \rightarrow$ no reaction; Hg^{2+} lies above Pt in the activity series

$2\ Al\ (s) + 3\ Zn^{2+}\ (aq) \rightarrow 2\ Al^{3+}\ (aq) + 3\ Zn\ (s)$
$Ni\ (s) + Mn^{2+}\ (aq) \rightarrow$ no reaction; Mn^{2+} lies above Ni in the activity series
$Ni\ (s) + 2\ HCl\ (aq) \rightarrow NiCl_2\ (aq) + H_2\ (g)$
$Pt\ (s) + HBr\ (aq) \rightarrow$ no reaction; H^+ lies above Pt in the activity series.

Workbook Problem 4.7

Using the activity series, write balanced chemical equations for the following reactions:

a. $Cu\ (s) + HCl\ (aq) \rightarrow$
b. $Mn\ (s) + ZnCl_2\ (aq) \rightarrow$
c. $Mg\ (s) + Al(NO_3)_3\ (aq) \rightarrow$

Strategy: Remember that any element higher in the activity series will react with the ion of any element lower in the activity series. Also remember, metals above the H^+ ion in the activity series will displace the hydrogen ion from an acid to form H_2 gas.

Step 1: Predict the outcome of these reactions.

4.9 Balancing Redox Reactions: The Half-Reaction Method

By the nature of redox reactions, there is a reduction going on at the same time as an oxidation. Separating the reaction into two **half-reactions** gives another way to balance these complex equations.

In this method, the two portions of the equation are separated, and balanced alone, just as the reactions above were balanced. To balance for charge, electrons will need to be added in.

As electrons do not appear in the final equation, the proportions of the two half-reactions relative to one another have to be adjusted so that when the equations are added back together, the electrons on the product side and the reactant side cancel out.

EXAMPLE:

Using the half-reaction method balance the following reaction which takes place in acidic solution:

$Mn^{2+}(aq) + NaBiO_3(s) \rightarrow Bi^{3+}(aq) + MnO_4^-(aq)$

SOLUTION:

1. Write two unbalanced half-reactions:

   Oxidation: $Mn^{2+}(aq) \rightarrow MnO_4^-(aq)$ (Mn goes from +2 to +7)

   Reduction: $NaBiO_3(s) \rightarrow Bi^{3+}(aq)$ (Bi goes from +5 to +3)

2. Balance each half reaction for atoms other than H and O.

   $Mn^{2+}(aq) \rightarrow MnO_4^-(aq)$

   $NaBiO_3(s) \rightarrow Bi^{3+}(aq) + Na^+(aq)$

3. Add H_2O to balance in oxygen, and H^+ to balance in hydrogen.

   $4\ H_2O(l) + Mn^{2+}(aq) \rightarrow MnO_4^-(aq) + 8\ H^+(aq)$

   $6\ H^+(aq) + NaBiO_3(s) \rightarrow Bi^{3+}(aq) + Na^+(aq) + 3\ H_2O(l)$

4. Balance each reaction for charge.

   $4\ H_2O(l) + Mn^{2+}(aq) \rightarrow MnO_4^-(aq) + 8\ H^+(aq) + 5e^-$

   $2\ e^- + 6\ H^+(aq) + NaBiO_3(s) \rightarrow Bi^{3+}(aq) + Na^+(aq) + 3\ H_2O(l)$

5. Make the electron count the same in both reactions.

   $(4\ H_2O(l) + Mn^{2+}(aq) \rightarrow MnO_4^-(aq) + 8\ H^+(aq) + 5e^-) \times 2$

   $(2\ e^- + 6\ H^+(aq) + NaBiO_3(s) \rightarrow Bi^{3+}(aq) + Na^+(aq) + 3\ H_2O(l)) \times 5$

   $8\ H_2O(l) + 2\ Mn^{2+}(aq) \rightarrow 2\ MnO_4^-(aq) + 16\ H^+(aq) + 10\ e^-$

$$10\ e^- + 30\ H^+\ (aq) + 5\ NaBiO_3(s) \rightarrow 5\ Bi^{3+}\ (aq) + 5\ Na^+\ (aq) + 15\ H_2O\ (l)$$

6. Add the two half-reactions together, canceling anything that appears on both sides of the equation.

$$\cancel{8\ H_2O\ (l)} + 2\ Mn^{2+}(aq) + \cancel{10\ e^-} + 14\ \cancel{30}H^+(aq) + 5\ NaBiO_3(s) \rightarrow 2\ MnO_4^-\ (aq) + \cancel{16\ H^+}\ (aq) + \cancel{10e^-} + 5Bi^{3+}\ (aq) + 5\ Na^+\ (aq) + 7\ \cancel{15}H_2O\ (l)$$

$$2\ Mn^{2+}(aq) + 14\ H^+\ (aq) + 5\ NaBiO_3(s) \rightarrow 2\ MnO_4^-\ (aq) + 5\ Bi^{3+}\ (aq) + 5\ Na^+\ (aq) + 7\ H_2O\ (l)$$

7. Check to make sure all atoms and charges are balanced.

Reactant side:	2Mn	14H	5Na	5Bi	15O	net charge: +18
Product Side:	2Mn	14H	5Na	5Bi	15O	net charge: +18

Workbook Problem 4.8

Using the half-reaction method, balance the following reaction which takes place in basic solution:

$KMnO_4 + Na_2SO_3 \rightarrow MnO_2 + Na_2SO_4$

Strategy: Follow the steps in the preceding worked example.

Step 1: Write two unbalanced half-reactions.

Step 2: Balance each half–reaction for atoms other than H and O.

Step 3: Add H_2O to balance in oxygen, and H^+ to balance in hydrogen.

Step 4: Balance each reaction for charge.

Step 5: Make the electron count the same in both reactions.

Step 6: Add the two half-reactions together, canceling anything that appears on both sides of the equation.

Step 7: Make the solution basic by adding 1 OH^- to each side for every H^+.

Step 8: Cancel water molecules that appear on both sides of the equation.

Step 9: Check your answer to make sure both atoms and charges are balanced.

4.10 Redox Stoichiometry

Just as the concentrations of acids and bases in solution can be determined by titration using an indicator, the concentration of an oxidizing or reducing agent can be similarly determined given that the following conditions are met:

- The reaction must have a 100% yield
- A color change in one of the reactants must signal the endpoint.

As many redox reagents are strongly colored, this can be a useful method. A balanced equation is of course required, as is a known quantity of one of the reactants.

Workbook Problem 4.9

A 50 mL solution of iodine and starch was titrated with an acidic 0.15 M thiosulfate solution. After 15.7 mL of the thiosulfate solution had been added, the blue color of the iodine:starch complex completely disappeared, indicating that the iodine had been completely reduced to I^- ions.

What was the original molar concentration of the iodine in solution?

The unbalanced equation is:

$$I_2 + S_2O_3^{2-} \rightarrow I^- + S_4O_6^{2-}$$

Strategy: Balance the equation, proceed as for an acid—base titration:

Step 1: Determine the number of moles of thiosulfate ion that were required to react with the iodine solution to completely reduce the iodine.

Step 2: Balance the equation for acid solution.

Step 3: Use mole ratios to determine the amount of iodine present in the reaction mixture.

Step 4: Determine the original concentration of I_2 in solution.

4.11 Some Applications of Redox Reactions

Although reactions such as the ones discussed here may seem rare, the reality is that electrons move around in many different types of reactions, even when they are not discussed this way. All of the following have a redox component:

- Combustion: the oxidation state of oxygen shifts from 0 when it is an element,to -2 after a reaction occurs. This causes a reduction in whatever is being burned.
- Bleaching: strong oxidizing agents such as peroxides, hypochlorite, or elemental chlorine can destroy colored molecules.
- Batteries: not surprisingly, the ability to generate a current, which is a flow of electrons, relies on this type of reaction.
- Metallurgy: most metals are found in the Earth's crust as oxides, and thus must be reduced to get their elemental forms.
- Corrosion, or rusting, is simply combustion writ small: it is a slower process, but if you set steel wool on fire, or leave it out in the rain, you will end up with the same product: rust.
- Cellular respiration: This is a combustion reaction that operates in an extremely controlled manner, but when reduced to its basics, glucose is combined to form carbon dioxide (which you then breathe out), water, and the heat that maintains your body temperature.

Self–Test

This section is intended to test your knowledge of the material covered in this chapter. Think through these problems, and make certain you understand what is going on. Ask yourself if your answer makes sense. Many of these questions are linked to the chapter learning goals. Therefore, successful completion of these problems indicates you have mastered the learning goals for this chapter. You will receive the greatest benefit from this section if you use it as a mock exam. You will then discover which topics you have mastered and which topics you need to study in more detail.

True–False

1. In acid—base reactions, an acid and a base react to form carbon dioxide and water.
2. Sodium nitrate will precipitate out of aqueous solution.
3. When electrons move in reactions, this is known as oxidation (if electrons are lost) and reduction (if electrons are gained).
4. Weak electrolytes do not dissociate very much.
5. Spectator ions do not participate in precipitation reactions.
6. Acids release nitronium ions into solution.
7. NaCl is a strong electrolyte.
8. Oxidations are always equal to ionic charge.

9. An activity series makes it possible to predict the outcome of precipitation reactions.

10. Redox reactions can be broken into half-reactions to make them easier to balance.

Matching

Reduction	a. dissociating only slightly
Net ionic equation	b. reactions involving transfer of electrons
Oxidation number	c. H_3O^+
Weak electrolyte	d. ranking of elements by reducing ability
Redox reaction	e. substance that gives up electrons
Dissociate	f. reaction description without spectators
Spectator	g. substance that accepts electrons
Hydroxide ion	h. gain of electrons
Activity series	i. loss of electrons
Electrolyte	j. ion that does not participate in a reaction
Reducing agent	k. reaction in which two soluble salts form an insoluble product
Precipitation reaction	l. measure of electron richness
Oxidation	m. substance that will conduct electricity when dissolved
Hydronium ion	n. OH^-
Oxidizing agent	o. separate into ions in solution
Strong electrolyte	p. substance that dissociates to produce hydroxide ions in solution
Base	q. substance that dissociates to produce hydronium ions in solution
Acid	r. dissociating completely

Fill-in-the-Blank

1. When _______________ reactants yield a(n) _________________ product, removing some of the dissolved ions, a(n)_________________________ reaction has occurred.
2. In neutralization reactions, a(n) ________ and a(n) ________ react to form __________ and ________.
3. In __________ reactions, electrons are transferred between reactants.
4. Acids, bases, or electrolytes that dissociate completely in solution are known as ____________, while those that dissociate incompletely are ________.
5. A(n) ________________________________ shows all reactants in their undissociated forms, a(n) ________________________________ shows all reactants dissociated, and a(n) ________________________________ shows only those reactants that have undergone a change.
6. In the Arrhenius definition, acids dissociate in aqueous solution to form an anion and a(n) __________ ion, while bases dissociate to form a cation and a(n) _______________ ion.
7. When a substance is oxidized, it __________ electrons, when it is reduced, it __________ electrons.

Problems

1. Classify each reaction as a precipitation, neutralization, or redox reaction, and justify your answer.

   a. $HNO_3\ (aq) + KOH\ (aq) \rightarrow KNO_3\ (aq) + H_2O\ (l)$

   b. $Cu(NO_3)_2\ (aq) + 2\ NaOH\ (aq) \rightarrow Cu(OH)_2\ (s) + 2\ NaNO_3\ (aq)$

   c. $Cu(NO_3)_2\ (aq) + Zn\ (s) \rightarrow Zn(NO_3)_2\ (aq) + Cu\ (s)$

2. If 15 g of Na_3PO_4, a strong electrolyte, are dissolved in 250 mL of solution, what is the total molar concentration of ions?

3. Write balanced molecular, ionic, and net ionic equations for the following reactions in aqueous solution:

   a. lead(II) nitrate + sodium chloride

   b. sulfuric acid + lithium hydroxide

   c. potassium chloride + silver acetate

4. Given 1.00 M solutions of $Ca(NO_3)_2$ and KOH, how much of each would need to be combined to form 1.8 g of $Ca(OH)_2$?

5. Predict reactions for each of the following. If a reaction occurs, write the net ionic equation. If no reaction occurs, list the possible soluble products.

a. $Cu(CH_3CO_2)_2 + Pb(NO_3)_2$

b. $NaNO_3 + (NH_4)_2SO_4$

c. $CaCl_2 + AgNO_3$

6. Assign oxidation numbers to all of the atoms in the following:

a. Cl_2

b. NaCl

c. CO_3^{2-}

d. Cr_2O_3

e. ClO_3^-

7. For each unbalanced equation given below, identify the oxidizing agent and the reducing agent.

a. $C_2H_6 + O_2 \rightarrow CO_2 + H_2O$

b. $Na + Cl_2 \rightarrow NaCl$

c. $Mg + Fe^{2+} \rightarrow Mg^{2+} + Fe$

d. $Ca + H^+ \rightarrow Ca^{2+} + H_2$

8. Given the activity series provided here, predict the outcome of the following reactions. When a reaction occurs, balance the equation.

a. chromium and copper(II) nitrate

b. tin and silver(I) nitrate

c. copper(II) nitrate and cobalt

d. silver and hydrochloric acid

Activity Series
$Li \rightarrow Li^+ + e^-$
$Ba \rightarrow Ba^{+2} + 2e^-$
$Na \rightarrow Na^+ + e^-$
$Cr \rightarrow Cr^{+3} + 3e^-$
$Co \rightarrow Co^{+2} + 2e^-$
$Sn \rightarrow Sn^{+2} + 2e^-$
$H_2 \rightarrow 2H^+ + 2e^-$
$Cu \rightarrow Cu^{+2} + 2e^-$
$Ag \rightarrow Ag^+ + e^-$

9. Balance the following reactions in acidic aqueous solution:

a. $Mg\ (s) + Cr(NO_3)_3\ (aq) \rightarrow Mg(NO_3)_2\ (aq) + Cr\ (s)$

b. $Al\ (s) + FeCl_2\ (aq) \rightarrow AlCl_3\ (aq) + Fe\ (s)$

c. $BrO_3^-\ (aq) + N_2H_4 \rightarrow Br^-\ (aq) + N_2\ (g)$

d. $MnO_4^-\ (aq) + Al\ (s) \rightarrow Mn^{2+}\ (aq) + Al^{3+}\ (aq)$

10. Balance the following reactions in basic aqueous solution:

a. NO_2^- (*aq*) + Al (*s*) → NH_3 (*aq*) + AlO_2^- (*aq*)

b. Cl_2 (*g*) → Cl^- (*aq*) + ClO^- (*aq*)

c. MnO_4^- (*aq*) + Br^- (*aq*) → MnO_2 (*s*) + BrO_3^- (*aq*)

d. MnO_4^- (*aq*) + I^- (*aq*) → I_2 + Mn^{2+} (*aq*)

11. Calculate the molarity of a Na_2SO_3 solution if 15.0 mL of the solution reacts completely with 23.7 mL of 0.124 M $K_2Cr_2O_7$ in acid solution according to the following unbalanced equation:

$$SO_3^{2-} + Cr_2O_7^{2-} \rightarrow SO_4^{2-} + 2\,Cr^{3+}$$

12. 5.3 g of potassium sulfate are combined with 4.9 g of lead nitrate in aqueous solution.
    a. Write the molecular, ionic, and net ionic equations for this reaction.
    b. How many grams of solid will be produced?
    c. How much of the excess reactant will be left over?
    d. If the experimental yield is 4.37 g, what is the percent yield?
    e. What will be the molar concentration of ions remaining in the solution if the reaction mixture is made up to 200 mL?

Challenge Problem

A 5.36 g sample that contains some elemental tin is dissolved in strong acid to give a solution of Sn^{2+} (*aq*). This is then titrated with a solution that is 0.0563 M in NO_3^-, which is reduced to NO (*g*). If the equivalence point is reached at 27.6 mL, what is the percentage of tin in the original sample?

Inquiry Based Problem

In the copper cycle experiment, a pre-1982, and therefore mostly copper penny is dissolved in nitric acid to produce copper(II) nitrate and nitrogen dioxide gas. Next, sodium hydroxide is added to the copper(II) nitrate solution. The product from this reaction is heated and dried. It undergoes dehydration (loss of water), producing copper oxide. This oxide is then reacted with sulfuric acid, which produces a light blue solution. Dipping in an aluminum foil strip until the reaction is complete (indicated by the cessation of bubbling) generates a metallic precipitate.

a. Write and balance equations for each of the steps described above.

b. If you begin with a mostly copper penny that weighs 3.01 g, how much copper is present if you use 36 mL of 5.0 M nitric acid to completely dissolve the penny into ions? What is the percent of copper in the penny?

c. If the final yield is 1.94 g, what is the overall percent yield of the procedure?

CHAPTER 5

PERIODICITY AND THE ELECTRONIC STRUCTURE OF ATOMS

Chapter Learning Goals

A. Electromagnetic Radiation – Characterization

1. Interconvert wavelength, frequency, and energy of electromagnetic radiation.
2. Using the Balmer-Rydberg equation, calculate the wavelength and energy of a photon absorbed or released when an electron changes orbitals.
3. Interconvert energy, frequency, and wavelength for photons.
4. Using the de Broglie equation, calculate the mass of an object knowing its wavelength and vice versa.

B. Wave Functions and Quantum Numbers

1. Relate a set of quantum numbers to a particular orbital.
2. Sketch and name each of the *s, p,* and *d* orbitals.

C. Electron Configurations

1. State the Pauli Exclusion Principle, Hund's Rule and the Aufbau Principle.
2. Predict ground-state electron configurations for elements.

D. Valence Electrons and the Periodic Table

1. Explain what is meant by effective nuclear charge, Z_{eff}.
2. Write the general valence-shell electron configuration for each group of the periodic table, and identify the blocks in which the elements are located.
3. Given a set of atoms, determine which atom is expected to have the largest radius.

Chapter in Brief

The birth of quantum mechanics at the turn of the last century greatly expanded the understanding of the atomic nature of matter, and blurred the lines between matter and energy, demonstrating that at the smallest levels, the two both share properties: matter is wave-like, and light is particle-like. The observation of line spectra and the development of the Balmer-Rydberg equation, though it was some years before it was fully explained, was the first step in understanding the quantization of energy: it led to the Bohr model of the hydrogen atom, which was soon greatly expanded by Schrödinger's wave functions. The variety of orbital shapes, and the development of the system of quantum numbers to identify each electron in an element only helped to solidify the remarkable depth of information contained in the periodic table, and helped to explain reactivity and periodic properties of the elements that seemed paradoxical, such as the decrease in atomic radius as the atomic number increases across the periodic table.

5.1 Light and the Electromagnetic Spectrum

Although we experience them quite differently, visible light, infrared and ultraviolet radiation, and X-rays are all forms of **electromagnetic energy**. Collectively, they make up the **electromagnetic spectrum**.

Electromagnetic energy travels in waves. These can be characterized using four properties:

- *Frequency*: the number of waves that go past a certain point in one second. Units will be "per second", abbreviated s^{-1} or /s, or given the name hertz (Hz). Frequency is indicated by the Greek lowercase *nu* (ν).

- *Wavelength*: the length of one complete wave. Usually measured from peak to peak. Units are meters or nanometers (1×10^{-9} m). Wavelength is indicated by the Greek lambda – λ.
- *Amplitude:* The height of a wave is proportional to its intensity.
- *Speed:* if wavelength (m) and frequency (/s) are multiplied together, the result is m/s. All electromagnetic radiation moves at the same speed: 3.00×10^8 m/s. This is indicated with the letter "c", which always means the speed of light.

Because all electromagnetic radiation has the same speed, it is possible to use the speed of light as a conversion factor to convert between wavelength and frequency:

$$c = \lambda \cdot \upsilon$$

which can also be rearranged to

$$\lambda = \frac{c}{\upsilon}$$

and

$$\upsilon = \frac{c}{\lambda}.$$

It is also helpful to remember the results of this relationship: a longer wavelength is lower in frequency, and a shorter wavelength is higher in frequency.

EXAMPLE:

Blu-Ray lasers typically operate at 473 nm. What is their frequency?

SOLUTION: Although it is not stated in the problem, you have a three-variable equation ($\upsilon = \frac{c}{\lambda}$) and two of the variables: wavelength and speed. Remember that c is a constant.

$$\upsilon = \frac{3.00\text{x}10^8\,\text{m/s}}{473\text{x}10^{-9}\,\text{m}} = 6.34\text{x}10^{14}\,\text{Hz}$$

Note: you will quickly notice that a frequency in this range (10^{14} Hz) is typical of visible light.

Workbook Problem 5.1

Your college radio station transmits at 88.5 MHz. What is the wavelength associated with this frequency?

Strategy: Use the relationship between speed, wavelength, and frequency.

Step 1: Choose the appropriate form of the equation, and solve:

5.2 Electromagnetic Energy and Atomic Line Spectra

Normal sunlight has a continuous spectrum. It contains all wavelengths of visible light, and even extends outside of the visible range to ultraviolet and infrared.

When individual elements are subjected to electricity or heat, they are said to be "excited." When this happens, they give off light only in specific wavelengths unique to that element–a **line spectrum**.

In 1885, Johannes Balmer was able to mathematically describe the pattern in the visible atomic line spectrum of the hydrogen atom. His equation was later extended by Johannes Rydberg to describe all the lines of the spectrum, including those outside the visible range.

$$\text{Balmer-Rydberg equation} = \frac{1}{\lambda} = R\left[\frac{1}{m^2} - \frac{1}{n^2}\right] \text{ or } \upsilon = R \bullet c\left[\frac{1}{m^2} - \frac{1}{n^2}\right]$$

R is the *Rydberg constant* – 1.097×10^{-2} /nm, and n and m are integers such that n > m.

EXAMPLE:

Calculate the wavelength of light emitted when an electron falls from the $n = 8$ to $n = 3$ levels in the hydrogen atom.

SOLUTION: We solve this problem by using the Balmer-Rydberg equation. Assign the higher energy level to be "n", and the lower level to be "m".

$$\frac{1}{\lambda} = R\left[\frac{1}{m^2} - \frac{1}{n^2}\right] = 1.097\text{x}10^{-2}\ \text{nm}^{-1}\left[\frac{1}{3^2} - \frac{1}{8^2}\right] = 0.00105\ \text{nm}^{-1}$$

Don't forget to invert the answer – nm^{-1} is not the desired unit!

$\lambda = 1/0.00105/\text{nm} = 955$ nm. This is in the infrared range.

Workbook Problem 5.2

The Pfund series of spectral bands has m = 5, and n > m. What are the longest and shortest wavelengths in this series?

Strategy: Determine the values of n that will make λ the longest and shortest. Remember that the value of λ is greatest when the value of n is smallest and the value λ is smallest when the value of n is greatest.

Step 1: Determine the value of n that will make λ the shortest and longest.

Step 2: Use the Balmer–Rydberg equation with $m = 5$ and solve for the shortest wavelength.

Step 3: Solve for λ using the Balmer–Rydberg equation and the values of m that make λ the longest.

5.3 Particlelike Properties of Electromagnetic Energy

The *photoelectric effect* is the fact that light of a specific frequency can eject electrons from the surface of a clean metal. In 1905, Albert Einstein proposed that this was due to light behaving like streams of small particles or **photons**. Below that specific threshold frequency, there is not enough energy per photon to eject an electron.

The relationship of energy to frequency can be expressed thus:

$$E = h\nu \text{ or, since } c = \lambda\nu,\ E = \frac{hc}{\lambda}$$

"h" is Planck's constant. It has a value of 6.626×10^{-36} J·s. This gives the energy of one photon of a specific frequency or wavelength.

Higher frequencies and shorter wavelength give higher energy. This is borne out in the way humans experience the electromagnetic spectrum: no one ever got a radio-wave burn, but UV radiation damages skin, and X-rays can do even deeper damage, as they have enough energy to pass right through the body.

Matter is quantized at the level of molecules and atoms, and electromagnetic energy is quantized in **photons.**

Using this information, Niels Bohr developed a model of the hydrogen atom as a proton with one electron that had discrete orbits available to it: the electron could be *here*, or *here*, but not in between. This explained the phenomenon of line spectra, and allowed the numbers n and m in the Balmer–Rydberg equation to be identified as energy levels.

The energy of photons emitted is related to the difference in energy levels in the atom.

EXAMPLE:

If one mol of photons have an energy of 53.7 kJ, what is the wavelength of the light from which they came?

SOLUTION: We need to rearrange the equation $E = \frac{hc}{\lambda}$ to calculate the wavelength. However, the given energy must first be converted from kJ/mol to J/photon.

$$53.7 \frac{\text{kJ}}{\text{mol}} \text{ x } \frac{1 \text{ mol}}{6.022 \text{ x } 10^{23} \text{ photons}} \text{ x } \frac{1000 \text{ J}}{\text{kJ}} = 8.92 \text{ x } 10^{-20} \text{ J/photon}$$

Now, rearrange the equation to solve for wavelength and put in the known values:

$$\lambda = \frac{hc}{E} = \frac{6.626 \text{ x } 10^{-34} \text{ J} \cdot \text{s x } 3.00 \text{ x } 10^{8} \text{ m/s}}{8.92 \text{ x } 10^{-20} \text{ J/photon}} = 2.23 \text{ x } 10^{-8} = 22.3 \text{ nm}$$

Workbook Problem 5.3

The threshold frequency of sodium metal is 4.4×10^{14} Hz. What is the wavelength and color of light to which this corresponds, and what is the energy of one mol of photons at this frequency?

Strategy: Use the equation for the energy of a photon, and multiply by Avogadro's number. Then convert frequency to wavelength.

Step 1: Determine the energy of one photon, then convert to one mole of photons.

Step 2: Convert the frequency to wavelength.

Step 3: Consult your textbook for the color of this wavelength of light.

5.4 Wavelike Properties of Matter

In 1925, Louis de Broglie suggested that if light can behave like a particle, it is only logical that particles can also behave like light: all matter will have wavelike properties. He supported this with a mathematical derivation.

Einstein proposed the following famous relationship between mass and energy: $E = mc^2$. Using the previous relationships between wavelength and frequency, and substituting simple velocity (v) for the speed of light, he came up with the following equation, that can be used to solve for the wavelength of any matter:

$$\lambda = \frac{h}{m\text{v}}$$

For this equation, h is also rearranged. A joule is kg•m^2/s or mass times velocity. For problems involving the deBroglie equation, h = 6.626×10^{-34} kg•m^2/s.

EXAMPLE:

The muzzle velocity of a 50 g sniper bullet is 860 m/s. The sniper must consider wind, movement of target, humidity, and altitude. What is the deBroglie wavelength of the ammunition at this speed? Could the deBroglie wavelength contribute to a missed shot?

SOLUTION:

Using the de Broglie equation, we can solve for wavelength:

$$\lambda = \frac{\text{h}}{\text{m} \bullet \text{v}} = \frac{6.626\text{x}10^{-34}\,\text{kg} \bullet \text{m}^2/\text{s}}{0.050\text{kg} \bullet 860\text{m/s}} = 1.54\text{x}10^{-35}\,\text{m}$$

The deBroglie wavelength of the bullet is less than the diameter of an atom. But if I'd missed my shot, I'd still give it a try.

5.5 Quantum Mechanics and the Heisenberg Uncertainty Principle

The Bohr model was an essential step in the understanding of the structure of matter, but, like the Balmer-Rydberg equation, it only works for hydrogen. Its most important insight was the idea that the electrons in an atom are confined to specific energy levels.

Ernest Schrödinger proposed the **quantum mechanical model** of the atom in 1926. In this model, the idea of electrons as particles is abandoned in favor of their wavelike properties.

Just one year later, Werner Heisenberg provided support for this idea by demonstrating that it is impossible to know both where an electron is, and what path it follows. Any attempt to "see" an electron's position would require that it interact with photons of light. As these photons would transfer energy to the electron, the speed of the electron would be altered by the interaction.

This is called the Heisenberg Uncertainty Principle, and can be expressed mathematically:

$$(\Delta x)(\Delta mv) \geq \frac{h}{4\pi}$$

Expressed in words, the uncertainty in the position of the object (x), multiplied by the uncertainty in its momentum (mass × velocity), is greater than or equal to Planck's constant divided by 4π.

EXAMPLE:

The mass of an electron is 9.11×10^{-31} kg, and its velocity, known to within 15%, is 2.5×10^{6} m/s. What is the uncertainty in its position?

SOLUTION:

The first thing to do is to determine the uncertainty in the velocity:

$0.15 \times 2.5 \times 10^{6}$ m/s = 3.8×10^{5} m/s

$(\Delta x)(\Delta mv) \geq \frac{h}{4\pi}$ can now be rearranged to solve for Δx: $(\Delta x) \geq \frac{h}{4\pi(\Delta mv)}$

$$(\Delta x) \geq \frac{6.626 \text{ x } 10^{-34} \text{ kg} \cdot \text{m}^2/\text{s}^2}{4 \text{ x } 3.14 \text{ (}9.11 \text{ x } 10^{-31} \text{kg x } 3.8 \text{ x } 10^{5} \text{m/s}} \geq 1.52 \text{ x } 10^{-10} \text{m or } 152 \text{ pm}$$

Although these numbers are quite small, so are the objects in question. Like deBroglie wavelengths, these calculations lose their relevance in the macroscopic world.

5.6 Wave Functions and Quantum Numbers

Although the position and momentum of an electron cannot be precisely known, its probable location can be calculated used differential equations known as Schrödinger's wave equation. The solutions to this equation are called wave functions or orbitals, and are represented with the Greek ψ (psi). ψ^2 defines the probability of finding the electron in a specific area around the nucleus.

Three parameters are required for wave functions: these are called the **quantum numbers**, and are defined using the following rules:

- The **principal quantum number** is abbreviated "*n*". It defines the energy level of the electron, and can be any positive integer. This also defines the *shell* of the electron. Electrons with $n = 5$, for example, are said to be in the fifth shell.

- The **angular momentum quantum number** is abbreviated "*l*". It defines the shape of the orbital, and can be any integer (including zero) *less than n*. These numbers also define the *subshell* of the electron by the following system:

    - $l = 0$, *s* subshell
    - $l = 1$, *p* subshell
    - $l = 2$, *d* subshell
    - $l = 3$, *f* subshell

- The **magnetic quantum number** is abbreviated "m_l". It defines the orientation of an orbital, and can be any integer from $-l$ to l. If the angular momentum quantum number is 2, for example, m_l can be -2, -1, 0, 1, or 2.

The energy levels for hydrogen electrons are dependent only on the principal quantum number. In multielectron atoms, things are complicated by the fact that electrons can interact with one another as well as with the nucleus. In general, as energy is required to separate positive and negative charges, an electron with a larger value of n will be higher in energy.

EXAMPLE:

Give the values of the quantum numbers of a 2*p* subshell.

SOLUTION:

For a 2*p* subshell, $n = 2$. The fact that it is a *p* orbital says that l must be 1. As m_l can be any number from $-l$ to l, it has possible values of -1, 0 and 1. There are three orbitals in a 2*p* subshell, as there are three possible orientations (as shown by the three magnetic quantum numbers).

Workbook Problem 5.4

Determine the subshell for an electron having $n = 3$, $l = 0$. How many orbitals are in this subshell?

***Strategy**:* Determine the subshell associated with a value of $l = 0$.

Step 1: Identify the subshell with the value of n and the letter designation for $l = 0$.

Step 2: Determine the number of orbitals in this subshell.

5.7 The Shapes of Orbitals

s orbitals are spherical, with the probability of finding an electron dependent only on the distance from the nucleus. The size increases in successively higher shells, with nodes – areas of zero probability – in between the shells. This is due to the wave nature of electrons: an area of zero probability corresponds to an area of zero amplitude in a wave. There is only one *s* orbital per shell ($l = 0$, so $m_l = 0$).

p orbitals are commonly described as dumbbell-shaped, with a node near the nucleus. As expected from the quantum number rules, there are three *p* orbitals per shell: $l = 1$, so $m_l = -1$, 0, or 1. These are defined as occurring along the *x*-axis, along the *y*-axis, and along the *z*-axis of a Cartesian coordinate system. The two lobes of a *p* orbital have different phases, which is important for covalent bonding: only electrons that are in the same phase can interact. As the principal quantum number increases, the size of the *p* orbitals also increases.

d orbitals have a more complex shape. With the angular-momentum quantum number of 2, there are 5 orbitals in each *d* subshell: $m_l = -2, -1, 0, 1, 2$. Four of the orbitals are shaped like paired *p* orbitals, 90 degrees rotated from one another; the fifth is shaped like a *p* orbital with a doughnut-shaped "skirt" in the *xy* plane. As in the *p* orbitals, the alternating lobes have different phases.

f orbitals have eight lobes, and three nodal planes through the nucleus. $l = 3$, so $m_l = -3, -2, -1, 0, 1, 2, 3$. This additional complexity makes them difficult to describe, and elements that contain them will not be much discussed at this level of study.

5.8 Quantum Mechanics and Atomic Line Spectra

The quantum-mechanical model provides an explanation for the line spectra of atoms, and for the Balmer-Rydberg equation that can be used to quantify their energies. When the electrons in an atom have energy added to them – when they are "excited" – they can jump up to energy levels with different principal quantum numbers.

Excited electrons are unstable: as a result, they will drop back down to a lower level at the earliest opportunity, and release the extra energy they gained as a photon. Since the energy levels are quantized, the energy emitted in the photons will be as well: all the transitions between those two energy levels will have the same wavelength.

The variables in the Balmer-Rydberg equation can now be seen to correspond to energy levels: "n" is an excited, higher-energy state, and "m" is the lower-energy state to which the electron returns. Thus, it is possible to measure the energy differences between orbitals for hydrogen electrons using this equation.

$$\frac{1}{\lambda} = R\left[\frac{1}{m^2} - \frac{1}{n^2}\right]$$

EXAMPLE:

Calculate the energy difference between the fourth and second energy levels of the hydrogen atom in kilojoules per mole.

SOLUTION:

For this problem, n = 4, and m = 2. First, solve for the wavelength of the energy:

$$\frac{1}{\lambda} = 1.097\text{x}10^{-2}\ \text{nm}^{-1}\left[\frac{1}{2^2} - \frac{1}{4^2}\right] = 0.00205\ \text{nm}^{-1}$$

$$\lambda = 486\ \text{nm}$$

Now solve for the energy of one photon:

$$E = \frac{hc}{\lambda} = \frac{6.626 \text{ x } 10^{-34}\ \text{J} \bullet \text{s x } 3.00 \text{ x } 10^{8}\ \text{m/s}}{486 \text{ x } 10^{-9}\ \text{m}} = 4.09 \text{ x } 10^{-19}\ \text{J}$$

This is now converted to kilojoules per mole fairly simply:

$$4.09 \text{ x } 10^{-19}\ \frac{\text{J}}{\text{photon}} \text{ x } 6.022 \text{ x } 10^{23}\ \frac{\text{photons}}{\text{mol}} \text{ x } \frac{1\ \text{kJ}}{1000\ \text{J}} = 246\frac{\text{kJ}}{\text{mol}}$$

Workbook Problem 5.5

Calculate the energy required (in kJ/mol) to remove an electron from the third shell of a hydrogen atom.

***Strategy**:* Identify n and m, solve for wavelength, then energy, and convert to kJ/mol.

Step 1: Identify n and m for this problem.

Step 2: Solve the Balmer-Rydberg equation to get the wavelength of the energy needed:

Step 3: Solve for the energy of one photon of this energy:

Step 4: Convert to kJ/mol:

5.9 Electron Spin and the Pauli Exclusion Principle

The three quantum numbers discussed earlier define the orbitals available in an atom, but do not allow for a unique description of the electrons in an orbital.

The property that distinguishes the individual electrons in an orbital is their "spin.". Electrons will spin either clockwise or counterclockwise, which causes them to have a tiny magnetic field. Each orbital can hold one each: an electron spinning clockwise and an electron spinning counterclockwise.

To describe this, a fourth quantum number, the spin quantum number – abbreviated m_s – is used. It can have values of +½ or -½, referred to as "up" and "down" spins, and is not constrained by the other quantum numbers.

Wolfgang Pauli proposed the **Pauli exclusion principle**, which states that no two electrons in the same atom can have the same set of quantum numbers. Each orbital has the ability to hold two paired electrons: ↑↓.

EXAMPLE:

State the four quantum numbers for each of the electrons in a 2*p* subshell.

SOLUTION:

A 2*p* orbital has a principal quantum number *n* of 2, an angular momentum quantum number *l* of 1, and magnetic quantum numbers m_l of -1, 0, and 1. For each of these orbitals, two electrons with +½ and -½ are possible. So the quantum numbers possible in this subshell are:

n	l	m_l	m_s
2	1	-1	+½
2	1	-1	-½
2	1	0	+½
2	1	0	-½
2	1	1	+½
2	1	1	-½

There are six unique sets of quantum numbers for a *p* subshell, so it can hold six electrons.

5.10 Orbital Energy Levels in Multielectron Atoms

In hydrogen atoms, only the distance from the nucleus determines the energy level of an electron. In multielectron atoms, electron-electron repulsions also play a part in the energy level of an electron.

Outer-shell electrons are further from the nucleus and held less tightly. The electrons that are closer to the nucleus also shield the outer electrons from the nuclear charge. These outer electrons are not feeling the full charge of the nucleus, but only an **effective nuclear charge, Z_{eff}.**

$$Z_{eff} = Z_{actual} - \text{Electron shielding}$$

This can be seen to relate to the shapes of the orbitals: a lower angular momentum quantum number will correspond to less electron shielding. For example, a 2*s* electron and a 2*p* electron are in the same shell, but because of the shape of the orbital, a 2*s* electron will spend more time close to the nucleus than a 2*p* electron. Therefore, the 2*s* electron is lower in energy than the 2*p* electron.

5.11 Electron Configurations of Multielectron Atoms

It is now possible to describe the **ground-state electron configuration**, or the lowest-energy configuration of any atom. Orbitals with the same energy level, such as the three orbitals in a *p* subshell are referred to as **degenerate**.

Three rules are used, collectively known as the **aufbau principle**:

- Lower-energy orbitals fill before higher-energy orbitals.
- An orbital can hold only two electrons with opposite spins.
- If degenerate orbitals are available, they will first fill singly with electrons with parallel spins. Only after there is an electron in each orbital will electrons begin to double up in orbitals. This is **Hund's rule**.

There are two ways of representing these configurations. The first gives the principal quantum number, *n*, followed by the letter designation of the subshell, with the number of electrons in the subshell represented by a superscript:

H: $1s^1$
He: $1s^2$
Li: $1s^2 2s^1$
Be: $1s^2 2s^2$

One of the consequences of the rules about quantum numbers is that there is no 1*p* subshell: with a principal quantum number of 1, the only angular momentum quantum number allowed is 0, as the number must be less than *n*.

B: $1s^22s^22p^1$

S subshells will contain two electrons, *p* subshells will hold six electrons, *d* subshells will hold ten electrons, *f* subshells will hold 14. These numbers can all be pulled straight from the periodic table, as will be explained in a later section. After argon, the energy levels will begin to cross one another, but these can still be taken from the periodic table, as will be explained below.

The other way that these configurations are written uses *orbital filling diagrams* with the electrons represented by arrows. This method better shows the functioning of Hund's Rule:

C: ↑↓ ↑↓ ↑ ↑ __ or $1s^22s^22p^2$
1*s* 2*s* 2*p*

N: ↑↓ ↑↓ ↑ ↑ ↑ or $1s^22s^22p^3$
1*s* 2*s* 2*p*

O: ↑↓ ↑↓ ↑↓ ↑ ↑ or $1s^22s^22p^3$
1*s* 2*s* 2*p*

EXAMPLE:
Give the electron configuration for neon.

SOLUTION: The electron configuration for neon, with 10 electrons, is $1s^22s^22p^6$.

Noble gases, which are known to be unreactive, all have completely full energy levels: there are no more subshells available in the $n = 2$ energy level. This pattern will continue with other noble gases as well.

EXAMPLE:
Give the orbital-filling diagram for magnesium.

SOLUTION: The orbital-filling diagram for magnesium would be:

↑↓ ↑↓ ↑↓ ↑↓ ↑↓ ↑↓
1*s* 2*s* 2*p* 3*s*

For elements that have more electrons than a noble gas, the configuration of the noble gas can be inserted in brackets. So, the electron configuration of magnesium can also be written [Ne] $3s^2$.

Workbook Problem 5.6
Give the ground-state electronic configuration, both complete and shorthand, for calcium. Also draw the orbital-filling diagram.

Strategy: Using the periodic table, determine the number of electrons in calcium.

Step 1: Determine the number of electrons in calcium.

Step 2: Use the Aufbau principle to determine the ground-state electronic configuration.

Step 3: Determine the noble gas in the previous row, and remove the electrons associated with it to a shorthand notation. Specifically name the electrons in unfilled subshells.

Step 4: Draw the orbital-filling diagram.

5.12 Some Anomalous Electron Configurations

The larger the atom, the closer together the energy levels are found: in the transition metals, a number of electron configurations are seen that are not accounted for by the previous rules.

Having completely filled and exactly half-filled subshells confers stability. As a result, when energy levels are very close together, electrons will move from lower-energy shells into higher energy shells to gain these configurations.

EXAMPLE:
Give the orbital-filling diagram of chromium.

SOLUTION: Chromium is the smallest element that will transfer electrons to gain stability. Its electron configuration is [Ar]$4s^1 3d^5$. The reason that this is more stable can be seen in the orbital-filling diagram:

↑↓	↑↓	↑↓ ↑↓ ↑↓	↑↓	↑↓ ↑↓ ↑↓	↑	↑ ↑ ↑ ↑ ↑
1*s*	2*s*	2*p*	3*s*	3*p*	4*s*	3*d*

5.13 Electron Configurations and the Periodic Table

The electrons that cannot be put into a noble-gas shorthand are **valence shell** electrons. They are the most loosely held electrons, and determine the chemical behavior of an element. All of the elements in a given group will have the same valence shell configuration: only the principal quantum number will change.

All alkali metals (group 1A) have a valence configuration of s^1. They are all reactive, and tend to lose one electron to form +1 ions. The electron configuration supports this observation: there is one loosely held electron that is the only electron in its energy level. The same pattern can be seen across the groups: metals have loosely held electrons that are easily lost, and non-metals have gaps in their valence shells that they fill by taking electrons to form negative ions.

The periodic table is divided into "blocks" based on what type of subshell the outermost valence electrons are in:

- *s*–block elements — Groups 1A and 2A (filling of an *s* orbital). n = row number.
- *p*–block elements — Groups 3A through 8A (filling of *p* orbitals; *ns* orbitals are filled). n = row number.
- *d*–block elements — transition metals. n = row number -1.
- *f*–block elements — lanthanide and actinide elements. n = row number -1

EXAMPLE:

Using only the periodic table, write the shorthand electron configuration for technetium (Tc).

SOLUTION:

Technetium is in the d-block, in the 5th row of the periodic table. The noble gas that is below it in electron configuration is krypton (the end of row 4). The electron configuration will begin with this. Because it is in the 5th row of the periodic table, the *s* electrons that will immediately follow will be 5*s*. This subshell is filled. The *d* electrons will be in the *n* - 1 energy level, so they will be 4*d* electrons. There will be 5 of them.

Tc = $[Kr]5s^2 4d^5$

5.14 Electron Configurations and Periodic Properties: Atomic Radii

Although atoms cannot be said to have a definite size, atomic radii can be assigned to atoms by measuring covalent bonds. When two identical atoms are covalently bonded together, half of the bond length is the atomic radius.

As *n* increases down through a group, atomic radius increases as larger valence-shell orbitals are being occupied.

As subshells fill across the periodic table from right to left, the effective nuclear charge felt by each electron increases: electrons within the same subshell only shield each other weakly, so the atomic radius decreases, even as the atomic number increases.

EXAMPLE:

Choose the smaller of the following pairs:

K and Fe N and Be S and Al Br and Ca

SOLUTION:

These electron pairs are all in the same row with one another, so in each case the element that is rightmost in the periodic table will be smaller.

K > Fe N < Be S < Al Br > Ca

Self-Test

This section is intended to test your knowledge of the material covered in this chapter. Think through these problems, and make certain you understand them. Ask yourself if your answer makes sense. Many of these questions are linked to the chapter learning goals. Therefore, successful completion of these problems indicates you have mastered the learning goals for this chapter. You will receive the greatest benefit from this section if you use it as a mock exam, as this will clarify which topics you have mastered and which topics you need to study in more detail. Some of the problems may require you to think slightly beyond the material presented in the text. Remember to use your units!

True-False

1. The amplitude of a wave is the number of crests that pass a certain point per second.

2. Each element has a unique spectral "signature."

3. The photoelectric effect provides evidence for the particle-like nature of light.

4. Even at the size of everyday objects, deBroglie wavelengths are relevant.

5. The Bohr model of the atom holds for all elements.

6. The energy of a photon can be calculated from its frequency.

7. 4 4 3 +½ is an allowed set of quantum numbers.

8. Electrons reside in *f* orbitals in common nonmetals.

9. Electrons in the same subshell do not effectively shield one another from nuclear charge.

10. Electrons that share an orbital must have parallel spins.

Multiple Choice

1. Electromagnetic energy is characterized by
   a. wavelength
   b. amplitude
   c. frequency
   d. all of the above

2. The Balmer-Rydberg equation accounts for the spectral lines of
   a. helium.
   b. hydrogen
   c. halfnium
   d. any element

3. For photons to knock electrons out of a metal requires that they
   a. have sufficient intensity
   b. have values close to Planck's constant
   c. be above the threshold frequency
   d. occur in large numbers

4. deBroglie determined that at the atomic level
   a. matter has a wavelike nature.
   b. electrons are very slow-moving
   c. elements are sometimes photons
   d. noble gases are mostly energy

5. Which of the following are permitted sets of quantum numbers?
   a. 3 4 -2 +½
   b. 4 3 -2 -½
   c. 4 2 2 1
   d. 4 2 2 +⅔

6. Frequency, wavelength, and energy are related such that:
   a. low-frequency electromagnetic energy has a short wavelength and high energy photons
   b. high-frequency electromagnetic energy has a long wavelength and high energy photons
   c. low-frequency electromagnetic energy has a long wavelength and low energy photons
   d. high-frequency electromagnetic energy has a short wavelength and low energy photons

7. The Heisenberg uncertainty principle states that
   a. matter and energy are the same thing
   b. the position of an electron or its velocity can be known, but not both

c. electrons need to fill degenerate orbitals with perpendicular spins first
d. everything with momentum has a wavelength

8. The alkali metals are
   a. *s* block elements
   b. *p* block elements
   c. *d* block elements
   d. *f* block elements

9. The electron configuration for F is
   a. $1s^2 1p^6 1d^7$
   b. $1s^2 2s^2 2p^6 3s^2 3p^5$
   c. $[Ar]3s^2 3p^5$
   d. $1s^2 2s^2 2p^5$

10. The size of elements *increases*
    a. down the groups and left to right along the periods
    b. up the groups and left to right along the periods
    c. down the groups and right to left along the periods
    d. up the groups and right to left along the periods

Matching

Degenerate	a. can hold two electrons with opposite spins
Wavelength	b. equal in energy
Frequency	c. unit of s^{-1} or /s
Amplitude	d. height of a wave from the midline
Hertz	e. number of crests that pass a certain point per second
Photon	f. set of rules that guide the filling of atomic orbitals
Orbital	g. distance from peak to peak of a wave
Aufbau principle	h. ejection of electrons from a clean metal surface by certain wavelengths of light
Photoelectric effect	i. packet of light energy

Fill-in-the-Blank

1. When discussing electromagnetic energy, the Greek letter λ indicates _______________, while the Greek letter υ indicates _________________.
2. When excited with electricity or heat, elements give off a characteristic ___________________.
3. The photoelectric effect was explained by Einstein and Planck as demonstrating that light contained _____________, or particles of light.

4. In the Bohr model of the atom, a transition from a higher energy level to a lower energy level corresponds with the ______________ of a photon with specific energy.
5. An electron can be specifically described using four ________________ ______________.
6. ___________________ are spherical. Their energy depends only on their distance from the nucleus.
7. ___________________ subshells are dumbbell shaped, with nodes at the nucleus. Each of the orbitals can hold _____ electrons, for a total of ______ electrons in the subshell.
8. Hund's rule states that when _________________ orbitals are being filled, they must first fill with electrons having _____________________________________.
9. The outermost electrons in an atom are called the ____________________. These determine the chemical reactivity of the element.
10. One-half of the bond length when two identical atoms are covalently bonded is known as the ___________ ____________________.

Problems

1. Cellular telephones commonly transmit at 835.6 MHz. What is the wavelength of this energy? How much energy is contained in 1 mole of cell phone photons?

2. Gamma rays have a wavelength of 2.15×10^{-5} nm. What is the frequency of this energy? How much energy is contained in one mole of gamma ray photons?

3. A 432 nm laser emits a quick pulse that contains 2.5 mJ of energy. How many photons were in the pulse?

4. The ionization energy of sodium metal is 495.8 kJ/mol. If this much energy is required to eject 1 mol of electrons from the surface, what frequency and wavelength of light corresponds to the energy of photons required? Will light with wavelength 540 nm dislodge electrons from sodium? Will light with wavelength 195 nm?

5. An electron has an uncertainty in its position of 327 pm. What is the uncertainty in its velocity?

6. If a hydrogen electron in the $n = 6$ energy level returns to ground state, what is the wavelength of the light emitted? What is the energy of the photon emitted?

7. Without using shorthand notation, write the electron structure for ytterbium.

8. Write the four quantum numbers for every electron in the ground state of neon.

9. Write the first two quantum numbers for the highest-energy electron in bromine.

10. When electrons are unpaired, the weak magnetic field they generate is reinforced by any other unpaired electrons in the same atom. This leads to a phenomenon called *paramagnetism*: the more unpaired electrons, the more the atom is attracted to a magnetic field. Draw the orbital-filling diagrams for Si, P, and Cl and rank them from most paramagnetic to least paramagnetic.

11. If an excited hydrogen atom falls to $n = 2$ and emits a photon of light with 4.58×10^{-19} J, at what energy level did it begin?

12. If an electron is accelerated to 2.57×10^6 m/s, what is its deBroglie wavelength? The mass of an electron is 9.109×10^{-31} kg.

13. Identify the atoms with the following electron configurations:
    $[Ar]4s^1 3d^{10}$
    $[Xe]6s^2 4f^{14} 5d^{10} 6p^5$
    $[Kr]\ 4d^{10}$
    $[Ar]4s^1 3d^5$

14. Group the following atoms in order of increasing atomic radius: Os, Rn, Fe, Ru

15. Group the following atoms in order of increasing atomic radius: S, Mg, Cl, Ne

Challenge Problem

1. Watts are units of power defined as J/s. If an incandescent bulb consumes 40 watts at 550 nm for 4 hours, how many photons would it have emitted if it were 100% efficient? If only 5% of the electricity consumed by the bulb was converted to light, and the rest was converted to heat, how long would the bulb need to be lit to heat 200 g of water from 20 °C to 95 °C? The specific heat of water is 4.184 J/g•°C. Use the units of specific heat to figure out how to determine the temperature change.

2. If a laser consumes 1500 W of power, and produces 2.73×10^{20} photons per second, what would the wavelength be if the efficiency of the laser were 100%? 80%? 20%? If the wavelength at which the laser operates is 1542 nm, how efficient is it?

CHAPTER 6

IONIC BONDS AND SOME MAIN-GROUP CHEMISTRY

Chapter Learning Goals

A. Formation of Ions

1. Predict the ground-state electron configuration for ions.
2. Given a set of ions, determine which ion is expected to have the largest radius.

B. Ionization Energies and Electron Affinities

1. For any two elements predict which has the higher first ionization energy.
2. For any two elements, predict which has the higher second, third, fourth, etc. ionization energy.
3. For any two elements, predict which has the more negative first electron affinity.

C. Ionic Solids and Lattice Energies

1. Identify the energies involved in a Born–Haber calculation of lattice energy. Know whether these energies are positive or negative, large or small, and use the Born–Haber cycle to calculate the lattice energy of an ionic compound.
2. On the basis of ionic charges and ionic radii, predict which of two ionic compounds should have the greater lattice energy.

D. Octet Rule

1. Use the octet rule to generalize the chemistry of each family of elements studied in this chapter. Know when to expect the octet rule to be valid and when it can fail.
2. Give the noble gas configuration of cations and anions in ionic compounds.

E. Main-Group Chemistry: Oxidation–Reduction Reactions

1. Know which alkali and alkaline earth metals form a) oxides, b) peroxides, and c) superoxides upon reaction with oxygen. Assign oxidation numbers to the oxygen atoms in these compounds.
2. Give the formulas of products formed when alkali metals react with halogens, hydrogen, nitrogen, oxygen, water, or ammonia. Balance the equations.
3. Give the formulas of products formed when alkaline earth metals react with halogens, hydrogen, oxygen, or water. Balance the equations.
4. Give the formulas of products formed when aluminum reacts with halogens, nitrogen, oxygen, acids, or bases. Balance the equations.
5. Give the formulas of products formed when halogens react with metals, hydrogen, or other halogens.
6. Balance redox equations representing reactions studied in this chapter. Identify which species are oxidized and which are reduced and which elements have undergone a change in oxidation state.

Chapter in Brief

Chemical bonds are the forces that hold atoms together. There are two types: ionic bonds and covalent bonds. This chapter concentrates on the reasons ionic bonds are formed and the energies involved in the formation of these bonds. You will learn that the underlying reason for the formation of ions and ionic compounds is the valence-shell configuration of the elements, which determines the amount of energy needed for losing or gaining electrons. Knowing the periodic trends in the energies for ion formation, you will learn how to predict which elements form cations or anions and if the formation of ionic compounds is energetically favorable. You will also learn about the Born–Haber cycle, a series of hypothetical steps used to calculate the overall energy involved in the formation of ionic compounds. Finally, you will learn about the chemistry of the groups 1A – 3A, 7A, and 8A elements and how the chemistry of these elements is governed by the octet rule.

6.1 Electron Configuration of Ions

Metal elements reside on the left-hand side of the periodic table, and tend to give up electrons to form cations. When they do so, they adopt the electron configurations of noble gases: alkali metals lose one electron to ionize and adopt the electron configurations of the noble gases one down from them on the periodic table, alkaline earth metals lose two and adopt the same configurations. In groups IA and 2A, the topmost energy level is emptied to leave behind a noble-gas configuration.

Nonmetals, on the right side of the periodic table, gain electrons and move up to the electron configurations of the noble gases above them. Halogens all gain one electron to ionize, the group 6A elements gain two electrons and adopt the same electron configurations. In these elements, the topmost energy level is filled to gain a noble-gas configuration.

Transition metals ionize so that they lose their *s* electrons first, then begin losing *d* electrons. This sometimes gives multiple stable ions, which accounts for the multiple ionization states of some transition metals.

EXAMPLE:

Write shorthand electron configurations for the following elements and their ions:

Mg^{2+} F^- S^{2-} Fe^{3+}

SOLUTION:

Mg =	$[Ne]3s^2$	Mg^{2+} - [Ne]	Noble-gas configuration
F =	$[He]2s^22p^5$	F^- - [Ne]	Noble-gas configuration
S =	$[Ne]3s^23p^4$	S^{2-} - [Ar]	Noble-gas configuration
Fe =	$[Ar]4s^23d^6$	Fe^{3+} - $[Ar]3d^5$	Half-filled *d* orbital provides stability

6.2 Ionic Radii

Cations are smaller than neutral atoms. When cations are formed, electrons are removed from valence-shell orbitals. In the case of groups 1A and 2A, the outermost energy levels are removed entirely. In addition, the effective nuclear charge (Z_{eff}) on all the remaining electrons increases, and those electrons are held more closely.

Anions are larger than neutral atoms. When anions are formed, valence-shell orbitals are completed. There are more electrons than protons, so the effective nuclear charge felt by the electrons will decrease. In addition, the additional electrons will add to the electron-electron repulsions felt by the outer electrons. The changes can be dramatic: a chlorine ion is almost twice the size of a neutral chlorine atom.

EXAMPLE:

Which atom or ion in the following pairs has the larger radius:

a. Cr^{2+} or Cr^{4+}
b. P or P^{3-}
c. S or S^{2-}
d. Na or Na^{+}

SOLUTION:

a. Cr^{2+}: The radius of Cr^{4+} is smaller, as both the *s* and *d* orbitals are gone, and the pull on the remaining electrons is correspondingly stronger.
b. P^{3-}: the addition of three electrons to a neutral phosphorus atom will greatly decrease the Z_{eff} felt by the electrons.
c. Se^{2-}: The addition of electrons causes a decrease in Z_{eff}: the outer electrons are more loosely held, and the radius is greater.
d. Na: The removal of electrons causes an increase in Z_{eff} creating a decrease in the radius.

6.3 Ionization Energy

Ionization energy (abbreviated E_i) is the amount of energy required to create a +1 cation from a gas-phase atom. It is always positive: energy must always be added to an atom to remove an ion, no matter how stable the resulting ion.

Ionization energy shows periodic trends:

- The lowest ionization energies are those for the alkali metals, as the loss of one electron empties the highest energy level.
- The highest ionization energies are those for the noble gases, as the loss of an electron breaks into a filled energy level.
- Ionization energy decreases down a group: it takes less energy to remove the outermost electron from a large atom than from a small atom, as it is farther from the nucleus.

The periodic trends are not perfectly smooth:

- group 3A ionization energies are lower than those for group 2A. When group 3A ionizes, a shielded *p* subshell is emptied, while when group 2A ionizes, a less-stable, half-filled *s* subshell is created.
- group 6A ionization energies are lower than those for group 5A, as the loss of one electron in this group creates a more-stable half-filled *p* subshell.

EXAMPLE:

Draw the electron-filling diagrams for Be, B, N, and O. Explain how these demonstrate the reasons behind the lower E_i for group 3A compared to group 2A, and group 6A compared to group 5A.

SOLUTION:

Be:	↿⇂ ↿⇂ 1*s* 2*s*	Ionization requires disruption of filled subshell
B:	↿⇂ ↿⇂ ↿ __ __	Ionization removes single, well-shielded electron

	1s	2s		2p		
N:	↑↓	↑↓	↑	↑	↑	Ionization requires disruption of half-filled subshell
	1s	2s		2p		
O:	↑↓	↑↓	↑↓	↑	↑	Ionization creates stable half-filled subshell
	1s	2s		2p		

6.4 Higher Ionization Energies

It is possible to remove more electrons sequentially from an atom, and these energies can also be measured. Each step will take more energy, as it is the energy required to remove an electron from an already positively charged ion. When all of the valence electrons have been removed, there will be a large jump (4x or more) in ionization energy.

EXAMPLE:
Identify the following fourth-period element from these ionization energies:

E_i = 590 kJ/mol
E_{i2} = 1145 kJ/mol
E_{i3} = 4912 kJ/mol
E_{i4} = 6491 kJ/mol
E_{i5} = 8153 kJ/mol

SOLUTION: To determine which group this element is in, it is necessary to look for a jump of 4x or more in the ionization energies. This occurs between E_{i2} and E_{i3}. This indicates that two electrons can be lost fairly easily, but that once two electrons are lost, the ion has a noble gas configuration and is difficult to ionize further. This describes the alkaline earth metals; the period four alkaline earth metal is calcium.

Workbook Problem 6.1
From the following list of ionization energies, identify the group of the following element:

E_{i1} = 1012 kJ/mol
E_{i2} = 1907 kJ/mol
E_{i3} = 2914 kJ/mol
E_{i4} = 4964 kJ/mol
E_{i5} = 21 267 kJ/mol
E_{i6} = 25 431 kJ/mol

Strategy: Look for a jump of 4x or more between ionization energies: this will show the number of valence electrons and thus the distance from the nearest noble gas.

Step 1: Find the large gap in ionization energies, identify the group number.

6.5 Electron Affinity

Electron affinity (abbreviated E_{ea}) is a measure of the energy released when an electron is added to a neutral, gas-phase atom. For an atom that will accept an electron, the value of electron affinity is therefore negative. The more stable the anion, the more energy is released. Many atoms, noble gases, for example, will not accept additional electrons, so their electron affinity is theoretical only. It would be positive, as energy would be required to hold an electron where it does not want to be, but it is impossible to measure a process that can't be made to occur.

Remember!

Positive energy change = energy required for a process (ionization)
Negative energy change = energy released as a result of a process (electron affinity)
This definition will recur throughout chemistry, and is not particularly intuitive. Learn it *now*.

Periodic trends in electron affinity are related to electron configurations. In the nonmetals, electron affinities generally become more favorable moving toward the halogens, with the exception that adding an electron to a somewhat stable half-filled *p* subshell will be less favorable. For example, nitrogen, with a half-filled *p* subshell, is more reluctant to accept an electron than is either of its neighbors, as that stability is disrupted in adding another electron.

For metals, those that can accept an electron into a partially filled *s* orbital (alkali metals, for example), will have a slightly negative electron affinity. Atoms that must accept electrons into an otherwise unoccupied subshell will be very reluctant to do so: alkaline earth metals need to add the electron into a *p* subshell that is empty, but in an energy level that already contains electrons.

Noble gases, which would need to add an electron not only to a new subshell, but to a new energy level, simply will not do so.

EXAMPLE:

Using electron filling diagrams, rank the E_{ea} of Si, P, S, and Cl

SOLUTION:

	1*s*	2*s*	2*p*	3*s*	3*p*	
Si:	↑↓	↑↓	↑↓ ↑↓ ↑↓	↑↓	↑ ↑ _	Ionization adds to stability by half-filling the 3*p* subshell.
P:	↑↓	↑↓	↑↓ ↑↓ ↑↓	↑↓	↑ ↑ ↑	Ionization destabilizes half-filled subshell.
S:	↑↓	↑↓	↑↓ ↑↓ ↑↓	↑↓	↑↓ ↑ ↑	Ionization moves toward noble gas configuration.
Cl:	↑↓	↑↓	↑↓ ↑↓ ↑↓	↑↓	↑↓ ↑↓ ↑	Ionization completes noble gas configuration.

Based on the electron-filling diagrams, and the general trend that the electron affinities increase moving toward the halogens, one would expect that Cl would have the largest negative E_{ea}, followed by S, then Si, with P having the smallest negative E_{ea} due to the destabilizing effect of adding to a half-filled *p* subshell.

6.6 The Octet Rule

From the past few sections it can be seen that:

- Alkali metals (group 1A) have low first ionization energies and so tend to lose their ns_1 electrons in reactions to gain a noble gas configuration.
- Alkaline earth metals (group 2A) have low first and second ionization energies, and so tend to lose all of their *ns* electrons in reactions so as to gain a noble gas configuration.
- Halogens (group 7A) have a large, negative electron affinity, so tend to ionize readily to obtain a noble gas configuration.

All of these observations can be summarized in the **octet rule**: Main-group elements tend to undergo reactions that leave them with eight outer-shell electrons.

Why does it work? The group 1A and 2A metals lose valence electrons easily, as the electrons they lose have a low Z_{eff}: they are the only electrons in the outermost *s* orbital. The Group 7A elements gain electrons easily, as the electrons have a high Z_{eff}: they are in *p* orbitals that do not shield one another very well. In both cases, the elements have gotten to the point where they have eight electrons in their outermost shell.

EXAMPLE:

Which noble gas configurations are adopted by the atoms in the following ionic compounds:

Li_2O AlN $CaBr_2$

SOLUTION:

In lithium oxide, the two lithium atoms each lose an electron and adopt the electron configuration of helium. The two lost electrons are gained by oxygen, which then has the electron configuration of neon.

In aluminum nitride, the aluminum metal gives up three valence electrons to adopt the electron configuration of neon, while the nitrogen takes on the three electrons and also has the electron structure of neon.

When calcium loses two electrons, it gains the electron structure of argon. The bromine atoms each take one of the lost electrons to take on the electron structure of krypton.

Workbook Problem 6.2

Will the following elements be more likely to gain or lose electrons when they form ions? How many electrons will they gain or lose?

Rb Sc Se As

Strategy: Find these elements on the periodic table, and locate their nearest noble gas. Remembering the general rule that metals lose electron and nonmetals gain them, and the knowledge that main-group elements ionize in predictable ways, determine the most likely ionized form of the element.

Step 1: Identify the group number of the element, and identify them as metals or nonmetals.

Step 2: Find the nearest noble gas.

Step 3: Identify the number of electrons that must be gained/lost to achieve the electron configuration of that noble gas.

Step 4: Write the ionized form of the element.

6.7 Ionic Bonds and the Formation of Ionic Solids

When a metal with a low ionization energy comes into contact with a nonmetal with a favorable electron affinity, a reaction can occur in which an electron is transferred from the metal to the nonmetal, forming a cation and an anion.

Energetically, the electron affinity of the nonmetal is unlikely to completely offset the energy required to remove an electron from a metal. But the reaction between sodium and chloride described in the text proceeds explosively, with a large release of energy. What accounts for this?

These reactions not only form two ions with very stable noble-gas configurations, but they also neutralize the charges that have been created in forming an ionic bond and, by extension, an **ionic solid**.

Ionic solids are networks of ions, in which the charge of one ion is neutralized by surrounding oppositely charged ions. These do not form molecules: we cannot assign one cation and one anion to one another: it is only possible to give the ratio required for the charges to be neutralized:

Na Cl Na Cl Na
Cl Na Cl Na Cl
Na Cl Na Cl Na
Cl Na Cl Na Cl
Na Cl Na Cl Na

If you look at this "salt crystal," you can see that there are four chlorine ions around each sodium ion, and four sodium ions around each chlorine ion. If this were extended into three dimensions, there would be even more ions surrounding and neutralizing each other. This provides exceptional stability.

These reactions proceed rapidly and cannot be observed to occur in stages. However, when dealing with reaction energetics, it is possible to break apart the reaction mathematically. In the case of ionic solid formation, this is done with what is known as a **Born-Haber cycle**. There are five common steps in this cycle, though there can be more or less:

	Step	
1.	Sublimation of the metal atoms: conversion of the metal to gas	Unfavorable
2.	Nonmetal molecules broken into individual atoms	Unfavorable
3.	Ionization of the gaseous metal atoms	Unfavorable
4.	Ionization of the gaseous nonmetal atoms	Favorable
5.	Formation of ionic bond	Very favorable

Each of these steps can be assigned numerical values. The energetically favorable steps will have negative values, as they are releasing energy, and the unfavorable steps will have positive values, as they are consuming energy.

EXAMPLE:

Write the Born-Haber cycle for the formation of lithium fluoride from lithium metal and gaseous fluorine.

SOLUTION:

	Step	Reaction	Energy
1.	Sublimation of lithium	$Li\ (s) \rightarrow Li\ (g)$	+159.4 kJ/mol
2.	Splitting of fluorine molecules	$\frac{1}{2}\ F_2\ (g) \rightarrow F\ (g)$	+158 kJ/mol
3.	Ionization of lithium	$Li\ (g) \rightarrow Li^+\ (g) + e^-$	+520 kJ/mol
4.	Ionization of fluorine	$F\ (g) + e^- \rightarrow F^-\ (g)$	-328 kJ/mol
5.	Formation of ionic solid	$Li^+\ (g) + F^-\ (g) \rightarrow LiF\ (s)$	-1036 kJ/mol
		Total:	-527 kJ/mol

Workbook Problem 6.3

Write the Born-Haber cycle for the formation of sodium bromide. The following information is provided:

Sublimation of Na:	+107.3 kJ/mol
Vaporization of $Br_2(l)$:	+ 30.9 kJ/mol
Breaking of Br_2 bond:	+224 kJ/mol
Ionization of Na:	+495.8 kJ/mol
Ionization of Br:	-325 kJ/mol
Formation of NaBr:	-747 kJ/mol

Strategy: Write and sum the energies of each of the steps in the reaction to calculate the overall energy change of the formation of rubidium chloride from its elements.

Step 1: Write each step of the reaction with its energy required or produced per mole.

Step 2: Sum the energy terms to find the overall energy change of the reaction.

6.8 Lattice Energies in Ionic Solids

Ionic solids have a structure that is referred to as a *crystal lattice*. The energy required to break this crystal lattice back into gaseous ions is called **lattice energy**, abbreviated *U*. It has the same magnitude but the opposite sign of the final step of the Born-Haber cycle: just as that last step, the formation of the ionic solid from gaseous ions, is always negative, the reverse will always be positive.

$$Li^+ (g) + F^- (g) \rightarrow LiF (s) \qquad -U = -1007 \text{ kJ/mol}$$

$$LiF (s) \rightarrow Li^+ (g) + F^- (g) \qquad U = +1007 \text{ kJ/mol}$$

The strength of ionic interactions is described by **Coulomb's law**, which has the mathematical form:

$$F = k\frac{z_1 \bullet z_2}{d^2}$$

k is a mathematical constant, z_1 and z_2 are the charges on the ions, and d is the distance between the centers of the ions.

The practical result of this is that the lattice energy is the largest, meaning the ions are most difficult to separate, when the charges are large, and the distance is small. Small, highly charged ions will have the strongest crystal lattice.

If one ion is changed, and the other held constant, the lattice energy will drop as the ion grows (assuming no change in the charge of the ion). This allows for the prediction of relative lattice energies.

EXAMPLE:

Rank the following compounds from strongest to weakest lattice energy:

LiCl CsCl KCl NaCl RbCl

SOLUTION:

The anion in each of these cases is chlorine, so only the size of the cation is varying. According to Coulomb's Law, the smaller the distance between the ions, the stronger the lattice energy, so the strongest lattice energy will occur with the smallest cation, and the weakest lattice energy with the largest cation.

$$LiCl > NaCl > KCl > RbCl > CsCl$$

EXAMPLE:

Rank the following compounds from strongest to weakest lattice energy:

$MgCl_2$ $TiCl_4$ $CaCl_2$ KCl $AlCl_3$

SOLUTION: This is a slightly more complex situation, but changes in charge will have a huge effect on the lattice energy. Ti, with a charge of +4, will have a stronger lattice energy than Al, with a charge of +3. Only when the charges are the same will size of ions play a role. Mg is smaller than Ca, so will form a stronger lattice.

$$TiCl_4 > AlCl_3 > MgCl_2 > CaCl_2 > KCl$$

Workbook Problem 6.4

Which of each of the following pairs of ions has the greatest lattice energy?

NaCl, KBr AlF_3, $MgCl_2$ $CuBr_2$, CuCl

Strategy: Determine the relative sizes and charges for the ions and assign relative strengths based on charge and size.

Step 1: Identify the relative charges in the two sets of ions and how they change from one set to the other.

Step 2: If necessary, identify the relative ion sizes in the two sets of ions and how they change from one to the other.

Step 3: Determine the relative lattice energies of the ion pairs.

6.9 Some Chemistry of the Alkali Metals

Description: Because of their electron configuration, have the lowest ionization energies of any elements, as they lose an *ns^1* electron to gain a noble gas configuration. As a result, they are among the most powerful **reducing agents** in the periodic table.

Alkali metals are *metallic*: they are shiny, malleable, and good conductors of electricity. In contrast to more common metal elements, they are soft enough to cut with a knife, have low melting points and densities, and are so reactive that they are stored under oil. They are not found in nature as metals, but only as salts.

Production: Lithium and sodium are produced by *electrolysis*: the chloride salt is melted, and an electric current passed through it while it is held at high temperature.

Potassium, rubidium, and cesium are produced by reduction at high temperatures, using sodium to reduce potassium, and calcium to reduce rubidium and cesium. Although this seems to contradict the activity series, the high temperatures allow tiny amounts of the desired metals to be produced: removing them from the reaction mixture shifts the equilibrium toward production of the desired metal. (This will be explained further in later chapters.)

Reactions: Alkali metals react with halogens to form *halides*: the reactivity of the alkali metals increases as the ionization energy decreases: cesium is the most reactive, lithium the least reactive.

Reactions with oxygen vary with the size of the ion: lithium produces a simple oxide, Li_2O. Sodium produces a peroxide, Na_2O_2, in which the anion is O_2^{2-}. Potassium produces a superoxide, KO_2, in which the anion is O_2^-. This last is particularly valuable in spacecraft for its ability to react with moisture and carbon dioxide to generate oxygen.

Alkali metals take their name from their reaction with water to yield hydrogen gas and an alkali metal hydroxide. These reactions are vigorous enough to split water molecules, and, as with the halide reactions above, increase in vigor with the decreasing ionization energy. The reaction of lithium and water generates bubbles, the reactions of other alkali metals result in flames as the heat generated burns the hydrogen gas that has been generated.

$$2\ M\ (s) + 2\ H_2O\ (l) \rightarrow 2\ MOH\ (aq) + H_2\ (g) + \text{heat}$$

A similar reaction occurs when alkali metals are added to liquid ammonia. These reactions produce hydrogen gas and a dissolved metal amide:

$$2\ M\ (s) + 2\ NH_3\ (l) \rightarrow 2\ MNH_2\ (aq) + H_2\ (g)$$

Liquid ammonia will dissolve alkali metals at temperatures below -33 °C to form metal ions and dissolved electrons. These solutions are very strong reducing agents.

6.10 Some Chemistry of the Alkaline-Earth Metals

Description: These elements are able to lose two ns^2 electrons to gain a noble gas configuration, but have slightly higher ionization energies than the alkali metals. However they do follow the same reactivity trend, in which the larger members of the group, with lower ionization energies, are the most reactive.

Alkaline earth metals are harder than the alkali metals, and have higher densities and melting points. Despite their lower reactivity, they are also found in nature only as salts.

Production: Pure alkaline-earth metals are produced by chemical or electrolytic reduction. Beryllium is produced by reduction of beryllium fluoride with magnesium, and magnesium is prepared by electrolysis of its melted chloride. Calcium, strontium, and barium are all made by high-temperature reduction with aluminum metal.

Reactions: Alkaline-earth metals react with halogens to form halides with form MX_2, and with oxygen to form oxides with the expected form MO, though strontium and barium will also form peroxides. Beryllium and magnesium are relatively unreactive, but will burn with a bright flame. The other alkaline-earth metals are all reactive enough that they are best protected from air.

With the exception of beryllium, the alkaline-earth metals will react with water to form hydroxides, but with much less vigor than the alkali metals. Magnesium will react only at temperatures exceeding 100 °C, calcium and strontium will react slowly at room temperature, and barium, with its low ionization energy, reacts vigorously at room temperature.

Workbook Problem 6.5

What would be the predicted outcome of the following reactions? Balance the reactions that occur.

Sr (s) + O_2 (g)

Ca (s) + H_2O (l)

Be (s) + H_2O (l)

Strategy: Based on the reactivity of the metals and the expected reactions:

Step 1: Determine what reaction, if any, will occur.

Step 2: Write balanced equations for the reactions that occur.

6.11 Some Chemistry of the Halogens

Description: These are diatomic nonmetals, characterized by gaining electrons in chemical reactions. They have large negative electron affinities, and large positive ionization energies. Halogens are too reactive to occur free in nature: their name itself is taken from this fact, coming from the Greek roots *hals* (salt) and *gennan* (to make).

Production: Halogens are produced by oxidation of their anions. Fluorine and chlorine are produced by electrolysis, while bromine and iodine are prepared by oxidation of their anions by chlorine.

Reactions: Halogens are among the most reactive elements in the periodic table, with fluorine being the most reactive: it will form compounds with every other element in the periodic table excepting only He, Ne, and Ar. Halogens become less reactive as they get larger, for exactly the same reason that metals gain reactivity as they get larger: outer electron shells are more weakly held, so elements that are gaining rather than losing electrons will become less reactive as their size increases.

Metal halides can be formed from every metal, predictably with main-group metals, and less-predictably with transition metals, some of which will form more than one halide.

Hydrogen and halogens will react to form hydrogen halides: fluorine will react explosively and spontaneously, chlorine will react in the presence of UV light or a spark, and bromine and iodine very slowly. These compounds will dissolve in water to form **acids**. Hydrofluoric acid, though it is the only weak acid among the hydrogen halides, is also one of the few substances that will etch glass.

6.12 Some Chemistry of the Noble Gases

Description: Noble gases are unreactive gases. They have complete outer valence shells, and complete energy levels. They become more reactive as they become larger, as outer electrons are less tightly held than those closer to the nucleus. Even the larger noble gases will react only with fluorine, the most electronegative of all the elements.

Putting It Together

It is possible to experimentally determine the empirical formula of an oxide in the laboratory. By placing magnesium in a weighed crucible and reweighing it, then heating it carefully, the magnesium will be slowly converted to a mixture of magnesium oxide and magnesium nitride. The magnesium nitride is then converted to magnesium oxide by reacting it with water. The mass is then the magnesium oxide alone.

Write and balance reactions for magnesium and oxygen, magnesium and nitrogen, and magnesium nitride and water.

If the experiment is done using 0.237 g of Mg, and the mixture of magnesium nitride and magnesium oxide weighs b. Reaction of magnesium ribbon in a weighed crucible produced 0.277 g of product. After weighing the product, water was added and the crucible was gently reheated. After cooling, the product in the crucible weighed 0.391 g. How much magnesium oxide was formed? What mass percent of magnesium nitride was formed in the first heating? (Assume a 100% yield at each step.)

Self-Test

This section is intended to test your knowledge of the material covered in this chapter. Think through these problems, and make certain you understand what is being asked, and that your answer makes sense. By using this section as a mock exam, you can discover which topics you have mastered and which topics you need to study in more detail.

True-False

1. Main-group metals tend to form cations that give them noble-gas electron configurations.
2. Nonmetals lose electrons to adopt lower energy levels.
3. Transition metals always ionize the same way.
4. When ions form, their radii increase if they are forming anions, and decrease if they are forming cations.
5. Ionization energy is always positive.
6. Ionization energies show no periodic properties.
7. Electron affinity applies primarily to metals, and is difficult to determine for nonmetals.
8. The octet rule can be used to predict how ions will form in reactions.
9. The Born-Haber cycle provides information about the strength of covalent bonds.
10. Lattice energy is related only to ionic size.

11. Alkali metals are powerful oxidizing agents.

12. Halogens are powerful oxidizing agents.

13. Noble gases participate in very few reactions, but the larger of them will react with fluorine.

14. Half-filled *d* orbitals provide a measure of stability.

Multiple Choice

1. Main-group elements form ions by:
   a. losing electrons
   b. gaining electrons
   c. working toward a half-filled *d* orbital
   d. losing or gaining electrons to obtain a noble-gas configuration

2. What +3 ion has the electron configuration $1s^22s^22p^6$?
   a. B^{3+}
   b. Al^{3+}
   c. P^{3+}
   d. Fe^{3+}

3. Rank the following elements in order of greatest to least first ionization energy (IE_1):
   Rb Li Cs K Na
   a. Na > Li > K > Cs > Rb
   b. Cs > Rb > K > Na > Li
   c. Li > Na > K > Rb > Cs
   d. Li > K > Na > Cs > Rb

4. Given the following ionization energies, which element is this most likely to be?
   IE_1: 738
   IE_2: 1451
   IE_3: 7733
   IE_4: 10 540

   a. Na
   b. Mg
   c. Al
   d. Si

5. Ionization energies are always positive, but electron affinities are generally negative because:
   a. energy is released when an atom achieves a more stable configuration
   b. an electron can only be added to a cation
   c. energy is required to hold an extra electron onto an atom
   d. atoms will not accept extra electrons

6. Born-Haber cycles do NOT require which of the following?
   a. ionization energy of the metal
   b. formation of the ionic solid
   c. liquefying the nonmetal
   d. sublimation of the metal

7. Lattice energies vary with
   a. ion charge
   b. electron affinity

c. ionic radius
d. A and C

8. Which of the following will have the highest lattice energy?
   a. AlN
   b. $MgCl_2$
   c. NCl_3
   d. LiCl

9. Alkali metals
   a. decrease in reactivity as they increase in size
   b. increase in reactivity as they increase in size
   c. increase in reactivity as they decrease in size
   d. are equally reactive going down the group

10. Alkaline-earth metals
    a. are the most reactive metals
    b. are commonly found pure in nature
    c. are powerful oxidizing agents
    d. will form MO oxides

11. Halogens
    a. readily form salts with alkali metals and alkaline earth metals
    b. are all gases
    c. are powerful reducing agents
    d. are found free in nature

12. Noble gases
    a. readily form salts with alkali metals and alkaline earth metals
    b. are found as either gases or liquids at room temperature
    c. are essentially unreactive
    d. can react under extreme conditions with alkali metals

Matching

Ionization energy	a. the energy change that occurs when an electron is added to an isolated atom in the gaseous state.
Coulomb's Law	b. sum of interaction energies of ions in a crystal.
Electron affinity	c. main-group elements tend to undergo reactions that leave them with eight valence electrons.
Lattice energy	d. mathematical relationship between charge interaction and energy
Born–Haber cycle	e. the amount of energy required to remove the outermost electron from an isolated atom in the gaseous state.

Core electron	f. a bond maintained by electrostatic interactions.
Ionic bond	g. electrons that form the inner shell of an element and do not participate in reactions.
Octet rule	h. a series of hypothetical steps, each of which contributes to the overall energy change during the formation of an ionic compound.

Fill-in-the-Blank

1. Chlorine and potassium will both ionize so as to have the electron configuration of __________.
2. When metals ionize to form cations, the atomic radius _________________.
3. Valence electrons are strongly shielded by inner, __________ electrons, making them relatively ____________ to remove.
4. Multiple ionization energies are possible with multielectron atoms. A very large _____________ in ionization energies indicates that all of the _______________ electrons have been removed.
5. Electron affinities cannot be determined for elements that do not form ____________.
6. The higher the ionic ________, the stronger the lattice energy is a use of _______________ Law.
7. Main-group elements tend to undergo reactions that leave them with ____________ outer shell electrons. This is the __________________.
8. When alkali metals react with halogens, the alkali metal is _______________, and the halogen is ______________.
9. The least reactive group on the periodic table is the _________________, group ________.
10. The larger an alkali metal, the ______________ reactive it is.

Problems

1. Write the electron configurations of the following elements first as neutral atoms and then as their most stable ion.

   P Ti Mn K Se

2. Rank these ions and atoms in order from smallest to largest.

   Cl^- Al Li^+ Cl Al^{3+} Li

3. Identify the following third-period element from these ionization energies:

   E_{i1} = 578 kJ/mol
   E_{i2} = 1817 kJ/mol
   E_{i3} = 2745 kJ/mol
   E_{i4} = 11 575 kJ/mol
   E_{i5} = 14 830 kJ/mol

4. Draw electron configurations for the following, and use them to rank these elements by electron affinity, lowest to highest. Justify your answer.

Ga Br Ge

5. Calculate the heat of sublimation for aluminum given that the net energy change for formation of $AlCl_3$ (*s*) from the elements, Al (*s*) + 3/2 Cl_2 (*g*) $\rightarrow$ $AlCl_3$ (*s*) is –704.2 kJ/mol. Use the following information:

Al: E_{i1} = 578 kJ/mol, E_{i2} = 1817 kJ/mol, E_{i3} = 2745 kJ/mol
Cl: E_{ea} = -348.6 kJ/mol
Bond dissociation energy for Cl_2 = 244 kJ/mol
Lattice energy for $AlCl_3$ (*s*) = 5492 kJ/mol

6. Using the octet rule, write and balance equations for the reactions of Mg, Li, and Al with Cl, O, and P.

7. Write the Born-Haber cycle for the formation of KF from its elements given the following information:

Sublimation of K:	+89.2 kJ/mol
$F_2 \rightarrow 2$ F (g):	+158 kJ/mol
Ionization of K:	+418.8 kJ/mol
Ionization of F:	-328 kJ/mol
Formation of KF:	-821 kJ/mol

8. Rank the following from strongest to weakest lattice energy, and justify your answers:

LiCl $MgCl_2$ KCl $AlCl_3$ CsCl

Challenge Problem

A 1.53 g sample of an alkaline earth metal was reacted with 4.75 L of chlorine gas (3.17 g/L). The resulting metal chloride was analyzed by dissolving a 5.25 g sample in water and adding an excess of silver nitrate. The analysis yielded 18.8 g of silver chloride.

a. What is the percent chlorine in the alkaline earth chloride?

b. What is the identity of the alkaline earth metal?

c. Write the balanced chemical equations for all reactions.

d. What was the limiting reactant in the reaction between the alkaline earth metal and the chlorine? If beryllium is in excess, how many grams did not react? If chlorine, how many liters remained?

CHAPTER 7

COVALENT BONDS AND MOLECULAR STRUCTURE

Chapter Learning Goals

A. Covalent Bonds

1. From a list of compounds, predict which are ionic and which are molecular.
2. Using only the periodic table, predict which of two elements is more electronegative.
3. Using only the periodic table, predict whether a given bond is ionic, polar covalent, or nonpolar covalent.
4. Using a table of electronegativities, predict which of two bonds is expected to be more polar.

B. Lewis Theory

1. Write Lewis symbols for atoms, and tell how many electrons must be shared to enable the atom to achieve a completed valence shell. Give the symbol of the noble gas with the same number of valence electrons.
2. For each atom in an electron-dot structure, give the number of bonded electron pairs and the number of nonbonded electron pairs.
3. For a given electron-dot structure, give the number of single bonds, double bonds, and triple bonds. Give the bond order of each bond.
4. Draw electron–dot structures of molecules and polyatomic ions, recognizing when multiple bonding and resonance structures are needed.
5. Determine the formal charge on each atom in a resonance structure, and use the formal charges to select the best resonance structure.

C. VSEPR Theory

1. Use the VSEPR model to predict the geometries of molecules and polyatomic ions, including those with more than one central atom.

D. Valence Bond Theory

1. For molecules and polyatomic ions, sketch and identify the orbitals used by each atom to form bonds. Show which orbital overlaps result in σ bonds and which result in π bonds.

E. Molecular Orbital Theory

1. Sketch a molecular orbital diagram for a diatomic molecule. Use the molecular orbital diagram to determine the number of unpaired electrons and to calculate the bond order of the molecule described.

Chapter in Brief

Covalent bond formation is the result of electron sharing between atoms. Covalent bonds have specific lengths and energies: energy is released when a bond is formed, absorbed when a bond is broken. Covalent compounds generally obey the octet rule, but atoms in the third period and below can form expanded octets as well. These structures can be represented with electron-dot structures, which can also be used to predict the existence of resonance hybrids. When dissimilar atoms participate in covalent bonding, differences in electronegativity can lead to a polar bond. These polarity differences can be predicted with the periodic table. Three-dimensional molecular structures are described using the VSEPR model, which takes both atoms and unbonded electrons into account. This is enhanced by valence bond theory, by which electrons are spread throughout hybrid orbitals that permit the electrons in pairs and bonds to maximize their distances from each other. Molecular orbital theory provides mathematical precision, and solves some problems that the previous models cannot.

7.1 The Covalent Bond

Hydrogen forms the simplest covalent molecules, and provide a good model for covalent bonding: as the atoms approach each other, the positive nuclei repel each other, and the electrons repel each other, but the nuclei pull on the electrons of both atoms. If the attractive forces are greater than the repulsive forces, a covalent bond is formed, with the electrons shared between the nuclei.

There is an optimum distance for atoms: too far away, and no bond can form, too close and the nuclei repel. Every covalent bond has a characteristic bond length that leads to maximum stability. These can be roughly predicted from atomic radii.

7.2 Strengths of Covalent Bonds

The energy of a covalently bonded molecule is lower than that of the individual molecules: energy is released when they come together. This same amount of energy has to be put in to the bond to break it. This is the **bond-dissociation energy (D)**.

Bond-dissociation energies are defined as the amount of energy needed to break a bond in an isolated, gaseous molecule. They are always positive, and equal in magnitude but opposite in sign to the energy released when the bond forms.

Although atoms in different molecules are in different environments, bond dissociation energies tend to be the same: A C-C bond always requires about the same energy to break no matter what molecule it is in.

7.3 A Comparison of Ionic and Covalent Compounds

Ionic compounds are solids with very high melting points, as the entire crystal lattice must be disrupted for the compound to melt.

Covalent compounds are low-melting solids, liquids, or gases. The atomic interactions within the molecules are strong, but the interactions between molecules are fairly weak.

7.4 Polar Covalent Bonds: Electronegativity

Chemical bonding exists on a continuum from purely covalent bonds such as those found in diatomic molecules, to purely ionic molecules. In between are molecules in which electrons are unequally shared: **polar covalent** molecules.

In polar covalent molecules, the electrons are unequally shared: one atom will pull harder on the electrons, and will therefore carry a slight negative charge. The other atom, deprived of its share of the electrons, will have a slight positive charge. These partial charges are represented with the lowercase Greek delta: $\delta-$ for the electron-rich atom, $\delta+$ for the electron-poor atom.

Bond polarity is the result of differences in **electronegativity** – the ability of an atom in a molecule to attract the shared electrons in a covalent bond. Fluorine is the most electronegative element, and is given a value (unitless) of 4. Electronegativity decreases down a group and left across the periodic table. Metals have very low electronegativities: cesium has such a low electronegativity that it is sometimes referred to as "electropositive"! Electronegativity is related to electron affinity and ionization energy.

Bond polarity is generally predicted using the following scale:

- Atoms with similar electronegativities ($\Delta EN \leq 0.4$) form nonpolar bonds.
- Atoms whose electronegativities differ by more than two ($\Delta EN > 2$) form ionic bonds.
- Atoms whose electronegativities differ by less than two ($\Delta EN < 2$) form polar covalent bonds.

EXAMPLE:

Using the electronegativity values in Figure 7.4 in the textbook, predict whether the following bonds are polar, nonpolar, or ionic: Cl-Cl, O-F, Mg-Se, C-H, Li-F

SOLUTION: Cl-Cl: nonpolar (3.0 – 3.0 = 0); O-F: polar (4.0 – 3.5 = 0.5); Mg-Se: polar (2.4 – 1.2 = 1.2); C-H: nonpolar (2.5 – 2.1 = 0.4); Li-F: ionic (4.0 – 1.0 = 3)

Workbook Problem 7.1

Predict whether the following are polar, nonpolar, or ionic: H – F, F – Cl, C – O, K – Cl, Ge – Br

Strategy: Predict based on periodic properties first, then confirm your answer with the numbers in Table 7.4 in the textbook.

Step One: Predict the polarity of the bonds based on their locations, then check with electronegativity values from Table 7.4 in the textbook:

Step Two: Put in the numbers to check.

7.5 Electron-Dot Structures

Electron-dot structures, or Lewis structures, represent a molecule's valence electrons using dots, and make it possible to simply model the formation of stable covalent compounds by combining atoms to fill valence shells.

The octet rule is the guiding principle for these structures, just as it was for ionic structures. The number of valence electrons an element has is the same as its group number: Group 7A elements (halogens) have 7 valence electrons, Group 5A elements have 5, and so on. Electrons are placed around the element singly first, then paired, until there are no more electrons to distribute. The single electrons are bonding electrons, and the paired electrons are non-bonding pairs or lone pairs.

EXAMPLE:

Draw electron-dot structures for the following neutral atoms:

N, Si, F, S

SOLUTION:

•N̈• •Ṡi• •F̈: •S̈•

The octet rule can be used to determine how bonds will form: only unpaired electrons will form bonds, and an atom will share electrons until it is surrounded by four pairs of electrons (an octet) or has no more electrons to share. Bonding pairs share electrons, and these shared electrons can be represented with a line.

EXAMPLE:

Draw the electron-dot structure for CH_4.

SOLUTION:

First draw the electron-dot structure for carbon. Carbon has four valence-shell electrons.

Each hydrogen has an electron that it can share with an unpaired electron on carbon.

H•

The hydrogens can be bonded to the carbon to complete the octet:

H
H:C:H
H

The paired electrons can also be represented as bonds:

H
|
H—C—H
|
H

Workbook Problem 7.2

Draw electron-dot structures for the following compounds:

HF, CH_2Br_2, Cl_2

Strategy: Determine valences of compounds, then combine them to form complete octets.

Step 1: Determine the number of electrons on each atom based on its position on the periodic table:

Step 2: Combine the atoms to complete octets, substituting bond lines for paired electrons.

It is possible for atoms to share more than one pair of electrons: when this happens, multiple bonds are formed. These are shorter and stronger than single bonds. Double bonds are formed when two pairs of electrons are shared, triple bonds are formed when three pairs of electrons are shared. The **bond order** is the number of electron pairs: a double bond has a bond order of 2, a triple bond has a bond order of 3.

Sometimes an atom donates a lone pair of electrons to a bond, so that both of the electrons in that bond come from one atom. These are coordinate covalent bonds: they are frequently formed by N, O, P, and S, which have lone-pairs to donate.

7.6 Electron-Dot Structures of Polyatomic Molecules

Many compounds contain only hydrogen and second-row elements. Structures for these are straightforward to draw, as the octet rule almost always applies, and the number of bonds formed by each element is easy to predict. The second-row atoms are bonded to one another to form the core of the molecule, and the hydrogens are found around the periphery.

EXAMPLE:

Draw the electron-dot structure for CH_3NH_2

SOLUTION:

To draw the electron-dot structure of CH_3NH_2:

1. First draw the electron-dot structure of the core atoms:

```
  •      ••
 •C•    •N•
  •      •
```

2. Form a bond between the two core elements by forming an electron pair:

```
  •  ••
 •C–N•
  •  •
```

3. Five hydrogens need to be placed: pair each one with a single electron to complete the octets:

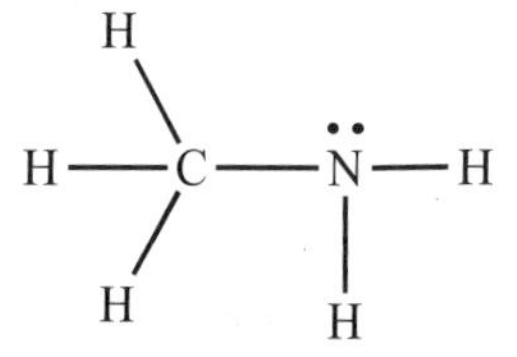

```
     H
      \       ••
 H —— C —— N —— H
      /       |
     H        H
```

Note that this compound can form coordinate covalent bonds: hydrogen ions can bond to the nitrogen to form a positive ion, among other possibilities.

Workbook Problem 7.3

Draw electron-dot structures for the following compounds:

$CH_3CH_2CH_3$, CH_3OH, C_2H_4

Strategy: Determine valences of compounds, then combine them to form complete octets.

Step 1: Determine the number of electrons on each atom based on its position on the periodic table:

Step 2: Combine the atoms to complete octets, substituting bond lines for paired electrons.

Compounds that contain elements beyond the second row can form expanded octets, and do not always abide by the octet rule: their larger size makes it possible for them to accommodate more electrons. All electron-dot structures can be drawn using the same basic rules:

- Find the total number of valence electrons in the molecule. Add electrons for any negative charges, and subtract them for positive charges.
- Determine how the atoms are connected: the most electronegative atom is generally central, halogens and hydrogens are generally peripheral.
- Place two electrons in between each pair of atoms.
- Distribute non-bonding electrons to terminal atoms, and assign any "left-over" atoms to the central atom, expanding its octet.
- When in doubt, choose the most symmetrical arrangement.

EXAMPLE:

Draw the electron-dot structure for SO_4^{2-}.

SOLUTION:

To draw the electron-dot structure of SO_4^{2-}:

1. Find the total number of valence electrons in the molecule:

   Each oxygen has six electrons, the sulfur atom has 6 valence electrons, and two more need to be added for the two negative charges: the total number of electrons is 32.

2. Determine how the atoms are connected:

   The sulfur can form an expanded octet, and placing it in the center gives the most symmetrical arrangement:

```
     O
  O  S  O
     O
```

3. Place electrons between each atom to form single bonds, and putting the correct number of valence electrons around the oxygen:

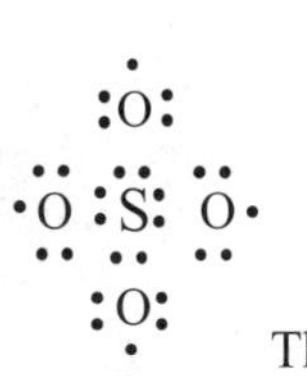

```
      •
     :O:
  ••  ••  ••
 •O  :S:  O•
  ••  ••  ••
     :O:
      •
```

This accounts for 28 electrons, leaving 4 to be distributed. Two of the remaining electrons will form the charges on the ion, but even after that, two more remain. Also, the octets of the oxygens are not satisfied.

4. Satisfy the octets of the oxygens with the four remaining electrons:

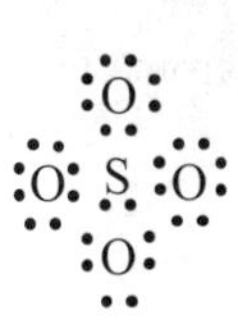

```
     ••
    :O:
 ••  ••  ••
:O: S :O:
 ••  ••  ••
    :O:
     ••
```

This accounts for all of the electrons, but oxygens typically form two bonds, not one. Two of the oxygens are ionized, making them stable with only one bond, but the other two will be happier double-bonded to the sulfur, giving it an expanded octet of 10 electrons.

```
        :O:
    ..  ||  ..
  -:O - S - O:-
    ..  ||  ..
        :O:
```

Workbook Problem 7.4

Draw the electron-dot structure for PF_5.

Strategy:

Step 1: Determine the total number of valence-shell electrons.

Step 2: Determine the connections.

Step 3: Draw the bonds and subtract the number of electrons used from the total number of valence electrons available.

Step 4: Complete the octets of the outer atoms.

Step 5: Place any remaining electrons on the central atom.

7.7 Electron-Dot Structures and Resonance

It is often the case, as in the sulfate ion above, that there is more than one possibility for an electron structure. When deciding which of the sulfur atoms was double-bonded to oxygen, there was no reason that up and down double bonds were any more or less valid than side-to-side double bonds. When this occurs, the "real" structure is an average of all possible structures. This going back and forth between equivalent structures is known as **resonance**. It is indicated by a single, straight, double-headed arrow:↔. Resonance structures differ only in placement of valence electrons.

EXAMPLE:

Draw the resonance structures for NO_3^-.

SOLUTION:

1. Total number of valence-shell electrons = 5 (from N) + 18 (from 3 O) + 1 (from the 1– charge) = 24

2. Determine the connections. In this anion, N is the central atom.

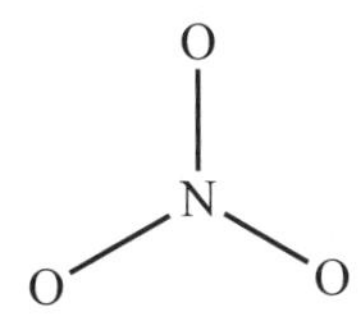

3. Six of the 24 electrons are used for forming the three N—O bonds, leaving 18. All of these electrons are used to form the octet around the three oxygen atoms.

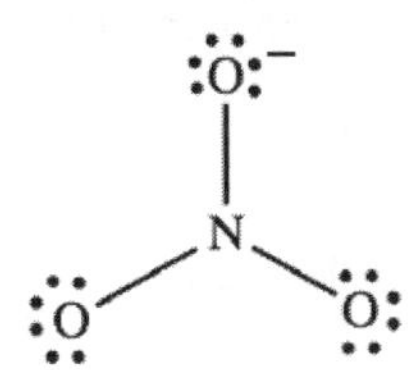

4. There are no more electrons to distribute; however, nitrogen does not have a completed octet. To form a completed octet, we can borrow a pair of lone electrons from one of the oxygen atoms. However, since all three N—O bonds are equal, we must draw three different resonance structures, each of which has a N=O bond between nitrogen and a different oxygen.

Workbook Problem 7.5

Draw as many resonance structures as possible for the carbonate (CO_3^{2-}) ion.

Strategy:

Step 1: Determine the total number of valence-shell electrons.

Step 2: Determine the connections.

Step 3: Draw the bonds and subtract the number of electrons used from the total number of valence electrons available.

Step 4: Complete the octets of the outer atoms.

Step 5: Place any remaining electrons on the central atom, looking for multiple possibilities.

7.8 Formal Charges

Sometimes it is difficult to distinguish which of a number of resonance possibilities is the "best." Formal charges make it possible to distinguish between structures and find the one that is the most stable.

To determine the formal charge on an atom, take the number of valence electrons in a neutral, unbonded atom, and subtract from it one-half the number of bonded electrons and all of the nonbonded electrons:

Formal charge = valence electrons - ½(bonding electrons) – nonbonding electrons

Structures with the lowest formal charges overall, and structures that isolate negative charges on electronegative atoms are the most stable.

For an ion, the overall formal charge of the molecule should be equal to the charge on the ion.

EXAMPLE:

Distinguish the best of these two structures for carbon dioxide by checking their formal charges.

:Ö—C≡O: ⟷ Ö=C=Ö

SOLUTION:

Symmetry favors the structure on the right, but both structures satisfy the octets of all the atoms. Formal charges make it possible to distinguish between them.

For the structure on the left:

Left oxygen	= 6 – 1 – 6 = -1
Carbon	= 4 – 4 – 0 = 0
Right oxygen	= 6 – 3 – 2 = 1

For the structure on the right:

Oxygen	= 6 – 2 – 4 = 0
Carbon	= 4 – 4 – 0 = 0

The formal charges confirm that the intuitive choice is correct: although both structures have an overall charge of 0, the asymmetrical structure has two of the same atom carrying opposite charges, an unstable situation.

Workbook Problem 7.6

Draw three resonance structures for the cyanate ion (CNO^-) and use formal charges to determine which is the most stable. Carbon is the central atom.

Strategy: Draw electron-dot structures, then check the formal charges on the atoms to determine which is the most stable.

Step 1: Determine the number of valence electrons in the ion

Step 2: Determine three reasonable electron dot structures with carbon central and this number of valence electrons.

Step 3: Calculate the formal charge on each of the atoms in each of the structures.

Step 4: Choose the most stable structure.

7.9 Molecular Shapes: The VSEPR Model

Molecules exist not only in two dimensions: their three-dimensional shapes are also very important for their biological functions, and can often be predicted using the **valence-shell electron-pair repulsion (VSEPR) model**. Electrons in lone pairs, or in bonds, try to repel each other, because of their negative charges.

Molecular shape is the result of these repulsions: it can be predicted by counting the number of electron clouds (bonds and lone pairs). Note that a double or triple bond counts as only one electron cloud:

- Two electron clouds (CO_2) – linear
- Three electrons clouds (BH_3) – trigonal planar. Molecular shapes (as opposed to electronic shapes) can vary: two bonds and one non-bonding pair is bent.
- Four electron clouds (NH_3) – tetrahedral. Molecular shapes can vary: three bonds, one non-bonding pair is trigonal pyramidal; two bonds, two non-bonding pairs is bent.
- Five charge clouds (PCl_5) – trigonal bipyramidal: three equatorial, two axial groups. Molecular shapes can vary: four bonds, one non-bonding pair is seesaw shaped; three bonds, two non-bonding pairs is T-shaped; two bonds, three non-bonding pairs is linear.
- Six charge clouds (SF_6) – octahedral: four equatorial, two axial groups. Molecular shapes can vary: five bonds, one non-bonding pair is square pyramidal; four bonds, two non-bonding pairs is square planar.

The shapes of larger molecules can be predicted by looking at the geometries around individual atoms.

EXAMPLE:

Predict the shape of XeF_2.

SOLUTION:

1. First, draw the electron-dot structure of the molecule. There are a total of 8 valence electrons from xenon, and 14 from fluorine, for a total of 22.

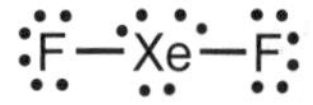

:F̤̈—Xe—F̤̈:

2. There are five charge clouds around xenon, two of which are attached to atoms. Referring back to the list above, five electron groups, two in bonds, and three non-bonding pairs, is linear.

EXAMPLE:

Ethanol has the formula CH_3CH_2OH. It is the component of alcoholic beverages that gives them their vivifying effects. The structural formula for this compound is

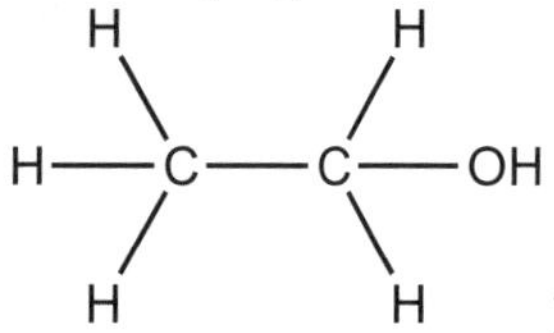

```
 H         H
  \       /
H——C——C——OH
  /       \
 H         H
```

Describe the geometry around the carbon and oxygen atoms.

SOLUTION:

The two carbons each have four bonds and no non-bonding pairs: the geometry around these atoms is tetrahedral. The oxygen atom has two bonds, and two non-bonding pairs: the geometry around this atom is bent.

Workbook Problem 7.7

Predict the molecular shape of CS_2 and SF_4.

Strategy: Draw the electron dot structure for each of the molecules and use the list above to determine the molecular shape.

Step 1: Draw the electron dot structure for the molecules, using the number of valence electrons and the most symmetrical structure.

Step 2: Determine the number of charge clouds around the central atom. How many are used for bonding and how many are nonbonding?

Step 3: Based on the chart, determine the shape.

7.10 Valence Bond Theory

VSEPR theory provides information about molecular shape, but says nothing about the nature of covalent bonds.

Valence bond theory states that when atoms approach one another, singly occupied electronic orbitals will overlap, allowing the now-paired electrons to be influenced by both nuclei. The orbitals must be in the same phase, and will contain electrons of opposite spin.

If anything other than *s* orbitals overlaps, there will be a directionality to the bond.

Bonds resulting from head-on overlap of orbitals are called **sigma (σ) bonds**.

7.11 Hybridization and sp^3 Hybrid Orbitals

For carbon in particular, the electron structure of the atom does not mesh easily with the VSEPR structure of molecules. Carbon has two unpaired *p* electrons, and no unpaired *s* electrons, but it forms four bonds.

Linus Pauling provided a mathematical solution, showing that the Schrödinger wave equations for *s* and *p* orbitals can be combined to form **hybrid orbitals**. In the case of carbon (and others) one *s* combines with three *p* to form four sp^3 orbitals.

Each sp^3 orbital has two lobes, a large and a small. The large lobes point toward the corners of a tetrahedron. Bonds formed by overlapping sp^3 orbitals are very strong: a tetrahedral arrangement of charge clouds always implies sp^3 hybridization.

7.12 Other Kinds of Hybrid Orbitals

Atoms with three electron clouds: these are formed by sp^2 hybridization: the combination of one *s* orbital and two *p* orbitals. The hybridized orbitals interact head-on to form σ bonds between the nuclei, but this leaves an empty *p* orbital above and below the plane of the hybrid orbitals. The unhybridized *p* orbitals can interact side-to-side, forming a π bond that shares electrons not between the nuclei, but above and below. The combination of σ and π bonds results in four electrons being shared: a double bond.

Atoms with two electron clouds: these are formed by *sp* hybridization: the combination of one *s* orbital and one *p* orbital. The two unhybridized *p* orbitals can interact side-to-side, forming two π bonds that share electrons above and below the nucleus. The combination of one σ and two π bonds results in six electrons being shared: a triple bond.

Atoms with five and six charge clouds: These have recently been found to be more complex than can be explained by valence bond theory.

EXAMPLE:
Determine the hybridization for the central atom in H_2CO (formaldehyde) and its bond types.

SOLUTION: This molecule has the following electron-dot structure:

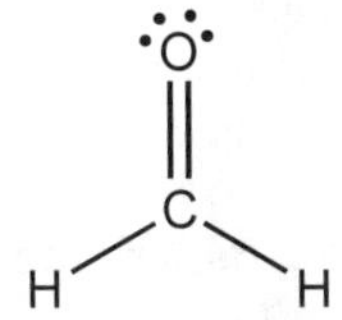

With three charge clouds, the central carbon atom is sp^2 hybridized. It has a σ bond to each hydrogen, and a σ and one π bond with the oxygen atom.

Workbook Problem 7.8
Given the following structure:

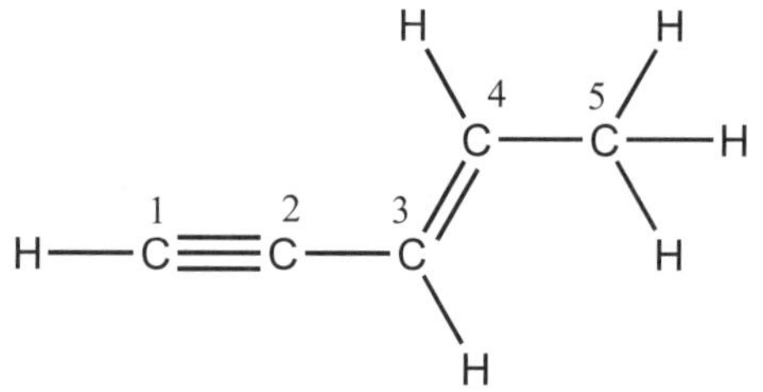

```
                      H         H
                       \ 4    5 /
                        C------C------H
        1     2     3  //       \
   H----C≡≡≡≡C-----C            H
                    \
                     H
```

Describe the hybridization and bonding of each of the carbon atoms.

Strategy: Determine the number of electron clouds around each carbon to determine hybridization, then identify the number of σ and π bonds.

Step 1: Determine the number of electron clouds around each carbon and its hybridization.

Step 2: Determine the bond types present.

7.13 Molecular Orbital Theory: The Hydrogen Molecule

The valence bond model does not always provide agreement with experimental observations. A more complex model known as **molecular orbital theory** provides closer results. In contrast to the previous ideas, which focused on atomic orbitals, this looks at the molecule as a whole.

When two hydrogen atoms interact, the wave portions of the electrons will interfere. They can interfere constructively, and add together, or interfere destructively, and subtract from one another.

If the H-atoms' *s* orbitals add together, they will form a single, egg-shaped *bonding* orbital. If they subtract, they will form two small orbitals with a node in between the nuclei – an *antibonding* orbital, denoted σ^*. The bonding orbital is stable and low in energy, the antibonding orbital is high-energy and unstable. If more electrons are added to an H_2 molecule, they will have to be added to the antibonding orbital, destabilizing the molecule.

Bond order is defined as ½(bonding electrons – antibonding electrons)

EXAMPLE:
What is the bond order of a He_2^{2+} ion? Is it stable or unstable?

SOLUTION: This molecule has two electrons, just as hydrogen does. The electrons will go first into the bonding orbital, leaving no electrons to go into the anti-bonding orbital. The bond order will be ½(2 – 0). Contrary to expectations, this will be a stable ion.

7.14 Molecular Orbital Theory of Other Diatomic Molecules

Experimental evidence shows that electron-dot structures fail to predict the properties of some molecules, particularly the diatomic gases in period 2. Of these three gases, nitrogen, oxygen, and fluorine, nitrogen and fluorine are diamagnetic, meaning that they are unaffected by magnetic fields, but oxygen is **paramagnetic**, meaning that it is attracted to a magnetic field. This phenomenon is only seen in atoms or molecules with unpaired electrons.

The electron-dot structures of these do not indicate that they have any unpaired electrons, so it is necessary to turn to molecular orbital theory to demonstrate why oxygen is paramagnetic: when a molecular orbital diagram is drawn, two electrons are found in antibonding π^* *2p* orbitals.

7.15 Combining Valence Bond Theory and Molecular Orbital Theory

Both of these theories are useful to chemists, but when valence bond theory fails, molecular orbital theory can save the day.

σ bonds are "local" bonds - they occur between the nuclei of two atoms, and go no further. Valence bond theory describes them well.

π bonds are not local, they occur outside of the internuclear space, and as a result, they can spread out over more than two atoms in resonance structures. When resonance structures occur, electrons are said to be "delocalized." Valence bond theory deals with this awkwardly at best: molecular orbital theory explains it perfectly.

The simplicity of valence bond theory makes it desirable whenever it can be useful, but where it fails, the mathematical precision of molecular orbital theory succeeds.

Putting It Together

Oxalic acid can be used in low concentrations to prevent mites in beehives. Its molar mass is 93.03 g/mol and its percent composition is 26.68% carbon, 71.09% oxygen, and 2.224% hydrogen. The structure for this molecule includes one C—C single bond, with two carbons sp^2 hybridized. The hydrogen atoms are bonded to oxygen. Determine the empirical formula, molecular formula, and structure.

Self-Test

This section is intended to test your knowledge of the material covered in this chapter. Think through these problems, and make certain you understand what they are asking. Make sure your answers make sense. Successful completion of these problems indicates that you have mastered the material in this chapter. You will receive the greatest benefit from this section if you use it as a mock exam, as this will allow you to determine which topics you need to study in more detail.

True-False

1. Covalent bonds involve transfer of electrons from one atom to another.

2. Bond dissociation energies are always positive.

3. Covalent compounds have lower melting points and weaker intermolecular interactions than ionic compounds.

4. Bond polarity is usually the result of differences in electronegativity.

5. Electron-dot modeling can predict stable covalent compounds.

6. Valences of main-group elements cannot be predicted based on their position on the periodic table alone.

7. Only the structures of binary compounds can be predicted with electron-dot modeling.

8. It is not possible to model charged compounds using electron-dot structures.

9. The more symmetrical an atomic structure is, the more likely it is to be stable.

10. The number of electron clouds in a molecule can be used to determine its three-dimensional shape.

11. Methane (CH_4) molecules have a pyramidal shape.

12. Molecular shape can be used to determine orbital hybridization.

13. When orbitals overlap head-on, π bonds are formed.

14. Molecular orbital theory can be used to explain molecular properties that cannot be explained by valence bond theory.

15. Multiple bonds can be explained by electron-dot and valence bond theories.

Fill-in-the-Blank

1. Covalent bonds have characteristic ________________. If the atoms are too far away, no ________ can form. Too close, and the ______________ repel each other.
2. The energy required to break a bond in an isolated gaseous atom is the ________ ____________ __________. It is always ______________, and equal in magnitude to the energy ______________ when the bond forms.
3. Chemical bonds exist on a continuum from pure covalent to _________ ______________, to pure ___________.
4. In electron-dot structures, single electrons are _____________ electrons, and paired electrons are ___________________.
5. Second-row elements are sometimes exceptions to the _____________ ________, as they can form ________________ ____________.
6. Sometimes multiple structures can be drawn for a compound. In this situation, ___________ ___________ can be used to distinguish the structure that is the most stable.
7. A molecule with four electron clouds will have a __________________ electronic shape, but the molecular shape may also be ___________, or _____________ ________________.
8. A trigonal planar arrangement of charge clouds always implies ________________ hybridization.
9. _____ bonds are always local, but ____ bonds can be delocalized over more than two atoms.
10. _______________ is a measure of the strength with which an element pulls on its electrons.

Matching

Bond length	a. a combined solution for the Schrödinger wave equations that gives a new set of values
Bond dissociation energy	b. attracted to a magnetic field as a result of unpaired electrons
Coordinate covalent bond	c. the energy required to break a chemical bond in an isolated molecule in the gaseous state
Polar covalent bond	d. a bond formed by head-on overlap of orbitals in which the electrons reside primarily between the two nuclei.
Lone pair	e. a bond in which the electrons are more strongly attracted to one atom
Electronegativity	f. a bond formed by side-to-side interactions with parallel orbitals in which the electrons do not reside between the nuclei
VSEPR model	g. an area in which the electrons involved in an atom or bond are most likely to be found
Hybrid orbitals	h. substance that is weakly repelled by a magnetic field as all of its electrons are spin paired
σ bond	i. an average of several valid electron-dot structures for a moleucle
π bond	j. the distance at which attractive forces and nuclear repulsions are in balance.
Molecular orbital	k. a pair of electrons not participating in a chemical bond.
Paramagnetic	l. "bookkeeping" charge assigned to an atom
Diamagnetic	m. pair of electrons participating in a chemical bond
Bonding pair	n. the strength with which an atom pulls on the electrons associated with it
Formal charge	o. a bond in which one atom donates a pair of electrons to an atom with a vacant orbital
Resonance hybrid	p. a method for determining the three-dimensional structures of molecules using electron repulsions

Problems

1. Identify each of the following compounds as primarily ionic or covalent:

$$LiBr \quad HCl \quad MgO \quad NF_3$$

2. Draw electron-dot structures for H_2O_2, NH_4^+, Cl_2CO, SF_4, and XeF_4. Use resonance structures if necessary.

3. Two white crystalline compounds are tested. The first has a melting point of 800 °C, a solubility of 36 g/100 mL of water, and conducts electricity in solution. The second has a melting point of 180 °C, a solubility of 200 g/100 mL of water, and does not conduct electricity in solution. What conclusions can you make about the chemical bonds in these compounds?

4. Draw all resonance structures for CO_3^{2-}, and assign formal charges to each atom.

5. Rank the following bonds from most to least polar, using only the periodic table:

C-H Al-F Li-Cl Cl-Cl H-Cl

Justify your answers.

6. Rank these atoms in order of increasing electronegativity:

Cs F Fe O

7. Given the following three structures for nitrous oxide, which is likely to contribute most to the structure? Use formal charges to justify your answer.

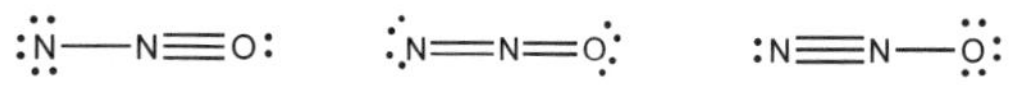

:N̈—N≡O: :N=N=O: :N≡N—Ö:

8. Determine the electron geometry and molecular geometry for the following molecules:

$$NF_3 \quad H_2S \quad IF_4^+ \quad PF_5$$

9. Predict the geometry and hybridization of the carbon atoms in acetaldehyde:

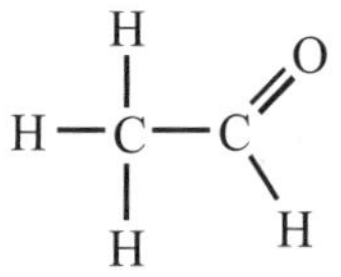

```
    H     O
    |   //
H — C — C
    |    \
    H     H
```

What types of bonds are found in this molecule? What atoms are linked by these bond types?

10. Sketch bonding and anti-bonding orbitals that come from the combination of 2 *s* orbitals. Indicate where they are interfering and how they are interfering (constructively or destructively).

Challenge Problem

360 mg of an unknown acid require 78.24 mL of 0.10 M NaOH to titrate to neutrality. Elemental analysis shows that the acid is 71.1% oxygen, 26.7% carbon, and 2.22% hydrogen. Determine the formula of the acid and draw its electron-dot structure. Using VSEPR theory, provide the molecular shape around each of the central atoms. Using valence-bond theory, how are the oxygen and carbon atoms hybridized, and what bond types – σ and π – are each participating in?

CHAPTER 8

THERMOCHEMISTRY: CHEMICAL ENERGY

Chapter Learning Goals

A. Heat Transfer

1. State the First Law of Thermodynamics, and understand its application to thermochemistry.
2. Differentiate between the concepts of heat and temperature.
3. Identify a state function.

B. Energy Change

1. Define and calculate PV work. Know whether work is being done by the system or on the system.
2. Differentiate between energy and enthalpy, and perform calculations interconverting the two. From ΔH or ΔE, tell whether energy is being lost from or gained by the system.
3. Given a balanced chemical equation and enthalpy change for a chemical reaction, calculate the enthalpy change per mole or per gram of each reactant and product.
4. Perform calculations involving specific heat (or molar heat capacity), heat flow, and temperature change.

C. Enthalpy Change

1. Perform calculations involving Hess's law.
2. Use standard heats of formation to calculate a standard heat of reaction.
3. Use bond dissociation energies to approximate a standard heat of reaction.

D. Spontaneous Reactions

1. Predict whether entropy increases or decreases for a chemical reaction or physical change.
2. Use the equation $\Delta G = \Delta H - T\Delta S$ to determine whether the forward reaction or the reverse reaction is favored.
3. Use ΔH and ΔS to determine the temperature at which a reversible system is at equilibrium.

Chapter in Brief

This chapter introduces you to the concept of thermochemistry: the heat changes that take place during reactions. You begin the study of this topic by learning the difference between heat and energy, and the types of energy changes that can take place. You are then introduced to the Law of Conservation of Energy, the First Law of Thermodynamics, and the concept of state functions. With this background, you will then learn how to calculate the internal energy of the system using $P\Delta V$ work and how the internal energy of the system is related to the enthalpy (ΔH) of the system. You will spend much of the rest of the chapter exploring how to use specific heat calculations in the laboratory and how to use Hess's law, standard heats of formation, and bond dissociation energies to calculate heats of reaction. Finally, you are introduced to the topics of entropy and free energy, topics that will be explored in more detail in later chapters.

8.1 Energy and Its Conservation

Energy is the capacity to do work or supply heat. It can be changed from one form to another, but cannot be created or destroyed. This is the **conservation of energy law**, also known as the **first law of thermodynamics**.

Kinetic energy (E_K) is the energy of motion, and is described by the following formula:

$$E_K = \frac{1}{2} mv^2$$

where m = mass and v = velocity in m/s. Thermal energy is the energy of molecular motion: fast-moving molecules have a higher temperature than slow-moving molecules. More collisions, higher temperature. When thermal energy is transferred from one object to another, it is transferred as heat.

Potential energy (E_P) is stored energy. This can be energy of position, as in a marble sitting at the top of a hill, or chemical potential energy, which is capacity to react and release energy. A bucket of gasoline has a great deal of chemical potential energy: a small spark will release a tremendous amount of energy as the hydrocarbons react with the oxygen in the air.

EXAMPLE:

Describe the energy changes as a rubber band is shot.

SOLUTION:

As the rubber band is stretched, potential energy is added to it. When the rubber band is released, the potential energy is converted to kinetic energy as it flies across the room.

8.2 Internal Energy and State Functions

When accounting for the energy changes in a reaction, the reaction is thought of as being isolated. The reactants and products are referred to as the **system**, while everything else is the **surroundings**.

In a completely isolated system – a theoretical situation – no energy is transferred to the surroundings, so the internal energy of the system is constant. (This is the first law of thermodynamics stated in a different way.)

It is not really possible to completely isolate a system, so energy will flow into and out of a system. When this is measured, it is expressed as $\Delta E = E_{final} - E_{initial}$.

Changes in energy are expressed in terms of the system. So if energy flows into the surroundings, the energy change is negative: the system has lost energy. If energy flows into the system, the energy change is positive: the system has gained energy. What makes this difficult to get a handle on is that generally, you are a part of the surroundings. When you hold a glass of cold water in your hand, the system of the water is gaining energy, while you are losing it. So a process that generates heat has a negative ΔE, while a process that requires heat has a positive ΔE.

Internal energy is a **state function**: it depends only on current conditions and not at all on how the system came to be in that condition. State functions are reversible: if you start at sea level and climb 2000 ft to the top of a mountain, then return to sea level, your overall change in altitude is zero: altitude is a state function.

EXAMPLE:

Which of the following are state functions:

Temperature Volume Mass Work

SOLUTION:

Temperature is a state function: if you warm a beaker in your hands, then put it back down on the lab bench, it will return to its previous temperature. The process is reversible, and you have a $\Delta T = 0$

A balloon carried outside on a cold day will shrink and look miserable, but when brought back indoors, it will resume its previous cheerful volume. Reversible process, $\Delta V = 0$, so volume is a state function.

Mass is a state function: how an object got its mass is irrelevant.

Work is *not* a state function. If an object has been moved from one side of a room to the other and back again, its overall position has not changed, but work was still expended moving it from one place to the other. The sofa is in the same place it was, but you are still tired!

8.3 Expansion Work

In physics (and by extension, chemistry), work is defined as force × distance ($w = F \times d$). In a system in which a cylinder is fitted with a piston, changes in the pressure of the gas can force the piston upwards, doing work. In this case, the equation used is

$$w = -(P \times \Delta V)$$

This is referred to as "PV work" and has the units L•atm. This is easily converted to joules, as 1 L•atm = 101 J. An expanding system is doing work, so is losing energy to the surroundings. This will have a negative value. A contracting system is having work done on it by the surroundings and so is gaining. In this case, the work will have a positive value. If the volume of the system does not change, $\Delta V = 0$, and no work is done.

EXAMPLE:

Calculate the work done when the volume of a reaction changes from 3.0 L to 12 L against a pressure of 15 atmospheres.

SOLUTION:

Using the equation $w = -(P \times \Delta V)$:

$$\mathrm{w} = -(15 \text{ atm x } [12 \text{ L} - 3 \text{ L}]) = 135 \text{ L} \bullet \text{atm}$$

$$135 \text{ L} \bullet \text{atm} \times 101 \text{ J/L} \bullet \text{atm} = 13{,}635 \text{ J or } 14 \text{ kJ}$$

Workbook Problem 8.1

If 7.4 L of hydrogen gas are reacted with 3.7 L of oxygen gas at 13.0 atm, 7.4 L of water vapor are produced. How many kilojoules of work have been done? Have they been done by the system, or to the system?

Strategy: Use the equation $w = -(P \times \Delta V)$, convert to kilojoules.

Step 1: Use $w = -(P \times \Delta V)$:

Step 2: Convert from L • atm to joules to kilojoules.

Step 3: What does the sign tell you? Was work done by or to the system?

8.4 Energy and Enthalpy

Reactions can transfer energy through heat (q) or PV work. These terms can be combined to give the overall energy change of a system:

$$\Delta E = q + (-P\Delta V)\text{, which can also be rearranged to } q = \Delta E + P\Delta V$$

If a reaction is run in a closed container, the volume is constant and no *PV* work can be done. In this case, $\Delta E = q$.

If instead the reaction is carried out at constant pressure (such as the atmospheric pressure in the lab), *PV* work can be done, so $q = \Delta E + P\Delta V$.

Because reactions at constant pressure are so common, the change of heat in these systems is called the **heat of reaction** or **enthalpy** and abbreviated ΔH. Enthalpy is also a state function: it depends only on the initial and final states of the reaction, and not the path the reaction takes. As with all changes, $\Delta H = H_{\text{products}} - H_{\text{reactants}}$.

When the PV work done by a reaction at constant pressure is calculated, it generally makes only a very small contribution to the overall energy change of a reaction. As a result, $\Delta H \approx \Delta E$. If there is no volume change as a result of the reaction, then $\Delta H = \Delta E$.

EXAMPLE:

The equation for the reaction of nitrogen and hydrogen to form ammonia is:

$$N_2\,(g) + 3\,H_2\,(g) \rightarrow 2\,NH_3\,(g)$$

If 2 liters of nitrogen and 6 liters of hydrogen are reacted together, 4 liters of ammonia are formed against 1 atm of pressure. At the same time, 8.23 kJ of heat are released by the reaction. What is the overall energy change of the reaction, and what is the % contribution of *PV* work?

SOLUTION:

Using the equation $w = -(P \times \Delta V)$:

$$\text{w} = -\left(1 \text{ atm x} \left[4 \text{ L} - 8 \text{ L}\right]\right) = 4 \text{ L} \bullet \text{atm}$$

$$4 \text{ L} \bullet \text{atm} \times 101 \text{ J/L} \bullet \text{atm} = 404 \text{ J or } 0.4 \text{ kJ}$$

Using the equation $\Delta E = q - P\Delta V$

$$\Delta E = -8.23 \text{ kJ} - 0.4 \text{ kJ} = -8.63 \text{ kJ}.$$

The percent contribution of the *PV* work is $0.4/8.63 \times 100 = 4.6\%$

Workbook Problem 8.2

When 7.4 L of hydrogen gas are reacted with 3.7 L of oxygen gas at 13.0 atm, 7.4 L of water vapor are produced, and 79.9 kJ of heat are released. What is the total energy change of the system?

Strategy: Use the answer to the previous workbook problem, substituting into $\Delta E = q - P\Delta V$.

Step 1: Use $\Delta E = q - P\Delta V$

8.5 The Thermodynamic Standard State

The value of the enthalpy change of a reaction is per mole of the reactant of interest. So if 200 kJ are released for one mole of the reactant, 400 kJ would be released if two moles reacted.

It is also essential to know the phases of all the reactants and products: for instance, the difference in energy change for producing liquid water is more than for producing gaseous water vapor.

To make sure that there is a common understanding regarding the amounts and phases of products and reactants, the **thermodynamic standard state** has been defined: a substance in its most stable form at 1 atm pressure and 25 °C is in its standard state. For solutions, the standard state is a solution under the above conditions and at 1 M concentration.

Measurements made at standard state are identified with a superscript (°) and can be used to calculate other quantities at standard state.

EXAMPLE:

Identify the standard state for each of the following:

bromine oxygen iron mercury sodium

SOLUTION:

At 25 °C and 1 atm,

- Bromine is a diatomic liquid Br_2 (*l*)
- Oxygen is a diatomic gas O_2 (*g*)
- Iron is a metallic solid Fe (*s*)
- Mercury is a metallic liquid Hg (*l*)
- Sodium is a metallic solid Na (*s*)

EXAMPLE:

What is the value of $\Delta E°$ if a reaction having $\Delta H° = 157.3$ kJ is carried out at a constant pressure of 12.3 atm and the volume change is 453 L?

SOLUTION:

$\Delta H° = \Delta E° + P\Delta V$

$157.3 \text{ kJ} = \Delta E° + (12.3 \text{ atm} \times 453 \text{ L})$

$157.3 \text{ kJ} = \Delta E° + (5572 \text{ L} \bullet \text{atm} \times 101 \text{ J/L} \bullet \text{atm}) = \Delta E° + 562.8 \text{ kJ}$

$\Delta E° = 157.3 \text{ kJ} - 562.8 \text{ kJ} = -405 \text{ kJ}$

Workbook Problem 8.3

When 0.5 mol of aqueous carbonic acid dissociates into water and carbon dioxide, the volume change at 1 atm is 11.2 L. If the $\Delta H°$ for this reaction is -20.7 kJ/mol, what is ΔE?

Strategy: Use the equation $\Delta H = \Delta E + P\Delta V$.

Step 1: Determine the ΔH for the reaction given the amount of carbonic acid that dissociated.

Step 2: Determine $P\Delta V$ and convert from L • atm to kJ.

Step 3: Determine ΔE from the equation: $\Delta H = \Delta E + P\Delta V$.

8.6 Enthalpies of Physical and Chemical Change

There are enthalpies associated with all physical changes or phase changes. Because all of these changes are reversible, it is important to realize that the energies associated with these changes are also reversible: the heat required to melt a certain amount of a substance is the same amount of heat that will be released when it freezes again. The two heats will be equal in magnitude, but opposite in sign. These heats are

- Heat of fusion ($\Delta H°_{fus}$) is the energy required to melt a substance at its normal melting/freezing point (for water, this is 32 °F or 0 °C). The name is a reference to the fact that when ice chips, for example, are melted, they will *fuse* together to form liquid with no air spaces. This is also the heat a substance releases when it freezes.

- Heat of vaporization ($\Delta H°_{vap}$) is the energy required to convert a substance from liquid to gas at its normal boiling point. This tends to be high, as all intermolecular interactions must be broken to vaporize a substance. This is also the energy released by a substance when it condenses from gas to liquid, which is why a steam burn can be so severe.

- Heat of sublimation/deposition ($\Delta H°_{sub}$): when a substance is converted straight from a solid to a gas, it is said to have sublimed. This is the sum of the heats of fusion and vaporization for a substance, as both of these phase changes are contained in sublimation.

Chemical changes also have associated enthalpies. At constant pressure, heat can flow into or out of a reaction system.

- Reactions that emit heat into the surroundings are **exothermic**: the enthalpy of the products is lower than that of the reactants, so $H_{products}$- $H_{reactants}$ is negative. $\Delta H < 0$ is exothermic.

- A reaction that absorbs heat from the surroundings is **endothermic**: in this case, the enthalpy of the products is higher than that of the reactants, so $H_{products}$- $H_{reactants}$ is positive. $\Delta H > 0$ is endothermic.

If an equation is balanced, and all substances are in their standard states, it is possible to calculate the enthalpy change of the reaction in either direction using the moles of a reactant. All reactions can be reversed, so $\Delta H°$ for the forward reaction = $-\Delta H°$ for the reverse reaction. It is also important to be careful in using balanced reactions: the coefficients are very important!

EXAMPLE:

The standard enthalpy for the combustion of propane (C_3H_8) is -2044 kJ. If a propane tank contains 3.5 kg of propane, how much energy will be released or absorbed when the propane is burned?

SOLUTION:

The $\Delta H°$ reported is for the combustion of 1 mol of propane. First, it is necessary to determine the number of moles of propane present. From the formula, the molecular weight of propane is 44 g/mol. 3.5 kg is 3500 grams.

$$3500 \text{ g propane x } \frac{1 \text{ mol propane}}{44 \text{ g propane}} = 79.5 \text{ mol propane}$$

The amount of heat involved in the combustion of this amount of propane is:

$$-2044 \text{ kJ/mol propane x } 79.5 \text{ mol propane} = 162{,}600 \text{ kJ}$$

This is an exothermic reaction. Two things make this clear: the first is that it is a combustion reaction: common sense says that heat is released when things are burned, and this is no exception. But the primary reason that it is obvious that this is an exothermic reaction is the sign on the enthalpy change. When the enthalpy change is negative, heat is being lost by the reactants and gained by the surroundings.

EXAMPLE:

The reaction of boron and hydrogen proceeds by the following reaction:

$$2\ B\ (s) + 3\ H_2\ (g) \rightarrow B_2H_6\ (g) \qquad \Delta H° = 36.4\ kJ/mol$$

If 25.4 g of boron react, what is the energy change? Is this reaction endothermic or exothermic?

SOLUTION:

The stoichiometric ratios here must be considered very carefully. The reported $\Delta H°$ is per mole of boron hydride, but two moles of boron are required per mole of boron hydride.

$$25.4\ g\ B \times \frac{1\ mol}{10.81\ g} \times \frac{1\ mol\ B_2H_6}{2\ mol\ B} \times \frac{36.4\ kJ}{mol\ B_2H_6} = 42.8\ kJ$$

No calculations are required to determine that this is an endothermic reaction – this can be seen immediately from the positive value of ΔH.

Workbook Problem 8.4

How much heat is released or required when 50 g of water are converted from liquid to gas at 100 °C? ΔH_{vap} for water is 40.7 kJ/mol.

Strategy: Use the heat of vaporization to determine the amount of heat involved.

Step 1: Convert the grams of water to moles.

Step 2: Multiply by the heat of vaporization.

Step 3: Make sure the sign of the heat transfer makes sense: the water is the system.

Workbook Problem 8.5

For the reaction

$$2\ Ca\ (s) + O_2\ (g) \rightarrow 2\ CaO\ (s) \qquad \Delta H = -634.9\ kJ/mol$$

How much heat is involved if 3.24 g of calcium metal react with an excess of oxygen gas? Is this an exothermic or an endothermic reaction?

Strategy: Convert to moles and use the stoichiometry of the reaction.

Step 1: Determine the number of moles of calcium.

Step 2: Using the stoichiometric ratio, determine the heat involved.

Step 3: Based on the sign of the heat transfer, determine whether the reaction is exothermic or endothermic.

8.7 Calorimetry and Heat Capacity

Calorimetry is an experimental method for measuring ΔH for a reaction. It can be carried out at either constant pressure or constant volume.

A constant-pressure **calorimeter** can be as simple as an insulated container with a loosely fitting lid to maintain the contents at atmospheric pressure, a stirrer, and a thermometer. This measures ΔH.

A constant-volume calorimeter - also called a **bomb calorimeter** – is more complex, and measures the ΔE of combustion reactions. A sample of a substance is placed in an oxygen environment, and the contents electrically ignited. The heat change is calculated from the increase in temperature of the surrounding water.

To calculate a heat change from a temperature change, it is necessary to know the relationship between the two, something that is distinctive for different materials.

The heat capacity (C) is calculated using the following formula:

$$C=\frac{q}{\Delta T}$$

Where q is heat, and ΔT is the temperature change, calculated, as always, final – initial. This is an extensive property: the amount of heat required will change based on the amount of the material present.

The specific heat is the amount of heat required to increase one gram of a material by 1 °C. For water, 4.184 J will raise the temperature of one gram of water 1 °C. The units of this heat capacity are therefore J/g •°C. Molar heat capacities are also sometimes used. The units for these are J/mol•°C. This is an intrinsic property, and can be used to identify unknown samples.

Specific heat can be used in calculations thus:

$$q = m \times c \times \Delta T$$

(If you look back, you will see that heat capacity (C) is simply mass × specific heat.) In calorimetry, the heat gained or lost by the water is the heat lost or gained by whatever is in it. In other words, the water in the calorimeter is the *surroundings*.

EXAMPLE:

Calculate the amount of heat lost when the temperature of 14.3 g of iron drops 17 °C. The specific heat of iron is 0.449 J/g·°C.

SOLUTION:

Use the formula $q = m \times c \times \Delta T$ to determine the amount of heat involved.

$$q = 0.449 \frac{J}{g\cdot{}^{o}C} \times 14.3\ g \times 17.0^{o}\ C = 109\ J$$

EXAMPLE:

If a 7.3 g sample of an unknown metal at 100 °C is dropped into 25 g of water at 20.0 °C, and the final temperature of the water is 22.1 °C, calculate the specific heat of the metal. The specific heat of water is 4.184 J/g·°C.

SOLUTION:

Use the formula $q = m \times c \times \Delta T$ to determine the amount of heat absorbed by the water. This is also the heat given off by the metal, which will allow the specific heat to be calculated.

$$q(\text{water}) = 4.184\,\frac{\text{J}}{\text{g}\cdot{}^{\circ}\text{C}} \times 25\text{ g} \times (22.1\ {}^{\circ}\text{C} - 20.0\ {}^{\circ}\text{C}) = 220\text{ J}$$

220 J are absorbed by the water, so 220 J were given off by the metal:

$$-220\text{ J} = c \times 7.3\text{ g} \times (22.1\ {}^{\circ}\text{C} - 100\ {}^{\circ}\text{C})$$

$$c = \frac{-220\text{ J}}{7.3\text{ g x }(22.1\ {}^{\circ}\text{C} - 100\ {}^{\circ}\text{C})} = 0.386\,\frac{\text{J}}{\text{g x }{}^{\circ}\text{C}}$$

If you were to consult a table, you would see that this indicates that the metal is copper.

Workbook Problem 8.6

A student mixes 1.50 g of NaOH with 50.0 mL of water at 22.73 °C in a coffee-cup calorimeter. The final temperature of the reaction is 30.70 °C. Assuming that the calorimeter absorbs a negligible amount of heat, and that the density of the solution is the same as that of water, calculate the amount of heat evolved in this dissociation reaction. What is the heat of dissociation of NaOH in kJ/mol?

Strategy: Determine the heat absorbed by the water, then reverse the sign to determine the number of joules evolved by the dissociation. Convert the grams of NaOH to moles, and divide the heat evolved by the number of moles of NaOH to determine the molar heat of dissociation.

Step 1: Determine the amount of heat gained by the water using the equation $q = mc\Delta T$.

Step 2: Determine the amount of heat released by the reaction.

Step 3: Determine the number of moles of NaOH involved.

Step 4: Determine the heat of dissociation per mole of NaOH.

8.8 Hess's Law

Reactions do not always go cleanly, and often produce a mixture of products. This makes direct measurement of the energetics of these reactions problematic. With enthalpy being a state function, the reactions can be broken down into steps and added together to determine the energetics of a difficult-to-measure step. This is **Hess's Law**: the overall enthalpy change for a reaction is equal to the sum of the enthalpy changes for the individual steps in the reaction.

Here are the rules for doing this:

- Reactants must all appear on the right. It is okay to reverse reactions to make this happen.
- Products must all appear on the left. Again, reverse reactions if necessary.
- Intermediates in the reaction must appear on both sides so that they cancel out.
- When a reaction is reversed, the sign on that reaction's ΔH must also be reversed.
- Reactions and their energies can be multiplied through by any necessary factor.
- Add all the reactions and their energies to get the desired reaction and its energy.

EXAMPLE:

Calculate $\Delta H°$ for the reaction $3\ C\ (s) + 4\ H_2\ (g) \rightarrow C_3H_8\ (g)$ given the following information:

$$C_3H_8\ (g) + 5\ O_2\ (g) \rightarrow 3\ CO_2\ (g) + 4\ H_2O\ (g) \qquad \Delta H° = -2043\ kJ$$
$$C\ (s) + O_2\ (g) \rightarrow CO_2\ (g) \qquad \Delta H° = -393.5\ kJ$$
$$2\ H_2\ (g) + O_2\ (g) \rightarrow 2\ H_2O\ (g) \qquad \Delta H° = -483.6\ kJ$$

SOLUTION:

Get the reactants on the right and the products on the left, flipping reactions around if needed, and multiplying them through if necessary to get the equation as written:

Reactants:

$$C\ (s) + O_2\ (g) \rightarrow CO_2\ (g) \qquad \Delta H° = -393.5\ kJ$$
$$2\ H_2\ (g) + O_2\ (g) \rightarrow 2\ H_2O\ (g) \qquad \Delta H° = -483.6\ kJ$$

These both need to be reacted through by coefficients, the first equation by 3, and the second equation by 2 to give them the coefficients needed:

$$3\ C\ (s) + 3\ O_2\ (g) \rightarrow 3\ CO_2\ (g) \qquad \Delta H° = -1180.5\ kJ$$
$$4\ H_2\ (g) + 2\ O_2\ (g) \rightarrow 4\ H_2O\ (g) \qquad \Delta H° = -967.2\ kJ$$

Products:

Take the first reaction in the given list, and turn it around, also reversing the sign on $\Delta H°$.

$$3\ CO_2\ (g) + 4\ H_2O\ (g) \rightarrow C_3H_8\ (g) + 5\ O_2\ (g) \qquad \Delta H° = 2043\ kJ$$

Stack up the rewritten reactions, and add them together:

$$3\ C\ (s) + 3\ O_2\ (g) \rightarrow 3\ CO_2\ (g) \qquad \Delta H° = -1180.5\ kJ$$
$$4\ H_2\ (g) + 2\ O_2\ (g) \rightarrow 4\ H_2O\ (g) \qquad \Delta H° = -967.2\ kJ$$
$$3\ CO_2\ (g) + 4\ H_2O\ (g) \rightarrow C_3H_8\ (g) + 5O_2\ (g) \qquad \Delta H° = 2043\ kJ$$

$$3\ C\ (s) + 3\ O_2\ (g) + 4\ H_2\ (g) + 2\ O_2\ (g) + 3\ CO_2\ (g) + 4\ H_2O\ (g) \rightarrow 3\ CO_2\ (g) + 4\ H_2O\ (g) + C_3H_8\ (g) + 5O_2\ (g)$$

Cancel out like terms on the reactant and product sides of the equation:

$$3\ C\ (s) + 4\ H_2\ (g) \rightarrow C_3H_8\ (g) \qquad \Delta H° = -1180.5\ kJ + (-967.2\ kJ) + 2043\ kJ$$

$$3\ C\ (s) + 4\ H_2\ (g) \rightarrow C_3H_8\ (g) \qquad \Delta H° = -104.7\ kJ$$

Workbook Problem 8.7

Calculate ΔH for the reaction:

$Fe_2O_3\ (s) + 3\ CO\ (g) \rightarrow 2\ Fe\ (s) + 3\ CO_2\ (g)$

Given the following information:

$4\ Fe\ (s) + 3\ O_2\ (g) \rightarrow 2\ Fe_2O_3\ (s)$ $\Delta H° = -1648.4\ kJ$

$2\ CO\ (g) + O_2\ (g) \rightarrow 2\ CO_2\ (g)$ $\Delta H° = -565.4\ kJ$

Strategy: Get the reactants on the right and the products on the left, flipping reactions around if needed, and multiplying them through if necessary to get the equation as written:

Step 1: Reactants on the left, multiplying through if necessary.

Step 2: Products on the right, multiplying through if necessary.

Step 3: Stack up the rewritten reactions, and add them together.

Step 4: Cancel terms that appear on both sides.

8.9 Standard Heats of Formation

To make it possible to determine the energy changes for any reaction, whether actual or theoretical, another method can be used. The **standard heat of formation** for a compound is the enthalpy change for the hypothetical formation of 1 mole of that compound in its standard state from its constituent elements in their standard states. The standard enthalpy of formation of an element in its standard state is zero.

EXAMPLE:

Write the equation for the reaction corresponding to the standard heat of formation of acetone: C_3H_6O.

SOLUTION:

The standard state of carbon is graphite, while hydrogen and oxygen are diatomic gases:

$3\ C\ (graphite) + 3\ H_2\ (g) + \frac{1}{2}\ O_2\ (g) \rightarrow C_3H_6O\ (l)$

It's as simple as that. You will see fractional coefficients in thermochemical equations, as they are written to provide a coefficient of "1" on the product.

There are extensive tables of heats of formation (one of which is in the appendix of your textbook), and they can be used to calculate heats of reaction in a variation on "final – initial": The sum of the heats of formation of the products, minus the sum of the heats of formation of the reactants will be the overall heat of reaction. This is summarized in the following formula:

$$\Delta H^\circ = \sum \Delta H^\circ_{\text{products}} - \sum \Delta H^\circ_{\text{reactants}}$$

It is important to note that the standard heats of formation must be multiplied by the coefficients in the balanced equation.

EXAMPLE:

Calculate ΔH° for the reaction $C_3H_6O\ (l) + 4\ O_2\ (g) \rightarrow 3\ CO_2\ (g)\ + 3\ H_2O\ (l)$ using the following information:

ΔH°_f for acetone = -248.4 kJ/mol

ΔH°_f for oxygen = 0 kJ/mol

ΔH°_f for CO_2 (*g*)= -393.5 kJ/mol

ΔH°_f for H_2O (*l*) = -285.8 kJ/mol

SOLUTION:

Subtract the total heats of formation of the reactants from the total heats of formation of the products.

$$\Delta H^\circ = [(3 \times \Delta H^\circ_f\ CO_2\ (g) + (3 \times \Delta H^\circ_f\ H_2O\ (l))] - [(1 \times \Delta H^\circ_f\ C_3H_6O\ (l)) + (\ 4 \times \Delta H^\circ_f\ O_2\ (g))]$$

$$\Delta H^\circ = [(3 \times -393.5\ \text{kJ/mol} + (3 \times -285.8\ \text{kJ/mol})] - [(1 \times -248.4\ \text{kJ/mol}) + (\ 4 \times 0\ \text{kJ/mol})]$$

$$= [(-1180.5\ \text{kJ}) + (-857.4\ \text{kJ})] - (-248.4\ \text{kJ})$$

$$= -1790\ \text{kJ}$$

Workbook Problem 8.8

During glycolysis, glucose ($C_6H_{12}O_6$) is taken through a series of reactions that combine it with oxygen and release carbon dioxide and water. How much energy is released through the slow combustion of 10.0 grams of glucose in this process? What would be the temperature change if this energy were used to heat 500 g of water?

You may use the following information:

ΔH°_f for glucose = -1273.3 kJ/mol

ΔH°_f for oxygen = 0 kJ/mol

ΔH°_f for CO_2 (*g*)= -393.5 kJ/mol

ΔH°_f for H_2O (*l*) = -285.8 kJ/mol

Strategy: Write a balanced equation for the reaction, and use the standard heats of formation to determine the overall heat of reaction per mole of glucose, then convert grams to moles and determine the heat released by 10 grams of glucose. Use this heat and the equation $q = mc\Delta T$ to determine the temperature change.

Step 1: Write and balance the equation.

Step 2: Determine the overall heat of reaction using $\Delta H^\circ = \sum \Delta H^\circ_{\text{products}} - \sum \Delta H^\circ_{\text{reactants}}$.

Step 3: Convert the grams of glucose to moles.

Step 4: Determine the energy released by 10 g of glucose.

Step 5: Using q = mcΔT, determine the temperature change of 500 g of water.

8.10 Bond Dissociation Energies

Heats of formation are not available for every compound. When there is not data for the heat of formation of a compound, it can be estimated using the known **bond dissociation energies** for every bond in the compound. The bond energies are positive: energy must always be put in to break a bond. The bond formation energies will be equal in magnitude and opposite in sign.

To determine the approximate enthalpy of any reaction, subtract the sum of the bond dissociation energies for the products from the bond dissociation energies in the reactants.

EXAMPLE:

Using bond-dissociation energies, estimate the heat of reaction of hydrogen and oxygen to form water.

	Bond Energy
H-H	436 kJ/mol
O=O	498 kJ/mol
O-H	464 kJ/mol

SOLUTION:

$2\ H_2\ (g) + O_2\ (g) \rightarrow 2\ H_2O\ (g)$

On the product side, 4 H-O bonds are formed:

$$4 \times 464\ \text{kJ/mol} = 1856\ \text{kJ/mol released}$$

On the reactant side, 2 H-H bonds are broken, and one O=O bond is broken:

$$(2 \times 436\ \text{kJ/mol}) + 498\ \text{kJ/mol} = 1370\ \text{kJ/mol absorbed}$$

Reactants – products = -486 kJ/mol

This is the answer for the problem as written – for the production of two moles of water. For one mole of water, the answer will be -486 kJ/mol x 0.5 = -243 kJ/mol. This is very close to the heat of formation of water vapor.

Workbook Problem 8.9

Predict the energy change for combustion of methane (CH_4) using bond dissociation energies.

Use the following information:

	Bond Energy
C-H	414 kJ/mol
O=O	498 kJ/mol

O-H	464 kJ/mol
C=O	799 kJ/mol

Strategy: Write a balanced equation for the reaction, and determine the bond energies for both products and reactants. Subtract products from reactants to determine the overall energy change of the reaction.

Step 1: Write and balance the equation.

Step 2: Determine the energy of the bonds on the reactant side.

Step 3: Determine the energy of the bonds on the product side.

Step 4: Determine the energy released by the reaction.

8.11 Fossil Fuels, Fuel Efficiency, and Heats of Combustion

Fuels all have standard heats of combustion (ΔH°_c) which is the energy released when they are burned.

Different applications have different fuel concerns. For example, mass is critical for a rocket, and a large volume of fuel is difficult in an automobile. So these energies can be converted to a variety of units for comparison: kJ/g where mass is a concern, kJ/mL where volume is a concern. These quantities of energy per mL or per g are referred to as **fuel efficiencies**.

EXAMPLE:

Toluene (C_7H_8) has a density of 0.8669 g/mL and a molar heat of combustion of -3910 kJ/mol. What is its heat of combustion in kJ/g and J/mL?

SOLUTION:

This is a pretty straightforward conversion problem:

$$\frac{-3910\ \text{kJ}}{\text{mol}} \times \frac{1\ \text{mol}}{92\ \text{g}} = -42.5\ \frac{\text{kJ}}{\text{g}}$$

$$\frac{-42.5\ \text{kJ}}{\text{g}} \times \frac{0.8669\ \text{g}}{1\ \text{mL}} = -36.8\ \frac{\text{kJ}}{\text{mL}}$$

With the exception of hydrogen gas, all common fuels are hydrocarbons. Fossil fuels gain their energy from photosynthesis – coal is primarily vegetable, petroleum is largely marine in origin, and natural gas is mostly methane. Coal and natural gas can be used as they are, but petroleum requires fractionation.

Alternative fuels are a burgeoning new industry. Ethanol can be produced by fermenting corn or cane sugar, but processes to produce it from waste wood are being investigated.

8.12 An Introduction to Entropy

Chemical reactions generally release energy in proceeding from higher-energy reactants to lower-energy products. But this is not always true: there are endothermic processes that occur even though energy must be added to them. For an ice cube to melt, it must spontaneously absorb energy from the surroundings. Release of energy cannot be the only factor in reactions.

Spontaneous processes occur on their own without external influence. Nonspontaneous processes require constant inputs of energy: to roll a marble down a hill simply requires that you get it started. Rolling it back up the hill requires a constant input of energy.

The unifying feature of spontaneous endothermic processes is an increase in the **entropy** (S) of the reaction mixtures. Entropy is an increase in molecular disorder or randomness. When an ice cube melts, the rigid crystalline structure of the solid water becomes the far less-orderly liquid water. The units for entropy are J/K: larger values indicate a greater degree of disorder.

Freedom to move and the ability to adopt numerous conformations are all related to entropy. Phase also plays a major role: solids have less entropy than liquids, with gases having the most entropy of all: solid iodine has an entropy of 116 J/K, while gaseous iodine has an entropy of 261 J/K.

As with all change measurements, $\Delta S = S_{\text{final}} - S_{\text{initial}}$. In contrast to ΔH, a positive ΔS is favorable for a reaction, indicating as it does, an increase in disorder.

EXAMPLE:

Predict whether ΔS° is likely to be positive or negative for the following processes:

a. $H_2O\ (s) \rightarrow H_2O\ (g)$
b. $Na\ (s) + \frac{1}{2}\ Cl_2\ (g) \rightarrow NaCl\ (s)$
c. $CH_3CH_2OH\ (l) + 3\ O_2\ (g) \rightarrow 2\ CO_2\ (g) + 3\ H_2O\ (g)$

SOLUTION:

a. Converting water from a crystalline solid to a gas will generate a large positive ΔS.
b. Going from a solid and a gas to a crystal lattice will cause a large negative ΔS.
c. Four moles of reactants (one a liquid) being converted to 5 moles of gas will increase the entropy, giving a positive ΔS.

8.13 An Introduction to Free Energy

A spontaneous process will be one where the overall energetics are favorable: ΔH and ΔS can both be favorable, with one negative and one positive, or one can be favorable and the other unfavorable, with the favorable outweighing the unfavorable. With entropy being temperature dependent, a reaction can be spontaneous at one temperature, and nonspontaneous at another.

All of this is bundled into the equation for the **Gibbs free-energy change** (ΔG):

$$\Delta G = \Delta H - T\Delta S.$$

The sign of ΔG determines whether or not a process is spontaneous: a negative ΔG indicates a spontaneous process in which free energy is being released. A positive ΔG indicates a process in which free energy is being absorbed. These are nonspontaneous. When $\Delta G = 0$, the reaction is at equilibrium, and no net change is occurring.

Entropy is temperature-dependent. If the free-energy equation is set to zero (equilibrium) it is possible to solve for the temperature at which the process will become spontaneous (or nonspontaneous):

$$T = \frac{\Delta H}{\Delta S}$$

EXAMPLE:

At what temperature is the vaporization of methanol at equilibrium, given the following values:

ΔH_f methanol (l) = -238.7 kJ/mol
ΔH_f methanol (g) = -201.2 kJ/mol
$S°$ methanol (l) = 127 J/K•mol
$S°$ methanol (g) = 238 J/K•mol

SOLUTION:

$$T = \frac{\Delta H}{\Delta S} = \frac{-201.2 \text{ kJ/mol} - (-238.7 \text{ kJ/mol})}{0.238 \text{ kJ/K} \bullet \text{mol} - 0.127 \text{kJ/K} \bullet \text{mol}} = \frac{37.5 \text{kJ/mol}}{0.111 \text{kJ/K} \bullet \text{mol}} = 338 \text{ K} = 65 \text{ °C}$$

Not at all surprisingly, this is the boiling point of methanol.

Workbook Problem 8.10

Given the following data, is the reaction of hydrogen and chlorine gas spontaneous at room temperature?

ΔH_f HCl (g) = -92.3 kJ/mol
$S°$ H_2 (g) = 130.6 J/K•mol
$S°$ Cl_2 (g) = 223 J/K•mol
$S°$ HCl (g) = 186.8 J/K•mol

Strategy: Write a balanced equation for the reaction, and determine ΔH and ΔS for the reaction. Using the Gibbs free energy equation, determine ΔG for the reaction at 25 °C, not forgetting to convert to Kelvin first.

Step 1: Write and balance the equation.

Step 2: Determine ΔH.

Step 3: Determine ΔS.

Step 4: Determine the free energy of the reaction at 25 °C, first converting the temperature to K.

Step 5: Determine the spontaneity of the reaction.

Putting It Together

Liquid hydrazine (N_2H_4) is used as the propellant in the maneuvering thrusters of spacecraft. When it is exposed to a catalyst, it undergoes a multistep decomposition that eventually results in hydrogen and nitrogen gas. If 3 g of hydrazine decompose, what is the free energy change of the system at 298 K? Use the following data:

ΔH_f hydrazine = + 95.4 kJ/mol.
S° hydrazine = + 121.2 J/mol•K
S° N_2 = + 191.5 J/mol•K
S° H_2 = + 130.6 J/mol•K

Self-Test

This section is intended to test your knowledge of the material covered in this chapter. Think through these problems, and make certain you understand what they are asking. Make sure your answers make sense. Successful completion of these problems indicates that you have mastered the material in this chapter. You will receive the greatest benefit from this section if you use it as a mock exam, as this will allow you to determine which topics you need to study in more detail.

True-False

1. Energy can be converted from one form to another.

2. Changes in energies are always calculated *initial – final.*

3. The energy changes in a state function depend on path.

4. When a system expands against pressure, it is doing work.

5. The thermodynamic standard state of an element is its most stable form at 0 °C.

6. A reaction can do work or transfer heat, but not both.

7. Melting is an exothermic process.

8. Specific heat is an extensive property, and cannot be used to identify a material.

9. Fractional coefficients can be used in thermochemical equations.

10. Standard heats of formation can be used to calculate reaction enthalpies.

11. Depending on the needs of the end user, different units can be used for heats of combustion of fuels.

12. A negative ΔS is favorable for a reaction.

13. Gases have lower entropy than liquids.

14. Free energy changes can be used to predict spontaneity of reactions.

15. Entropy is a temperature-dependent property.

Matching

Conservation of energy	a. heat released on reaction with oxygen
Energy	b. a process that proceeds without external influence
Temperature	c. going directly from solid to gas
System	d. a measure of molecular kinetic energy
Entropy	e. energy can be transferred between forms, but cannot be created or destroyed
State function	f. absorbing energy
Heat	g. a measure of randomness or disorder
Work	h. capacity to do work
Enthalpy	i. Heat required to convert a substance from solid to liquid
Heat of fusion	j. releasing energy
Sublimation	k. the amount of energy required to raise the temperature of one gram of a substance 1 °C
Specific heat	l. a measurement that is independent of path
Exothermic	m. the focus of a thermochemical process
Hess's Law	n. exerting a force over a distance
Heat of combustion	o. The overall enthalpy of a reaction can be determined by combining any number of theoretical steps
Spontaneous	p. $E + P\Delta V$
Endothermic	q. energy transferred as the result of a temperature difference

Fill-in-the-Blank

1. Energy can be either ________________________, the energy of motion, or __________________, stored energy.
2. Molecular kinetic energy is measured as __________________. When this property differs between objects, the energy transferred is called ____________.
3. Energy that flows out of a system is given a _____________________ sign. Reactions in which this occurs are called _______________________.

4. It is possible for state functions to return to their original condition. As a result, they are said to be completely ________________.
5. If there is no change in the ________________ of a system, no ________ is done.
6. The most stable form of a substance at 25 °C and 1 atm is referred to as its ____.
7. When calorimetry is used to measure heat flows, changes in the temperature of the water can be converted to changes in heat by using the mass of the water and its ____________________________.
8. During an endothermic process, heat flows from the ____________________ to the ______________ and ΔH is ______________.
9. The energy required to break a chemical bond is its ____. It is always ___________________________.
10. The greater the disorder in a system, the higher the ________________. If the temperature of a system is increased, this property will ____________________.
11. The ____________________________ can determine the spontaneity of a process. When it is zero, the system is at ______________________. When it is ___________ than zero, the reaction will proceed spontaneously. When it is ____________ than zero, the reaction will be nonspontaneous.

Problems

1. Kinetic energy is defined as $1/2\ mv^2$. How fast would a 2 gram marble need to be moving to have the same kinetic energy as a 1.3 kg bowling ball moving at 5 m/s?

2. If 950 mL of a gas is compressed to 550 mL under a constant external pressure of 8.00 atm and the gas absorbs 12 kJ, what are the values of q, w, and ΔE for the gas? Is work being done on the system or by the system? What is the value of ΔE for the surroundings?

3. If the specific heat of mercury is 0.140 J/g•°C, how much will the temperature of 15 grams of mercury change if 150 joules of heat are added to it?

4. If 2.00 L of chlorine gas react stoichiometrically with 1.00 L of oxygen at 1 atm to form 1.00 L of Cl_2O gas according to the following reaction, what is the overall energy change of the system (the density of chlorine gas is 3.17 g/L)? Is this an endothermic reaction or an exothermic reaction?

   $Cl_2\ (g) + \frac{1}{2}\ O_2\ (g) \rightarrow Cl_2O\ (g)$ $\qquad \Delta H = 80.3$ kJ/mol

5. Given the data above, and the following entropy values, is this reaction spontaneous at 25 °C? Will it be spontaneous at any temperature?

S° Cl_2	223.0 J/mol•K
S° O_2	205 J/mol•K
S° Cl_2O	266.1 J/mol•K

6. The ΔH of solution for $MgSO_4$•7 H_2O (Epsom salts) is +16.11 kJ/mol. If 15 g of Epsom salts are dissolved in 100 g of water at 24.36 °C, what is the final temperature of the solution? (Assume that the specific heat and density of the solution are the same as that of pure water.)

7. How much energy, in kJ, is required to take 50 g of water from -10 °C to steam at 100 °C? Use the following values:

c_{ice} = 2.108 J/g•°C
c_{water} = 4.184 J/g•°C
ΔH_{fus} = 6.01 kJ/mol
ΔH_{vap} = 40.7 kJ/mol

What is the largest contributor to the energy consumption? Why is this?

8. A 6.73 g chunk of metal known to be a mixture of copper (c = 0.385 J/g•°C) and gold (c = 0.129 J/g•°C) is taken from boiling water and dropped into 25 g of water at 25.00 °C. The final temperature is 26.65 °C. How many grams of copper and how many grams of gold are present in the sample?

9. Use Hess's Law to determine the value of ΔH for the following reaction:

$$2\ Fe\ (s) + 3/2\ O_2\ (g) \rightarrow Fe_2O_3\ (s)$$

from the following:

$Fe_2O_3\ (s) + 3\ CO\ (g) \rightarrow 2\ Fe\ (s) + 3\ CO_2(g)$ $\quad \Delta H° = -26.7$ kJ

$CO + ½\ O_2 \rightarrow 3\ CO_2\ (g)$ $\quad \Delta H° = -283.0$ kJ

10. Spraying water on a kitchen grease fire can spread it, so in commercial kitchens a large box of baking soda is often kept in case of fire. When it is thrown on the flames, they physically put the fire out, but the sodium bicarbonate also decomposes to form carbon dioxide and water by the following reaction:

$$2\ NaHCO_3\ (s) \rightarrow Na_2CO_3\ (s) + H_2O\ (l) + CO_2\ (g)$$

Given the following heats of formation, what is the approximate ΔH of this decomposition reaction?

$\Delta H_f^\circ\ NaHCO_3 = -947.7$ kJ/mol
$\Delta H_f^\circ\ Na_2CO_3 = -1131$ kJ/mol
$\Delta H_f^\circ\ H_2O = -285.9$ kJ/mol
$\Delta H_f^\circ\ CO_2 = -393.5$ kJ/mol

11. Using the bond dissociation energies given, calculate the enthalpy of the following reaction:

$2\ Cl_2\ (g) + CH_4\ (g) \rightarrow 2\ H_2\ (g) + CCl_4\ (g)$

Cl-Cl = 243 kJ/mol
H-H = 436 kJ/mol
C-H = 410 kJ/mol
C-Cl = 330 kJ/mol

12. Given the entropy values below, at approximately what temperature will the reverse of the above reaction become spontaneous?
$S°\ Cl_2$ = 223 J/mol•K
$S°\ H_2$ = 130.6 J/mol•K
$S°\ CH_4$ = 188 J/mol•K
$S°\ CCl_4$ = 214 J/mol•K

13. Using the following standard heats of formation, calculate the heat of combustion of pentane (C_5H_{12}) in kJ/mol and kJ/g. The density of pentane is 0.626 g/cm^3: what is the heat of combustion per mL?

ΔH°_f pentane $= -146.3$ kJ/mol

ΔH°_f H_2O $= -285.9$ kJ/mol

ΔH°_f CO_2 $= -393.5$ kJ/mol

Inquiry Based Problem

Anyone who has ever put Epsom salts in a bath is aware that the dissolving of a salt can alter the temperature of a solution. The heat of reaction can be measured in the laboratory by assembling a *calorimeter* consisting of two Styrofoam cups nested together and covered by a piece of corrugated cardboard through which is inserted a thermometer. Although some heat will be lost or gained by the apparatus itself, it is accurate enough to estimate the heats of dissociation reactions.

A student is given a 5.0 g sample of an unknown substance. Describe how to determine the approximate heat of dissociation. What units will this heat of dissociation have?

The possible identities of the unknowns are:

$CaCl_2$	ΔH_{sol} = -81.3 kJ/mol
NH_4NO_3	ΔH_{sol} = +25.8 kJ/mol
$NaCH_3COO$	ΔH_{sol} = -17.4 kJ/mol
NaCl	ΔH_{sol} = +3.88 kJ/mol

Which of these salts will increase the temperature of the solution? Which will decrease the temperature of the solution?

Convert these heats of solution into a form that is useful for the experiment at hand.

If the unknown substance causes a temperature change of -1.6 °C when dissolved in 50 g of water, which of these unknowns is it?

How many joules would be involved in the dissolution of 5 grams of each of the unknowns in 50 grams of water?

What would be the expected temperature change when each of these was dissolved in 50 g of water?

CHAPTER 9

GASES: THEIR PROPERTIES AND BEHAVIOR

Chapter Learning Goals

A. Gas Laws

1. Explain how the height of a liquid in a barometer depends on the density of the liquid.
2. Interconvert units of pressure.
3. Know how to determine the pressure of a gas using a manometer.
4. Use the ideal gas law to calculate pressure, volume, moles of gas, or temperature, given the other three variables.
5. Use the ideal gas law to calculate final pressure, volume, moles of gas, or temperature from initial pressure, volume, moles of gas, and temperature.
6. Perform stoichiometric calculations relating the mass of a reactant to the mass, moles, and volume or pressure of a gaseous product.
7. Use the ideal gas law to calculate the molar mass of a gas.
8. Use the ideal gas law to calculate the density of a gas.
9. Use Dalton's law to calculate the partial pressure of a gas in a mixture.

B. Behavior of Gases: Kinetic–Molecular Theory

1. Use the kinetic–molecular theory of gases to explain each of the gas laws.
2. Use Graham's law to calculate the relative rates of effusion of two different gases.
3. State the conditions under which a gas is expected to behave ideally or nonideally.

Chapter in Brief

Gases behave very differently than solids or liquids. They are compressible, as they are mostly empty space, and they exert pressure as a result of collisions between their rapidly moving particles and the walls of any container in which they are placed. Extensive experimental work in the early days of the science of chemistry worked out the gas laws: simple equations that can be used to predict how gases will behave when conditions are varied. The four variables that are relevant to gases are temperature, pressure, volume, and amount. Knowing the mathematical relationships between these variables makes it possible to perform stoichiometry with gas-generating equations. Mixtures of gases can also be dealt with using the gas laws, and in particular Dalton's law of partial pressures. Kinetic molecular theory provides a model for gas behavior that explains the gas laws, and a basis for understanding effusion and diffusion. Generally, it is assumed that gases will behave in an ideal way, that is, they will adhere closely to the tenets of kinetic molecular theory. At high pressures, this ceases to be true, so additional strategies are needed. The chapter ends with a brief look at the Earth's atmosphere.

9.1 Gases and Gas Pressure

The air in which we live is primarily nitrogen, 78% by volume, with the rest being made up primarily of oxygen and about 1% argon. All other gases, including carbon dioxide, exist only in very small amounts.

Gases have very consistent properties. All gases will mix with all other gases to form **homogeneous** mixtures. Gas molecules are very far apart, and interact very little, so the properties of one gas molecule compared to another do not really matter.

Gases are highly **compressible**, also as a result of the distance between molecules. The volume of a gas is 99.0% empty space.

Gases exert a measurable pressure on the walls of their container. Pressure is force per unit area, and force is mass x acceleration.

$$P = \frac{\text{force}}{\text{area}} = \frac{\text{kg} \bullet \frac{\text{m}}{\text{s}^2}}{\text{m}^2} = \frac{\text{N}}{\text{m}^2} = \text{Pa}$$

The S.I. unit for force is the Newton (N), and the unit for pressure is the pascal (Pa).

Pascals are very small, and there are a number of alternative pressure units, including **millimeters of mercury** (mmHg), also called **torr** after the inventor of the **barometer**, Evangelista Torricelli. In this piece of equipment, an open dish of mercury has in it a thin, glass tube full of mercury, closed at the top end, and open at the bottom. An increase in atmospheric pressure will force mercury higher into the tube, and a drop in atmospheric pressure will allow the mercury also to drop.

Atmospheric pressure is defined as 760 mmHg. This is about 101,000 Pa or 101 kPa. Because the pascal is inconveniently small for chemistry, it is rarely used. Instead, mmHg, atmospheres (atm), and bar (1 bar = 100,000 Pa) are more commonly used.

To measure the pressure of a gas, an open-end manometer can be used: this compares the pressure of the container to atmospheric pressure. If the liquid level in both is equal, the pressure of the contained gas is equal to atmospheric pressure. If the liquid is lower on the container side, the pressure in the container is higher than atmospheric pressure, and if the liquid is higher on the container side, the pressure is lower in the container. The differences can be measured to quantify the pressure in the container.

A barometer or manometer can be filled with a liquid other than mercury, of course. In that case, the density of the liquid can be used as a conversion factor to mmHg.

With problems involving a manometer, when in doubt, draw a picture!

EXAMPLE:

A barometer filled with water has a height of 9.563 m. What is the atmospheric pressure in atmospheres? The density of Hg is 13.6 g/mL, and the density of water is 1.00 g/mL.

SOLUTION:

The height of the barometer can be converted to mmHg by using the relative densities of the two liquids as a conversion factor. The pressure in mmHg can then be converted to atm.

$$9.563 \text{ m x} \frac{1000\text{mm}}{\text{m}} \text{ x} \frac{1.00 \text{ g/mL}}{13.6 \text{ g/mL}} = 703 \text{ mm Hg}$$

$$703 \text{ mm Hg x} \frac{1 \text{ atm}}{760 \text{ mm Hg}} = 0.925 \text{ atm}$$

EXAMPLE:

If a gas is contained in an open-end manometer, and the mercury on the side of the gas is 37 mm higher than the mercury in the side open to the air, what is the pressure of the contained gas?

SOLUTION:

The gas is at less than atmospheric pressure, as the pressure of the air on the open end of the tube pushes the mercury toward the contained gas. Atmospheric pressure is 760 mmHg, so the pressure of the gas must be 760 mmHg – 37 mmHg = 723 mmHg.

Workbook Problem 9.1

If the difference in height between the two sides of a manometer is 73 mmHg, with the mercury higher on the open side of the manometer, what is the pressure of the contained gas in atmospheres?

Strategy: Determine the pressure of the gas in mmHg, convert to atmospheres.

Step 1: Determine the pressure of the gas in mmHg.

Step 2: Convert from mmHg to atmospheres.

9.2 The Gas Laws

Although they have very different chemical properties, gases have very similar physical properties that can be defined by four variables: **pressure**, **temperature**, **volume,** and **number of moles** (amount). The relationships among these variables form the gas laws, and a substance that follows them exactly is known as an **ideal gas**.

The gas laws in this section appear almost absurdly simple, but they were worked out in the 17th and 18th centuries by extensive experimentation, and contributed tremendously to the understanding of matter that students today take for granted.

Boyle's law – When only volume and pressure are varying, they are inversely related: when volume increases, pressure decreases. Mathematically:

$$V \propto 1/P,\ PV = k \text{ (a constant), or, most usefully, } P_1V_1 = P_2V_2$$

Charles' law – When only volume and temperature are varying, they are directly related: when temperature (molecular kinetic energy, recall) drops, the volume drops as well. Temperatures *must* be expressed in Kelvin for the proportionality to be seen. Mathematically:

$$V \propto T,\ V/T = k \text{ (a constant) or, } \frac{V_1}{T_1} = \frac{V_2}{T_2}$$

Charles law is also the basis of the Kelvin scale: if volume is plotted against temperature, there is a point at which the volume drops (theoretically) to zero. This occurs at -273.15 °C, a temperature which is known as *absolute zero*. This is the lowest possible temperature, and so is 0 K.

Avogadro's Law – if only amount and volume are varying, they are directly related: more moles of gas will always mean more volume, with all else being equal. Mathematically:

$$V \propto n \text{ (}n\text{ = number of moles), } \frac{V}{n} = k \text{ (a constant), or } \frac{V_1}{n_1} = \frac{V_2}{n_2}$$

It can also be seen from Avogadro's Law that one mole of any gas will have the same volume at the same temperature and pressure. In fact, 1 mol of any gas at 0 °C and 1 atm will have a volume of 22.4 L. These conditions (0 °C and 1 atm) are referred to as STP – standard temperature and pressure – for gas law calculations.

EXAMPLE:

If 2 moles of a gas at 150 K and 1 atm occupy 24.6 L, how many liters will 4.3 moles of gas occupy under the same conditions?

SOLUTION:

There are a lot of numbers in this problem, making the primary challenge figuring out what is important and what is not. The temperature and pressure of the gas are given, but they do not vary, so need not be considered. The important numbers are the moles of gas, and the one volume given. This is enough information to solve for the other volume using Avogadro's law:

$$\frac{24.6 \text{ L}}{2 \text{ mol}} = \frac{V_2}{4.3 \text{ mol}}$$

$$V_2 = \frac{24.6 \text{ L x } 4.3 \text{ mol}}{2 \text{ mol}} = 52.9 \text{ L}$$

Workbook Problem 9.2

If a quantity of oxygen occupies 27 L at 37 °C, how many liters will it occupy at 350 °C?

Strategy: Determine which gas law is involved, convert temperatures, solve for the final volume.

Step 1: Determine the gas law to use.

Step 2: Convert the temperatures to K.

Step 3: Solve for the final volume.

9.3 The Ideal Gas Law

All three gas laws can be combined to form the ideal gas law, thus:

$$PV = nRT$$

R is the gas law constant = 0.08206 $\frac{\text{L} \cdot \text{atm}}{\text{K} \cdot \text{mol}}$. It can be calculated from the molar volume of a gas thus:

$$R = \frac{PV}{nT} = \frac{1 \text{ atm x } 22.414 \text{ L}}{1 \text{ mol x } 273.15 \text{ K}} = 0.08206 \frac{\text{L} \bullet \text{atm}}{\text{mol} \bullet \text{K}}$$

The ideal gas law can be used to calculate any one of the four variables that describe gases if the other three are known, and can also be rearranged to solve for any of the four variables, thus:

$$PV = nRT$$

$$P = \frac{nRT}{V}$$

$$V = \frac{nRT}{P}$$

$$n = \frac{PV}{RT}$$

$$T = \frac{PV}{Rn}$$

If the problem instead involves a change, and you are unsure of which gas law to use, try the following trick: set your two conditions equal to each other like this:

$$\frac{P_1V_1}{n_1RT_1} = \frac{P_2V_2}{n_2RT_2}$$

Then cancel out anything that is not changing. For instance, if the experiment is being done at constant pressure, $P_1 = P_2$ and they can be cancelled out. R is always the same on both sides, so can always be cancelled out.

Note: the standard state for gas laws is not the same as the thermodynamic standard state (apparently gas-law chemists like to be cold).

Also, there are no really "ideal" gases – all of them deviate from ideal behavior to a certain extent. However, the deviations are usually so small as to be negligible.

EXAMPLE:

What is the volume of 7.62 g of Cl_2 at 45.7 °C and 0.257 atm?

SOLUTION:

First, solve for the moles of chlorine, and the temperature in K:

$$n = 7.62 \text{ g } Cl_2 \text{ x } \frac{1 \text{ mol } Cl_2}{71.0 \text{ g}} = 0.107 \text{ mol } Cl_2$$

$$45.7 \text{ °C} + 273.15 = 318.85 \text{ K}$$

Then solve for the volume:

$$V = \frac{nRT}{P} = \frac{0.107 \text{ mol x } 0.08206 \frac{\text{L} \bullet \text{atm}}{\text{mol} \bullet \text{K}} \text{ x } 318.85 \text{ K}}{0.257 \text{ atm}} = 10.89 \text{ L}$$

Workbook Problem 9.3

If 6.7 g of fluorine were placed in a 3.0 L container at 25.0 °C, what pressure would there be in the container?

Strategy: Rearrange the ideal gas law to solve for pressure, convert grams to moles, and °C to K, solve.

Step 1: Rearrange the ideal gas law.

Step 2: Convert grams to moles and °C to K.

Step 3: Solve for the final pressure.

Workbook Problem 9.4

If 12.3 g of neon were taken from 3.2 L at 1 atm and 25 °C to 18 atm and 750° C, what would the final volume of the gas be?

Strategy: Determine what form of the gas laws is needed, convert temperatures and moles if necessary, and solve for final pressure.

Step 1: Determine the form of gas laws needed.

Step 2: Convert grams to moles and °C to K if needed.

Step 3: Solve for the final pressure.

9.4 Stoichiometric Relationships with Gases

Many chemical reactions involve gases as either reactants or products. The gas laws make it possible to calculate the volumes of these gases as well.

A corollary of Avogadro's Law is that the molar volumes of gases are equivalent. In other words, if you have 35 mL of one gas, and 35 mL of another gas, you have the same number of particles in each volume, and thus, the same number of moles.

The density of a gas can also be calculated using the ideal gas law: density is simply mass/volume. If either mass or number of moles is known, the density can be calculated.

If mass of a gas is known, the molar mass can also be calculated – solve for n in the ideal gas law, and then divide grams/mole to calculate the molar mass.

EXAMPLE:

If 2.52 g of zinc are dropped into concentrated HCl, how many liters of hydrogen will be produced at 33 °C and 1 atm?

$$\mathrm{Zn}\,(s) + 2\mathrm{HCl}\,(aq) \rightarrow \mathrm{ZnCl_2}\,(aq) + \mathrm{H_2}\,(g)$$

SOLUTION:

First, solve for the moles of hydrogen gas produced, then use the ideal gas law to calculate its volume:

$$2.52\text{ g Zn x } \frac{1\text{ mol}}{65.39\text{ g}} = 0.0385\text{ mol Zn x } \frac{1\text{ mol H}_2}{\text{mol Zn}} = 0.0385\text{ mol H}_2$$

$$V = \frac{nRT}{P} = \frac{0.0385\text{ mol}\cdot 0.08206\,\frac{\text{L}\cdot\text{atm}}{\text{mol}\cdot\text{K}}\cdot(33 + 273.15)\text{K}}{1\text{ atm}} = 0.967\text{ L}$$

EXAMPLE:

What is the density of chlorine gas at 85 °C and 1.5 atm?

SOLUTION:

It does not look like there is enough information here, but there is! Determine the volume of 1 mole of gas under the given conditions. The mass of 1 mole of Cl_2 is known, so the density can be calculated.

$$V = \frac{nRT}{P} = \frac{1\text{ mol}\bullet 0.08206\,\frac{\text{L}\bullet\text{atm}}{\text{mol}\bullet\text{K}}\bullet(85+273.15)\text{K}}{1.5\text{ atm}} = 19.59\text{ L}$$

1 mol of Cl_2 = 71 g, so the density of chlorine under these conditions is $\frac{71\text{g}}{19.59\text{L}} = 3.6\frac{\text{g}}{\text{L}}$

EXAMPLE:

If a sample of gas weighing 11.2 grams has a volume of 3.13 L at 300 K and 2.0 atmospheres, what is the molar mass of the gas? What is a possible identity for the gas?

SOLUTION:

To figure this out, first solve for the number of moles of gas present:

$$n = \frac{PV}{RT} = \frac{2.0\text{ atm x }3.13\text{ L}}{0.08206\frac{\text{L}\bullet\text{atm}}{\text{mol}\bullet\text{K}}\text{ x }300\text{ K}} = 0.254\text{ mol}$$

So 0.254 mol of the gas weighs 11.2 grams. This makes it possible to solve for the molar mass:

M.W. = 11.2 g/0.254 mol = 44.0 g/mol. It is likely that the gas is carbon dioxide.

Workbook Problem 9.5

In the following reaction, how many liters of oxygen will be needed to react with 7.45 L of NO? If the temperature of the reaction does not change, how many liters of NO_2 will be produced?

$$2\text{ NO }(g) + O_2\,(g) \rightarrow 2\text{ NO}_2\,(g)$$

Strategy: Determine which gas law is involved, solve for the volumes of oxygen and nitrogen dioxide.

Step 1: Determine the gas law to use.

Step 2: Solve for the volume of oxygen.

Step 3: Solve for the volume of nitrogen dioxide.

Workbook Problem 9.6

What is the density of xenon at 300 mm Hg and 100 °C?

Strategy: Convert temperatures and pressures to make it possible to solve for the volume of 1 mole under these conditions. Divide the mass of 1 mole of xenon by the volume to determine the density.

Step 1: Convert the temperature and pressure.

Step 2: Use the ideal gas law to calculate the volume of the gas.

Step 3: Determine the density of the gas.

Workbook Problem 9.7

The density of a gas was found to be 0.798 g/L at 700 °C and 3.75 atm. What is the molar mass of the gas? Can you suggest a possible identity?

Strategy: Decide on a volume of gas, then solve for n. Using the density makes it possible to solve for the mass for that volume, and then for the molar mass.

*Step 1:*Determine the number of moles in 1 L of gas.

Step 2: Solve for the mass in 1 L of gas using the density.

Step 3: Determine the molar mass by dividing g/moles.

9.5 Partial Pressure and Dalton's Law

Because gas particles do not interact much with one another, each gas in a mixture remains independent. This makes dealing with mixtures less complex than might be expected.

Dalton's Law of Partial Pressures states that the total pressure of a mixture of gases is the sum of the pressures of the individual gases:

$$P_{total} = P_1 + P_2 + P_3 + ... + P_n$$

Since all the gases in a mixture are by necessity at the same temperature and pressure, the pressure exerted by a mixture of gases is purely dependent on the number of moles in the mixture.

To determine the pressures of individual components of the mixture, the mole fraction is used. This is calculated by dividing the number of moles of the component of interest by the total number of moles in the mixture:

$$\text{Mole fraction } (X) = \frac{\text{moles of component}}{\text{total moles in mixture}}$$

The pressure of the individual component is then calculated as the total pressure of the gas times the mole fraction of the individual component:

$$P_1 = X_1 \times P_{total}$$

This equation also makes it possible to determine the mole fraction of a gas based on its partial pressure.

EXAMPLE:

A 3.0 L flask at 25 °C contains Ar at a partial pressure of 0.76 atm, He at a partial pressure of 0.32 atm, and Ne at a partial pressure of 0.42 atm. What is the total pressure of the mixture? What is the mole fraction of each gas?

SOLUTION:

P_{total} = 0.76 atm + 0.32 atm + 0.42 atm = 1.5 atm

The mole fraction of each gas can be calculated from the partial pressure:

$$X_{Ar} = \frac{0.76 \text{ atm}}{1.5 \text{ atm}} = 0.51;\ X_{He} = \frac{0.32 \text{ atm}}{1.5 \text{ atm}} = 0.21;\ X_{Ne} = \frac{0.42 \text{ atm}}{1.5 \text{ atm}} = 0.28$$

Note: Mole fractions are unitless, and will always add up to 1, just as percentages always add up to 100.

Workbook Problem 9.8

A 30 L tank is filled with the following: 18.3 g CO_2, 23.1 g Ne, and 5.26 g H_2. If the container is at 25 °C, what is the total pressure in the container, and what are the partial pressures of the individual gases?

Strategy: Determine the number of moles of each gas, the total number of moles, and the mole fraction of each component. Using these, calculate the total pressure and the partial pressures using the ideal gas law and Dalton's law.

Step 1: Determine the number of moles of each gas.

Step 2: Use the total number of moles to calculate the total pressure.

Step 3: Use the mole fraction (X) of each component to determine the partial pressures.

9.6 The Kinetic-Molecular Theory of Gases

The kinetic-molecular theory is a model developed over a century ago that describes the macroscopic behavior of gases based on a series of assumptions:

- Gas particles are tiny, and move randomly.
- The volume of the gas particles themselves is negligible.
- The gas particles are independent.
- Collisions of gas particles, with one another or with the walls of the container, are elastic. That is, no energy is lost in these collisions, so the total kinetic energy is constant at constant temperature.
- The average kinetic energy is proportional to Kelvin temperature.

These assumptions can be used to explain the individual gas laws quite effectively:

Boyle's Law: Pressure increases as volume decreases, because the smaller the space, the greater the number of collisions with the walls, and thus the greater the pressure.

Charles' Law: Temperature is a measure of kinetic energy. If the pressure is constant, particles moving faster will require more space.

Avogadro's Law: Volume increases as amount increases at constant pressure because adding more particles to the mixture will increase the number of collisions unless the space is also increased.

Dalton's Law: Because the identity of the particles is irrelevant and they do not interact, the pressures of individual component will depend only on the mole fraction of that component.

It is possible to derive from this the average speed of the particles of a gas:

$$u = \sqrt{\frac{3RT}{M}}$$

In this equation, u is speed in m/s, M is the molecular mass in kg/mol, and a different form of R must be used: $8.314 \frac{J}{K \bullet mol}$. All of these changes are required to make the units work out to m/s – remember from the last chapter that joules are derived from the definition of kinetic energy, and are $kg \text{ x } \frac{m^2}{s^2}$.

Gas particles move amazingly quickly. Larger molecules move more slowly than smaller molecules, and it important to remember that this is an average: there will be a distribution of actual speeds, but the total kinetic energy will remain the same. Also, gas particles do not move very far before colliding either with another gas particle or with something else. The *mean free path*, as it is known, is longer for a smaller, faster particle, and shorter for a larger, slower-moving particle. So gas molecules proceed in a zig-zag fashion.

EXAMPLE:

Calculate the average speed of a xenon atom at 398 K.

SOLUTION:

$$u = \sqrt{\frac{3RT}{M}} = \sqrt{\frac{3 \text{ x } 8.324 \frac{kg \cdot m^2/s^2}{K \cdot mol} \text{ x} 398K}{0.131 \text{ kg/mol}}} = \sqrt{76{,}000 \text{ m}^2/s^2} = 275 \text{ m/s}$$

Workbook Problem 9.9

What temperature is required to get gaseous iodine molecules moving at the same average speed achieved by hydrogen molecules at room temperature?

Strategy: Determine the speed of H_2 molecules at 25 °C, then solve the equation again for temperature needed for the much larger I_2 molecules.

Step 1: Determine the speed of H_2 at 25 °C

Step 2: Determine the temperature required for the iodine molecules to achieve the same speed.

9.7 Graham's Law: Diffusion and Effusion of Gases

Because of the high speeds at which gas molecules move, and their frequent collisions, gases mix rapidly with one another (this is why you can smell baking cookies from quite a distance). This rapid mixing of gases with frequent collisions is called **diffusion**: it is difficult to mathematically model diffusion because of the random nature of the collisions.

More readily characterized mathematically is **effusion**. This is the progress of gas particles through a tiny hole into a vacuum. Not surprisingly, this is inversely proportional to molar mass: a bigger, and thus, slower-moving, gas will effuse more slowly that a small, faster-moving gas.

Graham's law describes this property mathematically:

$$\text{Rate} \propto \frac{1}{\sqrt{M}}$$

This can be used most simply to calculate the relative rates of effusion in a mixture of gases:

$$\frac{\text{Rate}_1}{\text{Rate}_2} = \sqrt{\frac{M_2}{M_1}}.$$

EXAMPLE:

Calculate the ratio of effusion rates of Ne and Xe from the same container at the same temperature and pressure.

SOLUTION:

$$\frac{\text{Rate of effusion of Ne}}{\text{Rate of effusion of Xe}} = \sqrt{\frac{131.3 \text{ g Xe/mol}}{20.18 \text{ g Ne /mol}}} = 2.55$$

Workbook Problem 9.10

Oxygen-16 has an atomic mass of 15.995, and Oxygen-18 an atomic mass of 17.999. What will be their relative rates of effusion?

Strategy: Solve using Graham's law.

Step 1: Solve using Graham's law:

9.8 The Behavior of Real Gases

The ideal gas law is a good approximation of the behavior of gases under non-extreme conditions, but begins to fail when pressure is very high.

The actual volume of gas particles can no longer be neglected when they are very close together. As a result, the volume of a real gas is larger than that of an ideal gas at high pressure.

Attractive forces between gas particles will also become a factor at high pressures: these tend to reduce the volume of the gas.

The two factors negate each other at moderate pressures, but above 350 atm, the volume of the gas particles becomes the primary factor.

These factors have been accommodated mathematically in the **van der Waals equation**:

$$\left(P + {an^2}/{V^2}\right)\left(V - nb\right) = nRT$$

"a" is a correction for intermolecular attractions, "b" is a correction for molecular volume. Solving this equation for pressure can be done fairly simply, but solving for volume is non-trivial.

EXAMPLE:

Calculate the pressure of 1 mol of N_2 gas in 50 mL at 300 K using the ideal gas law and the van der Waals equation. The van del Waals constants for nitrogen are $a = 1.35(\mathrm{L^2 \bullet atm})/\mathrm{mol}^2$ and $b = 0.0387$ L/mol.

SOLUTION:

Ideal gas law:

$$\mathrm{P} = \frac{\mathrm{nRT}}{\mathrm{V}} = \frac{1\ \mathrm{mol\ x\ 0.08206}\ \dfrac{\mathrm{L \cdot atm}}{\mathrm{mol \cdot K}}\ \mathrm{x\ 300K}}{0.050\ \mathrm{L}} = 492\ \mathrm{atm}$$

van der Waals equation:

$$\mathrm{P} = \frac{\mathrm{nRT}}{\mathrm{V - nb}} - \frac{\mathrm{an^2}}{\mathrm{V^2}} = \frac{1\ \mathrm{mol\ x\ 0.08206}\ \dfrac{\mathrm{L \cdot atm}}{\mathrm{mol \cdot K}}\ \mathrm{x 300K}}{0.050\ \mathrm{L - (1\ mol\ x\ 0.0387\ L/mol)}} - \frac{(1\ \mathrm{mol})^2\ \mathrm{x}\ 1.35\ (\mathrm{L^2 \cdot atm})/\mathrm{mol}^2}{(0.050\ \mathrm{L})^2}$$

$$= \frac{24.6\ \mathrm{L \cdot atm}}{0.0113\ \mathrm{L}} - \frac{1.35\ \mathrm{L^2 \cdot atm}}{0.00250\ \mathrm{L^2}} = 2179\ \mathrm{atm}\ -\ 540\ \mathrm{atm}\ =\ 1639\ \mathrm{atm}$$

At very high pressures, the ideal gas law fails rather dramatically.

Workbook Problem 9.11

Calculate the pressure of 1 mol of O_2 gas in 100 mL at 300 K using the ideal gas law and the van der Waals equation. The van del Waals constants for oxygen are $a = 1.38(L^2 \bullet atm)/mol^2$ and $b = 0.0318$ L/mol. Does the ideal gas law approximate the pressure acceptably under these conditions?

Strategy: Solve using the ideal gas law and the van der Waals equation.

Step 1: Solve using the ideal gas law.

Step 2: Solve using the van der Waals equation.

Step 3: Compare the two values.

9.9 The Earth's Atmosphere

The atmosphere has four major regions. The *troposphere* is closest to the Earth's surface and both has the greatest effect on the surface and is most affected by human activity. Above the troposphere is the *stratosphere*, which extends from approximately 12 to 50 km above the earth's surface. The *mesosphere* extends another 35 km, and is below the final layer of the atmosphere, the *thermosphere,* which extends to 120 km.

In the troposphere, there are three major effects of human activity:

- Air pollution: as a side product of the Industrial Revolution, there has been an increase in hydrocarbon molecules and NO in the air. NO reacts to form NO_2, which is split by sunlight into NO and free radical oxygen molecules. These can attack oxygen molecules, forming ozone (O_3), a highly reactive molecule which further reacts to form *smog*.
- Acid rain: when high-sulfur coal is burned, SO_2 is released into the air. This is further oxidized to form SO_3, which can combine with water to form sulfuric acid. This has damaged forests and lakes, causing the extinction of fish in some areas. It also causes marble and limestone – two materials common in buildings and monuments – to slowly dissolve.
- Global warming: Some scientists worry that the increase in the levels of CO_2 in the atmosphere as a result of industrialization may reduce the amount of radiation that can be reflected back out into space. This has the potential to cause the warming of the troposphere.

The **ozone layer** is an atmospheric band that stretches from 20 to 40 km above the Earth's surface. Ozone is an irritating pollutant at low levels in the atmosphere, but absorbs ultraviolet light in the upper atmosphere, acting like sunscreen for the Earth below. Changes in the ozone layer, particularly thinning around the poles, are attributed to chlorofluorocarbons, which also react with ultraviolet light to form chlorine radicals that destroy ozone. These molecules have been banned internationally, but the ban is largely ignored in China and Russia. It is expected that the levels of CFCs will return to pre-1980 levels by mid-century.

Putting It Together

13.67 L of an unknown gas (kept at 3 atm and 500 K) weigh 197 g. What is the molecular formula of this gas if it is found to be 12.19% C, 0.51% H, 28.93% F, 18.02% Cl, and 40.56% Br?

Self-Test

This section is intended to test your knowledge of the material covered in this chapter. Think through these problems, and make certain you understand what they are asking. Make sure your answers make sense. Successful completion of these problems indicates that you have mastered the material in this chapter. You will receive the greatest benefit from this section if you use it as a mock exam, as this will allow you to determine which topics you need to study in more detail.

True-False

1. The atmosphere is about 70% oxygen.
2. Gases are more compressible than liquids.
3. Pascals, atmospheres, bars, and millimeters of mercury are all accepted units of pressure.
4. If the mercury in a manometer is lower on the side that is open to the air, then the pressure of the gas is lower than the pressure of the air.
5. Stoichiometric ratios can be determined by volume in gas reactions.
6. The smaller a gas particle, the slower it travels.
7. The ratio of effusion rates of two gases is directly proportional to the square roots of their masses.
8. The ideal gas law begins to fail at high temperatures.
9. When you inhale, you are breathing stratosphere.
10. Incomplete combustion of hydrocarbons is a major factor in air pollution.

Matching

Barometer	a. force per unit area
Pressure	b. the pressure of one component in a gas mixture is related to its mole fraction on the mixture
Atmospheric pressure	c. the rate at which a gas effuses is inversely proportional to the square root of its molar mass
Boyle's law	d. the escape of gas particles through a tiny hole in a membrane into a vacuum
Charles's law	e. the pressure exerted by one component in a gaseous mixture

Avogadro's law	f. the chaotic mixing of gases through random collisions
Ideal gas law	g. number of moles of a component divided by the total moles in a mixture.
Dalton's law	h. volume and temperature vary directly when pressure and amount are held constant
Partial pressure	i. volume and pressure are inversely proportional when temperature and amount are constant
Kinetic–molecular theory	j. a piece of equipment used to measure atmospheric pressure
Graham's law	k. equipment for measuring gas pressures
Diffusion	l. pressure generated by the atmosphere pressing down on the Earth's surface
Effusion	m. allows any of the four variables to be calculated - T, P, V or n – if the others are known
Ideal gas	n. a model that accounts for the behavior of gases
STP	o. moles and volume vary directly.
Standard molar volume	p. gas that follows precisely the tenets of kinetic molecular theory
Mole fraction	q. 22.4 L at STP
Manometer	r. 0 °C and 1 atmosphere pressure

Fill-in-the-Blank

1. Gases exert ________________ on the walls of their container as a result of __________________ with the walls.
2. The physical properties of a gas can be defined by four variables: ____________________, ____________________, ____________________, and ____________________.
3. The concentration of a component in a gas mixture is typically calculated as its ________________. This can also be used to determine the ________________ ________________ of the component.
4. In kinetic molecular theory, the _______________ of individual gas particles is said to be _________________. The assumption fails at high ______________________.

5. Effusion rates are proportional to the square root of molar mass. Gases with __________ molar masses move significantly _________________ than gases with high molar masses.

Problems

1. If atmospheric pressure is 0.983 atm, how high a column of silicone oil (density = 0.970 g/mL) will this pressure support (the density of mercury is 13.6 g/mL)?

2. If you are using an open–ended manometer filled with ethyl alcohol, rather than mercury, what is the gas pressure in mmHg if the level of ethyl alcohol in the arm connected to the bulb is a) 340 mm lower and P_{atm} = 760 mmHg; and b) 257 mm higher and P_{atm} = 760 mmHg? The density of the alcohol is 0.789 g/mL and the density of mercury is 13.6 g/mL.

3. How many moles of gas are present in a 50 mL flask at 5.78 atm and -20 °C? If the mass of the gas is 0.281 g, what is the gas?

4. What is the density of methane (CH_4) at 100 °C and 4 atmospheres?

5. A student carried out a reaction in the lab in which one of the products was a gas. She collected the gas for analysis and found that it contained 7.69% hydrogen and 92.3% carbon. She also observed that 250 mL of the gas at 35° C and 763 mmHg had a mass of 0.258 g. a) What is the empirical formula of the gas? b) What is the molar mass of the gas? c) What is its molecular formula?

6. What is the molar mass of a gas with a density of 14.04 g/L at 15 atmospheres and 300 °C?

7. When 25 mL of butane (C_4H_{10}, density 0.6014 g/mL) are burned, how many liters of oxygen are consumed, and how many liters of carbon dioxide and water vapor are produced at 30 °C and atmospheric pressure? What is the overall change in volume for the reaction? Is work being done as well as heat given off?

8. If 7.03 g Ne, 11.7 g N_2, and 15.7 g He are placed in a 5000 mL container at room temperature (25 °C), what is the total pressure, and what is the pressure of each component?

9. What are the relative effusion rates of H_2 and He?

10. If a 50 mL aerosol can at room temperature (25 °C) with an internal pressure of 7.8 atm is emptied into a 375 mL container at atmospheric pressure, what is the final temperature of the gas in Celcius?

Inquiry-Based Problem

As a lab assistant, you have been asked to develop parameters for a gas law experiment. Hydrogen gas will be generated by dropping solid zinc into hydrochloric acid: this is an exothermic reaction, so assume that the reaction will occur at an average of 32 °C and 1 atmosphere pressure. The gas collecting apparatus can measure between 200 and 800 mL of gas effectively. Less, and there are no gradations with which to estimate, more and the bottle will be overfilled and hydrogen gas will bubble out of the reaction (possibly splashing concentrated acid around at the same time). The students will be using mossy zinc: irregularly-shaped chunks of relatively pure metal.

The question you have been asked to answer is the following: what is the mass range of zinc chunks that the students should use?

CHAPTER 10

LIQUIDS, SOLIDS, AND PHASE CHANGES

Chapter Learning Goals

A. Intermolecular Forces

1. Using molecular geometries and electronegativity trends, determine whether molecules are polar or nonpolar.
2. Identify the major type of intermolecular force present in a substance, and determine which of two substances exhibits stronger intermolecular forces.

B. Relationships Between Phases

1. For a phase change, determine whether enthalpy is increasing or decreasing and whether entropy is increasing or decreasing.
2. Use the $\Delta G = \Delta H - T\Delta S$ equation to calculate the entropy change for a phase change or the temperature at which the phase change occurs.
3. Use the Clausius–Clapeyron equation to calculate vapor pressure or heat of vaporization.
4. Sketch a phase diagram, labeling the axes and each of the regions, and locate the triple point, critical point, normal melting point, and the normal boiling point. Use the phase diagram to describe physical changes.

C. Characterization of Solids

1. For metals crystallizing in one of the three cubic unit cells, determine the number of atoms, mass, volume, density, atomic radius, and packing efficiency.

Chapter in Brief

Unlike gases, liquids and solids have interactions between molecules. This chapter discusses the various types of intermolecular forces, and the effects they have on materials' properties and phase transitions. Free energy changes accompany all physical and chemical changes: the contributions of entropy and enthalpy to phase changes, as well as the contributions of changes in pressure and temperature will all be examined. Solids take a number of forms. Those that are crystalline can also be characterized by X-ray crystallography, which provides data about unit cells. This information can be used to gain insights into packing, and allows for calculations of atomic radius and density.

10.1 Polar Covalent Bonds and Dipole Moments

Polar covalent bonds result from differences in electronegativity between atoms: these differences lead to electrons being displaced toward one part of a molecule giving it a slight negative charge while the other end of the molecule is electron-poor and has a slight positive charge.

Graphically, the **dipole**, or charge separation, is indicated with a crossed arrow: +↓→. The crossed end is the "positive" end of the molecule. The point of the arrow indicates where the electrons are concentrated: the negative end of the molecule.

Molecules larger than two atoms can also have dipoles: the contributions of lone pairs and individual bond polarities can be summed to calculate a net dipole for a shaped molecule. In symmetrical molecules, the dipoles will sum to zero:

$$\longleftarrow\!\!+\!\!\downarrow\!\!\rightarrow$$

O=C=O

To determine the percent ionic character of a bond, the calculated dipole (full charges separated by the bond length) can be compared to the measured bond dipole. To calculate dipoles, the following formula is used:

$$\mu = Q \times r$$

Q is the charge (1.160×10^{-19} C), and r is the bond length. This is then converted to *debyes* (D). 1 debye is 3.336×10^{-30} C•m.

$$\%\text{ ionic character} = \frac{\text{measured dipole}}{\text{calculated dipole}} \times 100$$

EXAMPLE:

Determine the % ionic character of carbon monoxide, given the bond length of 112.8 pm and the measured dipole of 0.112 D.

SOLUTION:

Calculated dipole:

$$\mu = Q \times r = 1.160 \times 10^{-19} \times 112.8 \times 10^{-12}\,\text{m} \times \frac{1\text{ D}}{3.336 \times 10^{-30}\text{ C} \cdot \text{m}} = 3.92\text{ D}$$

$$\%\text{ ionic character} = \frac{0.112\text{ D}}{3.92\text{ D}} \times 100 = 2.86\%\text{ ionic}$$

Not surprisingly, carbon monoxide has very little ionic character.

EXAMPLE:

Which of the following would be expected to be polar, and which nonpolar:

CH_4 CH_3Cl HCl Cl_2

SOLUTION:

CH_4 and Cl_2 are both symmetrical molecules, which makes them non-polar. CH_3Cl and HCl are asymmetrical, and contain polar bonds. Drawing the electron-dot structures can help to confirm this initial determination:

```
    H              H
    |              |                H—:Cl:
H—C—H       H—C—:Cl:
    |              |               :Cl—:Cl:
    H              H
```

Workbook Problem 10.1

Indicate which of the following are likely to have dipole moments, and indicate the element with the highest electron density:

H_2CO- formaldehyde $N(CH_3)_3$ – trimethylamine   HBr – hydrogen bromide

Strategy: Look for asymmetrical molecules with electronegativity differences.

Step 1: Determine which molecules are asymmetrical:

Step 2: Determine which element is most electronegative, and thus the area of highest electron concentration.

Workbook Problem 10.2

BrF has a bond length of 176 pm and a measured dipole moment of 1.42 D. What is its percent ionic character?

Strategy: Calculate the dipole if it were purely ionic, then calculate the percent ionic character.

Step 1: Calculate ionic dipole.

Step 2: Determine percent ionic character.

10.2 Intermolecular Forces

Covalent and ionic bonds are *intramolecular* forces: they occur within molecules. **Intermolecular** forces occur *between* molecules.

Collectively, intermolecular forces are often called van der Waals forces. They are divided into four general categories:

- **Ion-dipole forces**: these occur between ions and polar molecules. The charge on the ion will attract the portion of a polar molecule that has an opposite partial charge. These are particularly important in aqueous solutions of ionic substances. Moderate strength: 10-50 kJ/mol.
- **Dipole-dipole forces**: dipoles will interact in liquids: the stronger the dipole, the stronger the interaction. Increasing strength of dipole moments correlates roughly with an increase in boiling points: stronger dipole-dipole interactions require more kinetic energy to disrupt. Weak: 3-4 kJ/mol.
- **London dispersion forces**: these forces are experienced by all molecules, but are the only ones that occur between nonpolar molecules. They are the result of varying electron densities: average electron densities are uniform, but a small variation in electron density can create a small dipole that can then induce an opposite dipole in another nearby molecule. *Polarizability* increases with molecular size, and varies with molecular shape: an extended molecule has more opportunity to create dipoles than one that is compact. Polarizability is related to boiling point: more dispersion forces mean a higher boiling point. Weak: 1-10 kJ/mol.
- **Hydrogen bonds**: very strong dipole-dipole forces between a hydrogen atom bonded to oxygen, nitrogen, or fluorine, and an unshared electron pair on another electronegative atom. Molecules that can hydrogen bond have boiling points that are unusually high for

their small size, as these interactions must be broken for the molecules to enter the gas phase. Moderate strength: 10-40 kJ/mol.

EXAMPLE:

Identify the intermolecular forces experienced by the following molecules, and rank them in order of increasing intermolecular forces:

NH_3 CH_4 HBr $CH_3CH_2CH_2CH_3$

SOLUTION:

NH_3 can hydrogen bond, which is the strongest intermolecular force. It will also experience dipole-dipole and dispersion forces.

CH_4 cannot hydrogen bond, and has no dipole. It will experience only dispersion forces.

HBr cannot hydrogen bond, but is polar and will have a dipole. This molecule will have both dipole-dipole and dispersion forces.

$CH_3CH_2CH_2CH_3$ cannot hydrogen bond, and has no dipole. It will experience only dispersion forces.

CH_4 will experience the weakest intermolecular forces. $CH_3CH_2CH_2CH_3$ will have stronger dispersion forces, as it is larger. HBr will have the next stronger, as it has both dipole-dipole and dispersion forces, and NH_3, as a hydrogen-bonder, will have the strongest intermolecular interactions. This will also be the ranking in order of increasing boiling point.

$$CH_4 < CH_3CH_2CH_2CH_3 < HBr < NH_3$$

Workbook Problem 10.3

Determine the intermolecular forces that are present in the following molecules and order by increasing strength:

CH_3OH Xe CH_3Cl

Strategy: Examine the structures of the molecules and determine what interactions are present, then rank in order of strength.

Step 1: List the interactions experienced by each molecule.

Step 2: Rank the molecules in order of increasing strength of interactions.

10.3 Some Properties of Liquids

Viscosity is the measure of a liquid's resistance to flow: the greater the intermolecular interactions, the more they hold on to one another: a long-chain molecule like those found in honey will not flow as readily as a nonpolar solvent held together only by dispersion forces.

Surface tension is also caused by intermolecular forces: at a liquid/gas interface, molecules with strong intermolecular forces will arrange themselves to maximize those interactions. It is this phenomenon that makes it possible for insects to walk on water, and causes water to bead up on a nonpolar surface: the nonpolar surface provides nothing for the water to hydrogen bond with, so it holds tightly together.

Both of these properties are temperature dependent: increasing kinetic energy will reduce both viscosity and surface tension.

10.4 Phase Changes

There are six possible phase changes, all associated with a change in free energy, ΔG.

- fusion (melting): solid $\rightarrow$ liquid
- freezing: liquid $\rightarrow$ solid
- evaporation: liquid $\rightarrow$ gas
- condensation: gas $\rightarrow$ liquid
- sublimation: solid $\rightarrow$ gas
- deposition: gas $\rightarrow$ solid

The free energy change can be described using the equation $\Delta G = \Delta H - T\Delta S$: and has two portions. The enthalpy change is related to the breaking or forming of intermolecular interactions, and the entropy change is associated with the increase or decrease in disorder associated with these changes.

For melting, sublimation, and vaporization, heat must be added, but disorder increases: ΔH and ΔS are both positive.

For freezing, condensation, and deposition, heat is released, but order increases: ΔH and ΔS are both negative.

At equilibrium, $\Delta G = 0$. This makes it possible to calculate freezing and boiling points, or, more usefully, to calculate entropy changes from measured freezing or boiling points:

$$T = \frac{\Delta H}{\Delta S}$$

$$\Delta S = \frac{\Delta H}{T}$$

EXAMPLE:

Methanol boils at 64.7 °C and has an enthalpy of vaporization of 35.3 kJ/mol. What is ΔS_{vap} for this phase transition?

SOLUTION:

$$\Delta S = \frac{\Delta H}{T} = \frac{35{,}300\frac{J}{mol}}{(64.7 + 273.15)K} = 104\frac{J}{mol \bullet K}$$

EXAMPLE:

A theoretical liquid has an enthalpy of fusion of -15.7 kJ/mol and an entropy of fusion of -142 J/mol•K. What is the freezing point of this liquid?

SOLUTION:

$$T = \frac{\Delta H}{\Delta S} = \frac{-15,700 J/mol}{-142 \frac{J}{mol \cdot K}} = 111 K = -163\ °C$$

Heating curves are graphical representations of phase changes. The vertical axis shows temperature, and the horizontal axis shows the heat added. What is seen in these curves, is that the temperature increases up to the normal melting point, then stops increasing as the phase change occurs. The same happens at the boiling point.

Heating curves have a number of notable features. The slopes when solids are being warmed to their melting point, liquids are being warmed to the boiling point, and gases are being heated are equivalent to the molar heat capacities of these phases. The number of kJ/mol that is added in the times that the temperature does not change are the heats of fusion (melting) and vaporization.

The heat of vaporization will always be greater than the heat of fusion: the flat line that occurs while the liquid is being vaporized is always longer than the flat portion of the line that corresponds to melting. This is due to the fact that in melting, intermolecular forces are weakened, while in vaporization they are completely eliminated.

Workbook Problem 10.4

If the enthalpy of fusion of ammonia is 5.97 kJ/mol, and the normal freezing point of ammonia is -107 °C, what is the change in entropy associated with the freezing of ammonia?

Strategy: Rearrange the equation $\Delta G = \Delta H - T\Delta S$ to solve for the entropy change.

Step 1: Rearrange to solve for ΔS.

Step 2: Solve for ΔS, being careful to maintain the correct sign.

10.5 Evaporation, Vapor Pressure, and Boiling Point

Liquid molecules at a certain temperature all have the same average kinetic energy, but some will have more energy than others. If a molecule with a high kinetic energy reaches the surface, it can escape, or evaporate.

In an open container, the molecules will drift away, and the liquid will evaporate completely. In a closed container, a dynamic equilibrium will develop: the same number of molecules will be escaping into the air as are returning to the liquid. As a result, at a specific temperature, a liquid will have a characteristic **vapor pressure**.

When the vapor pressure is equal to the external pressure, the liquid will boil. When the vapor pressure is 1 atm, the **normal boiling point** has been reached. If the external pressure is higher or lower than 1 atm, the boiling point will change correspondingly: the boiling point at lower pressures will be a lower temperature, the boiling point at a higher pressure will be a higher temperature.

When the natural log of the vapor pressure is plotted against the inverse of the temperature, they can be made to fit a linear relationship known as the Clausius-Clapeyron equation:

$$\ln P_{vap} = \left(\frac{-\Delta H_{vap}}{R}\right)\frac{1}{T} + C$$

"C" is a constant that is characteristic of each substance. This equation can also be rearranged so as to make it possible to solve for vapor pressures, temperatures, and ΔH_{vap}:

$$\ln P_1 + \frac{\Delta H_{vap}}{RT_1} = \ln P_2 + \frac{\Delta H_{vap}}{RT_2}$$

$$\Delta H_{vap} = \frac{(\ln P_2 - \ln P_1)R}{\left(\frac{1}{T_1} - \frac{1}{T_2}\right)}$$

In this situation, C can be disregarded, as it is the same on both sides of the equation. In all of these equations, R = 8.3145 J/K•mol.

EXAMPLE:

Methanol has a normal boiling point of 64.7 °C, and a vapor pressure of 400 mm Hg at 49.9 °C. What is ΔH_{vap} for methanol?

SOLUTION:

There is a great deal of information here, but some of it is hiding. What you have is two vapor pressures, and two temperatures with which to solve for ΔH. The vapor pressure of a liquid is always 760 mm Hg at its normal boiling point.

$$\Delta H_{vap} = \frac{(\ln P_2 - \ln P_1)R}{\left(\frac{1}{T_1} - \frac{1}{T_2}\right)} = \frac{(\ln 760 - \ln 400)8.314\frac{J}{mol \bullet K}}{\left(\left[\frac{1}{49.9 + 273.15}\right] - \left[\frac{1}{64.7 + 273.15}\right]\right)} = \frac{(6.63 - 5.99)8.314\frac{J}{mol \bullet K}}{0.000137\frac{1}{K}} = 38.8 kJ/mol$$

EXAMPLE:

Given the information above, what is the vapor pressure of methanol at 25 °C ?

SOLUTION:

With a variety of temperatures and vapor pressures to choose from, and having solved for ΔH_{vap} above, this is a fairly straightforward problem. Here, the equation above is used with 25 °C on one side, and the normal boiling point on the other:

$$\ln P_1 + \frac{38{,}800 J/mol}{8.314\frac{J}{mol \bullet K}(25 + 273.15)K} = \ln 760 + \frac{38{,}800 J/mol}{8.314\frac{J}{mol \bullet K}(64.7 + 273.15)}$$

$$\ln P_1 + 15.65 = 20.44$$
$$\ln P_1 = 4.787$$
$$P_1 = e^{4.787} = 120 \text{ mm Hg}$$

Workbook Problem 10.5

Ethanol has a normal boiling point of 78.4 °C, and a vapor pressure of 400 mm Hg at 63.5 °C. What is ΔH_{vap} for ethanol?

Strategy: Use the Clausius-Clapeyron equation to solve for ΔH.

Step 1: Convert temperatures to K.

Step 2: Solve for ΔH.

Workbook Problem 10.6

Given the information in the previous problem, what is the vapor pressure of ethanol at 25 °C ?

Strategy: Use the Clausius-Clapeyron equation to solve for the vapor pressure.

Step 1: Use the Clausius-Clapeyron equation to solve for the vapor pressure.

10.6 Kinds of Solids

Crystalline solids have an ordered, long-range structure. This order is visible macroscopically as sharp edges and flat faces. The four types are:

- **Ionic solids**: constituent particles are ions, held together by ionic bonds.
- **Molecular solids**: constituent particles are molecules, held together by intermolecular forces.
- **Covalent network solids**: no individual particles, but vast covalent networks.
- **Metallic solids**: individual metal atoms, but all associated in an "electron sea".

Amorphous solids have randomly-arranged constituent particles.

10.7 Probing the Structure of Solids: X–Ray Crystallography

Early in the 20th century, techniques were developed that made it possible to look into structures at the atomic level. These were – and are – based on diffraction of X-rays. It is a principle of optics that the wavelength of light must be no more than 2x the length of an object for it to be visible. As a result, the short wavelengths of X-rays are needed to visualize atoms.

When X-rays are directed at a crystal, a diffraction pattern develops as a result of the regular pattern of atoms: as the X-rays leave the crystal again, they interfere with one another both constructively and destructively. The diffraction pattern can then be interpreted to provide data about interatomic distances.

The basis for this interpretation is the Bragg equation, for which a father and son shared the Nobel Prize in physics in 1915:

$$d = \frac{n\lambda}{2\sin\theta}.$$

The wavelength, λ, is known, the angle of reflection, θ, can be measured, and n is a small , whole number, usually 1.

Computer-controlled diffractometers are now used for both data collection and analysis, making it possible to investigate the structures of even the largest macromolecules.

10.8 Unit Cells and the Packing of Spheres in Crystalline Solids

There are a few concepts that are unique to the idea of atomic packing. The **coordination number** is the number of other atoms touching the atom under discussion. Unit cells are the small repeating units of a crystal.

Four types of packing are possible:

- Simple cubic packing: atoms touch six others (coordination number of 6), 52% of space is used. The unit cell is primitive cubic: 1/8 of 8 atoms are found on the corners of the unit cell: 1 atom per unit cell, atomic diameter = edge length.
- Body-centered cubic packing: simple cubic packing arrangement with another central atom. Coordination number is 8 (four neighbors above, and four below), and 68% of space is used. 2 atoms per unit cell, atomic diameter = $\frac{2d}{\sqrt{3}}$, where d is the edge length of the unit cell.
- Cubic closest packing: two alternating layers simple cubic packing plus an additional atom halfway between the two unit cells – half in and half out, as it were. Four atoms per unit cell, in face-centered cubic unit cells: coordination number is 12, and 74% of the space is used. Atomic radius = $\sqrt{\frac{d^2}{8}}$, where d is the edge length of a unit cell.
- Hexagonal closest packing: this is a non-cubic arrangement, and thus does not have a cubic unit cell: Coordination number is 12, packing efficiency is 74%, just as in cubic closest packing.

EXAMPLE:

Aluminum has a face-centered cubic unit cell. If the edge of a unit cell is 404.5 pm, what is the density of aluminum in g/cm^3?

SOLUTION:

A face–centered cubic unit cell has a total of 4 atoms.

Based on this, it is possible to calculate the mass of a unit cell from the molar mass of aluminum and Avogadro's number:

$$4 \text{ atoms x } \frac{1 \text{ mol}}{6.022\text{x}10^{23} \text{ atoms}} \text{ x } 26.98 \frac{\text{g}}{\text{mol}} = 1.792\text{x}10^{-22} \text{ g}$$

The volume of a unit cell can be calculated from the edge length, being sure to convert from picometers to centimeters:

$$(404.5 \text{ x } 10^{-12} \text{ m x } 100 \text{ cm/m})^3 = 6.618 \text{ x } 10^{-23} \text{ cm}^3$$

Knowing mass and volume, density can be calculated:

$$d = \frac{m}{V} = \frac{1.792\text{x}10^{-22} \text{ g}}{6.618\text{x}10^{-23} \text{ cm}^3} = 2.71 \frac{\text{g}}{\text{cm}^3}$$

This agrees closely with the calculated density of aluminum.

EXAMPLE:

Nickel has a face-centered cubic unit cell. If the edge of a unit cell is 352.4 pm, what is the diameter of a nickel atom in pm?

SOLUTION:

The side of a face-centered cubic unit cell has 1full atom and 2 radii as the diagonal, making the diagonal equal to 4r or 2d.

Using the Pythagorean theorem:

$(352.4 \times 10^{-12}\ m)^2 + (352.4 \times 10^{-12}\ m)^2 = (2d)^2$

$2.478 \times 10^{-19}\ m = 4d^2$

$2.489 \times 10^{-10}\ m = d$

The diameter of a nickel atom is approximately 249 pm.

Workbook Problem 10.7

Polonium has a density of 9.3 g/cm^3 and a simple cubic structure. Estimate the radius of a polonium atom.

Strategy: Use the number of atoms per unit cell and the molecular weight to determine the mass of a unit cell, then use the density to solve for the volume of a cell. With the volume, determine the length of one side, then determine the atomic radius.

Step 1: Determine the mass of a unit cell.

Step 2: Solve for the volume of a unit cell.

Step 3: Determine the edge length of a unit cell.

Step 4: Determine the atomic radius.

10.9 Structures of Some Ionic Solids

When ionic solids crystallize, they adopt a unit cell structure that allows for the accommodation of ions of different sizes: cations are smaller than their neutral atoms, and anions are larger.

The unit cell of an ionic solid is always electrically neutral.

10.10 Structures of Some Covalent Network Solids

Carbon has more than 40 **allotropes** – different structural forms of the element with varying physical and chemical properties. Most are amorphous, but there are three famous exceptions:

- *Graphite*: the most common and most stable allotrope, graphite consists of two-dimensional sheets of fused, sp^2–hybridized, six-membered rings. Graphite is the "lead" in pencils, an electrode in batteries, and a lubricant for locks. All of these functions rely on the fact that the layers of graphite can slide over one another when air and water adsorb onto them. A closely related structure is carbon *nanotubes*, which are like sheets of graphite rolled into tubes. They are being investigated as structural composites, as they have 50-60x the tensile strength of steel.

- *Diamond*: a covalent network solid in which sp^3-hybridized carbon atoms bond to one another in a vast network. Diamond is the hardest known substance, and as a result is used in industrial saw blades and drill bits, as well as in jewelry. It is an electrical insulator, and has a melting point over 3550 °C.
- *Fullerene*: discovered in 1985 as a component of soot, fullerene is a soccer-ball-shaped molecule of 60 sp^2-hybridized carbon atoms in alternating hexagons and pentagons. When reacted with rubidium, a superconducting material called rubidium fulleride (Rb_3C_{60}) is formed.

Silicon has similar abilities to form network solids, but because it is larger than carbon, is less able to double bond with oxygen. As a result, it forms four single bonds with oxygen that are used to bridge to other silicon atoms. *Silica* has the empirical formula SiO_2. Most minerals and rocks are formed by silicates: these account for 75% of the Earth's crust by mass.

When silica is heated above 1600 °C, many of the Si-O bonds break, converting the silica into a viscous liquid. When it cools, the bonds reform into a random arrangement and an amorphous solid - *quartz glass* – is formed. Addition of transition metal ions lends color, and the addition of B_2O_3 adds heat resistance. This *borosilicate* glass is sold as Pyrex.

10.11 Phase Diagrams

Temperature can cause a spontaneous phase change, but pressure can as well. Graphs of temperature (horizontal axis) versus pressure (vertical axis) are called **phase diagrams**.

The lines on a phase diagram indicate interfaces between phases. It is fairly straightforward to remember which phase is which by thinking of water, a substance familiar in all three phases. As the temperature increases, the substance becomes first liquid, then gas.

A similar pattern is seen on the pressure axis: at the highest pressures, most substances are solids. The very few exceptions are substances, like water and gallium, in which the liquid form is more dense than the solid. These have a backward-sloping line between the solid and liquid phases that indicates that increasing pressure favors the liquid form.

There are a few features of phase diagrams that should be noted.

- Triple point: this is the temperature and pressure at which all three phases exist in equilibrium.
- Critical point: this is the temperature and pressure above which gases and liquids behave indistinguishably as *supercritical fluids*: gas and liquid phases have the same density and are miscible.
- Critical pressure: the pressure above which a liquid cannot be vaporized
- Critical temperature: the temperature above which a gas cannot be liquefied.
- Normal melting/boiling point: phase transition at 1 atm pressure.

Putting It Together

Sulfur dioxide is one of the compounds that causes acid rain. Acid rain is produced first by the oxidation of sulfur dioxide to sulfur trioxide in air. The sulfur trioxide then reacts with water, producing sulfuric acid. What is the molecular shape of sulfur dioxide? What intermolecular forces are present? Calculate the heat of reaction for the production of sulfur trioxide and sulfuric acid. (ΔH_f SO_2 = -296.8 kJ/mol, ΔH_f SO_3 = -395.7 kJ/mol, ΔH_f H_2O = -285.8 kJ/mol, and ΔH_f H_2SO_4 = -814.0 kJ/mol)

Self–Test

This section is intended to test your knowledge of the material covered in this chapter. Think through these problems, and make certain you understand what they are asking. Make sure your answers make sense. Successful completion of these problems indicates that you have mastered the material in this chapter. You will receive the greatest benefit from this section if you use it as a mock exam, as this will allow you to determine which topics you need to study in more detail.

True/False

1. If a molecule has polar bonds, it will always have a dipole.
2. Intermolecular forces all are equal in strength.
3. Intermolecular forces can be related to melting and boiling points.
4. Freezing causes a positive entropy change and a negative enthalpy change.
5. Vapor pressure does not vary with temperature.
6. Covalent network solids are extremely large molecules.
7. X-rays can be used to determine the strength of intermolecular forces in liquids.
8. The coordination number of a packing arrangement is the number of atoms in the unit cell.
9. Fullerene and graphite are both isotopes of carbon.
10. Increasing pressure increases the boiling point of a liquid.

Matching

Allotrope	a. temperature and pressure at which solid, liquid, and gas of the same substance are at equilibrium
Critical point	b. attractive force caused by temporary dipoles
Dipole	c. heat required for phase change from solid to liquid
Heat of fusion	d. point at which any increase in temperature or pressure produces a supercritical fluid
Molecular solid	e. resistance to flow
London dispersion force	f. solid in which the constituent particles are randomly arranged

Normal boiling point	g. different form of a pure element: can have different physical and chemical properties
Viscosity	h. temperature at which the phase spontaneously changes from liquid to gas at 1 atm pressure
Hydrogen bonding	i. solid held together by intermolecular forces
Vapor pressure	j. intermolecular force characterized by attraction between a hydrogen atom with a partial positive charge and an unbonded electron pair.
Triple point	k. a pair of separated partial charges
Amorphous solid	l. solid in which the constituent particles have long-range order
Crystalline solid	m. the partial pressure of a gas in equilibrium with a liquid in a closed container.

Multiple Choice

1. Interactions that result from the interactions between the δ– end of one molecule with the δ+ end of another molecule are
   a. dipole–dipole forces
   b. ion–dipole forces
   c. hydrogen bonding
   d. London dispersion forces

2. Small nonpolar molecules have fewer intermolecular forces than larger molecules. This is a result of their lower
   a. van der Waals forces
   b. polarizability
   c. density
   d. viscosity

3. The intermolecular forces present in CH_3OH are
   a. dipole–dipole forces
   b. ion–dipole forces, dipole-dipole forces
   c. London dispersion forces
   d. London dispersion forces, dipole-dipole forces, hydrogen bonding

4. Which of the following liquids will have the highest boiling point?
   a. $CH_3CH_2CH_3$
   b. $CH_3(CH_2)_2CH_3$
   c. $CH_3(CH_2)_4CH_3$
   d. $CH_3(CH_2)_6CH_3$

5. Which of the following liquids has the lowest vapor pressure?

```
   H H          H   H          H H
   | |          |   |          | |
 H-C-C-H      H-C-O-C-H      H-C-C-O-H
   | |          |   |          | |
   H H          H   H          H H
   (A)           (B)            (C)
```

a. A
b. B
c. C
d. both B and C will have comparable vapor pressures

6. If the external pressure is more than one atm, then a liquid boils at
   a. the normal boiling point
   b. a temperature below the normal boiling point
   c. a temperature above the normal boiling point
   d. there is not enough information to say

7. A solid whose constituent particles are held together by charge interactions is a(n)
   a. ionic solid
   b. molecular solid
   c. covalent network solid
   d. metallic solid

8. The type of packing present when the unit cell contains a complete central atom is
   a. simple cubic packing
   b. body–centered cubic
   c. hexagonal closest packed
   d. cubic closest packed

9. Most of the allotropes of carbon are amorphous, but diamond, fullerene, and graphite are all
   a. ionic solids
   b. covalent network solids
   c. forms of silica
   d. metallic

10. The point at which the phase transition from liquid to solid occurs at 1 atmosphere is
    a. the normal freezing point
    b. the triple point
    c. the critical point
    d. the critical pressure

Fill–in–the–Blank

1. There are several types of intermolecular forces. ______________________ result from the interactions of two polar molecules. ___________________________ occur in all molecules, and are the result of temporary charge imbalances. ___________________________ are the strongest intermolecular forces, and result from interactions between a H atom bonded to an N, O, or F, and a non-bonding pair on another atom.
2. When an ionic solid dissolves in water, _______________________ interactions occur.

3. Phase changes have an associated change in free energy. The _______________ portion of this change is involved in disrupting or forming intermolecular forces.
4. In a heating curve, the temperature increase stops during _________________________.
5. In a phase diagram, the phase change represented by the line between solid and gas is _________________________.
6. As temperature decreases, vapor pressure _________________________, and viscosity and surface tension ___________________.
7. A brittle, crystalline solid with sharp edges and a very high melting point is a(n) _________________________. A relatively soft, low-melting-point crystalline solid is most likely a _________________________, and an extremely hard, extremely high-melting-point solid is probably a _______________________________.

Problems

1. What is the most important intermolecular force in the following molecules?

   $CH_3CH_2CH_2OH$ $\quad$ $CH_3(CH_2)_4CH_3$ $\quad$ PCl_3

2. Which of the following would you expect to have the highest boiling point and why?
   a. Br_2 or Cl_2
   b. CH_3OH or CH_3CH_2OH
   c. CH_4 or CH_3OH

3. For platinum, ΔH_{vap} = 565.3 kJ/mol and ΔS_{vap} = 150.8 J/K·mol. What is the boiling point of platinum?

4. If 0.237 g of water condensed on a 50.0 g block of iron at 25.0 °C, what would the final temperature of the iron be? ΔH_{vap} = 40.7 kJ/mol, c_{iron} = 0.449 J/g•°C.

5. If a prankster were to take a 45 g ice cube out of a household freezer at -20 °C and drop it down the back of an unsuspecting sibling, how much heat would be absorbed from the hapless sibling if she decided to pretend she did not notice? Normal body temperature is 37 °C, the specific heat of ice is 36.57 J/mol•°C, ΔH_{fus} = 6.01 kJ/mol, and the specific heat of water is 4.184 J/g•°C.

6. The normal boiling point of benzene is 80.1 °C and ΔH_{vap} = 30.8 kJ/mol. What is the vapor pressure of benzene at 75.0 °C?

7. Gold has a density of 19.3 g/cm^3 and a face-centered cubic unit cell. What is the atomic radius of gold?

8. Palladium has a face-centered cubic unit cell. What is the density of palladium if the atomic radius is 137 pm?

9. For the phase diagram shown, identify the normal freezing point, the triple point, the critical point, and the lines that indicate boiling, sublimation, and melting. What does the slope of the central line indicate about the substance whose data is shown?

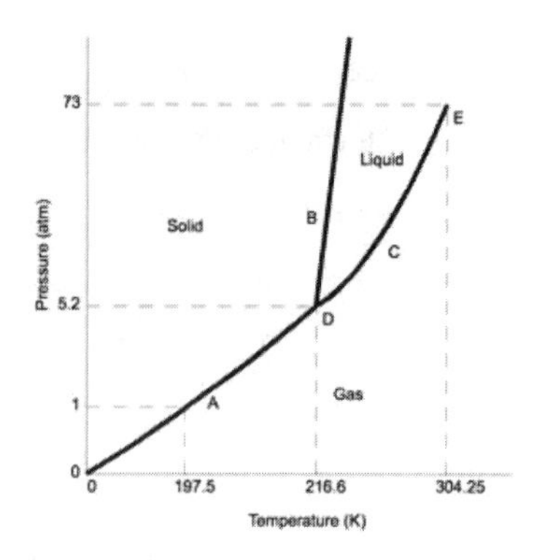

Challenge Problem

How many grams of water can be taken from -20 °C to 115 °C by burning 320 g of propane (C_3H_8)? Assume that all the energy of combustion goes into the water.

Necessary information:

ΔH_{comb} for propane	= 2,220 kJ/mol
Specific heat for ice	= 2.03 J/g•°C
Specific heat for water	= 4.184 J/g•°C
Specific heat of steam	= 2.08 J/g•°C
ΔH_{fus} for water	= 6.01 kJ/mol
ΔH_{vap} for water	= 40.67 kJ/mol

CHAPTER 11

SOLUTIONS AND THEIR PROPERTIES

Chapter Learning Goals

*A. **Formation of Solutions***
1. Explain the rule "like dissolves like" by analyzing the solution process in terms of forces overcome in the solute and solvent and forces formed between solute and solvent particles.
2. Use the equation $\Delta G = \Delta H - T\Delta S$ to calculate normal boiling point, heat of vaporization, or entropy of vaporization, given the other two.

*B. **Solutions and Concentration Units***
1. Define solution density, molarity, mole fraction, mass percent, parts per million, parts per billion, and molality, and perform calculations using these quantities.

*C. **Solutions and Solubilities***
1. Perform calculations using Henry's law.

*D. **Solutions and Colligative Properties***
1. Use Raoult's law to calculate the vapor pressure over a solution containing a nonvolatile solute and a solution containing two volatile liquids.
2. Perform calculations involving freezing–point depression and boiling–point elevation, and determine the molar mass of the solute.
3. Perform calculations involving the osmotic pressure equation, and determine the molar mass of the solute.
4. Describe fractional distillation with the aid of a liquid/vapor phase diagram.

Chapter in Brief

This chapter focuses on homogeneous mixtures, primarily solutions. The energetics of solution formation, the free energy changes involved, and factors that affect solubility are covered, as are the great variety of mass and mole-based concentration units. Calculations involving these units, and the ways that they are interconverted are explained and demonstrated. The effects of pressure and temperature on the solubility of gases, are explained and demonstrated, as are four colligative properties – lowering of vapor pressure, raising of boiling point, lowering of freezing point, and production of osmotic pressure are all discussed and methods of calculation explained. Finally, the practical uses of these properties are discussed.

11.1 Solutions

Mixtures can be classified as *homogeneous* or *heterogeneous*. In heterogeneous mixtures, the mixing of components is visually non-uniform: homogeneous mixtures are the same throughout.

Homogeneous mixtures are classified according to the size and behavior of the constituent particles:

- Solutions contain particles the size of an ion or small molecule. They are transparent, and do not separate on standing. An example is saltwater.
- Colloids contain particles with diameters of around 2-1000 nm. They scatter light, and may be murky or opaque, but do not separate on standing. An example is fog.

- Suspensions are only temporarily homogeneous mixtures: inputs of energy are required to keep them mixed. These contain the largest particles, and separate when left standing. Paint is an example of a suspension.

Any state of matter can form a solution with another state: seven possible solution types are known: gases can dissolve in solids, in liquids, or in other gases; liquids can dissolve in other liquids or in solids, and solids can dissolve in liquids or in one another.

When a gas or solid is dissolved in a liquid, the liquid is referred to as the **solvent**, and the dissolved substance is the **solute**.

When two liquids are dissolved in one another, the main component is the solvent and the lesser component is the solute.

EXAMPLE:

Drugstore "rubbing alcohol" can be purchased as 90% isopropyl alcohol or as 70% isopropyl alcohol For each of these, what is the solvent and what is the solute?

SOLUTION:

In both of these cases, the isopropyl alcohol is the solvent. The minor component, or solute is water.

Workbook Problem 11.1

A grocery-store bottle of hydrogen peroxide is labeled as a 3% solution. What is the solute and what is the solvent?

Strategy: Determine which is the major component and which is the minor component.

11.2 Energy Changes and the Solution Process

Except for mixtures of gases, intermolecular forces are at work in solutions as well as in pure substances, but the situation is made somewhat more complex by the fact that there are solvent-solvent interactions, solvent-solute interactions, and solute-solute interactions that all need to be considered.

When all three sets of interactions are similar, whether polar or nonpolar, dissolution will occur. This is often summarized simply in the expression "like dissolves like." Polar substances will dissolve in polar solvents, and nonpolar substances will dissolve in non-polar solvents.

When ionic substances dissolve in water, water molecules surround the ions and neutralize their charges. When this occurs, the ions are said to be *solvated* or, more specific to water, *hydrated.*

Free energy changes accompany solution formation: ΔS_{sol} will be positive, as disorder invariably increases when a solution is formed. ΔH_{sol} can be either positive or negative: some ionic compounds absorb heat when they dissolve, and some release it.

ΔH_{sol} can be broken into three components, the sum of which will be the overall enthalpy change:

- Solvent–solvent interactions: these will have a positive ΔH, as solvent molecules must be separated and their intermolecular forces disrupted to make room for solute particles.

- Solute–solute interactions: these will also have a positive ΔH, and for the same reason: the interactions between the solute particles will have to be overcome. For ionic solids, this is proportional to the lattice energy
- Solvent-solute interactions: these will have a negative ΔH as the solvent molecules solvate the solute particles. This portion of the energy has a larger contribution in the case of smaller ions and larger charges.

EXAMPLE:
Which of the following would you expect to be soluble in hexane, a nonpolar solvent? Rank them in order from most soluble to least soluble.

KCl $\quad C_2H_6 \quad$ CH_3OH

SOLUTION:
The polarity of the solutes will determine their solubility. Using "like dissolves like," the most non-polar solute will be the most soluble. The least polar solute here is C_2H_6, which has no dipole. Methanol (CH_3OH) is the next most soluble. It is a hydrogen bonder, but has a non-polar component that may provide some solubility, The most polar solute is KCl, which is unlikely to dissolve at all.

$$C_2H_6 > CH_3OH > KCl$$

Workbook Problem 11.2
Rank the following in order of increasing hydration energy (least negative to most negative):

Mg^{2+} $\quad Cl^- \quad$ Li^+

Strategy: Charge and molecular size determine the hydration energy.

11.3 Units of Concentration

There are four main methods for expressing the concentration of a solution, that is, the amount of solute per unit of solvent.

$$\text{Molarity} = \frac{\text{moles of solute}}{\text{Liter of solution}}; \text{ Units} = \text{mol/L} = \text{M}$$

This method has two main advantages: stoichiometry calculations and titrations are greatly simplified by using moles. Disadvantages are temperature dependence (volume of solvent will change slightly with increases and decreases in temperature) and inability to determine solvent quantities without density.

$$\text{Mole fraction } (X) = \frac{\text{Moles of component}}{\text{Total moles in the solution}}; \text{ Units} = \text{none}$$

The advantages of this method are temperature independence and convenience with gases. Rarely used for liquid solution as other methods are easier.

$$\text{Mass percent (mass \%)} = \frac{\text{Mass of component}}{\text{Total mass of solution}} \times 100\text{; units} = \%$$

This method is useful for small quantities, but density is needed to convert to molarity. Two variants on this method are often used for environmental calculations:

$$\text{Parts per} \quad = \frac{\text{Mas o componen}}{\text{Total}} \times 10^6$$

$$\text{Parts per billion (ppb)} = \frac{\text{Mass of component}}{\text{Total mass of solution}} \times 10^9$$

For such dilute solutions, 1 kg solution = 1 L solution.

$$1\ \text{ppm} = \frac{1\ \text{mg solute}}{1\ \text{L soln}}$$

$$1\ \text{ppb} = \frac{1\ \mu\text{g solute}}{1\ \text{L soln}}$$

$$\text{Molality} = \frac{\text{moles of solute}}{\text{kg of solution}}; \text{ Units = mol/kg or } m$$

The temperature independence of this method makes it particularly well suited to measuring properties of solutions at temperature extremes, but it is difficult to measure mass of liquids, and the density of the solution must be known to convert to molarity.

EXAMPLE:

A sample of water is found to have 65 ppm of arsenic. How many grams of arsenic will 1 liter contain? What is the molarity of this solution?

SOLUTION:

A concentration of 1 ppm means that each liter of an aqueous solution contains 1 mg of solute. Therefore, the number of grams of arsenic in 1 L of this water is 65 mg. To calculate the molarity of this solution, we need to convert milligrams to moles.

$$65\ \text{mg As} \times \frac{1\ \text{g}}{1 \times 10^3\ \text{mg}} \times \frac{1\ \text{mol}}{74.92\ \text{g As}} = 0.0000868\ \text{mol} = 867\ \mu\text{mol}$$

Since we have 1 liter of solution, the molarity is equal to 867 μM. Density is not needed to convert such dilute solutions from mass-based measures to molarity, as the solution will have the same density as water.

EXAMPLE:

A solution is made by dissolving 35.0 g of salt in 100 g of water. The resulting solution has a density of 1.16 g/mL. Determine the mass % and molarity of the solution.

SOLUTION:

The mass % of the solution is the mass of the solute divided by the total mass of the solution:

$$\text{mass \%} = \frac{35.0\ \text{g}}{135\ \text{g}} \text{x} 100 = 25.9\%\ \text{NaCl}$$

To convert to molarity, the mass of solute must be converted to moles, and the solution converted to volume using the density:

$$35.0 \text{ g} \times \frac{1 \text{ mole}}{58.44 \text{ g}} = 0.599 \text{ mol}$$

$$\frac{135 \text{ g}}{1.16 \frac{\text{g}}{\text{mL}}} = 116 \text{ mL}$$

$$M = \frac{0.599 \text{ mol}}{0.116 \text{ L}} = 5.15 \text{ M}$$

Workbook Problem 11.3

A solution is prepared by dissolving 15.43 grams of glucose ($C_6H_{12}O_6$) in 300 g of water. The density of this solution is 1.09 g/mL. Calculate the mass %, molality, and molarity of the solution.

Strategy: Calculate mass percent of solution, volume of solution, and moles of solute to provide all units.

Step 1: Determine mass %.

Step 2: Calculate the moles of solute, and the molality of the solution.

Step 3: Calculate the volume of the solution, and the molarity of the solution.

11.4 Some Factors Affecting Solubility

Solubility has natural limits: at some point, the solute molecules will be in equilibrium, meaning that they are going in and out of solution at an equal rate. When this occurs, the solution is said to be **saturated**.

Some solutes are more soluble at high temperatures. In this case, it is possible to produce a **supersaturated** solution with a greater-than-equilibrium amount of solute by creating a saturated solution at a high temperature and allowing it to cool undisturbed. These solutions are unstable.

The amount of solute that can be dissolved in a solvent to form a saturated solution at a given temperature is the **solubility** of that substance. A variation of solubility is miscibility: when the solution is a mixture of liquids, the liquids are sometimes soluble in one another regardless of proportions.

The temperature effect on solubility of solids is unpredictable, but gases become less soluble as the temperature of the solution increases. As a result, warm sodas quickly go flat, and less oxygen will dissolve in warmer water, making fish sensitive to thermal pollution.

Pressure does not affect the solubility of solids and liquids, but has a large effect on gases, which is described by **Henry's law**:

$$\text{Solubility} = k \bullet \text{P}$$

Where k is a constant related to the gas (with usual units of mol/L•atm), and P is the partial pressure of the gas. In contrast to other gas calculations, these measurements are reported at 25 °C.

EXAMPLE:

Determine the solubility of CO_2 (g) at 25 °C and a partial pressure of 33.5 mmHg.

$$k = 0.0032 \frac{\text{mol}}{\text{L} \cdot \text{atm}}$$

SOLUTION:

Use Henry's law, being careful to convert units as needed.

$$\text{Solubility} = 0.0032 \frac{\text{mol}}{\text{L} \cdot \text{atm}} \times 33.5 \text{ mm Hg} \times \frac{1 \text{ atm}}{760 \text{ mm Hg}} = 1.41 \times 10^{-4} \frac{\text{mol}}{\text{L}}$$

Workbook Problem 11.4

Given the Henry's law constant above (k = 0.0032 mol/(L•atm)), how many liters of CO_2 will bubble out of a 500 mL soda after it is opened at 25 °C if it was previously under 2.3 atmospheres of pressure with 100% CO_2 in the air, and comes to equilibrium with the atmosphere where the partial pressure of CO_2 is only 0.3 mmHg?

Strategy: Calculate the initial and final amounts of CO_2 in a saturated solution, determine the moles, and then the liters of CO_2.

Step 1: Calculate the moles of dissolved CO_2 under pressure.

Step 2: Calculate the moles of dissolved CO_2 after pressure is released and equilibrium with the atmosphere reached.

Step 3: Calculate the change in dissolved CO_2 – this is the CO_2 that has been released.

Step 4: Determine the volume of gas at 25 °C and 1 atm pressure.

11.5 Physical Behavior of Solutions: Colligative Properties

Solutions behave in some ways differently than pure solvents. Properties that depend on the amount of dissolved solute but not its chemical identity are called **colligative properties**. They include:

- Lowered vapor pressure
- Raised boiling point

- Lowered freezing point
- Osmosis: migration of solvent and small molecules through a membrane.

These properties will all be examined in detail in the following sections.

11.6 Vapor–Pressure Lowering of Solutions: Raoult's Law

In a solution with a non-volatile solvent (one that has no vapor pressure of its own), the vapor pressure is lower than it would be in the pure solvent: the solution already has a higher entropy than the pure solvent, so the entropy gain in evaporating is reduced, and the vapor pressure is reduced.

Mathematically, this is described by Raoult's law:

$$P_{soln} = P_{solv} \cdot X_{solv}$$

where *X* is the mole fraction of the solvent in the mixture.

When calculating the mole fraction, the total number of moles of particles must be used for the solvent. For a molecular solvent, like glucose, this will be the same as the moles of molecules. But for an ionic solvent, this will be the moles of ions generated.

Moles of ions generated are mathematically described by the **van't Hoff factor**: this is a multiplier that expresses the amount of dissociation of an ionic substance. Unfortunately, it is not as simple as 1 mol of NaCl breaking into 2 mol of ions: there is always some undissociated salt in a solution.

Just like the ideal gas law, Raoult's law applies to "ideal solutions." It works best at low solute concentrations.

EXAMPLE:

Determine the vapor pressure of a solution made by dissolving 25.7 g of LiBr in 1.00 kg of solution at 25 °C. The vapor pressure of pure water at 25 °C is 23.76 mm Hg. Assume complete dissociation.

SOLUTION:

Use Raoult's law.

Determine the mole fraction of water in the solution:

Moles LiBr = 25.7 g/86.85 g/mol = 0.2959 mol

Moles H_2O = 1000 g/18.0 g/mol = 55.56 mol

$$X_{H_2O} = \frac{55.56 \text{ mol}}{(55.56 \text{ mol} + 0.2959 \text{ mol})} = 0.995$$

Multiply by the vapor pressure of pure solvent:

$$P_{soln} = 23.76 \text{ mm Hg x } 0.995 = 23.64 \text{ mm Hg}$$

EXAMPLE:

How much LiBr would need to be dissolved in 1.00 kg water to lower the vapor pressure to 20 mm Hg, assuming complete dissociation?

First calculate the mole fraction of water:

$$20.0 \text{ mm Hg} = 23.76 \text{ mm Hg} \times X_{H_2O}$$

$$X_{H_2O} = 0.842$$

Next, determine the number of moles of solute needed for this mole fraction:

$$X_{H_2O} = \frac{55.56 \text{ mol}}{(55.56 \text{ mol} + x \text{ mol})} = 0.842$$

$$55.56 \text{ mol} = 0.842 \times (55.56 + x \text{ mol})$$

$$55.56 - 46.78 = 0.842x$$

$$x = 10.43 \text{ mol ions} = 5.21 \text{ mol LiBr}$$

$$5.21 \text{ mol x } 86.85 \text{ g/mol} = 453 \text{ g LiBr}$$

As you can see, these effects are rather small: if 454 g of LiBr were dissolved in 1 kg of water, the solution would no longer be dilute enough for Raoult's law to apply.

Workbook Problem 11.5

Assuming complete dissociation for both, calculate the quantity of $CaCl_2$ and LiCl that are needed to lower the vapor pressure of 500 g of water by 3 mm Hg at 70 °C. The vapor pressure of water at this temperature is 233.7 mm Hg.

Strategy: Determine the mole fraction of ions needed to lower the vapor-pressure by this amount, then convert to grams of each ionic substance.

Step 1: Calculate the mole fraction of water in a solution with this vapor pressure lowering.

Step 2: Calculate the moles of dissolved ions that will lead to this mole fraction.

Step 3: Convert to moles, then grams of $CaCl_2$.

Step 4: Convert to moles, then grams of LiCl.

When two liquids are mixed that each has its own vapor pressure, the total vapor pressure of the solution will be the combination of the vapor pressures of the two components. The vapor pressure of each component is calculated using Raoult's law, based on the mole fraction of each liquid:

$$P_{total} = P_A + P_B$$

The pressures of the individual components are calculated using Raoult's Law:

$$P_A = X_A(P_A^o);\ P_B = X_B(P_B^o)$$

$$P_{total} = X_A(P_A^o) + X_B(P_B^o)$$

EXAMPLE:

The vapor pressure of water at 25 °C is 23.8 mm Hg. The vapor pressure of ethyl alcohol (C_2H_5OH) at 25°C is 61.2 mm Hg. What is the vapor pressure of a solution prepared by combining 50 g of each liquid?

SOLUTION:

Calculate the moles of each to determine the mole fractions:

$$\text{mol } C_2H_5OH = 50 \text{ g } C_2H_5OH \times \frac{1 \text{ mol } C_2H_5OH}{46.0 \text{ g } C_2H_5OH} = 1.09 \text{ mol } C_2H_5OH$$

$$\text{mol } H_2O = 50 \text{ g } H_2O \times \frac{1 \text{ mol } H_2O}{18.0 \text{ g } H_2O} = 2.78 \text{ mol } H_2O$$

$$X_{C_2H_5OH} = \frac{1.09 \text{ mol } C_2H_5OH}{1.09 \text{ mol } C_2H_5OH + 2.78 \text{ mol } H_2O} = 0.282$$

$$X_{H_2O} = \frac{2.78 \text{ mol } H_2O}{1.09 \text{ mol } C_2H_5OH\ +\ 2.78\text{mol } H_2O} = 0.718$$

$$P_{total} = (0.282 \times 61.2 \text{ mm Hg})\ +\ (0.718\ \times 23.8 \text{ mm Hg})\ =\ 34.34 \text{ mm Hg}$$

Workbook Problem 11.6

The label on a bottle of wine states that it is 13% ethanol. Using the vapor pressures listed above, what is the pressure above a 150 g glass of wine, assuming (wrongly) that there are no other volatile components.

Strategy: Determine the mole fraction of each component, and calculate the partial pressure of each.

Step 1: Solve for the mass and moles of each component.

Step 2: Determine the mole fraction of each component.

Step 3: Determine the partial pressure of each component and the total pressure.

11.7 Boiling–Point Elevation and Freezing–Point Depression of Solutions

Another aspect of the reduced vapor pressure of a solution with a non-volatile solvent is the raised boiling point. If the vapor pressure is being held down, and the solution boils when the vapor pressure is equal to the external pressure, the vapor pressure will not reach the external pressure at the same temperature.

This is seen in the phase diagram of a solution, in which all of the lines are shifted down: the boiling point is raised, and the freezing point lowered by the addition of solutes. As in the case of all colligative properties, it is the number of particles that causes the effect.

The fundamental reason for this is the same as for vapor pressure depression: the increase in the entropy of the solution makes phase changes less favorable: more enthalpy needs to be added to get the liquid to boil, and more enthalpy needs to be removed to get the liquid to freeze.

Mathematically, this is described with the following equations:

$$\Delta T_b = K_b \bullet m$$
$$\Delta T_f = K_f \bullet m$$

For these equations, K_b is the **molal boiling-point elevation constant** and K_f is the **molal freezing-point depression constant**. Both are dependent on the *solvent*, not the identity of the particles. "*m*" is the molal concentration of solute particles, whether ions or molecules. Molality rather than molarity is used for these calculations, as temperature extremes affect the volume of solutions.

The actual dissociation of ionic compounds can also be taken into account by adding in the van't Hoff factor for the compound:

$$\Delta T_b = K_b \bullet m \bullet i$$
$$\Delta T_f = K_f \bullet m \bullet i$$

EXAMPLE:

What will be the freezing point and boiling point of an aqueous solution containing 55.0 g of KCl in 250 g of water? $K_b(H_2O) = 0.51\ ^{\circ}C/m$ and $K_f = 1.86\ ^{\circ}C/m$. Assume complete dissociation.

SOLUTION: Determine the molality of the solution, then calculate the boiling point elevation and freezing point depression:

$$55.0 \text{ g x} \frac{1 \text{ mol}}{74.55 \text{ g}} = 0.738 \text{ mol KCl x } \frac{2 \text{ mol ions}}{\text{mol KCl}} = 1.47 \text{ mol ions}$$

$$\frac{1.47 \text{ mol}}{0.250 \text{kg}} = 5.90\, m$$

$$\Delta T_b = 0.51^{\circ}\ C/m \text{ x } 5.90\ m = 3.01\ ^{\circ}C \quad 100\ ^{\circ}C + 3.01\ ^{\circ}C = 103\ ^{\circ}C$$

$$\Delta T_f = 1.86\ ^{\circ}C/m \text{ x } 5.90\ m = 11.0\ ^{\circ}C \quad 0\ ^{\circ}C - 11.0\ ^{\circ}C = -11\ ^{\circ}C$$

Workbook Problem 11.7

How many grams of $(NH_4)_3SO_4$ need to be added to 300 g of H_2O so that the freezing–point of the solution is lowered to –15.0 °C? K_f = 1.86 °C/*m*. Assume complete dissociation.

Strategy: Using the equation for freezing–point depression, solve for the molality of ions needed, then convert to moles and grams of ammonium sulfate.

Step 1: Determine the molality of ions needed to generate this freezing point depression.

Step 2: Determine the moles of $(NH_4)_3SO_4$ needed to generate this quantity of ions.

Step 3: Convert to grams of ammonium sulfate needed.

11.8 Osmosis and Osmotic Pressure

Cell membranes, along with some manufactured membranes are *semipermeable*, meaning that solvent molecules and some very small molecules can pass freely through the membrane, but large molecules and ions are blocked.

In the process of **osmosis**, solvent molecules pass preferentially from the side of low solution concentration to the side of high solution concentration. As a result, the level of the low concentration or pure solvent will drop, and the level of the high-concentration solution will rise.

This is an entropy-driven process: the dilution of a solution increases the disorder of the solution. The process will continue until the external pressure and the **osmotic pressure** are equal.

Osmotic pressure is calculated with the following equation:

$$\Pi = \text{MRT}$$

Π is the osmotic pressure, M is the molarity of solute particles in solution, and *R* and *T* are the gas law constant and the temperature in Kelvin. Molarity is used for these calculations as the temperature is a variable as well.

EXAMPLE:

Determine the osmotic pressure of a 0.050 M solution of sodium chloride at 24.3 °C.

SOLUTION:

$$\Pi = 0.050\text{M x } \frac{2 \text{ mol particles}}{\text{mol salt}} 0.08206 \frac{\text{L} \bullet \text{atm}}{\text{mol} \bullet \text{K}} \text{x}(273.15 + 24.3)\text{K} = 2.44 \text{ atm}$$

Workbook Problem 11.8

A solution containing an unknown substance has an osmotic pressure of 7.34 atm at 25 °C. What is the molarity of the solution?

Strategy: Using the equation for osmotic pressure, solve for the molarity of the solution.

11.9 Some Uses of Colligative Properties

Colligative properties are commonly used to melt ice: adding salt like calcium chloride that dissociates into three particles in an exothermic reaction is a particularly efficient way to deice a sidewalk and keep it clear.

The coolant added to an automobile engine both raises the boiling point of the coolant and lowers its freezing point, maintaining it as a liquid over a greater range.

The deicer sprayed on airplane wings operates on the same principle: maintaining the liquid range of water to prevent ice build-up.

Desalination of seawater can be accomplished through **reverse osmosis**: putting more than 30 atm of pressure on seawater on one side of a semi-permeable membrane will force pure water through the membrane until the osmotic pressure of the increasingly concentrated seawater is equal to the external pressure.

Any of the colligative properties can be used to determine molecular weights, but osmotic pressure, as the effect is so large, can be used most accurately.

Workbook Problem 11.9

A solution is prepared from 23 mg of a compound in 100 mL of aqueous solution. The osmotic pressure of the solution is 8.428 mm Hg at 25 °C. What is the approximate molecular weight of the compound?

Strategy: Using the equation for osmotic pressure, calculate the molarity of the solution, and from that the molecular weight of the compound.

Step 1: Calculate the molarity of the solution.

Step 2: Calculate the number of moles in the solution.

Step 3: Determine the molecular weight.

11.10 Fractional Distillation of Liquid Mixtures

Purification of hydrocarbons is accomplished using a method known as **fractional distillation**. When a group of volatile liquids is heated, the vapor component is enriched in the fraction with the lowest boiling point. Vapor is condensed by cooling, providing an enriched fraction of the lowest boiling-point component. Repeating the heating and cooling cycles allows for purification of the components of the volatile mixtures: this occurs naturally in a distillation column.

Putting It Together

Esters, organic compounds containing C, H, and O, often have pleasant odors. Butyl butanoate has the odor of pineapples. A 0.83 g sample of butyl butanoate was subjected to combustion analysis, which produced 2.026 g of CO_2 and 0.8288 g of H_2O. A 1.50 g sample was dissolved in enough solvent to make 250 mL of solution. The osmotic pressure of this solution at 25 °C was 1.02 atm. Determine the molecular formula of butyl butanoate.

Self–Test

This section is intended to test your knowledge of the material covered in this chapter. Think through these problems, and make certain you understand what they are asking. Make sure your answers make sense. Successful completion of these problems indicates that you have mastered the material in this chapter. You will receive the greatest benefit from this section if you use it as a mock exam, as this will allow you to determine which topics you need to study in more detail.

True–False

1. A solution of metals is called an *alloy*.

2. Entropy always increases when a solution is formed, but enthalpy can increase or decrease.

3. Molality is the most practical unit of concentration for environmental measurements.

4. Potassium nitrate will dissolve readily in oil.

5. The density of a solution is needed to convert from mass % to molarity.

6. Supersaturated solutions are stable at low temperatures.

7. Increasing pressure causes ionic solids to become more soluble.

8. The more concentrated a solution, the lower the vapor pressure.

9. The more concentrated a solution, the higher the freezing point.

10. Osmotic pressure can be overcome with high external pressures.

Multiple Choice

1. When a gas or solid is dissolved in a liquid, the solute is
   a. the dissolved substance
   b. the liquid
   c. the major component
   d. the minor component

2. ΔH_{soln} will be exothermic if
   a. ΔH for solvent–solvent interactions is negative
   b. ΔH for solute–solute interactions is negative
   c. ΔH for solute–solvent interactions is negative
   d. the sum of the three types of interactions leads to a negative ΔH

3. Benzene, a nonpolar organic compound, is most likely to dissolve in
   a. water

   b. CH_3CH_2OH
   c. CCl_4
   d. NH_3

4. When the temperature of a solvent is lowered, the solubility of gases
   a. decreases
   b. increases
   c. is not affected
   d. cannot be generally predicted

5. A mixture that contains particles large enough to be visible with a low–power microscope is a
   a. heterogeneous mixture
   b. solution
   c. suspension
   d. colloid

Matching

Colligative property	a. a technique for purifying hydrocarbon mixtures
Colloid	b. passage of solvent from a less concentrated to more concentrated solution.
Fractional distillation	c. a solution holding the maximum quantity of solute
Heat of solution	d. able to mix together in any proportions.
Miscible	e. the liquid portion of a solution
Molality	f. a homogeneous mixture containing large particles that do not settle out.
Osmosis	g. allows passage of small molecules only, not ions or large molecules.
Semipermeable membrane	h. a solid or gas that is dissolved in a liquid
Saturated	i. homogeneous mixture containing very small molecules or ions.
Solubility	j. a solution that has been heated to dissolve more solute then cooled. Unstable.
Solute	k. property that depends on the amount of dissolved solute, not its identity.
Solution	l. measure of concentration that is moles per kilogram of solvent.

Solvent	m. enthalpy change, positive or negative, that occurs when a solution forms.
Supersaturated	n. the amount of solute that a solution can contain at a specific temperature.

Fill–in–the–Blank

1. Solution formation has two endothermic elements: disrupting ____________________ interactions and disrupting ____________________ interactions both require energy.
2. ΔS_{soln} is ____________________ due to ______________________________ ____________________.
3. To convert between mass-based units and mole-based units requires that the ____________ of a solution be known.
4. Gases become ____________ soluble as the temperature decreases.
5. Adding a non-volatile solute to a solution causes the vapor pressure to ____________, the boiling point to ____________, the freezing point to ____________, and ____________ ____________ to develop across semi-permeable membranes.
6. Colligative properties can be used to determine the ____________ ____________ of an unknown solute.
7. Dissolved ions that are surrounded and stabilized by a shell of solvent molecules are said to be ____________. If water is the solvent, the term used is ____________.
8. The spontaneous dissolution of ionic solids can be either ____________ or ____________, depending on the energetics of solution formation.

Problems

1. Rank the following in order of increasing solubility in benzene, a nonpolar solvent.

   I_2 $\quad$ NaCl $\quad$ $CH_3(CH_2)_2CH_2OH$

2. If 28.4 g of sucrose ($C_{12}H_{22}O_{11}$) are dissolved in 350 g of water, the final volume is 374 mL. Calculate the concentration of the solution in terms of:

   a. mass percent
   b. mole fraction
   c. molality
   d. molarity

3. Household hydrogen peroxide is listed on the label as 3% H_2O_2. What is the molarity of this solution if the density of the solution is 1.01 g/mL?

4. A sample of industrial effluent is found to have an arsenic concentration of 15 ppb. What is the molarity of this solution? How many grams of arsenic would be found in 100,000 L of this effluent?

5. Goldfish prefer cold water to warm. If the mole fraction of oxygen in air is 0.21, what is the oxygen concentration in water at 25 °C? The Henry's law constant for oxygen is 1.32×10^{-3} mol/L·atm.

6. The Henry's law constant for nitrogen is 6.25×10^{-4} mol/L·atm. Calculate the solubility of nitrogen if its partial pressure above water is 976 mm Hg.

7. The vapor pressure of pure water is 42.175 mm Hg at 35 °C. Calculate the vapor pressure of a solution made with 4.5 g of glucose ($C_6H_{12}O_6$) and 55 g of H_2O.

8. The vapor pressure of water at 25 °C is 23.8 mm Hg. Calculate the vapor pressure of an aqueous solution that contains 5.8 g $FeCl_3$ and 72.1 g of water at this temperature. The van't Hoff factor for $FeCl_3$ is 3.4.

9. The vapor pressure of benzene, (C_6H_6) at 25°C is 93.4 mm Hg. The vapor pressure of toluene ($C_6H_5CH_3$) is 26.9 mm Hg at 25°C. Calculate the vapor pressure of a solution prepared from 10.0 g of benzene and 10.0 g of toluene.

10. The normal boiling point of carbon tetrachloride (CCl_4) is 76.7 °C. How many grams of naphthalene ($C_{10}H_8$) need to be added to 1.0 kg of carbon tetrachloride to raise the boiling point of the solution to 80.0 °C? K_b for CCl_4 is 5.03 °C/m.

11. Calculate the boiling point and freezing point of a solution prepared by mixing 10.0 g NaCl with 90.0 g of water. Assume complete dissociation. K_b = 0.51 °C/m, K_f = 1.86 °C/m

12. A solution was prepared by mixing 75.0 g of benzene (freezing point = 5.5°C) and 2.50 g of an unknown substance. The freezing point of the solution was 3.5 °C. Calculate the molar mass of the unknown substance. K_f = 5.12 °C/m.

13. Calculate the osmotic pressure of a solution at 15 °C containing 18.0 g of glucose ($C_6H_{12}O_6$) in 1.2 L of solution.

14. Calculate the molar mass of 0.250 g of a peptide in 1 L of water at 15 °C that has an osmotic pressure of 16.4 mm Hg.

Challenge Problem

The vapor above a room-temperature mixture of pentane and hexane is 43.2% pentane by mass. What is the mass percent composition of the solution if the vapor pressure of pure pentane is 425 mm Hg and the vapor pressure of hexane is 151 mm Hg at room temperature? What is the total pressure of the solution?

Inquiry-Based Problem

Your laboratory instructor gave you an unknown molecular solid that is soluble in water. An example of the experimental set-up – a test tube with a two-hole rubber stopper clamped to a ring stand over a burner – is also shown. A thermometer is inserted through one hole in the stopper and a stirrer is inserted through the other hole. With this experimental set-up, outline the procedure you will use to determine the molar mass of your unknown.

Consider the following:

1. What colligative property are you directed to use from the experimental set-up on display?
2. How should the formula for this colligative property be rearranged to determine the molar mass?
3. From the re-arranged formula for this colligative property, what data do you need to collect?
4. What physical property is involved? How does the phase change curve for this property influence the manner in which you should collect data?

CHAPTER 12

CHEMICAL KINETICS

Chapter Learning Goals

A. Concentration and Reaction Rates

1. Use a table of concentration versus time data to calculate an average rate of reaction over a period of time.
2. From the coefficients of a balanced chemical equation, express the relative rates of consumption of reactants and formation of products.
3. From a table of initial concentrations of reactants and initial rates, determine the order of reaction with respect to each reactant, the overall order of reaction, the rate law, the rate constant, and the initial rate for any other set of initial concentrations.
4. Use integrated first–, second–, and zeroth–order rate laws to find the value of one variable, given values of the other variables.
5. From plots of log concentrations versus time and 1/concentration versus time, determine the order of reaction.
6. Use the expression for half–life of a first– or second–order reaction to determine $t_{1/2}$ from k, or vice versa.
7. From a plot of concentration versus time, estimate the half–life of a first–order reaction.
8. Given the half life or the decay constant for an isotope, calculate the other variable. Given both decay constant and half life, calculate the quantity left undecayed after a specific time.

B. Reaction Mechanisms

1. Given a reaction mechanism and an experimental rate law, identify the reaction intermediates, determine the molecularity of each elementary reaction, and determine if the mechanism is consistent with the experimental rate law.
2. Given the rate-determining step, propose a rate law.

C. Temperature and Reaction Rates

1. Solve the Arrhenius equation for any variable, given the others.
2. Sketch a potential energy profile, showing the activation energies for the forward and reverse reactions and how they are affected by the addition of a catalyst.

Chapter in Brief

Chemical kinetics is the study of reaction rates and mechanisms. The appearance of reactants can be related to the disappearance of reactants, and reaction orders can be determined from experimental data and plots of concentrations versus time in both logarithmic and non-logarithmic forms. Integrated rate laws can be used to calculate the concentration of a reactant at any time, the fraction of reactant remaining, or the time required for the reaction to progress to a certain point. Radioactive decay follows first-order kinetics with a decay constant in place of a reaction constant. This makes it possible to determine how long an isotope has been decaying, how long it will take for a specific amount or proportion of decay to occur, or solve for either half-life or decay constant given experimental data. Kinetics make it possible to identify a rate-determining step and propose a reaction mechanism. You will also learn how to use plots of concentration versus time to determine the reaction order. Kinetic molecular theory and the Arrhenius equation make it possible to derive equations to determine activation energy and rate constants at different temperatures. Catalysts, both heterogeneous and homogeneous, and their mechanisms are also discussed.

12.1 Reaction Rates

Reaction rates are determined by how much the concentration of a reactant or product changes per unit of time.

$$\text{Rate} = \frac{\Delta\text{concentrat ion}}{\Delta\text{ time}}$$

The concentration of reactants will be decreasing, and the concentration of products increasing as time passes. As always, changes are calculated as final-initial, so for a disappearing reactant, the sign on the rate will be negative. For an appearing product, the rate will be positive.

The units associated with these rates are M/s or mol/(L • s) – this allows the calculated rate to be independent of the scale of the reaction. Part of the unit must be the reactant or product on which it is based.

The relative rates for different participants in the reaction will be related by the coefficients in the balanced equation. To use a very simple example:

$$2\ H_2\ (g) + O_2\ (g) \rightarrow 2\ H_2O\ (l)$$

In this reaction, hydrogen gas will disappear at the same rate that water is formed, but oxygen will disappear at half the rate of hydrogen disappearance and water formation. This leads to a number of possible rates. To avoid this ambiguity, chemists define a *general rate*:

$$\text{General rate of reaction} = \frac{\text{change in concentration}}{\text{coefficient}}$$

So, for the above reaction:

$$\text{General rate of reaction} = \frac{1}{2} \text{x} \frac{\Delta[H_2]}{\Delta t} = \frac{\Delta[O_2]}{\Delta t} = \frac{1}{2} \text{x} \frac{\Delta[H_2O]}{\Delta t}$$

When concentration is plotted versus time, average rates – those over a specific time period – can be seen, but *instantaneous rates* can also be seen. These are the rates defined by a tangent to the curve that touches at a specific time point. The instantaneous rate when time = 0, the very beginning of the reaction, is called the *initial rate*.

EXAMPLE:

In the following reaction:

$$4\ NH_3\ (g)\ +\ 3\ O_2\ (g)\ \rightarrow\ 2\ N_2\ (g)\ +\ 6\ H_2O\ (g)$$

the rate of formation of H_2O is 0.73 $M{\cdot}s^{-1}$ at 300 s. What is the rate of disappearance of NH_3?

SOLUTION: With the rate of formation of water being 0.73 $\frac{\text{mol } H_2O}{L \cdot s}$, the stoichiometry of the balanced equation can be used to determine the rate of disappearance of NH_3.

$$0.73\ \frac{\text{mol } H_2O}{L \cdot s} \times \frac{4\text{ mol } NH_3}{6\text{ mol } H_2O} = 0.49\ \frac{\text{mol } NH_3}{L \cdot s}$$

Because this is a rate of disappearance, it will be reported as –0.49 $M{\cdot}s^{-1}$ because we are reporting the rate of *disappearance* of NH_3.

Workbook Problem 12.1

Given the following unbalanced equation:

$$HBr\ (g) \rightarrow H_2\ (g) + Br_2\ (g)$$

Express the general rate of reaction with respect to all reactants and products. If the initial rate of disappearance of HBr is 3.26×10^{-3} M/s, what is the rate of appearance of bromine gas?

Strategy: The stoichiometry of the reaction allows relative reaction rates to be determined.

Step 1: Balance the equation.

Step 2: Write the reaction rates.

Step 3: Determine the rate of formation of bromine from the rate of disappearance of HBr.

12.2 Rate Laws and Reaction Order

The rate law for a reaction indicates its dependence on the concentration of each reactant. For a general reaction:

$$a\,A + b\,B \rightarrow \text{products,}$$

$$\text{rate} = k\,[A]^n[B]^m$$

In a **rate law**, “k” is a proportionality constant called the **rate constant**.

The exponents “m” and “n” must be experimentally determined – they are unrelated to the coefficients in the chemical reaction.

These coefficients determine **reaction order**. This is the sum of the exponents. So, a reaction in which m + n = 2 is said to be second-order, a reaction in which m + n = 3 is said to be third-order, etc.

The exponents themselves are usually small positive integers, but in complex reactions can be negative, zero, or even fractions.

- exponent = 1: rate depends linearly on the concentration of the reactant
- exponent = 0: rate is independent of the concentration of the reactant
- negative exponent: rate decreases as the concentration of the reactant increases

Workbook Problem 12.2

What are the units for zeroth-order, first order, and second-order rate constants if the rate is expressed as M/s?

Strategy: Determine the units needed to cancel out to the rate of each reaction.

Step 1: Zeroth-order reaction:

Step 2: First-order reaction:

Step 3: Second-order reaction:

12.3 Experimental Determination of a Rate Law

Determining the reaction order experimentally is accomplished by measuring the initial rate of a reaction with differing initial concentrations. This is done carefully, so as to make it possible to see what effect changing each reaction has independently.

The following are the most common results:

- Doubling the concentration has no effect: the reaction is zero-order in that reactant.
- Doubling the concentration doubles the rate: the reaction is first order in that reactant.
- Doubling the concentration quadruples the rate: the reaction is second-order in that reactant.
- Doubling the concentration causes rate to increase 8x: the reaction is third-order in that reactant.

Initial rates are always used for these determinations as the reverse reaction will begin to occur, even if only slightly, as the reaction continues.

Only reactants and catalysts appear in the rate law.

Once the rate law is determined, it can be used to solve for the rate constant. Rate constants are temperature-dependent, and are a characteristic of a reaction. The units of k, which can appear rather bizarre, depend on the order of the reaction.

EXAMPLE:

For the following theoretical reaction:

$$2\,A + 3\,B \rightarrow A_2B_3$$

the following measurements were made:

Initial [A] (M)	Initial [B] (M)	Rate of formation of A_2B_3
0.1	0.1	$1\ M{\cdot}s^{-1}$
0.2	0.1	$2\ M{\cdot}s^{-1}$
0.1	0.2	$1\ M{\cdot}s^{-1}$

Determine the rate law, reaction order, and rate constant.

SOLUTION:

What this shows is that doubling the concentration of A doubles the rate, while doubling the concentration of B does not affect the rate.

From this, the rate law can be determined to be:

$$\text{Rate} = k\,[A]^1[B]^0 = k\,[A]$$

This is a *first-order* reaction.

To solve for k, the rate expression must be rearranged:

$$k = \frac{\text{rate}}{[A]} = \frac{1\ M \cdot s^{-1}}{0.1\ M} = 10\ s^{-1}$$

The unit of s^{-1} is characteristic of a first-order reaction.

EXAMPLE:
The reaction:

$$2\,NO\,(g) + 2\,H_2\,(g) \rightarrow N_2\,(g) + 2\,H_2O\,(g)$$

is first-order in H_2 and second order in NO.

Determine the rate law and reaction order. How does the reaction rate change if the concentration of NO is doubled and the concentration of H_2 is held constant? How does the reaction rate change if the concentration of H_2 is cut in half and the concentration of NO is held constant?

SOLUTION:

Based on the information given, the rate law is:

$$\text{Rate} = k[NO]^2[H_2]$$

The reaction is third-order.

If the concentration of NO is doubled while the concentration of H_2 is held constant, the rate will quadruple. If the concentration of H_2 is halved, the rate will also be halved.

Workbook Problem 12.3
The following data was collected for the reaction

$$2\,NO_2\,(g) + F_2\,(g) \rightarrow 2NO_2F\,(g)$$

Exp	$[NO_2]$ (M)	$[F_2]$ (M)	Rate ($M \cdot s^{-1}$)
1	0.10	0.10	0.026
2	0.20	0.10	0.051
3	0.20	0.20	0.103

Determine the rate law from this data. What is the order of the reaction with respect to each reactant? What is the overall order of the reaction? Calculate the value of *k*.

Strategy: The rate law for the reaction is:

Rate = $k[NO_2]^m[F_2]^n$.

To find *m* and *n*, compare the change in concentration with the change in rate. Verify your answer by using the data from any of the three experiments.

Step 1: Determine the order of the reaction with respect to NO_2.

Step 2: Determine the order of the reaction with respect to F_2.

Step 3: Write the equation for the rate law.

Step 4: Calculate the value of *k*.

12.4 Integrated Rate Law for a First–Order Reaction

Using calculus, it is possible to convert a simple first-order rate law:

$$\text{Rate} = -\frac{\Delta[A]}{\Delta t} = k[A]$$

to an integrated form:

$$\ln\frac{[A]_t}{[A]_0} = -kt$$

This is enormously valuable, as this concentration-time format makes it possible to make determinations about reaction progress.

Graphically, a plot of ln [A] versus time gives a straight line if the reaction is first order in A. The slope of the line will be equal to -*k*. This is diagnostic of a first-order reaction: a concentration vs. time plot will be curved, but a plot of ln A vs. time will be straight.

EXAMPLE:

When sucrose reacts with water, glucose is formed according to the reaction:

$$C_{12}H_{22}O_{11} + H_2O \rightarrow 2\ C_6H_{12}O_6$$

This reaction is first-order in sucrose. follows first-order–kinetics with respect to the sucrose. Calculate the value of *k* if it takes 3.00 hours for the concentration of sucrose to decrease from 0.0500 M to 0.0442 M. Determine the amount of time required for the reaction to go to 95% completion.

SOLUTION: To calculate the value of *k*, we simply substitute the data given into the first-order integrated rate equation.

$$\ln\frac{0.0442}{0.0500} = -k(3.00\ \text{h}) \qquad k = 4.11 \times 10^{-2}\ \text{h}^{-1}$$

To determine the amount of time needed for the reaction to go to 95% completion, determine the concentration of sucrose at that point. The amount of sucrose remaining at 95% completion is 5% of the initial concentration:

$0.05 \times 0.0500\ \text{M} = 0.0025\ \text{M}$

This can be substituted into the integrated rate law with the rate constant previously calculated:

$$\ln\frac{0.0025}{0.0500} = -(4.11 \times 10^{-2}\ \text{h}^{-1})t \qquad t = 72.9\ \text{h}$$

Workbook Problem 12.4

In a first-order reaction, the concentration of the reactant goes from 0.100 M to 0.030 M over the course of 73 minutes. What is k, and how long will it take for the reaction to go to 80% completion?

Strategy: Substitute into the integrated rate law for a first-order reaction, then use k to solve for the time to 80% completion.

Step 1: Solve for k.

Step 2: Determine the concentration of the reactant at 80% completion.

Step 3: Solve for t at 80% completion.

12.5 Half–Life of a First–Order Reaction

The half-life of a reaction is the time it takes for the reactant concentration to drop to one-half its initial value. From the integrated rate law:

$$\ln\frac{1}{2} = -kt_{1/2}$$

$$-0.693 = -kt_{1/2}$$

$$t_{1/2} = \frac{0.693}{k}$$

For a first-order reaction, half-life is a constant.

EXAMPLE:

If the reaction of sucrose and water has a rate constant of -4.11×10^{-2}, what is the half-life for the reaction?

SOLUTION: Substitute the rate constant into the equation for the half-life of a first-order equation.

$$t_{1/2} = \frac{0.693}{4.11 \times 10^{-2}\ h^{-1}} = 16.9\ h$$

Workbook Problem 12.5

What is the half-life for the reaction described in workbook problem 12.4?

Strategy: Substitute into the half-life equation.

12.6 Radioactive Decay Rates

The decay of radioactive nuclei is a first-order process in which the number of radioactive nuclei in a sample can be used in lieu of the concentration of a reactant. The rate constant, *k*, is here called the **decay constant**.

The integrated rate law for radioactive decay is:

$$\ln\frac{N_t}{N_0} = -kt$$

The half-life and decay constant are related by the following equations:

$$t_{1/2} = \frac{\ln 2}{k} \qquad k = \frac{\ln 2}{t_{1/2}}$$

EXAMPLE:

If the decay constant of technetium-93 is 0.254 h^{-1}, what is its half-life?

SOLUTION: Substitute the decay constant into the above equation:

$$t_{1/2} = \frac{\ln 2}{0.254\ h^{-1}} = 2.73\ h$$

EXAMPLE:

If the half-life of technetium-96 is 4.3 days, what is its decay constant?

SOLUTION: Substitute the half-life into the above equation:

$$k = \frac{\ln 2}{4.3\ \text{days}} = 0.161\ \text{days}^{-1}$$

Workbook Problem 12.6

The half-life for the decay of C-14 is 5730 years. What is the decay constant? Since this is a first-order reaction, how old is a piece of wood in which 23% of the C-14 has decayed?

Strategy: Substitute into the equation relating half-life to the decay constant; use the integrated first-order rate law.

Step 1: Solve for *k*:

Step 2: Substitute into the integrated rate law and solve for *t* after 23% decay:

12.7 Second–Order Reactions

Second-order reactions are dependent either on the concentration of a single reactant raised to the second power – $[A]^2$ – or on two first-order reactants – [A][B].

The rate law for a reaction of the simpler type is

$$\text{Rate} = -\frac{\Delta[A]}{\Delta t} = k[A]^2$$

Integrated:

$$\frac{1}{[A]_t} = kt + \frac{1}{[A]_0}$$

Solved for $[A]_t = ½[A]_0$ (half-life):

$$t_{1/2} = \frac{1}{k[A]_0}$$

The half-life for a second-order reaction depends on both the rate constant and the initial concentration: as a result, each half-life for a second order reaction is twice as long as the previous half-life.

A second-order reaction can be identified by the fact that a plot of $\frac{1}{[A]_t}$ versus time is a straight line. The slope = *k*, and the intercept = $\frac{1}{[A]_0}$

EXAMPLE:

The reaction 2 NOBr (*g*) → 2 NO (*g*) + Br_2 (*g*) is a second-order reaction with respect to NOBr. The rate constant for this reaction is $k = 0.810\ M^{-1}\cdot s^{-1}$ when the reaction is carried out at a temperature of 10 °C. If the initial concentration of NOBr = 8.3×10^{-3} M, how much NOBr will be left after a reaction time of 50 minutes? What is the half-life of this reaction?

SOLUTION: We can solve for the amount of NOBr after 50 minutes by substituting into the integrated rate law for a second–order reaction.

$$\frac{1}{[\text{NOBr}]} = (0.810\ M^{-1} \bullet s^{-1}) \text{x } 50 \text{ min x } \frac{60 \text{ s}}{\text{min}} + \frac{1}{8.3\text{x}10^{-3}\text{M}} = 2550\ M^{-1}$$

$$[\text{NOBr}] = 3.92\text{x}10^{-4}\ M$$

To determine the half–life, again substitute into the equation for the half–life of a second–order reaction:

$$t_{1/2} = \frac{1}{0.810\ M^{-1} \bullet s^{-1}(8.3 \times 10^{-3})} = 149\ s$$

Workbook Problem 12.7

The following data were collected for the second-order reaction:

$$NO_2\ (g) \rightarrow NO\ (g) + O\ (g)$$

Time (s)	0	100	200	300	400
$[NO_2]$	0.100 0	0.079 7	0.066 2	0.056 7	0.049 5

Determine the rate constant, the time required for the reaction to reach 50% completion and the time required to reach 95% completion.

Strategy: Use the integrated second-order rate law and experimental data to determine the rate constant, then solve for the first half-life (50% completion) and 95% completion using the integrated second-order rate law.

Step 1: Solve for k

Step 2: Use the equation for half–life to determine the time required for 50% completion.

Step 3: Determine the concentration of NO_2 at 95% completion (5% of NO_2 remaining).

Step 4: Substitute the concentrations into the integrated rate law, and solve for t.

12.8 Zeroth–Order Reactions

A zeroth-order reaction is independent of the concentration of the reactants. Zeroth-order reactions are relatively uncommon, and have the equation:

$$\text{Rate} = -\frac{\Delta[A]}{\Delta t} = k[A]^0 = k$$

Integrated rate law:

$$[A] = -kt + [A]_0$$

A plot of [A] versus time is a straight line for a first-order reaction. Slope = -k, intercept = $[A]_0$

Workbook Problem 12.8

A general zeroth order reaction gave the following rate data:

Exp	[A] (M)	Rate ($M \cdot s^{-1}$)
1	0.10	0.026
2	0.20	0.026
3	0.40	0.026

Determine the rate constant and the amount of reactant remaining after 20 seconds with an initial concentration of 1.0 M.

Strategy: Substitute into the rate law for zeroth-order reactions, solve for [A] after 20 seconds.

Step 1: Solve for k:

Step 2: Substitute into the integrated rate law and solve for [A] after 20 seconds.

12.9 Reaction Mechanisms

The mechanism is the sequence of molecular events that leads from reactants to products. Each step is known as an **elementary reaction** or an **elementary step**. Knowing reaction mechanisms makes it possible to increase the efficiency of industrial reactions.

Elementary reactions provide information about individual molecular events, the overall reaction provides stoichiometry, but no mechanism information.

A species that is formed, then consumed in the course of reaction is known as a **reaction intermediate**. These do not appear in the overall equation.

Elementary reactions are classified based on their molecularity: the number of molecules or atoms on the reactant side. A unimolecular reaction involves only one reactant moleule. A bimolecular reaction involves energetic collisions between two reactant molecules, and a termolecular reaction (rare) involves collisions of three molecules or atoms.

The elementary steps must sum to give the overall equation.

12.10 Rate Laws for Elementary Reactions

Although overall rate laws must be determined experimentally, the rate law for an elementary reaction will contain the concentration of each reactant raised to an exponent equal to its coefficient in the chemical equation.

For a unimolecular reaction, the rate will always be first-order in the concentration of reactant, as the single molecule is being converted to products.

For a bimolecular reaction, the rate depends on the frequency of collisions between the reactant molecules, and thus on the concentration of those molecules.

This applies only to elementary reactions, not overall reactions.

EXAMPLE:

Write the rate law for the following elementary reaction:

$$2\ H_2\ (g) + O_2\ (g) \rightarrow 2\ H_2O$$

SOLUTION: The rate law for an elementary reaction is equal to the concentration of the reactants raised to their stoichiometric coefficients:

$$\text{Rate} = k[H_2]^2[O_2]$$

12.11 Rate Laws for Overall Reactions

The experimentally observed rate law for an overall reaction depends on the reaction mechanism.

When the overall reaction occurs in two or more steps, the slowest step will determine the rate of the overall reaction. This is known as the **rate-determining step**.

The rate-determining step will determine the rate law for the overall reaction, and will match the predicted rate law for the elementary reaction.

When a rate law for an overall reaction is proposed, the elementary reactions must sum to give the overall reaction, and the mechanism must be consistent with the observed rate law for the overall reaction.

To establish a reaction mechanism, the overall rate law must first be experimentally determined, then a series of elementary steps devised. The rate law can then be predicted based on the rate-determining step. If the observed and predicted rate laws agree, the proposed method is a possible one.

It is impossible to prove a mechanism correct, but easy to prove it incorrect.

EXAMPLE:

Given the following reaction mechanism:

$A_2 \rightarrow 2\ A$
$2\ A + 3\ H_2O \rightarrow 2\ AH_3 + 3/2\ O_2$
$2\ AH_3 + 4\ O_2 \rightarrow 2\ HAO_3 + 2\ H_2O$

a. Determine the overall reaction.
b. Identify the reaction intermediates, and determine the molecularity of each step.
c. Determine the rate law if the second step is the rate–determining step.

SOLUTION:

To determine the overall reaction, sum up the elementary reactions and cancel out the species that occur on both sides:

$$A_2 + 2\ A + 3\ H_2O + 2\ AH_3 + 4\ O_2 \rightarrow 2\ A + 2\ AH_3 + 3/2\ O_2 + 2\ HAO_3 + 2\ H_2O$$

Overall reaction: $A_2 + H_2O + 5/2\ O_2 \rightarrow 2\ HAO_3$

The reaction intermediates are the species that appear in the reaction mechanism, but not in the overall reaction: A, AH_3.

The molecularity of the first step = 1 (unimolecular).
The molecularity of the second step = 5.
The molecularity of the third step = 6.

Rate law = $k[A]^2[H_2O]^3$

Workbook Problem 12.9

What rate law and overall reaction would be expected from the following mechanism:

$2\ A \rightarrow A_2$ rate-determining step

$A_2 + B \rightarrow A_2B$ fast step

Strategy: Write the expected rate law based on the rate-determining step, and sum the reactions for the overall reaction.

Step 1: Write the expected rate law.

Step 2: Sum for the overall reaction.

12.12 Reaction Rates and Temperature: The Arrhenius Equation

Reaction rates tend to double when the temperature is increased by 10 °C. This is consistent with the **collision theory** model of reactions.

Increasing the kinetic energy of a reaction system will increase the energy with which the molecules collide, increasing the chances of a successful reaction.

Molecules must be correctly oriented at collision, and an intermediate must form that is an aggregate of all the reactants. This aggregate is known as the **transition state**, and the energy required to form it is known as the **activation energy**.

Evidence for the concept of an activation energy barrier comes from kinetic molecular theory: only a tiny proportion of gas-phase collisions result in reactions: the proportion of the collisions that do result in reactions is described by the formula:

$$f = e^{-E_a/RT}$$

As the temperature increases, there are more molecules available to collide at the energies necessary to generate a reaction. In fact, the distribution of these molecules from kinetic-molecular theory, and the dependence of reaction rate on temperature are both exponential.

Also involved is the orientation of the reactants when they collide: the proportion of molecules colliding with the proper orientation is the **stearic factor (*p*)**.

These factors are all brought together in the Arrhenius equation:

$$k = Ae^{\frac{-E_a}{RT}}$$

The parameter A is called the **frequency factor** or the pre-exponential factor. From this equation, the rate constant decreases as the activation energy (E_a) increases, and increases as T increases.

12.13 Using the Arrhenius Equation

If the rate constant is known at two temperatures, the natural log of the equation can be taken:

$$\ln k = \ln Ae^{\frac{-E_a}{RT}}$$

A plot of ln k versus $1/T$ gives a straight line, with a slope equal to $-E_a/R$. It can also be rearranged to give an estimate of the activation energy:

$$\ln\left(\frac{k_2}{k_1}\right) = \left(\frac{-E_a}{R}\right)\left(\frac{1}{T_2} - \frac{1}{T_1}\right)$$

EXAMPLE:

The activation energy for the reaction ClO_2F (*g*) → ClOF (*g*) + O (*g*) is 186 kJ/mol. If the value of k is $6.76 \times 10^{-4}\ s^{-1}$ at 322 °C, what is the value of k at 150 °C?

SOLUTION: Convert all temperatures to K, and substitute into the Arrhenius equation:

$$\ln\left(\frac{k_2}{6.76 \times 10^{-4}\ s^{-1}}\right) = \left(\frac{-1.86 \times 10^5\ J/mol}{8.314\ J/mol \bullet K}\right)\left(\frac{1}{423\ K} - \frac{1}{595\ K}\right)$$

$$\ln\left(\frac{k_2}{6.76 \times 10^{-4}\ s^{-1}}\right) = -15.3$$

Taking the antilog of both sides:

$$\frac{k_2}{6.76 \times 10^{-4}\ s^{-1}} = 2.29 \times 10^{-7}$$

$$k_2 = 1.53 \times 10^{-10}\ s^{-1}$$

Workbook Problem 12.10

If a reaction has a rate constant of $0.0123\ s^{-1}$ at 700 K and $0.00342\ s^{-1}$ at 250 K, what is the activation energy of the reaction, and what will the rate constant be at 500 K?

Strategy: Solve for E_a using the two-point Arrhenius equation, then use the same equation to solve for the rate constant at 500 K.

Step 1: Solve for E_a using the two-point Arrhenius equation.

Step 2: Solve for k at 500 K.

12.14 Catalysis

A catalyst is a substance that increases the rate of a reaction without being consumed. in the reaction. Catalysts are very important in the chemical industry, where they increase reaction efficiencies, and in living organisms where enzymes (biological catalysts) facilitate biologically essential reactions.

Catalysts make a lower-energy mechanism available, either by increasing the frequency factor (for instance, by ensuring that reactants are brought together in the proper configuration to react) or, more commonly, by reducing the activation energy.

12.15 Homogeneous and Heterogeneous Catalysts

Homogeneous catalysts exist in the same phase as the reactants: an aqueous enzyme, for example.

Heterogeneous catalysts are in a different phase than the reactants: a metal mesh that provides a reaction surface, for example. The mechanism for heterogeneous catalysis is not well understood, but seems to involve three steps:

- **Adsorption** of the reactants to the surface
- Facilitation of reaction on the surface
- **Desorption** of products from the surface.

Heterogeneous catalysts are the most important for industrial processes, in part because they are easily separated from the reactants and products. The catalytic converters in the exhaust systems of newer automobiles use heterogeneous catalysts to convert pollutants to water, carbon dioxide, nitrogen, and oxygen.

Putting It Together

The reaction mechanism for the reaction of triphenylphosphine, $P(C_6H_5)_3$, with $Ni(CO)_4$ is

$$Ni(CO)_4 \rightarrow Ni(CO)_3 + CO$$

$$Ni(CO)_3 + P(C_6H_5)_3 \rightarrow Ni(CO)_3[P(C_6H_5)_3]$$

a. What is the molecularity of each step?

b. Doubling the concentration of $Ni(CO)_4$ doubles the rate. However, doubling the concentration of $P(C_6H_5)_3$ does not affect the rate. What is the rate law for the reaction? Which step is the slow step in the reaction mechanism?

c. $k = 9.3 \times 10^{-3}\ s^{-1}$ at 20 °C. If the initial concentration of $Ni(CO)_4$ is 0.30 M, how long will it take the reaction to be 60% complete?

d. How many grams of $Ni(CO)_4[P(C_6H_5)_3]$ will be formed if you start with 0.237 g $Ni(CO)_4$ and allow the reaction to proceed for 2 minutes?

Self–Test

This section is intended to test your knowledge of the material covered in this chapter. Think through these problems, and make certain you understand what they are asking. Make sure your answers make sense. Successful completion of these problems indicates that you have mastered the material in this chapter. You will receive the greatest benefit from this section if you use it as a mock exam, as this will allow you to determine which topics you need to study in more detail.

True/False

1. The rate of a reaction can change as the reaction proceeds.

2. The exponents in rate laws cannot be experimentally determined.

3. The order of a reaction is determined by adding up the coefficients in the balanced equation.

4. The rate of a reaction is not the same as the rate constant unless the reaction is zeroth-order.

5. The rate of a first-order reaction is dependent on the concentration of a singe reactant raised to the first power – $[A]^1$.

6. Radioactive decay is a first-order process that is independent of initial concentration.

7. The half-life of a reaction is half of the original amount of a reactant.

8. The half-life of a second-order reaction is dependent on initial concentration.

9. Zeroth-order reactions are fairly common.

10. Reactions can occur in more than one step, and all the steps occur at the same rate.

11. High temperatures increase the likelihood that reactants will collide with sufficient energy to react.

12. Activation energy can be negative for exothermic reactions.

13. A catalyst is unchanged at the end of a reaction.

14. Biological reactions rarely involve catalysts.

15. Catalysts can be in a different phase than the rest of the reactant mixture.

Multiple Choice

1. The overall reaction order for the rate law, rate = $k[A]^2[B]$ is
   a. two
   b. three
   c. five
   d. six

2. A reaction has the rate law, rate = $k[A]^2$. If the initial rate for this reaction is 0.15 mol/L·sec, and the concentration of A is doubled, the new initial rate for this reaction will be
   a. 0.45 mol/L·sec
   b. 1.20 mol/L·sec

c. 0.60 mol/L·sec
d. 1.35 mol/L·sec

3. If a plot of ln [A] versus time gives a straight line for a particular reaction, that reaction is
   a. first-order
   b. ½ order
   c. second-order
   d. zeroth-order

4. The reaction order for the elementary reaction step, $A + B \rightarrow$ products, is
   a. second-order
   b. third-order
   c. cannot be determined from the balanced equation
   d. first-order

5. For every collision that occurs in a reaction
   a. products are formed
   b. the particles have enough kinetic energy to overcome the potential energy barrier
   c. a small fraction of molecules will overcome the potential energy barrier to reaction
   d. none of the above

6. The exponents in a rate law indicate the
   a. dependence of the rate on time
   b. sum of the coefficients in a balanced equation
   c. time required for a certain amount of reactant to disappear
   d. order of the reaction

7. The half–life of a first–order reaction with $k = 3.23 \times 10^{-3}\ s^{-1}$ is
   a. cannot be determined
   b. 1.62×10^{-3} s
   c. 3.10 x 10^2
   d. 2.15×10^2 s

8. The half–life of a second–order reaction depends on
   a. the initial concentration
   b. only the rate constant
   c. the rate constant and the initial concentration
   d. none of the above

9. Reaction rates tend to double when the temperature is
   a. doubled
   b. increased by 10 °C
   c. increased by 100 K
   d. increased by 100 °C

10. The activated complex
    a. has more energy than either the reactants or products
    b. is the configuration of the atoms at the maximum of the potential energy barrier
    c. is a weak linking of all the atoms involved in reaction
    d. all of the above

Matching

Activation energy	a. biological catalyst
Bimolecular reaction	b. study of rates and mechanisms of chemical reactions
Catalyst	c. having a different phase than the reaction
Kinetics	d. involving two molecules
Decay constant	e. not dependent on the concentration of reactants
Enzyme	f. first-order rate constant for radiodecay
Half-life	g. slowest elementary reaction in an overall reaction mechanism
Heterogeneous catalyst	h. factor related to orientation of molecules in collisions and their ability to react
Homogeneous catalyst	i. occurring when the concentration of products is zero.
Initial rate	j. involving three particles
Molecularity	k. the potential energy barrier to reaction
Rate-determining step	l. reaction intermediate with the maximum potential energy
Steric factor	m. time required for the concentration of a reactant to drop to one-half its initial value
Termolecular reaction	n. species that lowers the activation energy without being consumed in the reaction
Transition state	o. in the same phase as the reaction
Zeroth-order reaction	p. the number of particles, whether atoms or molecules, involved in a reaction

Fill–in–the–Blank

1. The ______________ ___________ is defined as either the ____________________ in concentration of a reactant, or the __________________ in the concentration of a __________________.

2. Rate laws depend on the concentration(s) of reactants and/or a proportionality constant known as the _____________ ________________, abbreviated ___.

3. In a first-order reaction, doubling the concentration of a reactant multiplies the rate by a factor of ___. In a second-order reaction, doubling the concentration of a reactant multiplies the rate by a factor of ___. In a third-order reaction, doubling the concentration of a reactant multiplies the rate by a factor of ___. In a zeroth-order reaction, doubling the concentration of a reactant multiplies the rate by a factor of ___.
4. In experimental determinations of rate laws, initial rates of reaction are used because when the products reach a certain concentration, it is likely that the ________________ ________________ will begin to occur.
5. If a plot of log [A] versus time gives a straight line, the reaction is ___________________ with respect to A.
6. Reaction mechanisms can be made up of individual ___________________ ________________, each of which will have its own rate. The overall rate of the reaction will be determined by the ________ ___________________ __________, which is the ______________ step.
7. The Arrhenius equation can be used to calculate the __________________ _____________ of a reaction, the energy required to form a ________________ __________ and from there to form products.
8. A catalyst can either increase the efficiency of collisions, or make a ______________ _________ reaction mechanism available.
9. When the temperature increases by 10 °C, reaction rates tend to ___________________.
10. Heterogeneous catalysts _____________ reactants, facilitate reactions, then _____________ products.

Problems

1. The following data for the reaction A + B $\rightarrow$ C was collected. Calculate an estimate of the initial rate, and the rate between 30 and 75 seconds.

[C] (mol/L)	**Time (s)**
0	0
0.005	15
0.010	30
0.015	45
0.017	60
0.019	75
0.020	90
0.021	105
0.022	120

2. The rate of disappearance of PH_3 for the reaction

   $4\ PH_3(g) \rightarrow P_4(g) + 6\ H_2\ (g)$

   is 1.67×10^{-2} mol/L·sec. What is the rate of appearance of H_2?

3. The following data were collected for the reaction

   $2\ NO\ (g) + Cl_2\ (g) \rightarrow 2\ NOCl\ (g)$

[NO] (mol/L)	[Cl_2] (mol/L)	Initial Rate mol/(L·sec)
0.025	0.025	15.9
0.050	0.025	63.6
0.025	0.050	31.8

   Determine the rate law from this data. What is the order of the reaction with respect to each reactant? What is the overall order of the reaction? Provide the rate law, and solve for *k*.

4. The following data were collected for the reaction

   3 A + 2 B → 2 D

[A] (mol/L)	[B] (mol/L)	Initial Rate mol/(L·sec)
0.150	0.150	6.70×10^{-2}
0.300	0.150	0.134
0.150	0.300	6.70×10^{-2}

   Determine the rate law from this data. What is the order of the reaction with respect to each reactant? What is the overall order of the reaction? What is the rate law? Calculate the value of *k*. If the initial concentration of A = 0.542 M and the initial concentration of B = 0.830 M, what would the initial rate be?

5. Plutonium-239 has a half-life of 24,000 years. How long would it take for one mole of plutonium to decay until only 200 atoms remained?

6. The rate law for the decomposition of O_3 to O_2 is second-order in ozone. The rate constant for this reaction equals 1.40×10^{-2} L/(mol·sec). If the initial concentration of ozone is 2.37 M, how much O_3 will be present after 12 hours?

7. The following data were collected for the reaction $2\ HI\ (g) \rightarrow H_2\ (g) + I_2\ (g)$

[HI] (mol/L)	Time (min)
2.50	0
1.45	3
1.02	6
0.788	9
0.641	12
0.541	15
0.468	18
0.412	21
0.368	24

[HI] (mol/L)	Time (min)
0.332	27
0.303	30

Determine the order of this reaction. Calculate the value of k and the half-life from the initial reaction.

8. The proposed reaction mechanism for a reaction is

$NO_2\ (g) + Cl_2\ (g) \rightarrow NO_2Cl\ (g) + Cl\ (g)$

$Cl\ (g) + NO_2\ (g) \rightarrow NO_2Cl\ (g)$

What is the overall reaction? What is the molecularity of each step? What intermediates are there? If the first step is the slow step in the mechanism, write the expected rate law. What will the order of the reaction be?

10. If the rate constant for a reaction is $k = 3.79 \times 10^{-5}$ L/mol·sec at 108 °C and the activation energy for the reaction is 11.8 kJ/mol, what is the value of A?

11. The activation energy is $E_a = 98.62$ kJ/mol for the reaction

$2\ NOCl \rightarrow 2\ NO + Cl_2$

If $k = 7.2 \times 10^{-5}\ s^{-1}$ at 67 °C, at what temperature will $k = 3.2 \times 10^{-3}\ s^{-1}$?

Challenge Problem

For the reaction

$C_2H_5I + OH^- \rightarrow C_2H_5OH + I^-$

$E_a = 79.7$ kJ/mol and $A = 1.09 \times 10^9\ M^{-1}s^{-1}$at 45°C. If the concentration of C_2H_5I is doubled while $[OH^-]$ remains constant, the rate doubles. If the $[OH^-]$ is doubled while $[C_2H_5I]$ remains constant, the rate doubles. If a 250 mL solution of 0.475 g KOH in ethanol is mixed with a 250 mL solution of 1.378 g C_2H_5I in ethanol, what is the initial rate at 35 °C?

Inquiry-Based Problem

You arrive in lab to learn that you will be investigating the rate law for the reaction between iodite and sulfite ions. Your laboratory instructor has written the steps for this reaction on the board:

$IO_3^-\ (aq) + 3\ SO_3^{2-}\ (aq) \rightarrow I^-\ (aq) + 3\ SO_4^{2-}\ (aq)$

$5\ I^-\ (aq) + 6\ H^+\ (aq) + IO_3^-\ (aq) \rightarrow 3\ H_2O + 3\ I_2\ (s)$

$3\ I_2\ (s) + 3\ SO_3^{2-}\ (aq) + 3\ H_2O \rightarrow 6\ I^-\ (aq) + 3\ SO_4^{2-}\ (aq) + 6\ H^+\ (aq)$

$2\ IO_3^-\ (aq) + 6\ SO_3^{2-}\ (aq) \rightarrow 2\ I^-\ (aq) + 6\ SO_4^{2-}\ (aq)$

I_2, which is produced in the second step and consumed in the third step, reacts with starch to produce a deep blue color due to the formation of an I_2•starch complex. A 0.10 M solution of HIO_3, a 0.05 M solution of H_2SO_3, a solution of starch, and a graduated pipet are provided. Develop an experimental procedure to determine the rate law for the overall reaction, and the activation energy for the reaction.

Consider the following:

1. Determine the form of the rate law based on the overall reaction.
2. How can you determine the impact of each individual reactant on the overall rate?
3. How are you going to monitor the progress of the reaction? What does this tell you about the consumption of reactants?
4. How are you going to define rate? What unit will you use?
5. How are you going to determine the activation energy for this reaction once you determine the rate law?

CHAPTER 13

CHEMICAL EQUILIBRIUM

Chapter Learning Goals

A. Extent of Chemical Reactions

1. Given a balanced chemical equation representing a homogenous or heterogeneous equilibrium, write the equilibrium equation.
2. From the equilibrium concentrations of products and reactants, calculate the equilibrium constant K_c.
3. Given the equilibrium partial pressures of reactants and products, calculate the equilibrium constant K_p.
4. From a value of K_c and a balanced equation, calculate K_p. From a value of K_p and a balanced equation, calculate K_c.

B. Equilibrium Mixture Composition

1. From the value of K_c or K_p, determine whether mainly products or mainly reactants exist at equilibrium.
2. For a given mixture of reactants and products, determine whether a system is at equilibrium. If not, determine the direction the reaction will go to achieve equilibrium.
3. Given K_c and initial concentrations of reactants and/or products, calculate the final concentrations of reactants and/or products.

C. Systems Under Stress

1. Determine the reaction direction when a system at equilibrium reacts to a stress applied to the system, including changes in concentrations, pressure and volume, or temperature.
2. Describe the effect of adding a catalyst to a system at equilibrium.

D. Chemical Kinetics and Chemical Equilibrium

1. Describe the relationship between the equilibrium constant and the ratio of the rate constants for the forward and reverse reactions. Solve problems involving this relationship.

Chapter in Brief

This chapter introduces the concept of chemical equilibrium, the state reached when the concentration of reactants and products remains constant over time despite both forward and backwards reactions continuing to occur. Equilibrium constants can be related to either concentrations or partial pressures, and equations calculated in different ways can be related to one another. These constants can be used to determine the final concentrations of reactants and products in an equilibrium mixture. When equilibria are stressed through changing conditions, the equilibrium will shift to minimize the stress. Changes in temperature, pressure/volume, and concentrations will all shift the equilibrium mixture; adding a catalyst will speed the achievement of equilibrium, but does not change the equilibrium concentrations. Reaction conditions must be chosen carefully along with the use of a catalyst. Kinetics and equilibrium are linked, and their constants can be compared.

13.1 The Equilibrium State

Many reactions do not go to completion, but reach an *equilibrium* at which the concentrations of reactants and products are no longer changing.

To avoid confusion, species on the left of an equation will always be called "reactants" and those on the right will always be called "products," even though the direction of the reaction may be unclear. To indicate that the reaction can proceed in either direction, a double arrow is used.

As a reaction proceeds, there comes a point when the reverse reaction (right to left) proceeds at the same rate as the forward reaction. This occurs in all reactions, though some proceed so far toward completion that they are called irreversible: the equilibrium mixture will contain almost all products and almost no reactants.

When equilibrium is reached, the reactions, forward and backward, continue to occur, but have the same rates: there is no net conversion of reactants to products.

13.2 The Equilibrium Constant K_c

When reactions are permitted to go to equilibrium, the concentrations of reactants and products are related by the equilibrium constant, K_c.

For the generic reaction $a\,A + b\,B \leftrightarrows c\,C + d\,D$ the equilibrium constant is:

$$K_c = \frac{[C]^c[D]^d}{[A]^a[B]^b}$$

The equilibrium constant is temperature-dependent, and unitless: all of the concentrations are considered to be divided by their standard state concentrations (1 M), which eliminates the units from the concentration values.

For the reverse reaction, the equilibrium constant will be the inverse of the constant for the forward reaction. This makes it important that the form of the equation is given.

EXAMPLE:

For the reaction $CO\,(g) + 2\,H_2\,(g) \leftrightarrows CH_3OH\,(g)$, the equilibrium concentrations are $[CO]_e = 2.38$ M; $[H_2]_e = 0.260$ M; $[CH_3OH]_e = 3.15$ M at a temperature of 110 °C. Calculate the equilibrium constant, K_c, for this reaction.

SOLUTION: Write an equilibrium equation for the balanced chemical reaction:

$$K_c = \frac{[CH_3OH]}{[CO][H_2]^2} = 19.6$$

Substitute the equilibrium concentrations and solve for K_c.

$$K_c = \frac{(3.15)}{(2.38)(0.260)^2} = 19.6$$

Workbook Problem 13.1

A weak acid dissociates according to the following equation:

$$HA\ (aq) \leftrightarrows H^+\ (aq) + A\ (aq)$$

What is the equilibrium constant at 20 °C if a 0.250 M solution is 1.73% ionized?

Strategy: Write the equilibrium equation for K_c, calculate concentrations of all species and solve:

Step 1: Write the equation for the equilibrium constant.

Step 2: Determine the concentration of each species.

Step 3: Solve for K_c.

13.3 The Equilibrium Constant K_p

For gas-phase reactions, partial pressures can be used instead of molar concentrations. The ideal gas law can be rearranged to show the relationship between concentration (n/V) and pressure:

$$P_A = [A]RT$$

When these terms are substituted into the equation for an equilibrium constant, the relationship between K_c and K_p can be shown to be:

$$K_p = K_c(RT)^{\Delta n}$$

Where Δn is the sum of the coefficients of the gaseous products minus the sum of the coefficients of the gaseous reactants.

EXAMPLE:

Calculate K_p for the reaction in the previous example.

SOLUTION: The coefficients of the balanced equation can be used to calculate Δn. The sum of the coefficients on the products side is 1, while the sum of the coefficients on the reactant side is 3. Δn = products – reactants = –2.

Substitute Δn and K_c into the equation relating K_p to K_c:

$$K_p = (19.6)\left[\left(0.0821\ \frac{L \cdot atm}{mol \cdot K}\right)(383\ K)\right]^{-2} = 1.98 \times 10^{-2}$$

Workbook Problem 13.2

Determine the value of K_c and K_p for the reaction:

$$PCl_5\ (g) \leftrightarrows PCl_3\ (g) + Cl_2\ (g)$$

At equilibrium at 250 K, the concentrations measured for each reactant are $[PCl_5]_e = 2.35$ M and $[PCl_3]_e = [Cl]_e = 1.56$ M.

Strategy: Write the equilibrium equation for K_c.

Step 1: Solve for K_c using the molar concentrations given.

Step 2: Determine Δn.

Step 3: Solve for K_p, using the equation that expresses the relationship between K_c and K_p.

13.4 Heterogeneous Equilibria

Homogeneous equilibria have all reactants and products in the same phase. But it is also possible to have reactants in a mixture of phases: this is **heterogeneous equilibrium**.

Because the concentrations of solids and liquids are constant, they are not included in the equilibrium expression.

EXAMPLE:

Write the equilibrium expression for the reaction $H_2CO_3\ (s) \leftrightarrows CO_2\ (g) + H_2O\ (l)$.

SOLUTION: This reaction generates only one product that is not a pure solid or liquid. As a result, the equilibrium constant will be:

$$K_c = [CO_2]$$

As this is a gas, K_p should be used:

$$K_p = K_c RT$$

13.5 Using the Equilibrium Constant

Because of the form in which the equilibrium constant is written, its value provides information about the way the reaction proceeds at a specific temperature.

If K_c is large, over 10^3, the reaction proceeds almost to completion. Products predominate over reactants.

If K_c is small, under 10^{-3}, the reaction proceeds very little. Reactants will predominate over products in an equilibrium mixture.

With an intermediate value of K_c, there will be appreciable quantities of both reactants and products in an equilibrium mixture.

The equilibrium constant can also be used to predict the direction of reaction given concentrations of reactants and products.

The reaction quotient, Q_c, is calculated in exactly the same way as the equilibrium constant expression, but uses concentrations that may or may not be equilibrium values. Comparing Q and K makes it possible to predict which way the reaction will go:

- $Q_c < K_c$; reaction goes from left to right, more products will be formed.
- $Q_c > K_c$; reaction goes from right to left, products will be converted to reactants.
- $Q_c = K_c$; reaction is at equilibrium, there will be no net change.

If all but one of the equilibrium concentrations is known, the last can be calculated algebraically using the equilibrium constant.

It is also possible to calculate equilibrium concentrations from initial concentrations using the equilibrium coefficient and the balanced chemical equation. To do this, the following steps are required:

- Write and balance the chemical equation.
- Make a table in which to list the initial concentrations of all reactants and products, and the change expected. For the change, define the concentration of one of the species as x and use the balanced equation to define the changes in the other species in terms of x.
- Substitute the algebraic statements of final concentration into the equilibrium equation and solve for x. Should this require solving a quadratic equation, choose the answer that makes chemical sense (there are no negative concentrations).

EXAMPLE:

$K_c = 4.18 \times 10^{-9}$ at 425 °C for the following reaction:

$2\ HBr\ (g) \leftrightarrows H_2\ (g) + Br_2\ (g)$

What is the position of equilibrium? If the concentrations of all species present are [HBr] = 0.35 M; $[H_2] = [Br_2] = 3.2 \times 10^{-3}$ M is the reaction mixture at equilibrium? If not, in which direction will the reaction proceed?

SOLUTION: We can determine the position of equilibrium by looking at the value of K_c. Since $K_c < 10^{-3}$, we know that reactants predominate over products and that the position of equilibrium is to the left. This is reflected in the concentrations provided, but to determine if the reaction mixture is at equilibrium, we need to calculate Q_c and compare that value to K_c.

$$Q_c = \frac{[H_2][Br_2]}{[HBr]^2} = \frac{(3.2 \text{ x } 10^{-3})^2}{(0.35)^2} = 8.4 \times 10^{-5}$$

Since $Q_c > K_c$, the reaction is not at equilibrium and will proceed from right to left.

EXAMPLE:

The reaction $PCl_5\ (g) \leftrightarrows PCl_3\ (g) + Cl_2\ (g)$ has $K_c = 85.0$ at a temperature of 760 °C. Calculate the equilibrium concentrations of PCl_5, PCl_3, and Cl_2 if the initial concentration of PCl_5 is 3.50 M. Assume a volume of 1.00 L

To solve this problem, follow the procedure outlined above:

The balanced equation is given.

The initial concentration of PCl_5 is 3.50 M. Let x be the concentration of PCl_5 that reacts on going to the equilibrium state. Since the mole to mole ratios are 1:1:1, if x amount of PCl_5 reacts, then x mol/L of PCl_3 and Cl_2 will be present at equilibrium. This can be summarized in the following table:

	$PCl_5\,(g) \leftrightarrows$	$PCl_3\,(g)$ +	$Cl_2\,(g)$
Initial Concentration (M)	3.5	0	0
Change (M)	$-x$	$+x$	$+x$
Equilibrium Concentration (M)	$3.5 - x$	x	x

The equilibrium expression for the reaction is:

$$K_{eq} = \frac{[PCl_3][Cl_2]}{[PCl_5]}$$

The equilibrium concentrations from the above table can be substituted into this expression, and the equation rearranged to facilitate the use of the quadratic equation.

$$85.0 = \frac{x^2}{3.5 - x}$$

$$x^2 + 85x - 298 = 0$$

Use the quadratic equation to solve for x:

$$x = \frac{-85 \pm \sqrt{(85)^2 - 4(-298)}}{2} = 3.37, -88.3$$

A negative number does not make physical sense: the answer is 3.37.

Calculate the equilibrium concentrations:

$$[PCl_5] = 3.5 - 3.37 = 0.13\ M$$
$$[PCl_3] = [Cl_2] = 3.37\ M$$

Workbook Problem 13.3

The value of K_c for the reaction

$$C\,(s) + H_2O\,(g) \leftrightarrows CO\,(g) + H_2\,(g)$$

is 3.0×10^{-2}. Determine the equilibrium concentration if the initial concentration of water is 11.75 M.

Strategy: Follow the previously outlined steps:

Step 1: Write the balanced equation for the reaction.

Step 2: Make a table listing the initial concentration, the change in concentration, and the equilibrium concentration. (Let x = the amount of substance that reacts.)

Step 3: From the balanced equation, write the equilibrium equation. Substitute the algebraic expressions for the equilibrium concentrations into the equilibrium equation, and solve for x. Use the quadratic equation if necessary.

13.6 Factors That Alter the Composition of an Equilibrium Mixture: LeChâtelier's Principle

If a reaction does not proceed at room temperature and atmospheric pressure, the reaction conditions can be altered to force the reaction to proceed. There are four factors that alter the composition of an equilibrium mixture:

- Concentration of reactants or products
- Pressure and volume
- Temperature
- Catalysis

The fourth, addition of a catalyst, only affects the rate at which equilibrium is reached, not the composition of the equilibrium mixture.

If a stress – change in concentration, pressure, volume and/or temperature – is applied to a reaction mixture at equilibrium, the equilibrium will shift to minimize the stress. This is known as **Le Châtelier's principle**.

13.7 Altering an Equilibrium Mixture: Changes in Concentration

When equilibrium is disturbed by adding or removing a reactant or product, reaction will occur to *remove* an added substance, and *replenish* a removed substance.

Adding a reactant will drive the reaction towards products (to the right); adding a product will drive the reaction towards reactants (to the left).

Removing a reactant will drive the reaction towards reactants (to the left); removing a product will drive the reaction towards products (to the right).

Mathematically, this works by changing Q_c: when a reactant is added or a product removed, Q_c becomes smaller: the reaction must shift right towards products, increasing the numerator again. When a reactant is removed, or a product added, Q_c becomes larger. To return to equilibrium, the reaction must shift left: products will be converted to reactants.

EXAMPLE:

Consider the following reaction:

$$CaCO_3\,(s) \leftrightarrows CaO\,(s) + CO_2\,(g)$$

Predict the results of adding $CaCO_3$, adding CO_2, removing CO_2.

SOLUTION:

Adding $CaCO_3$ will not affect the equilibrium: solids do not appear in either K or Q.

Adding CO_2 will push the reaction to shift left towards reactants, thus reducing the amount of CO_2.

Removing CO_2 will push the reaction to shift right towards products, thereby increasing the amount of CO_2.

Workbook Problem 13.4

The following reaction is at equilibrium:

$$2\ BrNO\ (g) \leftrightarrows 2\ NO\ (g) + Br_2\ (g)$$

What would be the effect on the equilibrium if BrNO was added, BrNO was removed, NO was added, Br_2 was removed.

Strategy: Use Le Châtelier's principle

Step 1: Determine the direction of reaction needed to minimize each stress.

13.8 Altering an Equilibrium Mixture: Changes in Pressure and Volume

When the moles of products and moles of reactants are different, changing the volume and/or pressure can shift the reaction direction. An increase in the pressure will shift the reaction to the side with fewer moles of gas. A decrease in the pressure will shift the reaction in the direction of more moles of gas.

Increases in pressure (due to decrease in volume) are akin to increases in concentration for gases, as n/V increases.

If a reaction has the same number of moles of gaseous products and reactants, changing the pressure will not affect the reaction.

In heterogeneous equilibria, changes in pressure do not affect solids or liquids, as their volume is essentially independent of pressure.

These effects only are seen when pressure is changed as a result of changing volume. If pressure is increased by adding a non-reacting (inert) gas to the mixture, the equilibrium will not change.

Workbook Problem 13.5

The following reaction is at equilibrium:

$$N_2O_4\ (g) \leftrightarrows 2\ NO_2\ (g)$$

What would be the effect on the equilibrium if the pressure was increased? If the pressure was decreased? If 10 L of neon were added to the reaction mixture at constant volume?

Strategy: Use Le Châtelier's principle

Step 1: Determine the direction of reaction needed to minimize each stress.

13.9 Altering an Equilibrium Mixture: Changes in Temperature

So long as the temperature remains constant, the value of the equilibrium constant remains the same. Changes in pressure of concentration only change *Q*.

When the temperature changes, the equilibrium constant itself is altered:

An exothermic reaction will be less favored as the temperature increases: the equilibrium constant will decrease with increasing temperature: the heat added by the reaction will increase, not decrease, the stress on the equilibrium. As the temperature decreases, the equilibrium constant will decrease: the reaction will be contributing heat, reducing the stress on the reaction.

An endothermic reaction will be more favored as the temperature increases: the equilibrium constant will increase with increasing temperature, as the reaction absorbs some of the temperature increase. As the temperature decreases, the equilibrium coefficient will decrease: the heat absorbed by the reaction increases, not decreases, the stress on the equilibrium.

EXAMPLE:

How will the following changes alter the equilibrium for the reaction?

$$C(graphite) + CO_2(g) \leftrightarrows 2CO(g) \quad \Delta H^{o} = +172.5\ \text{kJ}$$

a. CO is removed from the system.
b. CO_2 is removed from the system.
c. The volume of the container is increased.
d. Graphite is added to the system.
e. The temperature is raised.

SOLUTION:

a. When CO is removed from the system, the equilibrium shifts right.
b. When CO_2 is removed from the system, the equilibrium shifts left.
c. Increasing the volume of the container decreases the pressure in the container. The equilibrium will shift to the right to minimize the change.
d. Adding a solid to the system does not affect the equilibrium since solids are not included in the equilibrium expression.
e. Raising the temperature causes the equilibrium to shift right, as the reaction absorbs temperature.

13.10 The Effect of a Catalyst

Adding a catalyst lowers the activation energy for both the forward and reverse reactions. As a result, the reaction to reach equilibrium faster, but the equilibrium constant will be the same, and the equilibrium concentrations will be unaffected.

Although a catalyst does not affect the equilibrium, it can affect the reaction conditions chosen: if good yields are obtained slowly at a certain temperature, finding a catalyst can make the reaction more practical.

Workbook Problem 13.6

Consider the following reaction, which occurs in a catalytic converter in the presence of a palladium catalyst:

$$2CO\ (g) + O_2\ (g) \xrightarrow{Pd} 2CO_2\ (g) \qquad \Delta H° = -566\ kJ$$

What would be the effect on the concentration of CO_2 if the pressure was increased? If the pressure was decreased? If the palladium catalyst was removed? If the temperature was lowered? If the concentration of O_2 was increased?

Strategy: Use Le Châtelier's principle

Step 1: Determine the direction of reaction needed to minimize each stress, and its effect on $[CO_2]$.

13.11 The Link Between Chemical Equilibrium and Chemical Kinetics

If the reaction A + B ⇆ C + D is a simple reaction with an elementary forward and reverse reaction, it has two rate constants:

Rate of forward reaction = $k_f[A][B]$
Rate of reverse reaction = $k_r[C][D]$

At equilibrium, these rates are equal to one another, and can be related to the equilibrium constant:

$$k_f[A][B] = k_r[C][D] \text{ or } \frac{k_f}{k_r} = \frac{[C][D]}{[A][B]} = K_c$$

It is the relative values of k_f and k_r that determine the composition of the equilibrium mixture. If the rate constant for the forward reaction is much larger than that of the reverse reaction, K_c is very large, and the reaction goes to completion. With the reverse reaction being too slow to be detected, this is an irreversible reaction.

If instead the values of k_f and k_r are close to one another, there will be an equilibrium in which all species – reactant and product – will be present.

The temperature-dependence of equilibrium constants is also explained by the relationship derived above:

$$\frac{k_f}{k_r} = K_c$$

Because the kinetic constants are temperature-dependent, and increase by different amounts when the temperature increases (as described by the Arrhenius equation), equilibrium is also temperature-dependent.

EXAMPLE:

Calculate the equilibrium constant K_c for the reaction

$H_2O\ (l) \leftrightarrows H^+\ (aq) + OH^-\ (aq)$

given that $k_f = 2.4 \times 10^{-5}\ s^{-1}$ and $k_r = 1.3 \times 10^{11}\ M^{-1}{\cdot}s^{-1}$ and that K is for the expression $[H^+][OH^-]$

SOLUTION: To calculate K_c, use the equation

$$K_c = \frac{k_f}{k_r} = \frac{2.4 \times 10^{-5}\ s^{-1}}{1.3 \times 10^{11}\ s^{-1}} = 1.8 \times 10^{-16}$$

Putting It Together

Calculate the pressure of all species at equilibrium for the reaction

$$2\ NO\ (g) + Br_2\ (g) \leftrightarrows 2\ NOBr\ (g)$$

given an initial pressure of NO = 78.4 mm Hg and the initial pressure of Br_2 = 51.3 mm Hg. The total pressure at equilibrium = 120.5 mm Hg. Determine the value of K_p.

Self–Test

This section is intended to test your knowledge of the material covered in this chapter. Think through these problems, and make certain you understand what they are asking. Make sure your answers make sense. Successful completion of these problems indicates that you have mastered the material in this chapter. You will receive the greatest benefit from this section if you use it as a mock exam, as this will allow you to determine which topics you need to study in more detail.

True–False

1. All chemical reactions are reversible to some extent.
2. The equilibrium constant can be determined by running the same reaction at different temperatures.
3. The equilibrium constant has units of M^{-1}.
4. Equilibrium constants can be related to concentrations or to partial pressures.
5. A large equilibrium constant indicates a reaction with more reactants than products at equilibrium.
6. Increasing the pressure on a gas-phase reaction will drive the reaction in the direction of more moles of gas.
7. Adding more of a reactant pushes the reaction to the right toward products.
8. Catalysts speed the rate at which equilibrium is achieved.
9. The relative values of k_f and k_r determine the composition of the equilibrium mixture.
10. The value of K makes it possible to predict the direction of a reaction.

Multiple Choice

1. A state of chemical equilibrium is reached when
   a. the rate of the forward reaction is the same as the rate of the reverse reaction
   b. $Q = K$
   c. the concentrations of the products and reactants have reached constant value
   d. all of the above

2. The equilibrium constant is calculated as:
   a. concentration of products over concentration of reactants
   b. concentration of products raised to stoichiometric coefficients over concentration of reactants raised to stoichiometric coefficients
   c. concentration of reactants over concentration of products
   d. concentration of reactants raised to stoichiometric coefficients over concentration of products raised to stoichiometric coefficients

3. In heterogeneous equilibria, the following are included in the equilibrium constant
   a. solids, liquids, solutions, and gases
   b. liquids, solutions, and gases
   c. solutions and gases
   d. only gases

4. An equilibrium constant in the range of 10^{-3} to 10^{3} indicates that
   a. both products and reactants will be appreciably present at equilibrium
   b. only products will be appreciably present at equilibrium
   c. only reactants will be appreciably present at equilibrium
   d. no information about final concentrations can be gained from this

5. The following stresses will result in a change in equilibrium concentrations:
   a. change of temperature
   b. change of concentrations
   c. change in volume and pressure
   d. all of the above

6. Which of the following is an expected response to stress on a reaction?
   a. Lowering the temperature on an exothermic reaction increases the concentration of products.
   b. Lowering the temperature on an endothermic reaction increases the concentration of products.
   c. Raising the temperature on an exothermic reaction increases the concentration of products.
   d. Raising the temperature on an endothermic reaction has no effect on the concentration of products.

7. A catalyst:
   a. increases the concentration of products
   b. increases the percent yield of a reaction
   c. makes a reaction go to completion
   d. increases the speed at which equilibrium is obtained

8. K_p
   a. is the same as K_c
   b. is the reciprocal of K_c
   c. is the equilibrium constant calculated using partial pressures
   d. includes the concentration of pure liquids in the equilibrium expression

9. The reaction quotient, Q_c, is
   a. calculated in the same manner as K_c
   b. less than K_c at equilibrium
   c. greater than K_c at equilibrium
   d. equal to K_p

10. When a reactant is removed from a system at equilibrium
    a. the reaction shifts right

b. the reaction shifts left
c. the reactants are converted to products
d. the denominator in the K_c expression becomes larger

Matching

Chemical equilibrium	a. constant in the equilibrium equation for gas-phase reactions
Equilibrium constant, K_c	b. equilibrium in which all reactants and products are in the same phase
Equilibrium constant, K_p	c. calculated like K_{eq}, but with non-equilibrium concentrations.
Equilibrium mixture	d. constant in the equilibrium equation for solutions
Homogeneous equilibria	e. if a stress is applied to a reaction, the reaction occurs in the direction that relieves the stress
Heterogeneous equilibria	f. any state in which change in forward and reverse directions is equal in rate, leading to no net change
Reaction quotient, Q_c	g. a state in which the concentration of products and reactants in a reaction no longer shows any net change
Reversible reaction	h. equilibrium in which reactants and products are in different phases
Dynamic state	i. a reaction that can go forward toward products or back to reactants, depending on reaction conditions
Le Châtelier's principle	j. combination of reactants and products with characteristic proportions

Fill–in–the–Blank

1. Chemical equilibrium is a ____________________ state in which the ____________________ of reactants and products is ___________________.
2. The equilibrium equation for a ________________________ equilibrium does not include concentrations of pure ______________ or _____________.
3. When an equilibrium mixture is subjected to a stress, it responds to __________________ that stress. Adding more reactant will shift a reaction toward ________________, while adding more products will shift a reaction toward ____________________. Changing the __________________ affects the mixture by changing the _____________________ constants. Adding a

____________________ does not alter the equilibrium mixture, simply ________________ the rate at which it is achieved.

4. Comparing K and Q makes it possible to determine the __________________ of a reaction. When $K = Q$, the reaction is at ___________________________. When $K < Q$, the reaction will proceed from ____________________ to __________________________. When $K < Q$, the reaction will proceed from ____________________ to __________________________.
5. Increasing the pressure on a gas-phase reaction will shift the equilibrium away from small molecules toward larger molecules. This has the result of ______________________ the pressure.

Problems

1. Write equilibrium expressions for the following balanced equations.

   a. $O_3\,(g) + Cl\,(g) \leftrightarrows O_2\,(g) + ClO\,(g)$

   b. $2\,NO\,(g) + Br_2\,(g) \leftrightarrows 2\,NOBr\,(g)$

   c. $2\,Cu\,(s) + O_2\,(g) \leftrightarrows 2\,CuO\,(s)$

   d. $2\,N_2O\,(g) \leftrightarrows 2\,N_2\,(g) + O_2\,(g)$

2. The partial equilibrium pressures for N_2, O_2, and NO in the reaction

   $N_2\,(g) + O_2\,(g) \leftrightarrows 2\,NO\,(g)$

   are $p(N_2) = 0.27$ atm, $p(O_2) = 0.187$ atm, and $p(NO) = 0.045$ atm. Calculate K_p for this reaction.
3. Calculate K_c for the reaction in Question 2.

4. From the value of K_c, determine whether mainly products or mainly reactants exist at equilibrium for the following balanced equations:

   a. $CH_3Cl\,(aq) + OH^-\,(aq) \leftrightarrows CH_3OH\,(aq) + Cl^-\,(aq)$ $\quad K_c = 1 \times 10^{16}$

   b. $2SO_2\,(g) + O_2\,(g) \leftrightarrows 2\,SO_3\,(g)$ $\quad K_p = 3.3$

   c. $2\,HBr\,(g) \leftrightarrows H_2\,(g) + Br_2\,(g)$ $\quad K = 2 \times 10^{-19}$

5. For the reaction

   $PCl_5\,(g) \leftrightarrows PCl_3\,(g) + Cl_2\,(g)$

   $K_c = 33.3$ at 760 °C. Is this system at equilibrium if $[PCl_5] = 2.43 \times 10^{-3}$ M, $[PCl_3] = [Cl_2] = 0.830$ M? If not, determine the direction the reaction must go to reach equilibrium.

6. Nitrogen dioxide is produced from the reaction of dinitrogen oxide and oxygen according to the equation

   $2\,N_2O\,(g) + 3\,O_2\,(g) \leftrightarrows 4\,NO_2\,(g)$

At 25 °C, 0.0342 moles of N_2O and 0.0415 moles of O_2 are placed in a 1.00 L container and allowed to react. The $[NO_2]$ at equilibrium is 0.0298 moles. What are the equilibrium concentrations of N_2O and O_2? What is the value of K_c?

7. The equilibrium constant for the reaction

   $SO_2\ (g) + NO_2\ (g) \leftrightarrows NO\ (g) + SO_3\ (g)$
   is $K_c = 85.0$ at 460 °C. Calculate the concentrations of all species at equilibrium when the initial reaction mixture contains 0.100 M SO_2 and 0.100 M NO_2.

8. Consider the reaction

   $2\ SO_2\ (g) + O_2\ (g) \leftrightarrows 2\ SO_3\ (g)$ $\Delta H° = +197$ kJ

   What will be the direction of the reaction when the following stress is applied?

   a. addition of SO_3
   b. decrease in temperature
   c. increase in volume
   d. addition of N_2 gas
   e. addition of SO_2 gas
9. For the reaction $A + B \leftrightarrows 2\ C$, $k_f = 6.8 \times 10^{-5}\ s^{-1}$ and $k_r = 7.2 \times 10^{-7}\ s^{-1}$. Calculate K.
10. Calculate the equilibrium concentration for all species present in the reaction

    $HCONH_2\ (g) \leftrightarrows NH_3\ (g) + CO\ (g)$

    if the initial concentration of formamide ($HCONH_2$) is 0.125 M. $K_c = 4.84$ at 127 °C.

11. For the reaction

    $2\ H_2S\ (g) + CH_4\ (g) \leftrightarrows 4\ H_2\ (g) + CS_2\ (g)$

    $K_c = 5.27 \times 10^{-8}$ at 700 °C. What is the position of equilibrium for this reaction? Is the reaction mixture at equilibrium when the concentrations of the reactants and products are $[CH_4] = 1.75$ M, $[H_2S] = 2.00$ M, $[H_2] = 0.450$ M, and $[CS_2] = 0.085$ M? If not, in which direction will the reaction proceed?

Challenge Problem

The following reaction was carried out in a 4.00 L vessel:

$NO_2\ (g) + NO\ (g) \leftrightarrows N_2O\ (g) + O_2\ (g)$

Determine the number of moles of reactants and products present at equilibrium if the initial number of moles are 0.100 mol NO_2, 0.050 mol NO, 0.0125 mol N_2O, and 0.00125 mol O_2. Use a value of K_c = 0.914.

CHAPTER 14

AQUEOUS EQUILIBRIA: ACIDS AND BASES

Learning Goals

A. Arrhenius Acids and Bases

1. Define acids and bases according to the Arrhenius, Brønsted–Lowry, and Lewis theories.

B. Dissociation of Water

1. Calculate H_3O^+ concentration from OH^- concentration and *vice versa*. Determine whether a solution is acidic, neutral, or basic based on these concentrations.
2. Interconvert pH and $[H_3O^+]$. Use pH values to classify solutions as acidic, neutral, or basic.
3. Given the molar concentration of a strong acid or a strong base, determine the pH of the solution.

C. Brønsted–Lowry Acids and Bases: Proton–Transfer Reactions

1. From Lewis structures, determine which chemical species can act as Brønsted–Lowry acids, Brønsted–Lowry bases, or both.
2. From the chemical equation for a proton–transfer reaction, identify the conjugate acid–base pair.
3. Given the extent of dissociation of an acid in water, determine whether the acid is a stronger or weaker acid than water and whether the conjugate base of the acid is a stronger or weaker base than water.
4. Given a proton–transfer reaction and the relative strengths of the acid and base involved, determine which direction the reaction will proceed.
5. Interconvert K_a and K_b.
6. Classify salt solutions as acidic, neutral, or basic. Calculate the pH of these solutions.

D. Brønsted–Lowry Acids and Bases: Dissociation in Water

1. Given the pH of a weak acid solution, determine the K_a of the acid.
2. Given the K_a value and the initial concentration of a weak monoprotic acid, calculate the concentrations of all species at equilibrium, the pH of the solution, and the percent dissociation of the acid.
3. Given the K_a value and the initial concentration of a weak diprotic acid, calculate the concentrations of all species at equilibrium and the pH of the solution.
4. Given the K_b value and the initial concentration of a weak base, calculate the concentrations of all species at equilibrium and the pH of the solution.
5. Identify which of two substances is more acidic.

E. Lewis Acids and Bases

1. Identify Lewis acids and bases.

Chapter in Brief

The study of equilibrium carries over into the study of acids and bases. The three definitions of acids and bases are discussed, as well as conjugate acid and base pairs. The reactions of strong and weak acids and bases are discussed, and the concepts of pH and pK_a introduced. Polyprotic acids and their reactions are explained, as are the identifications of acidic salts and basic salts, and the pH of their solutions. Periodic variations in acid strength are explained, and can be used to predict the strength of various acids.

14.1 Acid–Base Concepts: The Brønsted–Lowry Theory

The Arrhenius definition of acids and bases is the simplest, and the one that has been used thus far. In this definition, acids dissociate to form hydrogen ions, and bases dissociate to form hydroxide ions. This theory is limited to aqueous solutions, and is unable to account for the basicity of substances – like ammonia (NH_3) – that are bases without having any hydroxide ions to contribute.

The **Brønsted–Lowry theory** of acids and bases solves these problems by describing an acid as a proton donor, and a base as a proton acceptor, with acid-base reactions being proton transfers.

Chemical species that differ only by a proton are conjugate acid-base pairs.

- A is the acid, and A^- the conjugate base. Conjugate base has one less proton than the acid.
- BH^+ is the conjugate acid of B: the conjugate acid has one more proton than the base.

Acid dissociation: proton transferred to solvent:

$$HA + H_2O\ (l) \leftrightarrows A^-\ (aq) + H_3O^+\ (aq)$$

Base dissociation: proton taken from solvent:

$$B + H_2O\ (l) \leftrightarrows BH^+\ (aq) + OH^-\ (aq)$$

All Brønsted–Lowry bases have one or more lone pairs of electrons. This makes it possible for them to pull protons off solvent molecules.

EXAMPLE:

Write the proton–transfer equilibria for the following acids or bases in aqueous solution, and identify the conjugate acid–base pairs in each one: CH_3COOH (acetic acid), PO_4^{3-} (phosphate ion – base)

SOLUTION:

$$CH_3COOH\ (l) + H_2O\ (l) \leftrightarrows CH_3COO^-\ (aq) + H_3O^+\ (aq)$$

Acid = CH_3COOH
conjugate base = CH_3COO^-

$$PO_4^{3-}(aq) + H_2O\ (l) \leftrightarrows HPO_4^{2-}\ (aq) + OH^-\ (aq)$$

Base = PO_4^{3-}
conjugate base = HPO_4^{2-}

Workbook Problem 14.1

Write the proton-transfer reactions for nitrous acid (HNO_2) and basic sulfate (SO_4^{-2}). Identify the acid or base and its conjugate.

Strategy: Write the reaction with water, in which the acid protonates the water, and the base deprotonates the water.

Step 1: Write the reactions in the forward direction.

Step 2: Identify the conjugate acid and conjugate base.

14.2 Acid Strength and Base Strength

In an acid-dissociation reaction, the acid and water molecules compete for the loosely-attached hydrogen ion that defines the acid. If the water molecules are more basic than the acid ion, they will pull the hydrogen ion off the acid, and the acid will dissociate. If the acid ion (A^-) is more basic than the water molecules, the acid will not dissociate.

A **strong acid** dissociates completely in aqueous solution. It is thus also defined as a **strong electrolyte**. The remaining ion, A^-, is a weak base with only a negligible tendency to combine with a proton in aqueous solution. Strong acids include perchloric ($HClO_4$), hydrochloric (HCl), sulfuric (H_2SO_4), and nitric (HNO_3).

A **weak acid** only partially dissociates in aqueous solution, and is therefore a **weak electrolyte**. The remaining ion, A^-, is a relatively strong conjugate base, and will thus tend to pull protons back out of solution. Weak acids include acetic acid (CH_3COOH), nitrous acid (HNO_2) and hydrofluoric acid (HF).

The weaker the acid, the stronger its conjugate base. The stronger an acid, the weaker its conjugate base.

Protons will always be transferred from a weaker base to a stronger base, so protons will transfer from a strong acid to the conjugate base of a weak acid.

A table of acids and their conjugate bases, arranged by strength, is on page 543 of the textbook.

EXAMPLE:

Determine the direction of reactions involving hydrochloric acid, Cl^-, acetic acid, and acetate ion.

SOLUTION: Hydrochloric acid is a strong acid, making Cl^- a very weak base. Acetic acid is a weak acid, so acetate ions are a relatively strong conjugate base. As a result, hydrochloric acid will transfer protons to acetate ions. The reaction will be:

$$HCl + CH_3COO^- (aq) \leftrightarrows CH_3COOH (aq) + Cl^- (aq)$$

Workbook Problem 14.2

Write the proton–transfer reaction between HCl and CO_3^{2-} and the proton–transfer reaction between CN^- and H_3PO_4.

Strategy: Determine which species is the stronger acid and which is the stronger base, then transfer protons accordingly.

Step 1: Write the reactions, transferring protons from the stronger acid to the stronger base.

14.3 Hydrated Protons and Hydronium Ions

When acids dissociate, the acid releases a hydrogen ion – H^+. This is too reactive ot exist loose in solution, so it associates with a water molecule to form a hydronium ion - H_3O^+.

This can be considered to be the simplest hydrate of the proton - $[H(H_2O)]^+$- but others are also formed with the general formula $[H(H_2O)_n]^+$.

For practical purposes, H^+ (*aq*) and H_3O^+ (*aq*) are used interchangeably to represent a proton hydrated by an unspecified number of water molecules.

14.4 Dissociation of Water

Water can act both as an acid and as a base, and will dissociate in a reaction in which water acts as both acid and base:

$$H_2O\,(l) + H_2O\,(l) \leftrightarrows H_3O^+\,(aq) + OH^-\,(aq)$$

Water has a characteristic dissociation constant, abbreviated K_w. Water is omitted from the equilibrium expression as it is a pure liquid.

$$K_w = [H_3O^+]\,[OH^-] = 1.0 \times 10^{-14}$$

This equilibrium lies very far to the left: in pure water, only about two out of 10^9 molecules are dissociated.

The forward and reverse reactions are rapid, and in pure water at 25 °C,

$$[H_3O^+] = [OH^-] = 1.0 \times 10^{-7}\ M$$

In aqueous solutions, the relative values of $[H_3O^+]$ and $[OH^-]$ determine whether the solution is acidic, basic, or neutral:

$[H_3O^+] = [OH^-]$, neutral solution; $[H_3O^+] = 1.0 \times 10^{-7}$ M.
$[H_3O^+] > [OH^-]$, acid solution; $[H_3O^+] > 1.0 \times 10^{-7}$ M.
$[H_3O^+] < [OH^-]$, basic solution; $[H_3O^+] < 1.0 \times 10^{-7}$ M

Given the concentration of one ion, it is possible to calculate the concentration of the other using the equilibrium constant:

$$[H_3O^+] = \frac{1 \text{ x } 10^{-14}}{[OH^-]}$$

$$[OH^-] = \frac{1 \text{ x } 10^{-14}}{[H_3O^+]}$$

EXAMPLE:

Calculate the molarity of OH^- in a solution with an H_3O^+ concentration of 0.0042 M. Is this solution acidic, basic, or neutral? What is the $[H_3O^+]$ in a solution with $[OH^-]$ of 9.35×10^{-7} M acidic, basic, or neutral?

SOLUTION: If we know either the $[H_3O^+]$ or $[OH^-]$, we can calculate the other from the relationship $K_w = [H_3O^+][OH^-]$.

$$[OH^-] = \frac{1 \times 10^{-14}}{0.0042} = 2.4 \times 10^{-12}; \quad [H_3O^+] > [OH^-], \text{ so this solution is acidic.}$$

$$[H_3O^+] = \frac{1.0 \times 10^{-14}}{9.35 \times 10^{-7}} = 1.07 \times 10^{-8}; \ [H_3O^+] < [OH^-]; \text{ this solution is basic.}$$

Workbook Problem 14.3

Determine the $[OH^-]$ in a solution with $[H_3O^+] = 2.3 \times 10^{-5}$ M, and the $[H_3O^+]$ of a solution with $[OH^-] = 5.73 \times 10^{-9}$ M. Are these solutions acidic, basic, or neutral?

Strategy: Use the water dissociation constant to determine the concentrations. A basic solution will have $[OH^-] > [H_3O^+]$, and an acid solution will have $[H_3O^+] > [OH^-]$.

Step 1: Use K_w to calculate the ion concentrations:

14.5 The pH Scale

The pH scale is a logarithmic scale used to express the hydronium ion concentration:

$$pH = -\log[H_3O^+].$$

$$[H_3O^+] = \text{antilog}(-pH) = 10^{-pH}$$

Note that only the digits to the right of the decimal point are significant in a logarithm: the number to the left of the decimal point is an exact number having to do with the integral power of 10.

As $[H_3O^+]$ increases, pH decreases:

Acidic solutions, pH < 7.
Basic solutions, pH > 7.
Neutral solutions, pH = 7.

A less-often-used scale is pOH: this is $-\log[OH^-]$. Just as $[H_3O^+][OH^-]$ always equals 10^{-14}, pH + pOH always equals 14.

EXAMPLE:

Calculate the $[H_3O^+]$ and $[OH^-]$ for orange juice, which has a pH of 3.87.

SOLUTION:

$$[H_3O^+] = \text{antilog}(-pH) = \text{antilog}(-3.80) = 1.35 \times 10^{-4} \text{ M}$$

$$[OH^-] = \frac{1 \text{ x } 10^{-14}}{1.35 \text{ x } 10^{-4}} = 7.41 \text{ x } 10^{-10} \text{ M}$$

EXAMPLE:

Calculate the pH of a solution with a $[H_3O^+]$ of 3.56×10^{-7} M. Is this solution acidic, basic, or neutral?

SOLUTION:

$pH = -\log 3.56 \times 10^{-7} = 6.45$

The pH is less than 7: this solution is slightly acidic.

Workbook Problem 14.4

Calculate the pH of a solution whose $[OH^-] = 2.35 \times 10^{-6}$ M. Is the solution acidic, basic, or neutral?

Strategy: Calculate the $[H_3O^+]$, determine pH.

Step 1: Calculate the $[H_3O^+]$, using the expression for the dissociation of water.

Step 2: Calculate the pH.

Step 3: Determine if the solution is acidic, basic, or neutral.

14.6 Measuring pH

To measure pH, there are a number of acid-base indicators, compounds that change color in a specific pH range. They are weak acids with different colors in their acid and conjugate base forms, and change color over a range of 2 pH units.

Particularly useful is a mixture of indicators that change color over the entire range of pH: this is known as a *universal indicator.*

For greater precision than is possible with indicators, pH meters are electronic instruments that measure the pH–dependent electrical potential of the test solution.

14.7 The pH in Solutions of Strong Acids and Strong Bases

Three of the strong acids are mono-protic, meaning that they dissociate completely to form one proton. Because these dissociate completely, $[H_3O^+] = [A^-]$ = initial concentration of the acid, and $[HA] = 0$. This makes the pH of a strong acid solution very straightforward to calculate:

$$pH = -\log [\text{acid}]$$

The most common strong bases are alkali metal hydroxides with formula MOH. These also dissociate completely to form metal ions and hydroxide ions. The pH can be calculated from the $[OH^-]$.

Alkaline earth hydroxides have the formula $M(OH)_2$. When dissolved, they dissociate completely to form one metal ion and two hydroxide ions, but they are not as soluble as the alkali metal hydroxides.

Alkaline earth oxides are very strong bases because the O^{2-} ion is an even stronger base than the hydroxide ion. An O^{2-} ion will immediately pull a proton off a water molecule, forming two hydroxide ions: one from the oxide ion, nowbound to a hydrogen, and the other from the split water molecule that is now missing a hydrogen:

$$O^{2-} + H_2O \rightarrow 2\ OH^-$$

EXAMPLE:

Calculate the pH of a 2.37×10^{-2} M HNO_3 solution.

SOLUTION: HNO_3 is a strong acid that completely dissociates in water. Therefore, $[H_3O^+]$ = initial concentration of the undissociated acid.

$$pH = -\log [H_3O^+] = -\log (2.37 \times 10^{-2}) = 1.63$$

EXAMPLE:

Calculate the pH of a 2.37×10^{-6} M $Mg(OH)_2$ solution.

SOLUTION: $Mg(OH)_2$ is a strong base that completely dissociates in water to form a metal cation and two hydroxide anions. Therefore, $[OH^-]$ = 2 x initial concentration of the undissociated base.

$$[OH^-] = 2 \text{ x } 2.37 \times 10^{-6} \text{ M} = 4.74 \times 10^{-6} \text{ M}$$

in dilute aqueous solution, $[OH^-][H_3O^+] = 1 \text{ x } 10^{-14}$

$$[H_3O^+] = \frac{1 \text{ x } 10^{-14}}{4.74 \text{ x } 10^{-6}} = 2.11 \text{ x } 10^{-9} \text{M}$$

$$pH = -\log [H_3O^+] = -\log (2.11 \times 10^{-9}) = 8.67$$

Workbook Problem 14.5

Calculate the pH of a 7.2×10^{-3} M KOH solution.

Strategy: Determine the $[H^+]$ from the K_w expression and then calculate pH.

Step 1: Determine the concentration of $[OH^-]$.

Step 2: Calculate the $[H_3O^+]$.

Step 3: Calculate the pH.

What key concept did you use?

14.8 Equilibria in Solutions of Weak Acids

A weak acid only partially dissociates in solution, allowing for an equilibrium expression, abbreviated K_a:

$$HA\ (aq) + H_2O\ (l) \leftrightarrows H_3O^+\ (aq) + A^-\ (aq)$$

$$K_a = \frac{[H^+][A^-]}{[HA]}$$

As it is a pure liquid, water is omitted from the equilibrium expression.

K_a values are experimentally determined: the higher the value, the stronger the acid.

Also sometimes used are pK_a values. p$K_a = -\log K_a$. The higher the K_a is, the lower pK_a will be.

EXAMPLE:
The pH of a 0.400 M solution of nicotinic acid ($HC_6H_4NO_2$) is 2.63. Determine the value of K_a for this acid.

SOLUTION: To determine K_a, first write the balanced equation for the dissociation equilibrium and the equilibrium equation that defines K_a.

$$HC_6H_4NO_2\ (aq) + H_2O\ (l) \leftrightarrows H_3O^+(aq) + C_6H_4NO_2^-\ (aq)$$

$$K_a = \frac{[H_3O][C_6H_4NO_2^-]}{[HC_6H_4NO_2]}$$

To determine K_a, we need to know the concentrations of the species in the equilibrium mixture. We can determine the concentration of H_3O^+ from the pH.

$[H_3O^+]$ = antilog (–pH) = antilog (–2.63) = 2.34×10^{-3} M.

Dissociation of one $HC_6H_4NO_2$ molecule produces one H_3O^+ and $C_6H_4NO_2^-$ ion; the H_3O^+ and $C_6H_4NO_2^-$ concentrations are equal.

$[H_3O^+] = [C_6H_4NO_2^-] = 2.34 \times 10^{-3}$ M

The $[HC_6H_4NO_2]$ concentration at equilibrium is equal to the initial concentration minus the amount of $HC_6H_4NO_2$ that dissociates.

$[HC_6H_4NO_2] = 0.400 - (2.34 \times 10^{-3}) = 0.398$ M

$$K_a = \frac{[H_3O^+][C_6H_4NO_2^-]}{[HC_6H_4NO]} = \frac{(2.34 \times 10^{-3})(2.34 \times 10^{-3})}{(0.398)} = 1.38 \times 10^{-5}$$

14.9 Calculating Equilibrium Concentrations in Solutions of Weak Acid

Both equilibrium concentrations and pH can be calculated using the K_a or pK_a of a weak acid.

Although there are two reactions occurring, the autoionization of water and the partial dissociation of the weak acid, the autoionization of water can be disregarded when $K_a >> K_w$, and the $[H_3O^+]$ attributed only to the acid. When these values are close together, however, both reactions must be considered.

When equilibrium problems involving acids are worked, it is often possible to make the simplifying assumption that so little of the acid dissociates that the initial and final concentrations are the same.

For the equation:

$$HWA + H_2O \leftrightarrows WA^- + H_3O^+$$

the equilibrium expression is:

$$K_a = \frac{[H_3O^+][WA^-]}{[HWA]}$$

When solving such an equation, it is frequently the case that the amount of dissociation is so small that [HWA] before and after dissociation is essentially the same, which greatly simplifies the solution of the equilibrium expression. Proceed with caution, however! The assumption needs to be checked every time a problem is worked, by comparing the value obtained for x with the concentration of the acid. If $[HWA] - x \neq [HWA]$, the problem will need to be solved using the quadratic equation without simplifying assumptions.

EXAMPLE:

Calculate the pH and the concentration of all species present (H_3O^+, $C_4H_7O_2^-$, $HC_4H_7O_2$, and OH^-) and the pH in 0.355 M butyric acid ($HC_4H_7O_2$), which has $K_a = 1.5 \times 10^{-5}$.

SOLUTION:

Because $K_a >> K_w$, it is possible to disregard the contribution of autoionization of water. The dissociation reaction is:

$$HC_4H_7O_2\,(aq) + H_2O\,(l) \leftrightarrows H_3O^+\,(aq) + C_4H_7O_2^-\,(aq)$$

Principal reaction	$HC_4H_7O_2\,(aq)$	$\leftrightarrows$	H_3O^+	+	$C_4H_7O_2^-$
Initial concentration (M)	0.355		0		0
Change (M)	$-x$		$+x$		$+x$
Equilibrium concentration (M)	$0.355 - x$		$+x$		$+x$

Substitute the equilibrium concentrations into the equilibrium expression.

$$K_a = 1.5 \times 10^{-5} = \frac{[H_3O^+][C_4H_7O_2^-]}{[HC_4H_7O_2]} = \frac{(x)(x)}{(0.355 - x)}$$

Assume that x is negligible compared with the initial concentration of the acid, so $0.355 - x \approx 0.355$. Using this value in the denominator, solve for x.

$$x^2 \approx (1.5\times10^{-5})(0.355)$$

$$x^2 \approx 5.33 \times 10^{-6}\,M$$

$$x \approx 2.31 \times 10^{-3}\,M$$

The equilibrium concentrations are $[HC_4H_7O_2] = 0.355 - 0.00231 = 0.353$ M; $[H_3O^+] = [C_4H_7O_2^-] = 2.31 \times 10^{-3}$ M

$[OH^-]$ is obtained from the dissociation of water.

$$[OH^-] = \frac{K_w}{[H_3O^+]} = \frac{1.0\times10^{-14}}{2.31\times10^{-3}} = 4.33\times10^{-12}\,M$$

$\text{pH} = -\log\ (2.31 \times 10^{-3}) = 2.64$

Workbook Problem 14.6

Determine the pH of a 1.63 M lactic acid ($HC_3H_5O_3$) solution. ($K_a = 1.4 \times 10^{-4}$)

Strategy: Follow the steps outlined previously.

Step 1: Determine the species present initially.

Step 2: Write the proton–transfer reaction.

Step 3: Make a table showing the principal reaction and the initial and equilibrium concentrations.

Step 4: Substitute the equilibrium concentrations into the equilibrium equation, and solve for x. (Since $K_a < 1.0 \times 10^{-3}$, assume that x is negligible and that $1.63 - x \cong 1.63$.)

Step 5: Calculate the equilibrium concentrations.

Step 6: Calculate the pH.

14.10 Percent Dissociation in Solutions of Weak Acids

Another way of expressing acid strength is in terms of % dissociation. This is calculated as

$$\text{Percent Dissociation} = \frac{[A^-]}{[HA]} \times 100$$

The value depends on the concentration and type of acid: in general, the percent dissociation increases with an increasing K_a, and increases in more dilute solutions.

EXAMPLE:

Calculate the percent dissociation of 0.100 and 1.00 M vitamin C solutions (ascorbic acid = $C_6H_8O_6$, $K_a = 8.0 \times 10^{-5}$).

SOLUTION:

$$K_a = \frac{[H_3O^+][C_6H_7O_6^-]}{[C_6H_8O_6]}$$

0.100 M solution:

$$K_a = \frac{[H_3O^+][C_6H_7O_6^-]}{[C_6H_8O_6]} = \frac{x^2}{0.100}$$

$$x^2 = 0.100 \times 8.0 \times 10^{-5} = 8.0 \times 10^{-6}$$

$$x = 0.00283 \text{ M}$$

Check the assumption that concentration of the dissociated ion can be neglected in calculation:

0.100 - 0.00283 = 0.09717 (The answer is within 1% of being correct.)

$$\% \text{ dissociation} = \frac{0.00283}{0.100} \text{x}100 = 2.8\%$$

1.00 M solution:

$$K_a = \frac{[H_3O^+][C_6H_7O_6^-]}{[C_6H_8O_6]} = \frac{x^2}{1.00}$$

$$x^2 = 1.00 \times 8.0 \times 10^{-5} = 8.0 \times 10^{-5}$$

$$x = 0.00894 \text{ M}$$

Check the assumption that concentration of the dissociated ion can be neglected in calculation:

1.00 - 0.00894 = 0.991 (The answer is within 1% of being correct.)

$$\% \text{ dissociation} = \frac{0.00894}{1.00} \text{x}100 = 0.89\%$$

As expected, the more dilute solution shows more dissociation.

Workbook Problem 14.7

Calculate the percent dissociation of a 0.79 M benzoic acid (C_6H_5COOH) solution. ($K_a = 6.5 \times 10^{-5}$)

Strategy: To determine the percent dissociation we need to first determine the concentration of the dissociated HA. This requires a determination of the equilibrium concentration of H_3O^+ and

$C_6H_5COO^-$.

Step 1: Determine the species present initially, and write the proton-transfer reaction.

Step 2: Make a table showing the principal reaction, initial concentration, change in concentration, and the equilibrium concentration.

Step 3: Substitute the equilibrium concentrations into the equilibrium expression, and solve for x.

Step 4: Calculate the concentrations of the major species present.

Step 5: Calculate the percent dissociation.

14.11 Polyprotic Acids

Some acids, such as sulfuric acid (strong) and phosphoric acid (weak) have more than one proton than can dissociate. As a result, they have a stepwise dissociation, and each step has its own characteristic K_a.

It is more difficult to pull a hydrogen off of an already negatively charged ion, so the dissociation constants decrease dramatically:

$$K_{a1} > K_{a2} > K_{a3}$$

A diprotic acid solution contains a mixture of acids: H_2A, HA^-, and H_2O. The diprotic species is by far the strongest, and as a rule the first dissociation step is the only one that needs to be considered, as the hydronium ions contributed by the second dissociation (or the third in the case of phosphoric acid) are too few to change the pH calculated from the first dissociation.

The exception to the above rule is sulfuric acid, the only strong polyprotic acid. Sulfuric acid dissociates completely at the first dissociation step, and has a $K_{a2} = 1.2 \times 10^{-2}$, stronger than the K_a of many weak acids. In this case, it is necessary to calculate the $[H_3O^+]$ generated by both dissociations.

EXAMPLE:

Calculate the pH of a 0.145 M sulfuric acid solution. K_{a1} = complete, $K_{a2} = 1.2 \times 10^{-2}$.

SOLUTION:

First dissociation:

$[H_3O^+] = [HSO_4^-] = 0.145$ M

Second dissociation:

$$K_{a2} = \frac{[H_3O^+][HSO_4^-]}{[SO_4^{2-}]} = \frac{x^2}{0.145 - x} = 1.2 \times 10^{-2}$$

With a Ka_2 of this magnitude, it is unlikely that the value of x will be negligible. Rearrange and solve using the quadratic equation:

$$x^2 + 0.012x - 0.00174 = 0$$

$$x = \frac{-0.012 \pm \sqrt{(0.012)^2 - 4(-0.001740)}}{2}$$

$$x = 0.036 \text{ M}$$

The total $[H_3O^+] = 0.145 \text{ M} + 0.036 \text{ M} = 0.181 \text{ M}$

$pH = -\log(0.181) = 0.74$

Workbook Problem 14.8

Write the stepwise dissociation for arsenic acid (H_3AsO_4) and determine the pH of a 1.69×10^{-3} M solution. ($K_{a1} = 5.62 \times 10^{-3}$; $K_{a2} = 1.70 \times 10^{-7}$; $K_{a3} = 3.95 \times 10^{-12}$)

Step 1: Write the stepwise dissociation for arsenic acid.

Step 2: Determine the principal reaction.

Step 3: Make a table showing the principal reaction, initial concentration, change in concentration and equilibrium concentration of the reactant and products.

Step 4: Substitute the equilibrium concentrations into the equilibrium expression.

Step 5: Calculate the pH of the solution.

14.12 Equilibria in Solutions of Weak Bases

Weak bases will accept a proton from water, leaving OH^- ions and the protonated conjugate acid of the base:

$$B\,(aq) + H_2O\,(l) \leftrightharpoons BH^+\,(aq) + OH^-\,(aq)$$

The position of the equilibrium is characterized by the base–dissociation constant, K_b.

$$K_b = \frac{[OH^-][BH^+]}{[B]}$$

Many weak bases are amines, ammonia derivatives in which one of the hydrogen atoms is replaced by another chemical group: the basicity of these compounds is due to the lone-pair of electrons on the nitrogen atom, which can be used to bind a proton.

Problems involving weak bases are solved just like those involving weak acids: the only difference is that $[OH^-]$ is being solved for in the equilibrium equation. To obtain the $[H^+]$ and the pH of the solution requires using the autoionization constant of water.

EXAMPLE:

Calculate the pH of a 0.10 M solution of methylamine, CH_3NH_2. $K_b = 3.7 \times 10^{-4}$.

SOLUTION:

$$K_b = 3.7x10^{-4} = \frac{[CH_3NH_3^+][OH^-]}{[CH_3NH_2]} = \frac{x^2}{0.10 - x}$$

K_b is within 3% of the concentration of the base: use the quadratic equation:

$$x^2 + 3.7x10^{-4}x - 3.7x10^{-5} = 0$$

$$x = \frac{-3.7x10^{-4} \pm \sqrt{(3.7x10^{-4})^2 - 4(-3.7x10^{-5})}}{2}$$

$$x = 0.0059 \text{ M}$$

To solve for pH, calculate the concentration of H^+ using the autoionization constant of water:

$$[H^+] = \frac{1 \times 10^{-14}}{[OH^-]} = \frac{1 \times 10^{-14}}{[0.0059]} = 1.69 \times 10^{-12}$$

$$pH = -\log(1.69 \times 10^{-12}) = 11.8$$

Workbook Problem 14.9

Determine the pH of a 0.975 M trimethylamine, $(CH_3)_3N$, solution. $K_b = 6.5 \times 10^{-5}$.

Strategy: Solve using a table of equilibrium values.

Step 1: Write the principal reaction.

Step 2: Construct a table with concentrations of the reactant and products.

Step 3: Substitute the equilibrium concentrations into the equilibrium expression, and solve for *x*.

Step 4: Determine the equilibrium concentrations of the species present.

Step 5: Calculate the pH of the solution.

14.13 Relation Between K_a and K_b

For a conjugate acid-base pair, K_a and K_b can be related through the ionization constant of water:

$$K_a \times K_b = K_w$$

As the strength of an acid increases, the strength of the conjugate base decreases. As the strength of a base increases, the strength of the conjugate acid decreases, always proportional to K_w.

EXAMPLE:
Calculate K_a for methylamine, $K_b = 1.4 \times 10^{-4}$.

SOLUTION:

$$K_a = \frac{K_w}{K_b} = \frac{1 x 10^{-14}}{1.4 x 10^{-4}} = 7.14 x 10^{-11}$$

14.14 Acid–Base Properties of Salts

The pH of a salt solution is determined by the acid-base properties of the constituent cations and anions. The salts formed in an acid-base reaction will contain the conjugate acid and the conjugate base of the reactants. The stronger of the two will predominate:

Strong acid + Strong base → Neutral solution
Strong acid + Weak base → Acidic solution
Weak acid + Strong base → Basic solution

When a strong acid and a strong base react, they form a neutral salt. As an example, the reaction of sodium hydroxide and hydrochloric acid forms sodium chloride. These are all strong electrolytes, meaning that they have no tendency to associate with anything else: the Cl^- ions in solution are not looking to pull hydrogens off water: they do not even combine with their Na^+ counterparts.

These form neutral cations: Group 1A, Group 2A (except beryllium)

These form neutral anions: NO_3^-, Cl^-, ClO_4^-, Br^-, I^-

When a weak base reacts with a strong acid, the anion is neutral, as above, but the cation will be the conjugate acid of a weak base. This weak acid can react with water to form H_3O^+ ions, making the solution slightly acidic.

Small, highly charged metal cations are also acid in solution: the oxygens in the water molecules bind to them so tightly that the hydrogens can dissociate easily.

When a strong base reacts with a weak acid, the opposite occurs: the salts will have a strong, neutral anion that does not wish to associate with anything, and a weak base – the conjugate base of the weak acid.

When a weak base and a weak acid react, the pH of the solution will tend toward the stronger of the two compounds:

a. K_a (for the cation) > K_b (for the anion); acidic solution
b. K_a (for the cation) < K_b (for the anion); basic solution
c. K_a (for the cation) ≈ K_b (for the anion); neutral solution

EXAMPLE:

Determine whether aqueous solutions of the following salts are acidic, neutral, or basic. Write the hydrolysis reaction for those solutions that are acidic or basic. NH_4ClO_4, RbCl, $NaCH_3CO_2$, NH_4CO_3

SOLUTION: To determine the acidity of an aqueous solution of a salt, we must determine if the salt is derived from a 1) strong acid/strong base reaction, 2) weak acid/strong base reaction, 3) strong acid/weak base reaction, or 4) weak acid/weak base reaction.

NH_4ClO_4: derived from the weak base NH_3 (*aq*) and the strong acid $HClO_4$ (*aq*). Acidic (NH_4^+ is the weak conjugate acid of NH_3).

$$NH_4^+ (aq) + H_2O \leftrightarrows NH_3 (aq) + H_3O^+ (aq)$$

RbCl: derived from the strong base RbOH (*aq*) and the strong acid HCl (*aq*). Neutral.

$NaCH_3CO_2$: derived from the strong base NaOH and the weak acid CH_3COOH. Basic. ($CH_3CO_2^-$ is the weak conjugate base of CH_3COOH.)

$$CH_3CO_2^- (aq) + H_2O \leftrightarrows CH_3COOH (aq) + OH(aq)$$

NH_4CO_3: derived from the weak base NH_3 and the weak acid $HCO_3^-(aq)$. To determine the acidity of this solution we must calculate the K_a of the cation and the K_b of the anion of the salt and compare the two.

$$NH_4^+ \; K_a = \frac{1 \text{ x } 10^{-14}}{K_b \; NH_3} = \frac{1 \text{ x } 10^{-14}}{1.8 \text{ x } 10^{-5}} = 5.6 \text{ x } 10^{-10}$$

$$CO_3^{2-} \; K_b = \frac{1 \text{ x } 10^{-14}}{K_a \; HCO_3^-} = \frac{1 \text{ x } 10^{-14}}{5.6 \text{ x } 10^{-11}} = 1.8 \text{ x } 10^{-4}$$

$K_b > K_a$; therefore, the solution is basic. The hydrolysis reaction is

$$CO_3^{2-} (aq) + H_2O (l) \leftrightarrows HCO_3^- (aq) + OH^- (aq)$$

EXAMPLE:

Determine the pH of a 0.150 M solution of NH_4Cl.

SOLUTION: To determine the acidity of an aqueous solution of a salt, the K_a for the acidic portion must be used:

From above, K_a of the ammonium cation is 5.6 x 10^{-10}:

$$5.6 \text{ x } 10^{-10} = \frac{[NH_3][H_3O^+]}{[NH_4^+] - x} = \frac{x^2}{0.150}$$

$$x = 9.16 \text{ x } 10^{-6}$$

$$pH = -\log(9.16 \text{ x } 10^{-6}) = 5.04$$

Workbook Problem 14.10

Calculate the pH of a 0.137 M KOCl solution. (K_a for HOCl = 3.5×10^{-8})

Strategy: Determine if KOCl produces an acidic, basic, or neutral aqueous solution, write the hydrolysis reaction for this salt, and determine pH.

Step 1: Write the hydrolysis reaction for this salt.

Step 2: Write an equilibrium expression for this reaction.

Step 3: Construct a table.

Step 4: Substitute the equilibrium concentrations into the equilibrium expression.

Step 5: Calculate the pH of the solution.

14.15 Factors that Affect Acid Strength

The extent of dissociation of an acid is often determined by the strength and polarity of the H–A bond: the weaker or more polar the H-A bond is, the stronger the acid.

For binary acids, as the size of A increases down the periodic table, the strength of the HA bond is weakened, so acidity increases down the groups. Within a row, the acidity will increase with increasing polarity. As A becomes more electronegative, acidity increases.

Oxoacids have the general formula H_nYO_m, where Y is a nonmetallic atom, *n* and *m* are integers. Y is always bonded to one or more hydroxyl (OH) groups, and may be bound to one or more oxygen atoms as well. The dissociation of an oxoacid involves the breaking of an O-H bond, so anything that weakens these bonds or increases their polarity will increase the strength of the acid.

As Y becomes more electronegative, the strength of oxoacids increases. As the number of oxygens changes in an oxoacid (sulfuric acid to sulfurous acid, for example) an increase in the oxidation number of Y will increase acid strength. In other words, the more oxygens the central atom is trying to hold on to, the stronger the acid will be. H_2SO_4 is stronger than H_2SO_3, HNO_3 is stronger than HNO_2.

EXAMPLE:

Order the following by increasing acid strength: a) HF, NH_3, H_2O; b) NH_3, AsH_3, PH_3; c) HIO_3, HIO, HIO_2, HIO_4; d) H_2SeO_4, H_2TeO_4, H_2SO_4.

SOLUTION:

a) Across a period, an increase in electronegativity gives rise to an increase in acidity: $NH_3 < H_2O < HF$.

b) Down a group, an increase in size gives rise to an increase in acid strength: $NH_3 < PH_3 < AsH_3$.

c) For oxoacids, acid strength increases with an increasing number of oxygen atoms: $HIO < HIO_2 < HIO_3 < HIO_4$.

d) For oxoacids that contain the same number of OH groups and the same number of O atoms, acid strength increases with increasing electronegativity of Y: $H_2TeO_4 < H_2SeO_4 < H_2SO_4$.

14.16 Lewis Acids and Bases

An even more general definition of acids and bases was proposed by G. N. Lewis. In this definition, a base is an electron–pair donor, and an acid is an electron-pair acceptor.

For bases, the Lewis and Brønsted–Lowry definitions agree, but for acids, the Lewis definition accommodates several situations that are not addressed by the Brønsted–Lowry definition, including cations and neutral molecules with vacant valence orbitals. This explains the acidity of Al^{3+} and CO_2, for example.

Putting It Together

Sulfanilic acid, a compound that is used in making dyes, is produced from the reaction of aniline and sulfuric acid in aqueous solution.

$$C_6H_5NH_2\ (aq) + H_2SO_4\ (aq) \rightarrow C_6H_4NH_2SO_3H\ (aq)$$

If you prepare 125 g of sulfanilic acid with an experimental yield of 90%, how much aniline will you use? ($d_{\text{aniline}} = 1.02$ g/mL) The K_a of sulfanilic acid is 5.9×10^{-4}. What is the pH of an aqueous solution prepared by dissolving 5.73 g of $NaC_6H_4NH_2SO_3H$ in 250 mL of water?

Self–Test

This section is intended to test your knowledge of the material covered in this chapter. Think through these problems, and make certain you understand what they are asking. Make sure your answers make sense. Successful completion of these problems indicates that you have mastered the material in this chapter. You will receive the greatest benefit from this section if you use it as a mock exam, as this will allow you to determine which topics you need to study in more detail.

True–False

1. The Brønsted-Lowry definition states that acids are electron donors, and bases are electron acceptors.

2. A strong acid is more corrosive than a weak acid.

3. H^+ is unstable in solution and attaches to a water molecule to form a hydronium (H_3O^+) ion.

4. A solution with a pH of 7.9 is slightly basic.

5. The strength of an acid and the strength of its conjugate base are directly related: the stronger the acid, the stronger its conjugate base.

6. Dilute solutions of weak acids will have a higher percent dissociation.

7. For most polyprotic acids, two or more dissociations must be considered.

8. Salts can be acidic, basic, or neutral.

9. K_a can be used to calculate the K_b of a weak base.

10. $Mg(OH)_2$ is a weak base.

Multiple Choice

1. In the Arrhenius, Brønsted-Lowry, and Lewis definitions of an acid, acids are
   a. hydrogen ion generators, proton acceptors, and electron acceptors
   b. hydroxide acceptors, proton donors, and electron acceptors
   c. hydrogen ion generators, proton donors, and electron acceptors
   d. hydroxide acceptors, proton acceptors, and electron donors

2. The pH of a salt solution
   a. = 7 for all salts
   b. > 7 for salts derived from a strong acid and a strong base
   c. = 7 for salts derived from a weak acid and a weak base
   d. < 7 for salts derived from a strong acid and a weak base

3. A weak acid or base
   a. dissociates incompletely in solution
   b. has a pH close to 7
   c. dissociates fully in solution
   d. is always less corrosive than a strong acid or base

4. The concentrations of H_3O^+ and OH^- in pure water at 25 °C are
   a. 1×10^{-7} M, 1×10^{-14} M
   b. 1×10^{-7} M, 1×10^{-7} M
   c. 1×10^{-14} M, 1×10^{-7} M
   d. 1×10^{-14} M, 1×10^{-7} M

5. Solutions with $[H_3O^+] = 1 \times 10^{-5}$ M have a pH of
   a. 1×10^{-5}
   b. -5

c. 5
d. 9

6. The strength of a binary acid is affected by
   a. polarity of the H-A bond
   b. the size of A
   c. the strength of the H-A bond
   d. all of the above

7. Acid-base indicators
   a. are strong acids
   b. are substances that change color in a specific pH range
   c. have the same colors in their acid (HIn) and conjugate base (In^-) forms
   d. can be used to determine the exact pH of a solution

8. Which of the following is the strongest acid?
   a. $HClO_4$
   b. $HClO_3$
   c. $HClO_2$
   d. HClO

Matching

Acid-base indicator	a. K_a
Acid dissociation constant	b. H^+ donor
Arrhenius acid	c. chemical species who differ only by one proton
Arrhenius base	d. molecule that dissociates in water to form OH^- ions
Base dissociation constant	e. electron acceptor
Brønsted-Lowry acid	f. a molecule with different colored forms at different pH values
Brønsted-Lowry base	g. K_b
Conjugate acid-base pair	h. dissociated/undissociated x 100
Hydronium ion	i. $-\log[H_3O^+]$
Ion–product for water	j. molecule that dissociates in water to release H^+ ions
Lewis acid	k. H^+ acceptor
Lewis base	l. acid that is capable of more than one dissociation
Percent dissociation	m. ion in which H^+ is bonded to the oxygen atom of a solvent water molecule

pH	n. electron donor
Polyprotic acid	o. given by $K_w = [H_3O^+][OH^-]$
Strong acid	p. acid that dissociates incompletely
Weak acid	q. acid that dissociates completely

Fill–in–the–Blank

1. Only a small fraction of __________ acid molecules transfer a ____________ to water.
2. In acid solution, the concentration of ______________ ions is greater than the concentration of ______________ ions. In _____________ solutions, these concentrations are _______________.
3. In general, the percent dissociation of an acid _______________ with increasing K_a, and _______________ with increasing concentration.
4. Polyprotic acids have a _________________ dissociation in which each step has a much ______________ K_a.
5. Amines are weak bases. They have ____________________ that can be donated to a bond with ________________.

Problems

1. What is the $[OH^-]$ for a solution of $Ca(OH)_2$ whose pH = 11.78?
2. Calculate the pH of a solution prepared by dissolving 1.83 g of $Ba(OH)_2$ in 150 mL of water.
3. The percent dissociation for a 7.35×10^{-3} M weak acid solution is 0.51%. Calculate the pH of this solution and determine the K_a for the acid.
4. Determine the K_a of histidine, a weak organic acid, if the pH of a 1.30×10^{-3} M solution is 6.03.
5. What is the pH of a 1.24 M solution of pyridine, a weak base whose formula is C_6H_5N? $K_b = 1.8 \times 10^{-9}$?
6. Calculate the concentration of all species present and the pH of a 0.125 M oxalic acid ($H_2C_2O_4$) solution. $K_{a1} = 5.9 \times 10^{-2}$, $K_{a2} = 6.4 \times 10^{-5}$.
7. Calculate the pH of a 0.350 M solution of benzylamine. ($K_b = 2.14 \times 10^{-5}$)
8. Determine a) K_b for the conjugate base of ascorbic acid ($K_a = 8.0 \times 10^{-5}$), and b) the K_a for the conjugate acid of hydrazine ($K_b = 8.9 \times 10^{-7}$).
9. 1.75 g of KCN is dissolved in 150 mL of water. What is the concentration of HCN (*aq*) at equilibrium? What is the pH of the solution? ($K_a = 4.9 \times 10^{-10}$)

10. Determine which acid is stronger, and explain why.
    a. HBr (*aq*) or HI (*aq*)
    b. H_3PO_4 or H_3PO_3

Challenge Problem

A 25.0% by mass solution of H_3PO_4 has a density of 1.1667 g/mL. Calculate the pH and concentrations of all phosphate–containing species present.

CHAPTER 15

APPLICATIONS OF AQUEOUS EQUILIBRIA

Chapter Learning Goals

A. Neutralization Equilibria

1. From the relative strengths of the acid and base in a neutralization reaction, predict whether the pH will be equal to, greater than, or less than 7.00 at the equivalence point.
2. Write balanced net ionic equations for the four types of neutralization reactions.
3. Calculate pH values for a strong acid–strong base titration.
4. Calculate pH values for a weak acid–strong base titration.
5. Given a titration curve, select which indicator(s) could be used to detect the equivalence point.
6. Calculate pH values for a weak base–strong acid titration.
7. Calculate pH values for a diprotic acid–strong base titration.

B. Common–Ion Effect and Buffer Solutions

1. Describe the effect on pH when the conjugate base of a weak acid is added to a solution of the weak acid and the conjugate acid of a weak base is added to a solution of the weak base. Calculate the concentrations of all species present at equilibrium.
2. Given the initial concentration of weak acid (or weak base) and its conjugate base (or weak acid), determine the equilibrium concentrations of all species and the pH of a buffer solution.
3. Calculate the pH of a buffer after the addition of OH^- or H_3O^+.
4. Use the Henderson-Hasselbalch equation to calculate the pH of a buffer.
5. From a table of weak acids and their K_a values, select the weak acid/conjugate base pair that would make the best buffer at a given pH.

C. Solubility Equilibria

1. Write the solubility product expression for a given ionic compound.
2. Given the K_{sp} of an ionic compound, calculate its solubility and vice versa.
3. Given the K_{sp} of an ionic compound, calculate its solubility in the presence of a common ion.
4. Identify ionic compounds that have an enhanced solubility at low pH. Write chemical equations showing why the solubility increases as $[H_3O^+]$ increases.
5. From K_{sp} values, determine whether a precipitate will form on mixing solutions of ionic compounds.
6. Determine which metal sulfides will precipitate from a solution of metal ions on addition of H_2S at a specified pH.

D. Complex Equilibria

1. Given K_f, calculate the concentrations of the species present in a complex–ion equilibrium. Given K_{sp} and K_f, calculate the solubility of a slightly soluble ionic compound in an excess of the complexing agent.
2. Write chemical equations showing how a given oxide or hydroxide exhibits amphoteric behavior.

Chapter in Brief

Concepts of equilibrium can be applied to neutralization reactions between strong and weak acids and bases. The titration curves of each of these reactions have distinctive features that make it possible to determine the relative strength of the reactants. The common-ion effect also affects equilibrium mixtures: it is this corollary of Le Châtelier's principle that explains the functioning of buffers: solutions of weak acids and their conjugate bases that resist changes in pH. The pH of a buffer system will be close to the pK_a of the weak acid, and can be calculated using the Henderson- Hasselbalch equation. The solubility product of an ionic compound, K_{sp}, is the equilibrium constant for its dissolution. This can be converted to molar solubility, and can be altered by changing the conditions surrounding the compound. Using knowledge of these constants and how they can be changed allows for the separations of ions in solution through selective precipitation.

15.1 Neutralization Reactions

When a **strong acid and a strong base** react, the net ionic equation for the reaction is:

$$H_3O^+\,(aq) \;+\; OH^-\,(aq) \rightarrow 2\,H_2O\,(l)$$

When equal numbers of moles of acid and base are mixed together, $[H_3O^+]$ and $[OH^-] = 1 \times 10^{-7}$ M, and the solution is neutral. The reaction proceeds essentially to completion. The equilibrium constant is the reciprocal of the ion–product constant for water. The reaction leaves behind a neutral salt, and the pH is 7.

When a **weak acid and a strong base** react, the net ionic reaction involves proton transfer from HA to the strong base:

$$HA\,(aq) \;+\; OH^-\,(aq) \rightarrow H_2O \;+\; A^-\,(aq)$$

The equilibrium constant for this reaction is obtained by combining the reactions for the dissociation of the weak acid and formation of water (reverse of the dissociation of water). As a result, the equilibrium constant for these reactions is:

$$K_n = K_a(1/K_w)$$

The neutralization reaction will go to completion, but will leave behind the conjugate base of the weak acid, therefore the pH will be greater than 7.

When a **strong acid and a weak base** react, the net ionic equation involves proton transfer from the strong acid to the weak base:

$$H_3O^+\,(aq) \;+\; B\,(aq) \rightarrow H_2O\,(l) \;+\; BH^+\,(aq)$$

The equilibrium constant for this reaction is obtained by multiplying the equilibrium constant for the dissociation of the base with the formation of water (reverse of dissociation of water).

$$K_n = K_b(1/K_w)$$

The neutralization reaction will go to completion, but will leave behind the conjugate acid of the base. As a result, the pH will be less than 7.

When a **weak acid and a weak base** react, the reaction involves proton transfer from the weak acid to the weak base:

$$HA\,(aq) \;+\; B\,(aq) \leftrightarrows BH^+\,(aq) \;+\; A^-\,(aq)$$

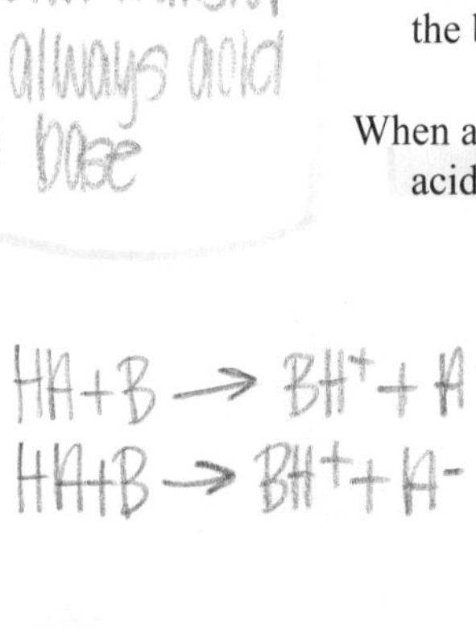

The equilibrium constant for the overall reaction will be the combination of the dissociation of the weak acid, the protonation of the weak base, and the formation of water:

$$K_n = \frac{K_a K_b}{K_w}$$

This type of reaction has far less tendency to proceed to completion than neutralizations involving strong acids or strong bases.

EXAMPLE:

Write the net ionic equation, and predict the pH for the following reactions:

HF (*aq*) + KOH HCl (*aq*) + CH_3NH_2 (a weak base)

SOLUTION:

The first equation is the reaction of a weak acid with a strong base. The net ionic equation is

$$HF\ (aq) + OH^-\ (aq) \leftrightarrows H_2O\ (l) + F^-\ (aq)$$

The pH of this solution will be higher than 7 (basic), as a weak base is left in solution.

The second equation is the reaction of a strong acid with a weak base. The net ionic equation is

$$H_3O^+\ (aq) + CH_3NH_2 \leftrightarrows H_2O\ (l) + CH_3NH_3^+\ (aq)$$

The pH of this solution is expected to be less than 7 (acidic) since the conjugate acid of a weak base is a weak acid.

Workbook Problem 15.1

Write a balanced net ionic equation for the neutralization of nitrous acid by sodium hydroxide. Determine K_n and the position of equilibrium for this neutralization reaction. Predict the pH of the solution. K_a for nitrous acid is 7.24 x 10^{-4}.

Strategy: Write the net reaction, determine K_n.

Step 1: Write the net neutralization reaction by writing individual reactions:

Step 2: Calculate K_n based on the equilibrium constants for the individual reactions in step 1.

Step 3: Determine the position of equilibrium from the value of K_n.

Step 4: Predict the pH:

15.2 The Common–Ion Effect

A solution of a weak acid and a salt of its conjugate base is an important mixture, as these mixtures regulate the pH of biological systems.

The common-ion effect is a shift in equilibrium caused by adding more of an ion that is involved in an equilibrium. It is an example of Le Châtelier's principle: adding the salt of the conjugate base of an acid will push the equilibrium toward less dissociation of the acid.

To determine the properties of a solution prepared from a weak acid and a salt of its conjugate base, the properties of the various species must be determined. The principal reaction is the dissociation of the weak acid. The equilibrium calculations must be adjusted for the initial concentration of the conjugate base – this provides an initial concentration of A^-, while the dissociation of the weak acid will provide $[H_3O^+]$ and the change in $[A^-]$.

EXAMPLE:

Calculate the concentration of all species present, the pH, and the percent dissociation of nitrous acid in a solution that is 0.150 M HNO_2 and 0.075 M $NaNO_2$.

SOLUTION: Because the salt is 100% dissociated, the species present initially are HNO_2, NO_2^-, Na^+, and H_2O. Na^+ is inert; HNO_2 is a weak acid ($K_a = 4.5 \times 10^{-4}$); NO_2^- is the conjugate base of a weak acid; and H_2O can be either an acid or a base. The principal reaction is transfer of a proton from HNO_2 to H_2O.

Principal Reaction	$HNO_2\ (aq)\ +\ H_2O\ (l) \leftrightarrows$	$H_3O^+\ (aq)$	$NO_2^-\ (aq)$
Initial Concentration (M)	0.150	0	0.075
Change (M)	$-x$	$+x$	$+x$
Eq. Concentration (M)	$0.150 - x$	$+x$	$0.075 + x$

The common ion in this problem is NO_2^-. The equilibrium equation for the principal reaction is

$$K_a = 4.5 \times 10^{-4} = \frac{[H_3O^+][NO_2^-]}{[HNO_2]} = \frac{(x)(0.075 + x)}{(0.150 - x)} \approx \frac{(x)(0.075)}{(0.150)}$$

x is assumed to be negligible because K_a is so small. The equilibrium is shifted to the left due to the common–ion effect.

$$x = [H_3O^+] = \frac{(4.5 \times 10^{-4})(0.150)}{(0.075)} = 9.0 \times 10^{-4}$$

Note that the assumption concerning the size of x is justified.

$pH = -\log (9.0 \times 10^{-4}) = 3.05$

The percent dissociation of HNO_2 is

$$\text{Percent dissociation} = \frac{[HNO_2]_{dissoc.}}{[HNO_2]_{initial}} \times 100 = \frac{9.0 \times 10^{-4}}{0.150} \times 100 = 0.60\%$$

Workbook Problem 15.2

Calculate the pH, and the percent dissociation of phosphorous acid in a solution that is 0.250 M H_3PO_3 and 0.175 M NaH_2PO_3. ($K_a = 1.0 \times 10^{-2}$)

Strategy: Write the reaction, determine the concentration of H_3O^+, $H_2PO_3^-$.

Step 1: Write the overall reaction.

Step 2: Make a chart of initial and equilibrium concentrations.

Step 3: Solve for the equilibrium concentrations of H_3O^+, $H_2PO_2^-$.

Step 4: Determine the pH of the solution.

Step 5: Determine the % dissociation of the acid.

15.3 Buffer Solutions

Solutions like these, that contain a weak acid and its conjugate base or a weak base and its conjugate acid are called **buffer solutions**. They are important to biological systems as they are able to resist changes in pH.

If a small amount of a base is added to one of these solutions, it will be neutralized by the weak acid. If a small amount of acid is added, it will be neutralized by the conjugate base.

For a buffer prepared from a weak acid, HA, and its conjugate base, B:

$$\left[H_3O^+\right] = K_a \frac{[HA]}{[B]}$$

Because the dissociations involved in these reactions are so small, these calculations can be done using initial concentrations.

Buffer capacity is a measure of how much acid or base a solution can absorb before the pH begins to be seriously affected. The higher the concentration of the acid and base, the greater the buffer capacity. For the same concentration, a larger volume of solution will also have a greater buffer capacity.

EXAMPLE:

Which of these solutions will have the greatest buffer capacity? 1 M acetic acid, 0.1 M sodium acetate; 0.1M acetic acid, 1 M sodium acetate; 1 M acetic acid, 1 M sodium acetate; or 0.1 M acetic acid and 0.1 M sodium acetate?

SOLUTION: The higher concentrations will have the greatest buffer capacities, so the solution that is 1 M in both species will have the highest buffer capacity.

Workbook Problem 15.3

Calculate the pH of a 1.0 L buffer solution containing 0.35 M HCOOH and 0.25 M HCO_2Na. Determine the pH of this solution after the addition of 0.10 mol HCl (assume no volume change). K_a for formic acid equals 1.8×10^{-4}

Strategy: The pH of the initial solution is determined from K_a and the equilibrium expression. The pH of the solution after addition of HCl is determined knowing the changes in concentration that occur after the addition of $[H_3O^+]$.

Step 1: Determine the principal reaction and equilibrium concentrations.

Step 2: Using the equilibrium equation, solve for $[H_3O^+]$.

Step 3: Solve for the pH.

Step 4: Write the neutralization reaction that occurs upon addition of HCl and set up a table showing the number of moles present before and after the addition of HCl.

Step 5: Determine the concentrations of the buffer components after neutralization occurs.

Step 6: Substitute the concentrations of the buffer components into the equilibrium expression and calculate the pH.

15.4 The Henderson–Hasselbalch Equation

The equation developed above can be rearranged in logarithmic form called the Henderson-Hasselbalch equation:

$$\text{pH} = \text{pK}_a + \log\frac{[\text{base}]}{[\text{acid}]}$$

When the ratio of base to acid is 1, the pH will be equal to the pK_a. When the pH is not equal to the p K_a. this equation can be used to determine the percent dissociation of the acid.

$$\log\frac{[\text{base}]}{[\text{acid}]} = \text{pH} - \text{pK}_a$$

If the pH of a solution is 2 units above the pK_a of the weak acid, pH = $pK_a + 2$.

$$\log \frac{[\text{base}]}{[\text{acid}]} = 2$$

$$\frac{[\text{base}]}{[\text{acid}]} = 1 \times 10^2 = \frac{100}{1}$$

This indicates that in a sample of molecules, there will be 100 base molecules for every one acid molecule:

$$\%\ \text{dissociation} = \frac{100 \text{ base molecules}}{101 \text{ total molecules}} \times 100 = 99\%$$

This equation is also extremely useful for preparing buffer solutions. A weak acid with a pK_a within 1 pH unit of the desired pH of the solution can be used to make a solution that will hold the desired pH by adjusting the amount of base.

The pH of the final solution will be determined only by the mole ration of base to acid, not by the volume, but the volume will also determine the buffer capacity.

EXAMPLE:

You are performing an experiment that requires your solution be buffered at a pH of 4.32. Benzoic acid has a K_a of 6.5×10^{-5}. What proportion of benzoic acid to sodium benzoate would give a pH of 4.32?

SOLUTION:

The pK_a of benzoic acid is $-\log(6.5 \times 10^{-5}) = 4.19$.

$$\log \frac{[\text{base}]}{[\text{acid}]} = \text{pH - pK}_a = 4.32 - 4.19 = 0.13$$

$$\frac{[\text{base}]}{[\text{acid}]} = 10^{.13} = 1.35$$

A good buffer for this pH would contain 1.35 moles of sodium benzoate for every mole of benzoic acid.

Workbook Problem 15.4

Determine the [sodium acetate]/[acetic acid] ratio for a buffer system with a pH = 5.75. The K_a of acetic acid is 1.8×10^{-5}. How much sodium acetate would need to be added to a 1 L of a 0.01 M solution of acetic acid to make a buffer at this pH?

Strategy: Use the Henderson–Hasselbalch equation to calculate the [base]/[acid] ratio. This information provides the mole ratio of acetate to acetic acid.

Step 1: Calculate the pK_a of acetic acid.

Step 2: Use the Henderson-Hasselbalch equation to calculate the [base]/[acid] ratio.

Step 3: Determine the mass of sodium acetate needed.

Workbook Problem 15.5

The pK_a of asparagine is 8.8. At what pH is asparagine 25% dissociated?

Strategy: Using the Henderson–Hasselbalch equation, calculate the log of the [base]/[acid] ratio, then determine the pH at which this value is obtained.

15.5 pH Titration Curves

In an acid-base **titration**, a solution of a known concentration of base (or acid), is slowly added from a buret to a solution containing an unknown concentration of acid (or base).

This allows for the determination of the **equivalence point** — the point at which stoichiometrically equivalent quantities of acid and base have been mixed together.

Titrations can be graphically represented as titration curves: a plot of the pH of the solution versus the volume of added titrant. These are useful in determining the equivalence point, which allows for the solution of a suitable indicator.

15.6 Strong Acid–Strong Base Titrations

Before any base is added, the $[H_3O^+]$ = concentration of strong acid. As base is added, it will neutralize acid, leading to a slow rise in pH:

$$\text{mmol } H_3O^+ \text{ after neutralization} = \text{mmol } H_3O^+ \text{ initial} - \text{mmol } OH^- \text{ added}$$

$$[H_3O^+] \text{ after neutralization} = \frac{\text{mmol } H_3O^+ \text{ after neutralization}}{\text{total volume of acid and base}}$$

At the equivalence point, all of the initial acid is neutralized, and the solution contains a neutral salt. The pH will be 7.00.

After the equivalence point, the pH will increase dramatically, as there is an excess of OH^- present.

$$\text{mmol of excess } OH^- = \text{mmol of } OH^- \text{ added} - \text{mmol of acid initially present}$$

$$[OH^-] \text{ after neutralization} = \frac{\text{mmol excess } OH^-}{\text{total volume of acid and base}}$$

Workbook Problem 15.6

What is the pH that results when 50 mL of 0.1 M NaOH are added to 15 mL of 1.0 M HCl?

Strategy: Determine the number of moles of H_3O^+ and OH^- present in the mixture and determine the pH based on the excess.

Step 1: Calculate the moles of OH^- and the moles of H_3O^+.

Step 2: Determine the excess quantity and the resultant $[H_3O^+]$.

Step 3: Determine the pH of the solution.

15.7 Weak Acid–Strong Base Titrations

Before any base is added, the pH can be calculated as usual for a solution of weak acid.

As base is added, but before the equivalence point is reached, the amount of conjugate base in the solution is equal to the amount of strong base added: the reaction between the weak acid and the strong base will proceed to completion.

Amount of base present $[A^-]$ = mL strong base added x [strong base]

$$[A^-] = \frac{\text{mmol of } A^-}{\text{total mL of acid and base}}$$

Amount of acid, HA, after neutralization = mmol of HA initially – mmol of A^-

$$[HA] \text{ after neutralization} = \frac{\text{mmol HA after neutralization}}{\text{total volume of acid and base}}$$

At the equivalence point, all the weak acid has been neutralized, leaving behind a basic salt solution: the conjugate base of the weak acid.

Mmol A^- at equivalence point = initial mmol weak acid

$$[A^-] = \frac{\text{initial mmol of HA}}{\text{total mL of acid and base}}$$

pH at the equivalence point will be the pH of the basic salt solution, and will be greater than 7.

After the equivalence point, the pH is determined by the excess OH^- from the strong base.

mmol OH^- added = mL of strong base × [strong base]

mmol A^- present = mmol A^- at the equivalence point

mmol OH^- present = mmol OH^- added – mmol A^- present

$$[H_3O^+] = \frac{K_w}{[OH^-]}$$

The pH titration curve for a strong base and a weak acid has a different appearance than that of a strong base and a strong acid. Because of the buffering action of the weak acid/conjugate base mixture that develops before the equivalence point, the pH initially rises more quickly, then levels off . Where pH = pK_a, the curve is flattest. Near the equivalence point, the pH increase is smaller than in a strong acid/strong base reaction, and the pH at the equivalence point is greater than 7.

As the K_a of a weak acid increases, the equivalence point gets more difficult to detect: in other words, the weaker the acid, the flatter the curve, as the initial pH is so close to neutral.

EXAMPLE:

25.0 mL of 0.200 M HF is titrated with 0.100 M NaOH. How many mL of base are required to reach the equivalence point? Calculate the pH at each of the following points: a) after addition of 10.0 mL of base, b) halfway to the equivalence point, c) at the equivalence point, and d) after addition of 65.0 mL of base.

SOLUTION: To determine the number of mL of base needed to reach the equivalence point, we need to first calculate the number of mmol of HF present.

$$\text{mmol HF} = 25.0 \text{ mL} \times \frac{0.200 \text{ mmol HF}}{1 \text{ mL}} = 5.0 \text{ mmol}$$

We need 5.0 mmol of NaOH to reach the equivalence point, which means we need 50 mL of 0.100 M NaOH.

To calculate the pH at the different points in the titration, we need to use the steps that are outlined above.

After addition of 10.0 mL of NaOH

mmol F^- = 10.0 mL × 0.100 mmol/mL = 1.00 mmol
$[F^-]$ = 1.00 mmol/35.0 mL = 2.86×10^{-2} M
mmol HF = 5.0 mmol – 1.00 mmol = 4.0 mmol
[HF] = 4.0 mmol/35.0 mL = 0.114 M

$$pH = pK_a + \log\frac{[F^-]}{[HF]} \qquad K_a = 3.5 \times 10^{-4} \quad pK_a = 3.46$$

$$pH = 3.46 + \log\frac{(2.82 \times 10^{-2})}{(0.114)} = 2.85$$

Halfway to the equivalence point, we have added 25.0 mL of NaOH (since 50 mL is required to reach the equivalence point).

mmol F^- = 25.0 mL × 0.100 mmol/mL = 2.50 mmol
$[F^-]$ = 2.50 mmol/50 mL = 0.050 M
mmol HF = 5.0 mmol – 2.50 mmol = 2.50 mmol
[HF] = 2.50 mmol/50 mL = 0.0500 M

$$\text{pH} = 3.46 + \log\frac{(0.050)}{(0.050)} = 3.46$$

At the equivalence point, all the HF has been neutralized, and the pH is determined by the concentration of F^-.

mmol F^- = 50 mL × 0.100 mmol/mL = 5.0 mmol

$[F^-]$ = 5.0 mmol/75 mL = 6.67 x 10^{-2} M

F^- is the anion of a weak acid and therefore gives a basic solution.

$$K_b = \frac{K_w}{K_a} = \frac{1.0 \times 10^{-14}}{3.5 \times 10^{-4}} = 2.9 \times 10^{-11}$$

Principal reaction: F^- (*aq*) + H_2O (*aq*) ⇆ HF (*aq*) + OH^- (*aq*)

$$K_b = 2.9\text{x}10^{-11} = \frac{[\text{HF}][\text{OH}^-]}{[\text{F}^-]} = \frac{x^2}{6.67 \text{ x } 10^{-2}}$$

$x = 1.4 \times 10^{-6} = [OH^-]$

$$\left[H_3O^+\right] = \frac{1.0 \times 10^{-14}}{1.4 \times 10^{-6}} = 7.1 \times 10^{-9} \quad \text{pH} = -\log(7.1 \times 10^{-9}) = 8.15$$

After addition of 65.0 mL of NaOH mmol OH^- added = 65.0 mL × 0.100 mmol/.mL = 6.5 mmol mmol F^- = 5.0 mmol mmol OH^- present = 6.5 mmol – 5.0 mmol = 1.5 mmol

$[OH^-]$ = 1.5 mmol/90.0 mL = 1.67 × 10^{-2} M

$$\left[H_3O^+\right] = \frac{1.0 \times 10^{-14}}{1.67 \times 10^{-2}} = 6.0 \times 10^{-13} \, \text{pH} = -\log(6.0 \times 10^{-13}) = 12.2$$

Workbook Problem 15.7

What quantity of 0.10 M NaOH is required to titrate 100 mL of 0.050 M phenol (K_a = 1.3 x 10^{-10}) to the equivalence point? Calculate the pH at the equivalence point.

Strategy: Determine the volume of NaOH required to provide equimolar amounts of base and acid, determine the concentration of conjugate base and pH.

Step 1: Calculate the moles of phenol present and thus the moles of OH^- needed.

Step 2: Determine the volume of sodium hydroxide solution required.

Step 3: Determine the total volume of the resultant solution.

Step 4: Solve for the concentration of salt.

Step 5: Set up an equilibrium table.

Step 6: Solve for K_b and [OH].

Step 7: Solve for [H_3O] and pH.

15.8 Weak Base–Strong Acid Titrations

Before any acid is added, the pH can be calculated as usual for a solution of weak base.

As acid is added, but before the equivalence point is reached, the amount of conjugate acid in the solution is equal to the amount of strong acid added: the reaction between the weak base and the strong acid will proceed to completion.

Amount of acid present [BH] = mL strong acid added x [strong acid]

$$[\mathrm{BH}] = \frac{\text{mmol of BH}}{\text{total mL of acid and base}}$$

Amount of weak base after neutralization = mmol of B initially – mmol of HA added

$$[\mathrm{B}] \text{ after neutralization} = \frac{\text{mmol B after neutralization}}{\text{total volume of acid and base}}$$

At the equivalence point, all the weak base has been neutralized, leaving behind an acid salt solution: the conjugate acid of the weak base.

Mmol BH at equivalence point = initial mmol weak base

$$[\mathrm{A}^-] = \frac{\text{initial mmol of HA}}{\text{total mL of acid and base}}$$

pH at the equivalence point will be the pH of the acidic salt solution, and will be less than 7.

After the equivalence point, the pH is determined by the excess H_3O^+ from the strong acid.

The pH titration curve for a weak base and a strong acid has a different appearance than that of a strong base and a strong acid. Because of the buffering action of the weak acid/conjugate base mixture that develops before the equivalence point, the pH initially drops more quickly, then levels off . Where pH = pK_a, the curve is flattest. Near the equivalence point, the pH increase is smaller than in a strong acid/strong base reaction, and the pH at the equivalence point is less than 7.

Workbook Problem 15.8

40.00 mL of 0.375 M trimethylamine, $(CH_3)_3N$, is titrated with 0.100 M HCl. a) How many mL of acid are required to reach the equivalence point? b) Calculate the initial pH, and the pH after

c) the addition of 10.00 mL of acid, d) halfway to the equivalence point, e) at the equivalence point, and f) after addition of 100.00 mL of acid. $K_b = 6.5 \times 10^{-5}$.

Strategy: Use the method outlined above to calculate the pH at various points in the titration.

a) mL of acid required to reach the equivalence point:
Step 1: Determine the number of mmoles of $(CH_3)_3N$ in the initial solution.

Step 2: Determine the mmol of HCl needed to react with the $(CH_3)_3N$ and the volume of 0.100 M HCl needed.

b) Initial pH of the $(CH_3)_3N$ solution:
Step 1: Determine the initial pH of the solution.

c) pH of the solution after addition of 10.00 mL of acid.
Step 1: Calculate the concentration of $(CH_3)_3N$ and its conjugate acid after the addition of 10.00 mL of HCl.

Step 2: Determine the K_a and pK_a of the conjugate acid of $(CH_3)_3N$.

Step 3: Using the Henderson–Hasselbalch equation, calculate the pH of the solution.

d) pH halfway to the equivalence point:
Step 1: Use the Henderson–Hasselbalch equation to calculate the pH. Halfway to the equivalence point, [base] = [acid].

e) pH at the equivalence point:
Step 1: Calculate the concentration of the conjugate acid, using the initial moles of base, and the total volume of solution.

Step 2: Determine the pH of the solution.

f) pH after the addition of 200.00 mL of 0.100 M HCl:

Step 1: Determine the number of mmoles of excess HCl.

Step 2: Determine the concentration of excess HCl, and calculate the pH.

15.9 Polyprotic Acid–Strong Base Titrations

These calculations share the same features of other calculations except that there are two equivalence points. Amino acids have two dissociable protons, and so will react with two molar amounts of strong base. These titrations are done with protonated amino acids, so they begin at a low pH with the form H_2A^+.

Before the addition of any base, the pH is calculated as an equilibrium problem, The principal reaction is the dissociation of H_2A^+:

$$H_2A^+\,(aq) \leftrightarrows HA\,(aq) + H_3O^+\,(aq)$$

Adding strong base before the first equivalence point generates an H_2A^+/HA buffer solution. The concentrations can be calculated as in any weak acid–strong base titration, using the Henderson–Hasselbalch equation to calculate pH. Halfway to the first equivalence point, pH = K_{a1}.

At the first equivalence point, all the H_2A^+ is converted to HA. At this point, the principal reaction will be proton transfer between HA molecules:

$$2\,HA\,(aq) \leftrightarrows H_2A^+\,(aq) + A^-\,(aq) \qquad K = K_{a2}/K_{a1}$$

At this point, the pH is the average of pK_{a1} and pK_{a2}.

$$pH = \frac{pK_{a1} + pK_{a2}}{2}$$

For an amino acid, this is also the *isoelectric point* — the neutral form [HA] is at a maximum and the $[H_2A^+]$ and $[A^-]$ are very small and equal. In biochemistry, the isoelectric point is useful for separating amino acids and proteins.

When strong base is added between the first and second equivalence points, the final proton is removed:

$$HA\,(aq) \leftrightarrows H_3O^+\,(aq) + A^-\,(aq)$$

This generates an HA/A^- buffer solution. The Henderson–Hasselbalch equation can be used to calculate pH.

At the second equivalence point, all of the HA is converted to A^-. $[A^-]$ is equal to the initial concentration of amino acid - $[H_2A^+]$. This is now the solution of a basic salt:

$$A^-\,(aq) + H_2O\,(l) \leftrightarrows HA\,(aq) + OH^-\,(aq) \qquad K_b = K_w/K_{a2}$$

Strong base that is added beyond the second equivalence point will determine the pH.

Workbook Problem 15.9

What would the pH be if 25 mg of NaOH were added to 100 mL of a 0.050 M solution of the amino acid proline? ($pK_{a1} = 1.952$, $pK_{a2} = 10.64$) What is the isoelectric point of proline?

Strategy: Determine how far this amount of sodium hydroxide takes the titration: use the Henderson-Hasselbalch equation to determine pH. Calculate isoelectric point from pK_{a1} and pK_{a2}.

Step 1: Determine how many moles of proline and how many moles of sodium hydroxide are present.

Step 2: Use the Henderson-Hasselbalch equation to determine the pH.

Step 3: Calculate the isoelectric point.

15.10 Solubility Equilibria

When a sparingly-soluble ionic compound dissolves, an equilibrium is set up between the solid, and the dissolved ions in solution.

The equilibrium constant for solubility is called the solubility product constant (K_{sp})

For the generic reaction

$$M_aX_b\,(s) \leftrightarrows a\,M^{b+}\,(aq) + b\,X^{a-}\,(aq)$$

$$K_{sp} = [M^{b+}]^a[X^{a-}]^b$$

15.11 Measuring K_{sp} and Calculating Solubility from K_{sp}

The solubility constant is measured experimentally: the concentrations of an equilibrium mixture of species are measured, either by allowing the solid to come to equilibrium, or by generating a solution by adding together soluble components to create a precipitate.

The solubility constant is temperature-dependent, and can be used to calculate molar solubility.

Molar solubility can also be calculated from K_{sp}.

EXAMPLE:

K_{sp} for AgCl is 1.8×10^{-10}. Calculate the molar solubility of AgCl.

SOLUTION: The solubility equilibrium for AgCl is

$$AgCl \leftrightarrows Ag^+(aq) + Cl^-(aq)$$

$$K_{sp} = [Ag^+][Cl^-]$$

$$1.8 \times 10^{-10} = x^2$$

$$x = 1.34 \times 10^{-5}$$

The molar solubility of AgCl is 1.34×10^{-5} M

EXAMPLE:
The molar solubility of copper (I) chloride is 1.095×10^{-3} M. Calculate K_{sp} for CuCl.

SOLUTION: The solubility equilibrium for AgCl is

$$CuCl \leftrightarrows Cu^+(aq) + Cl^-(aq)$$

$$K_{sp} = [Cu^+][Cl^-]$$

$$K_{sp} = (1.095 \times 10^{-3})^2$$

$$K_{sp} = 1.20 \times 10^{-6}$$

Workbook Problem 15.10
Calculate the K_{sp} for a solution of $CdCl_2$ prepared in pure water, given that the $[Cd^{2+}] = 1.10 \times 10^{-5}$ M.

Strategy: Write the balanced chemical equation for the solubility of $CdCl_2$, and use the stoichiometry and information given to calculate K_{sp}.

Step 1: Write the solubility equilibrium expression for $CdCl_2$.

Step 2: Determine the concentration of Cd^{2+} and Cl^- and calculate the K_{sp}.

Workbook Problem 15.11
The K_{sp} of $Cu(OH)_2$ is 2.18×10^{-20}. Calculate the molar solubility of $Cu(OH)_2$.

Strategy: Write the solubility equilibrium for $Cu(OH)_2$ and the equilibrium expression. Calculate the molar solubility of $Cu(OH)_2$.

Step 1: Let x be the number of mol/L of $Cu(OH)_2$. that dissolves. The saturated solution then contains x mol/L of Cu^{2+} and $2x$ mol/L of OH^-. Solve for K_{sp}.

15.12 Factors That Affect Solubility

The common ion effect will work to decrease the solubility of a slightly soluble ionic compound: the presence of an ion that is present in the compound will tend to cause ions that make it into solution to immediately precipitate back out.

If the compound contains a basic anion, the solubility will increase as the pH decreases.

The formation of complex ions, in which a Lewis base forms a coordinate covalent bond to the metal cation, will push the equilibrium in the direction of increased solubility by removing dissolved metal cations into complex ions.

The formation of complex ions is a stepwise process, but the equilibrium constant for the ion formation incorporates all of the steps. A high K_f indicates the formation of a stable complex ion.

Amphoteric oxides are metal hydroxides that will dissolve in both strongly basic and strongly acidic solutions. This occurs in basic solutions when excess OH^- ions convert the hydroxide into a complex ion in basic solution. In acid, the OH^- is neutralized, and the metal cation becomes soluble.

Only oxides that can be converted to more soluble complex ions exhibit this behavior.

EXAMPLE:

Determine the molar solubility of $Cu(C_2O_4)$ ($K_{sp} = 2.87 \times 10^{-8}$ at 25 °C) in a 0.25 M $CuCl_2$ solution.

SOLUTION: The solubility equilibrium expression is

$$Cu(C_2O_4) \leftrightarrows Cu^{2+}\ (aq) + C_2O_4^{2-}\ (aq)$$

The equilibrium expression for this reaction is

$$K_{sp} = [Cu^{2+}][C_2O_4^{2-}]$$

Let x be the number of mol/L of $Cu(C_2O_4)$ that dissolves. We can now construct a table showing the equilibrium concentrations:

Solubility Equilibrium	$Cu(C_2O_4)_2\ (s)$	$\leftrightarrows$	$Cu^{2+}\ (aq)$ +	$C_2O_4^{2-}\ (aq)$
Initial Concentration (M)			0.25	0
Equilibrium Concentration (M)			$0.25 + x$	$+x$

Given the value of K_{sp}, we can assume that x is negligible. Therefore, the equilibrium concentration of Cu^{2+} is approximately 0.25 M. Substituting these values into the equilibrium expression gives

$2.87 \times 10^{-8} = (0.25)(x)$ $\qquad$ $x = 1.15 \times 10^{-7}$ M

Workbook Problem 15.12

The initial pH of a solution containing $Sn(OH)_2$ was 9.45 after addition of excess NH_3. Determine the molar solubility of $Sn(OH)_2$ in this solution. For $Sn(OH)_2$, $K_{sp} = 5.4 \times 10^{-27}$.

Step 1: Write the solubility equation and solubility expression.

Step 2: Using the pH, determine the $[OH^-]$ concentration.

Step 3: Let x be the number of mol/L of $Sn(OH)_2$ that dissolves. Construct a table showing the equilibrium concentrations.

Step 4: Calculate the solubility of $Sn(OH)_2$.

EXAMPLE:

Write a balanced net ionic equation for the dissolution reaction between AgBr and $Na_2S_2O_3$, and calculate the equilibrium constant given K_{sp} (AgBr) = 5.4×10^{-13} and K_f $\{Ag(S_2O_3)_2^{3-}\} = 4.7 \times 10^{13}$.

SOLUTION:

The net ionic equation for the dissolution of AgBr in $Na_2S_2O_3$ is obtained by combining the solubility expression for AgBr and the formation expression for $Ag(S_2O_3)_2^{3-}$.

$AgBr\ (s) \leftrightarrows Ag^+\ (aq) + Br^-\ (aq)$ $\qquad$ $K_{sp} = 5.4 \times 10^{-13}$

$Ag^+\ (aq) + 2\ S_2O_3^{2-} \leftrightarrows Ag(S_2O_3)_2^{3-}$ $\qquad$ $K_f = 4.7 \times 10^{13}$

Adding the two equations together and canceling out what is common on both the reactant and product sides gives

$AgBr\ (s) \leftrightarrows \cancel{Ag^+}\ (aq) + Br^-\ (aq)$ $\qquad$ $K_{sp} = 5.4 \times 10^{-13}$

$\cancel{Ag^+}\ (aq) + 2\ S_2O_3^{2-}\ (aq) \leftrightarrows Ag(S_2O_3)_2^{3-}\ (aq)$ $\qquad$ $K_f = 4.7 \times 10^{13}$

$AgBr\ (s) + 2\ S_2O_3^{2-}\ (aq) \leftrightarrows Ag(S_2O_3)_2^{3-}\ (aq) + Br^-\ (aq)$ $\quad$ $K = K_{sp} \times K_f = 25.4$

15.13 Precipitation of Ionic Compounds

Solubility guidelines begin to describe the behavior of ions in solution, but a far more accurate method is the use of the **ion product**.

For the salt, M_aX_b; IP = $[M^{b+}]^a[X^{a-}]^b$.

The ion product is calculated using initial concentration, not equilibrium concentrations: this is a reaction quotient, and will determine which direction the reaction will go using the following guidelines:

If IP > K_{sp}, the solution is supersaturated, and precipitation occurs.

If IP = K_{sp}, the solution is saturated and at equilibrium.

If IP < K_{sp}, the solution is unsaturated and precipitation will not occur.

Workbook Problem 15.13

Will a precipitate form when 100 mL of 0.75 M $Zn(NO_3)_2$ is mixed with 250 mL of 1.50 M Na_2CO_3?

Strategy: Write a metathesis (exchange) reaction for the reaction between $Zn(NO_3)_2$ and Na_2CO_3, and apply the solubility rules in Chapter 4 to determine if a precipitate will form.

Step 1: Write the solubility equation and IP expression for the precipitate.

Step 2: Calculate the IP, and compare its value to the K_{sp}. Will a precipitate form?

15.14 Separation of Ions by Selective Precipitation

It is possible to separate a mixture of ions in solution by adding a reagent that will precipitate some of the ions but not others.

Using an acid solution, it is possible to separate insoluble and soluble metal sulfides. The equilibrium constant for this reaction is K_{spa} - the solubility constant in acid solution.

For the general equation

$$MS\,(s) + 2\,H_3O^+\,(aq) \leftrightharpoons M^{2+}\,(aq) + H_2S\,(aq) + 2\,H_2O\,(l)$$

$$K_{spa} = \frac{[M^{2+}][H_2S]}{[H_3O^+]^2}$$

Using a reaction quotient:

$$Q_c = \frac{[M^{2+}][H_2S]}{[H_3O^+]}$$

it is possible to adjust $[H_3O^+]$ so that $Q_c > K_{spa}$ for the more insoluble metal sulfide. The more insoluble sulfide will precipitate out, and the more soluble will stay in solution.

15.15 Qualitative Analysis

This is a procedure for identifying the ions present in an unknown solution.

In the traditional scheme of analysis for metal cations, 20 cations into five groups by selective precipitation.

Aqueous HCl will remove Ag, Hg, and Pb as insoluble chlorides (some Pb will remain).

Bubbling H_2S through the acid solution will precipitate out the remaining Pb, as well as Cu, Hg, Cd, Bi, and Sn.

Addition of NH_3 neutralizes the solution, precipitating out the remaining sulfates: Mn, Fe, Co, Ni, and Zn. Insoluble hydroxides of Al and Cr also precipitate out.

$(NH_4)_2CO_3$ is now added to precipitate out Ca and Ba.

$(NH_4)_2HPO_4$ will precipitate out Mg.

All that can remain in solution is K and Na. These can be identified using flame testing: Na imparts a persistent yellow color to a flame, while K imparts a transient violet flame.

The ion in each group can be separated with further analysis. Although there are now more sophisticated techniques for the analysis of metal ions in solution, this method is still an excellent technique for learning laboratory skills and learning about acid–base, solubility, and complex–ion equilibria.

Putting It Together

Aluminum phosphate is formed from the reaction of aluminum chloride and phosphoric acid. a) Write a balanced chemical equation for this reaction. b) If you begin with 95 g of aluminum chloride and 1.50 L of 0.75 M phosphoric acid, how many grams of aluminum phosphate will be produced? c) If you place 25.0 g of aluminum phosphate in water to produce a solution with a volume of 1.00 L, what are the equilibrium concentrations of Al^{3+} and PO_4^{3-}? ($K_{sp} = 1.3 \times 10^{-20}$) d) How does the addition of HCl affect the solubility of aluminum phosphate? e) If you mix 0.75 L of 4.00×10^{-3} M $AlCl_3$ with 1.25 L of 0.700 M Na_3PO_4, will a precipitate of aluminum phosphate form? If so, how many grams of $AlPO_4$ will form?

Self–Test

This section is intended to test your knowledge of the material covered in this chapter. Think through these problems, and make certain you understand what they are asking. Make sure your answers make sense. Successful completion of these problems indicates that you have mastered the material in this chapter. You will receive the greatest benefit from this section if you use it as a mock exam, as this will allow you to determine which topics you need to study in more detail.

True–False

1. Mixing 100 mL of 0.1 M HCl with 100 mL of 0.1 M NaOH will yield a solution with pH 7.

2. Mixing 100 mL of 0.1 M acetic acid with 100 mL of 0.1 M NaOH will yield a solution with pH < 7.

3. Adding a common ion will reduce the dissociation of a weak acid.

4. When OH^- ions are added to a buffer system, the equilibrium shifts toward undissociated acid molecules.

5. When pH = pK_a, all of the weak acid in a solution is dissociated.

6. A strong acid/strong base titration has an equivalence point at pH 7.

7. A strong acid/weak base titration has an equivalence point above pH 7.

8. Molar solubility and K_{sp} must be separately experimentally determined.

9. Metal ions in solution can be separated based on reactivity.

10. Na and K ions in solution can be separated by adding HCl to solution.

Matching

Amphoteric	a. metal cation bonded to one or more small molecules
Buffer capacity	b. solution of weak acid and conjugate acid that resists changes in pH
Buffer solution	c. point in a titration at which molar equivalents have been added to solution
Common-ion effect	d. exhibiting both acidic and basic properties
Complex ion	e. relationship between pH and pK_a of a weak acid, calculated using the amount of dissociation.
Equivalence point	f. procedure for identifying ions in solution
Formation constant	g. shift in equilibrium caused by adding an ion already involved in the equilibrium
Henderson-Hasselbalch equation	h. K_{eq} for complex ions
Ion product	i. The amount of acid or base a buffer can absorb without a significant change in pH
Qualitative analysis	j. Q for K_{sp}

Fill–in–the–Blank

1. The neutralization of a weak acid with a ___________ base, will go to _______________ because OH^- has a great affinity for ________________.

2. Solutions which contain a ____________ acid and ________________ base resist changes in ____________ through shifts in ____________________ . They are called _________________ .

3. Complex ions are formed when ___________________ cations form __________________ covalent bonds with small _________________ like ammonia or water.

Problems

1. Write balanced net ionic equations, and predict the pH for reactions between
   a. HNO_3 and NaOH
   b. HCl and NH_2NH_2
   c. CH_3COOH (acetic acid) and NaOH

2. Calculate the pH of a solution containing 0.10 mol of NH_3 and 0.25 mol of NH_4Cl in 1.0 L. For ammonia, $K_b = 1.8 \times 10^{-5}$.

3. Determine the pH and concentration of all species present in a buffer solution containing 0.45 mol HClO and 0.25 mol NaClO in 1.0 L of solution. Determine the change in pH upon addition of 0.10 mol NaOH. Determine the change in pH on addition of 0.10 mol HCl. K_a for HClO is 3.5×10^{-8}

4. Determine the ratio of lactic acid to lactate ion required for preparing a buffer solution whose pH is 4.75. The K_a for lactic acid is 1.4×10^{-4}. If the buffer is to be overall 0.750 M in these compounds, how many moles of each will be added per liter of solution?

5. Determine the change in pH that will occur on addition of 0.300 mol HCl (assuming no volume change) to a 1 L buffer solution prepared from 0.50 M HCOOH and 0.25 M HCO_2Na. The pK_a of formic acid is 3.74.

6. Calculate the pH of 75.0 mL of 0.0400 M HCl after addition of the following volumes of 0.1000M NaOH: a) 0.0 mL; b) 10.0 mL; c) 30.0 mL; d) 50.0 mL.

7. Calculate the pH of 50.0 mL of 0.0500 M acetic acid ($K_a = 1.8 \times 10^{-5}$) after addition of the following volumes of 0.1000 M NaOH: a) 0.0 mL; b) 10.0 mL; c) 25.0 mL; d) 30.0 mL.

8. The molar solubility of $CaCO_3$ is 7.07×10^{-5} M. Calculate the value of K_{sp}.

9. Calculate the molar solubility for $Ca_3(PO_4)_2$, given that the $K_{sp} = 2.1 \times 10^{-33}$.

10. Calculate the solubility of a solution of $CaSO_4$ that contains 0.50 M Na_2SO_4. ($K_{sp} = 7.1 \times 10^{-5}$)

11. Which of the following compounds are more soluble in acidic solution than in pure water? Why?
    a. AgBr b. Na_2S c. LiCN d. CaF_2

12. What are the concentrations of Zn^{2+} and $Zn(NH_3)_4^{2+}$ in a solution prepared by adding 0.50 mol of $Zn(NO_3)_2$ to 1.0 L of 4.0 M NH_3? ($K_f = 7.8 \times 10^8$)

13. Will a precipitate form when 350 mL of 0.75 M $CaCl_2$ is mixed with 200 mL of 1.50 M Na_3PO_4? $K_{sp} = 2.1 \times 10^{-33}$ for calcium phosphate.

14. Is it possible to separate Cu^{2+} from Fe^{2+} by bubbling H_2S through a 0.20 M HCl solution that contains 0.007 M Cu^{2+} and 0.007 M Fe^{2+}? K_{spa} for CuS $= 6 \times 10^{-16}$, K_{spa} for FeS $= 6 \times 10^2$, and $[H_2S] = 0.10$ M.

Challenge Problem

CuS has $K_{spa} = 6.0 \times 10^{-16}$. Determine the solubility of CuS in a buffer prepared from 0.45 M formic acid and 0.25 M sodium formate. $K_a = 1.8 \times 10^{-4}$ for formic acid.

CHAPTER 16

THERMODYNAMICS: ENTROPY, FREE ENERGY, AND EQUILIBRIUM

Chapter Learning Goals

A. Molecular Randomness and Chemical and Physical Changes

1. Qualitatively determine whether simple chemical or physical changes are spontaneous.
2. Qualitatively predict whether the sign of ΔS is positive or negative for a chemical or physical change.
3. On the basis of probability, determine which of two states has higher entropy.
4. Calculate the standard entropy of reaction from the standard molar entropies of products and reactants.
5. Determine whether a reaction is spontaneous by determining the sign of ΔS_{total}.

B. Free Energy Change

1. Use the equation $\Delta G = \Delta H - T\Delta S$ to calculate the free energy of reaction and to determine the temperature at which a nonspontaneous reaction becomes spontaneous.
2. Calculate the standard free energy of reaction from standard free energies of formation.
3. Calculate the free energy of reaction for a system having nonstandard pressures and concentrations.
4. From the standard free energy of reaction, calculate the equilibrium constant.

Chapter in Brief

Spontaneous processes are those that proceed toward equilibrium without any outside influence. Spontaneity is related to entropy, the amount of molecular randomness. The standard molar entropy of substances can be used to calculate the standard entropy of reaction, and the second law of thermodynamics states that in any spontaneous process, the total entropy of the system and surroundings increases. Free energy is related to entropy through the equation $\Delta G = \Delta H - T\Delta S$. It is a state function that determines the spontaneity of a reaction. The relative importance of the enthalpy (ΔH) and entropy (ΔS) terms is determined by the temperature of the reaction. Free energy of reaction can be calculated either using the equation above, or free energies of formation. The free energy of a reaction is also mathematically related to the equilibrium constant, K, through the equation $\Delta G° = -RT \ln K$.

16.1 Spontaneous Processes

A **spontaneous process** is one that proceeds on its own without any external influence or input of energy.

Spontaneous reactions always move toward equilibrium. Because of this, their spontaneity is dependent on the same things that the equilibrium is dependent on: temperature, pressure, and the composition of the reaction mixture.

When $Q < K$, reaction proceeds in the forward direction. When $Q > K$, reaction proceeds in the reverse direction.

Spontaneity of a reaction is no indication of the speed of the reaction. The speed is a function of kinetics and the height of the **activation energy** barrier.

EXAMPLE:

Determine which of the following processes are spontaneous and which are nonspontaneous.

a. The cooling of a cup of tea.
b. The smell of baking cookies being detectable by sensors throughout a house.
c. The decomposition of table salt into sodium metal and chlorine gas.
d. The reaction of S and H_2 to produce SH_2 if the concentration of H_2 is 1 M and the concentration of SH_2 is 5 M. $K_c = 7.8 \times 10^5$.

SOLUTION: Both a and b are spontaneous processes. C is a nonspontaneous process. To determine if d is spontaneous, determine Q for the reaction. If $Q < K$, then the reaction is spontaneous. If $Q > K$, then the reaction is nonspontaneous in the forward direction, but will be spontaneous in the reverse direction. The equilibrium expression for this reaction is

$$S(s) + H_2(g) \leftrightarrows SH_2(g)$$

$$Q = \frac{[SH_2]}{[H_2]} = \frac{(5)}{(1)} = 5$$

$Q < K$; the reaction is spontaneous in the forward direction.

16.2 Enthalpy, Entropy, and Spontaneous Processes: A Brief Review

Most spontaneous processes release heat, but not all do. For example, ice melts spontaneously even though this is an endothermic process.

Enthalpy alone cannot account for spontaneity. The second factor that is involved is **entropy**: the tendency for a system to move toward maximum randomness.

Entropy, abbreviated S, is a state function, meaning that it is independent of path. When a solid melts, a liquid vaporizes, a gas is heated or a solute dissolves, the freedom of movement of the individual particles, as well as the number of possible arrangements of those particles both increase; when this occurs entropy will also increase.

Dissolution of ionic solids is a slightly more complex case, which is why all ionic solids are not soluble. When ions are released into solution, they are also **hydrated** – surrounded by an orderly collection of water molecules that neutralizes their charge. This reduces the entropy of the water molecules. As a result, most ionic solids with +1 and -1 charges, such as NaCl, are soluble, while those with higher charges, and subsequently with more water molecules required for hydration, are not.

An increase in entropy is favorable. As with all state functions the change is calculated:

$$\Delta S = S_{\text{final}} - S_{\text{initial}}$$

When ΔS is positive, entropy is favorable as disorder has increased. When ΔS is negative, entropy is unfavorable, as order has increased.

EXAMPLE:

Predict the sign of ΔS for

a. $Br_2(l) \rightarrow Br_2(g)$
b. $H_2O(l) \rightarrow H_2O(s)$
c. $NaCl(aq) \rightarrow NaCl(s)$
d. $Zn(s) + 2\,HCl(aq) \rightarrow H_2(g) + ZnCl_2(aq)$

SOLUTION:

a. ΔS is positive; gas molecules are less orderly than liquid molecules.
b. ΔS is negative; liquid molecules decrease in randomness when they freeze into a solid.
c. ΔS is negative; aqueous ions lose their randomness when they condense into a crystal.
d. ΔS is positive; gas is generated, which greatly increases the disorder of the system.

16.3 Entropy and Probability

A random state is more likely, and can be achieved in a large number of ways. Although the discussions here are at the molecular level, this is something that can be seen in everyday life as well. It is a rare thing indeed, for instance, to open a clothes dryer and find that the clothes have spontaneously folded themselves. It is much more likely that they are found in one of the multitude of disordered states that are available to them.

Ludwig Boltzmann proposed that the entropy of a particular state is related to the number of ways that this state can be achieved by the following formula:

$$S = k \ln W$$

$k = R/N_A = 1.38 \times 10^{-23}$ J/K, and $\ln W$ = the number of available states.

For a perfect crystal, entropy will be 0 – there is only one way to achieve that state, and ln 1 = 0. Molecules with strong **dipole moments** are more likely to form more orderly crystals. As a result, their entropy will be lower.

Gas expands spontaneously because the state of greater volume is more probable: the larger the volume, the higher the number of disordered states available to the gas particles.

For an ideal gas it is possible to derive the following corollaries to the Boltzmann equation:

$$\Delta S = nR \ln \frac{V_{final}}{V_{initial}}$$

$$\Delta S = nR \ln \frac{P_{initial}}{P_{final}}$$

From this, it is possible to see that increasing the volume or decreasing the pressure both increase the entropy of a gas.

Workbook Problem 16.1

What is the entropy change if the volume of 3 g of hydrogen gas increases from 5 L to 20 L at a constant temperature? (R = 8.314 J/K)

Strategy: Determine the number of moles of hydrogen, calculate ΔS.

Step 1: Calculate the moles of hydrogen.

Step 2: Calculate ΔS.

16.4 Entropy and Temperature

As the temperature of a substance increases, the overall kinetic energy of the system also increases.

This affects entropy in several ways. When temperature increases, molecular motion increases, randomness increases, and individual molecular energies occur across a wider spectrum, introducing another area of disorder. All of these things increase entropy.

A plot of entropy versus temperature will show a steady increase in entropy with temperature, with large jumps for phase changes.

Third law of thermodynamics — The entropy of a perfectly ordered crystalline substance at 0 K is zero.

16.5 Standard Molar Entropies and Standard Entropies of Reaction

Standard molar entropies, $S°$ are available for many materials. This is the entropy of 1 mole of the pure substance at 1 atm pressure and a specified temperature, usually 25 °C. The units of these values are in J/(K•mol).

For the same substance, the gas-phase form will have the highest entropy, the solid phase the lowest entropy.

Within the same phase, larger molecules – which have the ability to adopt different conformations and thus can have internal entropy – have higher entropy than smaller molecules.

Entropies of reaction can be calculated from the standard entropies of the reactants and products, by subtracting the sum of the entropies of the reactants from the sum of the entropies of the products. Just as with enthalpies of reaction, the individual values must be multiplied by the stoichiometric coefficients.

EXAMPLE:

Calculate the $\Delta S°_{rxn}$ for

$$2\ NO\ (g)\ +\ O_2\ (g)\ \rightarrow\ 2\ NO_2\ (g)$$

SOLUTION: $\Delta S°_{rxn} = \Delta S°_{products} - \Delta S°_{reactants}$

$S°$ (NO) = 210.7 J/mol·K
$S°$ (O_2) = 205.0 J/mol·K
$S°$ (NO_2) = 240.0 J/mol·K

$\Delta S°_{rxn}$ = [2(240.0 J/mol·K)] – [(205.0 J/mol·K) + 2(210.7 J/mol·K)] = -146.4 J/mol•K

Workbook Problem 16.2

Calculate the $\Delta S°_{rxn}$ for the combustion of hydrogen to form water vapor.

$S°$ (H_2) = 130.6 J/mol·K
$S°$ (O_2) = 205.0 J/mol·K
$S°$ (H_2O) = 188.7 J/mol·K

Strategy: Write the formula for the reaction, calculate ΔS.

Step 1: Write the formula.

Step 2: Calculate ΔS.

16.6 Entropy and the Second Law of Thermodynamics

Free energy changes determine the spontaneity of a reaction, and are calculated:

$$\Delta G = \Delta H - T\Delta S$$

When $\Delta G > 0$, the reaction is nonspontaneous, when $\Delta G = 0$ the reaction is at equilibrium, and when $\Delta G < 0$ the reaction will occur spontaneously.

First law of thermodynamics — In any process, spontaneous or nonspontaneous, the total energy of a system and its surroundings is constant. This is a restatement of the law of conservation of energy. Exothermic reactions have energy flowing from system to surroundings, while endothermic reactions have energy flowing from surroundings into the system. This law says nothing about spontaneity.

Second law of thermodynamics — In any spontaneous process, the total entropy of a system and its surroundings always increases. This provides a clear-cut criterion of spontaneity using the definition:

$$\Delta S_{\text{total}} = \Delta S_{\text{system}} + \Delta S_{\text{surroundings}}$$

When ΔS_{total} is positive, the reaction is spontaneous. When $\Delta S_{\text{total}} = 0$, the reaction is at equilibrium, and when $\Delta S_{\text{total}} < 0$ the reaction is nonspontaneous. All reactions proceed spontaneously in the direction that increases the total entropy of the system plus surroundings.

At constant pressure, $\Delta S_{\text{surr}} = \frac{-\Delta H_{\text{rxn}}}{T}$. This makes it possible to rewrite the equation above:

$$\Delta S_{\text{total}} = \Delta S^{\circ}_{\text{rxn}} - \frac{\Delta H_{\text{rxn}}}{T}$$

EXAMPLE:

Determine whether the following reaction is spontaneous at 25 °C by determining ΔS_{total}:

$$H_2\,(g) + Cl_2\,(g) \rightarrow 2\,HCl\,(g)$$

SOLUTION: Calculate ΔS°, ΔH_{rxn}, and ΔS_{total}

S° (H_2) = 130.6 J/mol·K
S° (Cl_2) = 223.0 J/mol·K
S° (HCl) = 186.8 J/mol·K
ΔH_f°(HCl) = -92.3 kJ/mol

$\Delta S^\circ_{rxn} = [2(186.8 \text{ J/mol·K})] - [(130.6 \text{ J/mol·K}) + (223.0 \text{ J/mol·K})] = 20.0 \text{ J/mol•K}$

$\Delta H^\circ_{rxn} = 2(-92.3 \text{ kJ/mol}) - (0 \text{ kJ/mol} + 0 \text{ kJ/mol}) = -184.6 \text{ kJ/mol}$

$$\Delta S_{total} = \Delta S^o_{rxn} - \frac{\Delta H_{rxn}}{T} = 20.0 \text{ J/(mol• K)} - \frac{-184{,}600\text{J/mol}}{298\text{K}} = 639 \text{ J/(mol• K)}$$

The overall entropy change is positive, so the reaction is spontaneous.

Workbook Problem 16.3

Calculate ΔS_{total} for the following reaction at 298 K, and determine if the reaction is spontaneous.

$AgNO_3\ (aq) + NaBr\ (aq) \rightarrow AgBr\ (s) + NaNO_3\ (aq)$

$S^\circ (Ag^+)$ = 72.7 J/mol·K
$S^\circ (Br^-)$ = 82.4 J/mol·K
S° (AgBr) = 107 J/mol·K
$\Delta H^\circ (Ag^+)$ = 105.6 kJ/mol
$\Delta H^\circ (Br^-)$ = -121.5 kJ/mol
ΔH°(AgBr) = -100.4 kJ/mol

Strategy: Calculate the values of ΔS_{system} and ΔS_{surr}.

Step 1: Write the net ionic equation for this reaction.

Step 2: Calculate $\Delta S_{system.}$

Step 3: Calculate $\Delta H_{rxn.}$

Step 4: Calculate $\Delta S_{total.}$

16.7 Free Energy

The previously described equation.

$$\Delta G = \Delta H - T\Delta S$$

Can be rewritten:

$$-T\,\Delta S_{total} = \Delta G$$

In any spontaneous process at constant temperature and pressure, the free energy of the system decreases. Temperature can determine the relative importance of the enthalpy and entropy changes.

Using the free-energy equation, it is possible to determine the temperature at which a reaction becomes spontaneous, by finding the temperature at which the reaction is at equilibrium. Remember that at equilibrium, $\Delta G = 0$.

This allows the following rearrangement:

$$\Delta G° = \Delta H° - T\Delta S° = 0.$$

$$T = \frac{\Delta H°}{\Delta S°}$$

EXAMPLE:

Pure chromium is obtained by reducing Cr_2O_3 with aluminum.

$Cr_2O_3\ (s)\ +\ 2\ Al\ (s) \rightarrow 2\ Cr\ (s)\ +\ Al_2O_3\ (s)$

Determine $\Delta H°$ and $\Delta S°$ for the reaction. Is the reaction spontaneous at 350 °C?

SOLUTION:

Calculate ΔH°_{rxn} and ΔS°_{rxn}.

$$\Delta H^\circ_{rxn} = \Delta H^\circ_f(Al_2O_3) - \Delta H^\circ_f(Cr_2O_3)$$

$$\Delta H^\circ_{rxn} = [1\text{ mol } Al_2O_3 \times (-1676\text{ kJ/mol})] - [1\text{ mol } Cr_2O_3 \times (-1140\text{ kJ/mol})]$$

$$\Delta H^\circ_{rxn} = -536\text{ kJ}$$

$$\Delta S° = S^\circ_f(Al_2O_3) - S^\circ_f(Cr_2O_3)$$

$$\Delta S^o_{rxn} = [(2\text{ mol Cr} \times 23.8\text{ J/mol·K}) + (1\text{ mol } Al_2O_3 \times 50.9\text{ J/mol·K})] - [(2\text{ mol Al} \times 28.3\text{ J/mol·K}) + (1\text{ mol } Cr_2O_3 \times (-1058\text{ J/mol·K})]$$

$$\Delta S^\circ_{rxn} = 1099.9\text{ J/K}$$

To determine if the reaction is spontaneous, we need to calculate ΔG for the reaction.

$$\Delta G° = \Delta H° - T\Delta S°$$

$$\Delta G° = -536\text{ kJ} - (623\text{ K})(1.0999\text{ kJ/K}) = -1221\text{ kJ}$$

Since the value for $\Delta G°$ is negative, the reaction is spontaneous.

Workbook Problem 16.4

Given the following data, what is the normal boiling point of methanol?

$$CH_3OH\ (l) \rightarrow CH_3OH\ (g)$$

$S°\ (l)$	= 127 J/mol·K
$S°\ (g)$	= 238 J/mol·K
$\Delta H_f°\ (l)$	= -238.7 kJ/mol

ΔH_f° (*g*) = -201.2 kJ/mol

Strategy: Solve for ΔS and ΔH, solve for the temperature at which $\Delta G = 0$.

Step 1: Calculate ΔS.

Step 2: Calculate ΔH.

Step 3: Calculate T.

16.8 Standard Free–Energy Changes for Reactions

The **standard free–energy change**, ΔG°, is the change in free energy that occurs when reactants in their standard states are converted to products in their standard states. It is an extensive property, and refers to the number of moles in the equation.

Standard state is the pure form of a solid, liquid or gas, or a solute at 1 M, at 1 atm pressure and 25 °C.

ΔG° can be calculated from the standard enthalpy change, ΔH°, and the standard entropy change, ΔS° using the equation $\Delta G^\circ = \Delta H^\circ - T\Delta S^\circ$.

Workbook Problem 16.5

Determine ΔG° for the reaction:

$NaCl\ (aq) + AgNO_3\ (aq) \rightarrow NaNO_3\ (aq) + AgCl\ (s)$

given the following information:

	ΔH° (kJ/mol)	S° (J/mol·K)
AgCl (*s*)	–127.1	96.2
$AgNO_3$ (*aq*)	–101.8	219.1
$NaNO_3$ (*aq*)	–447.5	205.4
NaCl (*aq*)	–407.3	115.5

Strategy: From the information given, you can determine ΔG° from the equation $\Delta G^\circ = \Delta H^\circ - T\Delta S^\circ$.

Step 1: Determine ΔH°_{rxn}.

Step 2: Determine ΔS°_{rxn}.

Step 3: Calculate ΔG°_{rxn}.

16.9 Standard Free Energies of Formation

The **standard free energy of formation,** ΔG°_f, of a substance is the free–energy change for formation of 1 mole of the substance in its standard state from the most stable form of its constituent elements in their standard states. For an element in its most stable form at 25°C, $\Delta G^\circ_f = 0$.

If ΔG°_f is negative, the substance is stable and will not decompose back into its elements. If ΔG°_f is positive, the substance is unstable and prone to decompose. This says nothing about the rate of this decomposition, however.

ΔG°_f can be used to calculate the standard free–energy changes for reactions:

$$\Delta G^\circ = \Delta G^\circ_f(\text{products}) - \Delta G^\circ_f(\text{reactants})$$

EXAMPLE:

Calculate ΔG°_{rxn} for the following reaction:

$C_2H_4\,(g) + Cl_2\,(g) \rightarrow C_2H_3Cl\,(g) + HCl\,(g)$

given the following information:

	ΔG°_f (kJ/mol)
$C_2H_4\,(g)$	68.1
$Cl_2\,(g)$	0
$HCl\,(g)$	–95.3
C_2H_3Cl	51.9

SOLUTION:

$\Delta G^\circ_{rxn} = [51.9+(-95.3)] - [(68.1) + 0] = -111.5\text{ kJ}$

Workbook Problem 16.6

Determine ΔG°_{rxn} from ΔG°_f for the following reaction. Is the reaction spontaneous? If not, determine the temperature at which the reaction becomes spontaneous.

$CH_4\,(g) + 2\,Cl_2\,(g) \leftrightarrows CCl_4\,(l) + 2H_2\,(g)$

Strategy: Determine ΔG°_{rxn}, then calculate ΔH° and ΔS° to determine the temperature at which the reaction becomes spontaneous. Will the reaction be spontaneous at 400 °C? 45 °C? -20 °C?

	ΔG° (kJ/mol)	ΔH° (kJ/mol)	S° (J/mol·K)
$CH_4\,(g)$	-50.8	–74.8	186.2
$Cl_2\,(g)$	0	0	223.0
$CCl_4\,(l)$	-65.3	–135.4	216.4
$H_2\,(g)$	0	0	130.6

Step 1: Calculate ΔG°_{rxn}, determine if the reaction is spontaneous.

Step 2: Determine the temperature at which the reaction will be spontaneous.

Step 3: Determine the spontaneity at the temperatures listed.

16.10 Free–Energy Changes and Composition of the Reaction Mixture

When reactants and products are present at nonstandard state pressures and concentrations:

$$\Delta G = \Delta G^\circ + RT \ln Q$$

Where Q is the reaction quotient, and takes the same form as the equilibrium constant expression.

EXAMPLE:

Calculate ΔG for the formation of acetylene (C_2H_2) if the pressure of hydrogen is 415 atm, the pressure of acetylene is 0.1 atm, and the temperature is 25 °C.

SOLUTION:

The formula for this reaction is:

$$2\,C\,(s) + H_2\,(g) \rightarrow C_2H_4\,(g)$$

$$\Delta G^\circ = 209.2 \text{ kJ/mol}$$

$$\Delta G = 209,200 \frac{\text{J}}{\text{mol}} + 8.314 \frac{\text{J}}{\text{mol} \cdot \text{K}} (298 \text{ K}) \ln \frac{(0.1)}{(415)} = 188,600 \frac{\text{J}}{\text{mol}} = 189 \frac{\text{kJ}}{\text{mol}}$$

Workbook Problem 16.7

Consider the following reaction:

$$2\,NO\,(g) + O_2\,(g) \leftrightarrows 2NO_2\,(g)$$

Determine the spontaneity under standard conditions, and again under the following conditions: $P_{NO} = 0.150 \text{ atm}$, $P_{O_2} = 0.250 \text{ atm}$, $P_{NO_2} = 0.001 \text{ atm}$. Under which conditions is the reaction more spontaneous?

Strategy: Determine ΔG°_{rxn}, then calculate ΔG_{rxn} and determine which reaction is more spontaneous.

	ΔG° (kJ/mol)
$NO\,(g)$	86.6
$O_2\,(g)$	0
$NO_2\,(g)$	51.3

Step 1: Calculate ΔG°_{rxn}, determine if the reaction is spontaneous.

Step 2: Calculate ΔG.

16.11 Free Energy and Chemical Equilibrium

As the reaction proceeds toward equilibrium, the value of Q changes: when there are no products, $Q = 0$, and ln Q is infinite. As the reaction proceeds, the total free energy decreases. When the reaction is mostly products, Q >>, ln Q is greater than 0, and the reaction will proceed spontaneously in the reverse direction.

Free energy can be plotted against reaction progress: there will be a minimum between pure reactants and pure products: this minimum is the equilibrium mixture.

At equilibrium, $\Delta G° = 0$, and $Q = K$, allowing the following relationship to be derived:

$$\Delta G^\circ = -RT \ln K$$

EXAMPLE:

Calculate K for the reaction below at 25 °C:

$C_2H_6\,(g) + H_2\,(g) \rightarrow 2\,CH_4\,(g)$

ΔG° for CH_4 = -50.8 kJ/mol
ΔG° for C_2H_6 = -32.9 kJ/mol

SOLUTION:

Calculate ΔG°_{rxn}:

$$\Delta G^\circ_{rxn} = 2(-50.8) - (-32.9) = -68.7 \text{ kJ/mol}$$

Calculate K:

$$\ln K = \frac{\Delta G^\circ}{-RT} \frac{-68{,}700 \dfrac{\text{J}}{\text{mol}}}{-\left[8.314 \dfrac{\text{J}}{\text{mol}\cdot\text{K}}(298\text{ K})\right]} = 27.7$$

$K = 1.07 \times 10^{12}$

Workbook Problem 16.8

Consider the following reaction:

$$I_2\,(g) + Cl_2\,(g) \leftrightarrows 2ICl\,(g)$$

The equilibrium constant, K_p, for this reaction is 81.9 at 25 °C. What is the standard free-energy change for this reaction?

Strategy: Use the equation $\Delta G^\circ = -RT \ln K$

Step 1: Solve for the free energy:

Putting It Together

At 37 °C (normal body temperature), $\Delta G° = -14.0$ kJ for the reaction

$Hb\text{–}O_2\,(aq) + CO\,(g) \leftrightarrows Hb\text{–}CO\,(aq) + O_2\,(g)$

When a person is exposed to carbon monoxide gas, it binds preferentially to hemoglobin molecules – abbreviated Hb – in the blood, which normally operates as an oxygen carrier. The free-energy change above shows that this binding is favored, making even a small quantity of carbon monoxide dangerous. If the concentration of carbon monoxide is 10% of the concentration of oxygen, what percentage of hemoglobin molecules are bound to carbon monoxide?

Self–Test

This section is intended to test your knowledge of the material covered in this chapter. Think through these problems, and make certain you understand what they are asking. Make sure your answers make sense. Successful completion of these problems indicates that you have mastered the material in this chapter. You will receive the greatest benefit from this section if you use it as a mock exam, as this will allow you to determine which topics you need to study in more detail.

True–False

1. A spontaneous process occurs without any external energy and occurs quickly.
2. A reaction that is spontaneous and endothermic will have a large positive ΔS.
3. An increase in temperature always leads to an increase in entropy.
4. Standard entropies of reaction cannot be calculated without knowing the enthalpy of reaction.
5. A spontaneous reaction always increases the total entropy of the system plus the surroundings.
6. If ΔH is negative and ΔS is also negative, the reaction will be spontaneous at low temperatures.
7. The temperature of a reaction system determines the relative importance of the enthalpy and entropy terms.
8. The standard enthalpy of formation of a pure element is always negative.
9. ΔH determines whether a reaction is spontaneous or nonspontaneous.
10. The standard molar entropy of a pure substance is always positive.

Matching

Entropy	a. the total energy of a system and its surroundings is constant
First law of thermodynamics	b. the study of energy changes in chemical reactions
Free energy	c. the amount of molecular randomness in a system
Second law of thermodynamics	d. a process that proceeds on its own without any external influence
Spontaneous process	e. in any spontaneous process, the total entropy of a system and its surroundings always increases
Standard state	f. the capacity of a system to continue a chemical reaction.
Thermodynamics	g. the entropy of a perfectly ordered crystalline substance at 0 K is zero
Third law of thermodynamics	h. most stable form at 25 °C and 1 atm, 1 M for solutions

Fill–in–the–Blank

1. The expansion of a gas is a ____________________ process that __________________ the entropy of the gas.
2. When an ionic solid dissolves, the entropy of the solid ________________, but the entropy of the solvent _______________.
3. As the ________________ of a system increases, molecular __________________ and ______________ increase.
4. Standard molar entropies show that _____________ molecules have higher entropy than ________ molecules, _____________ have higher entropy than liquids, and liquids have higher entropy than ___________.
5. Reactions can be made spontaneous by the ____________ term, the ___________, or both. What matters is their relative contributions to ______ _________. The relative contributions of the two terms is determined by the _____________.
6. If a substance has a ________ free energy of formation, it is thermodynamically _____________. However, it may decompose imperceptibly slowly.
7. A reaction with a positive free energy will be spontaneous in the __________________ direction.

Problems

1. Which has the higher entropy?
   a. An orderly dishwasher, or a toy box
   b. Sugar crystals in a sugar bowl, or sweetened tea
   c. 32 g O_2 gas at STP or 1 mol of O_2 gas at 273 K in a volume of 35.5 L

2. Predict the sign of ΔS for the following processes or reactions:
   a. an increase in the volume of a gas at constant temperature
   b. formation of solid products from aqueous reactants
   c. $MgO\ (s) + 2\ NH_4Cl\ (s) \rightarrow 2\ NH_3\ (g) + MgCl_2\ (s) + H_2O\ (l)$
   d. $N_2\ (g) + H_2\ (g) \rightarrow N_2H_4\ (l)$
   e. separation of ${}^{235}_{92}U$ from a mixture of ${}^{235}_{92}U$ and ${}^{238}_{92}U$

3. Calculate the standard entropy of reaction for

   a. $2\ NH_4NO_3\ (s) \rightarrow 2\ N_2\ (g) + O_2\ (g) + 4\ H_2O\ (g)$

   $S°\ NH_4NO_3\ (s)$ = 151.1 J/(mol•K)
   $S°\ N_2\ (g)$ = 191.5 J/(mol•K)
   $S°\ O_2\ (g)$ = 205.0 J/(mol•K)
   $S°\ H_2O\ (g)$ = 188.7 J/(mol•K)

   b. $2\ S\ (s) + 3\ O_2\ (g) \rightarrow 2\ SO_3\ (g)$

   $S°\ S\ (s)$ = 31.8 J/(mol•K)
   $S°\ O_2\ (g)$ = 205.0 J/(mol•K)
   $S°\ SO_3\ (g)$ = 256.6 J/(mol•K)

4. For a given reaction at room temperature, if $\Delta S^\circ_{total} = 1.957 \times 10^3$ J/K and $\Delta S^\circ_{rxn} = 515.7$ J/K, calculate ΔH°_{rxn}.

5. Calculate ΔS if 1 mol of hydrogen gas at STP expands to a volume of 45.3 L at 273 K and 1 atm of pressure.

6. Determine the normal boiling point of ethanol given the following information:

   $S°\ (l)$ = 161 J/mol·K
   $S°\ (g)$ = 282.6 J/mol·K
   $\Delta H_f^\circ\ (l)$ = -277.7 kJ/mol
   $\Delta H_f^\circ\ (g)$ = -235.1 kJ/mol

7. Calculate ΔS_{total}, ΔS_{sys}, and ΔS_{surr} for the following reaction at 25 °C:

   $H_2\ (g) + Cl_2\ (g) \rightarrow 2\ HCl\ (g)$

$S°\ H_2\ (g)$	= 130.6 J/(mol•K)	$\Delta H_f^\circ\ H_2\ (g)$	= 0 kJ/mol
$S°\ Cl_2\ (g)$	= 223.0 J/(mol•K)	$\Delta H_f^\circ\ Cl_2\ (g)$	= 0 kJ/mol
$S°HCl\ (g)$	= 186.8 J/(mol•K)	$\Delta H_f^\circ\ HCl\ (g)$	= -92.3 kJ/mol

8. Calculate the standard free energy of reaction from the standard free energies of formation for the following reaction:

$C_2H_4\,(g) + 3\,O_2\,(g) \rightarrow 2\,CO_2\,(g) + 2\,H_2O\,(l)$

$\Delta G_f^\circ\ C_2H_4\,(g)$ = 68.1 kJ/mol
$\Delta G_f^\circ\ O_2\,(g)$ = 0 kJ/mol
$\Delta G_f^\circ\ CO_2\,(g)$ = -394.4 kJ/mol
$\Delta G_f^\circ\ H_2O\,(l)$ = -237.2 kJ/mol

9. Calculate the free–energy change for the following reaction if the partial pressures are 0.5 atm for CO, 15 atm for CO_2, and 0.1 atm for O_2.

$2\,CO\,(g) + O_2\,(g) \leftrightarrows 2\,CO_2(g)$

$\Delta G_f^\circ\ CO\,(g)$ = -137.2 kJ/mol
$\Delta G_f^\circ\ O_2\,(g)$ = 0 kJ/mol
$\Delta G_f^\circ\ CO_2\,(g)$ = -394.4 kJ/mol

10. Determine the equilibrium constant for the following reaction at 37 °C.

$H_2CO_3\,(aq) \leftrightarrows CO_2\,(aq) + H_2O\,(l)$

$\Delta G_f^\circ\ H_2CO_3\,(aq)$)= -623 kJ/mol
$\Delta G_f^\circ\ CO_2\,(aq)$ = -386.0 kJ/mol
$\Delta G_f^\circ\ H_2O\,(l)$ = -237.2 kJ/mol

11. Alkali metals are refined at high temperatures. What is ΔG_{rxn} for this process? At what temperature does it become spontaneous? What is the equilibrium constant for the reaction at this temperature?

$2\,NaCl\,(s) \leftrightarrows 2\,Na\,(s) + Cl_2\,(g)$

	ΔG° (kJ/mol)	ΔH° (kJ/mol)	S° (J/mol·K)
$NaCl\,(s)$	-384.2	-411.2	72.1
$Cl_2\,(g)$	0	0	223.0
$Na\,(s)$	0	0	51.2

Challenge Problem

Determine the value of K at 25 °C for the reaction

$N_2\,(g) + O_2\,(g) \leftrightarrows 2\,NO\,(g)$

	ΔG° (kJ/mol)	ΔH° (kJ/mol)	S° (J/mol·K)
$N_2\,(g)$	0	0	191.5
$O_2\,(g)$	0	0	205.0
$NO\,(g)$	86.6	90.2	210.7

CHAPTER 17

ELECTROCHEMISTRY

Chapter Learning Goals

A. Galvanic Cells: Spontaneous Oxidation–Reduction Reactions

1. Sketch a galvanic cell, identifying the anode and cathode half–reactions, the sign of each electrode, and the direction of electron and ion flow.
2. Write balanced chemical equations for reactions occurring in galvanic cells.
3. Write and interpret shorthand notation for galvanic cells.
4. Write balanced chemical equations for reactions occurring in common batteries.
5. Describe the reactions that occur when iron rusts.

B. Galvanic Cells: Cell Potentials

1. Use a table of standard reduction potentials to calculate standard cell potentials.
2. Use a table of standard reduction potentials to rank substances in order of increasing oxidizing strength or reducing strength and to determine whether a reaction is spontaneous.
3. Use the Nernst equation to calculate cell potentials for reactions occurring under non-standard–state conditions.
4. From a measured cell potential for a reaction involving hydrogen ion and a reference cell potential, calculate the pH of the solution.

C. Galvanic Cells: Free–Energy Changes

1. Interconvert cell potential and free–energy change for a reaction.
2. Calculate equilibrium constants from standard cell potentials and vice versa.

D. Electrolytic Cells

1. Describe half–cell and overall reactions occurring in electrolytic processes.
2. Perform electrolytic cell calculations interconverting current and time, charge, moles of electrons, and moles (or grams) of product.

Chapter in Brief

Electrochemistry is the area of chemistry concerned with the interconversion of chemical and electrical energy. Galvanic cells convert chemical energy to electrical energy. The standard cell potential of these reactions can be calculated from tables of standard half-reactions, and potentials at nonstandard-state conditions can be calculated using the Nernst equation. These potentials can also be used to calculate the free-energy changes and equilibrium constants of these reactions. The most common applications of these reactions are in batteries and fuel cells, with the same processes causing the much less-productive process of corrosion. When the process is reversed, electrical energy can be used to drive nonspontaneous chemical changes. This process, known as electrolysis, is used to produce sodium, chlorine, sodium hydroxide and is also used in electrorefining and electroplating. The products obtained during electrolysis depend on the overvoltage required to drive the processes, not only on the standard voltages of the reactions involved. Quantitative calculations involving electrolytic processes require that the amount of current, and thus the number of electrons that flow into the reaction, be known.

17.1 Galvanic Cells

In galvanic cells, a spontaneous chemical reaction generates an electric current. In electrolytic cells, an electric current drives a nonspontaneous reaction.

In redox reactions, recall that oxidation is a loss of electrons (an increase in oxidation number), and reduction is a gain of electrons (a decrease in oxidation number).

To make it easier to see what is happening in these types of reactions, they are typically broken into half-reactions: one for the oxidation portion (with electrons as products) and another for the reduction portion (with electrons as reactants). The species that causes the oxidation and accepts the electrons is referred to as the oxidizing agent, while the species that gives up the electrons is referred to as the reducing agent.

These reactions can be carried out directly, as when a strip of zinc is placed in a beaker that contains copper ions –in this case, the copper is reduced by taking electrons directly off the zinc metal, while the zinc is oxidized by giving electrons to the copper ions. Any enthalpy of reaction is lost into the solution.

Redox reactions can also be separated and run in a galvanic or Daniell cell. In this situation, the oxidation and reduction portions of the reaction are run separately: the metals are immersed in a solution that contains a common anion, and connected by a wire, through which electrons flow from the reducing agent to the oxidizing agent.

Specific names are given to the two metals, which are known as **electrodes**. The **anode** is the electrode at which oxidation takes place. It is defined as the negative (-) electrode and is the electrode where electrons are produced. The cathode is the electrode at which reduction takes place. It is defined as the positive (+) electrode, and it consumes electrons.

The solutions are connected by a salt bridge, a U-shaped tube that contains a gel permeated with an electrolyte that does not participate in the reaction. Neutrality of the solutions is maintained by the electrolyte in the salt bridge. Anions move toward the anode, and cations move toward the cathode.

The half-reaction at the anode and the half-reaction at the cathode must add together to give the overall reaction.

EXAMPLE:

Describe the construction of a galvanic cell based on the following reaction:

$$Ni^{2+}\,(aq) + Mg\,(s) \rightarrow Ni\,(s) + Mg^{2+}\,(aq)$$

SOLUTION: Start by taking the overall cell reaction and breaking it into two half–reactions.

$$Ni^{2+}\,(aq) + 2\,e^{-} \rightarrow Ni\,(s)$$

$$Mg\,(s) \rightarrow Mg^{2+}\,(aq) + 2\,e^{-}$$

Looking at the two half-reactions, we find that the Ni^{2+} is being reduced, and the Mg is being oxidized. Therefore, the anode compartment of our cell would consist of a strip of magnesium metal immersed in a solution containing Mg^{2+} ions (such as magnesium nitrate). The cathode compartment would consist of a strip of nickel immersed in a solution containing Ni^{2+} ions (such as nickel(II) nitrate). The two half–cells would be connected to each other with a salt bridge and an external wire. Electrons flow through the wire from the magnesium anode to the nickel cathode. Anions move from the cathode compartment toward the anode, while cations migrate from the anode compartment toward the cathode.

Workbook Problem 17.1
Describe the construction of a galvanic cell based on the following reaction:

$$Pb^{2+}(aq) + Cd(s) \rightarrow Pb(s) + Cd^{2+}(aq)$$

Strategy: Break the overall reaction into half-reactions and design a galvanic cell, identifying the cathode and the anode.

Step 1: Write the half–reactions, identifying the oxidizing agent and the reducing agent.

Step 2: Identify solutions for each half-cell, and a suitable inert electrolyte for the salt bridge.

17.2 Shorthand Notation for Galvanic Cells

To make it easier to see what is going on in a galvanic cell, a shorthand notation is used. For the example above, the equation

$$Ni^{2+}(aq) + Mg(s) \rightarrow Ni(s) + Mg^{2+}(aq)$$

would be written:

$$Mg \mid Mg^{2+}(aq) \parallel Ni^{2+}(aq) \mid Ni$$

In this shorthand, the single vertical line, |, represents a phase boundary – the solid electrode and the aqueous solution – while the double vertical line, ||, represents the salt bridge.

The anode half-cell (oxidation) is always written on the left, and the electrons flow to the right into the cathode half cell. Reactants are written first, followed by products.

If a gas is involved (as when hydrogen is reduced, or chlorine oxidized), an additional line will be present to identify the electrode:

$$Zn \mid Zn^{2+}(aq) \parallel H^{+}(aq) \mid H_2(g) \mid Pt(s)$$

A very detailed version of the shorthand notation also includes ion concentrations and gas pressures.

EXAMPLE:
Give the shorthand notation for a galvanic cell that employs the overall reaction

$$Ni(NO_3)_2(aq) + Cd(s) \rightarrow Ni(s) + Cd(NO_3)_2(aq)$$

Give a brief description of the cell.

SOLUTION: The two half–reactions for this overall reaction are

Ni^{2+} (*aq*) + 2 e^- → Ni (*s*)

Cd (*s*) → Cd^{2+} (*aq*) + 2 e^-

From these half–reactions, we know that nickel is being reduced and cadmium is being oxidized. Therefore, Cd is the anode and Ni is the cathode. The cell notation is

Cd (*s*) | Cd^{2+} (*aq*) || Ni^{2+} (*aq*) | Ni (*s*)

This cell would consist of a strip of cadmium metal as the anode dipping into an aqueous solution of $Cd(NO_3)_2$ and a strip of nickel metal as the cathode dipping into an aqueous solution of $Ni(NO_3)_2$. The two half–cells would be connected by a salt bridge and a wire.

Workbook Problem 17.2

Given the shorthand notation below, determine the half–reactions occurring at the anode and cathode. Write the overall reaction for the cell and give a brief description of the cell's construction.

Mg (*s*) | Mg^{2+} (*aq*) | | Cr^{3+} (*aq*) | Cr (*s*)

Strategy: Based on the shorthand notation above, determine which species is the cathode and which is the anode, and write the half–reaction that occurs at each electrode.

Step 1: Write the half–reaction occurring at the anode. Remember, the reactant in each half–cell is written first, followed by the product.

Step 2: Write the half–reaction occurring at the cathode.

Step 3: Combine the two half–reactions to obtain the overall cell reaction (recall that the equation must be balanced in electrons).

Step 4: Describe the cell indicated in the shorthand notation.

17.3 Cell Potentials and Free–Energy Changes for Cell Reactions

The chemical potential that pushes electrons away from the anode and toward the cathode is known as the **electromotive force**, the **cell potential (*E*)** or the **cell voltage**. The potential of a galvanic cell is defined as a positive quantity with units of volts (V).

The following relationships are found among the units of energy, electrical charge, and current:

$$1 \text{ watt} = \frac{1 \text{ J}}{\text{s}} = \frac{1 \text{ coulomb x } 1 \text{ volt}}{\text{s}} = 1 \text{ ampere x } 1 \text{ volt}$$

Watts are a measure of power, coulombs are a measure of charge, volts are a measure of potential, and amperes are a measure of current.

Cell potential is measured with a voltmeter: when the + and - terminals of the voltmeter are connected to cathode (+) and anode (-), a positive value will be obtained for the voltage. This makes it possible to use a voltmeter to determine which electrode is the cathode and which is the anode.

Cell potential and free energy are both driving forces for chemical reactions. They are related by the following equation:

$$\Delta G = -nFE$$

In this equation, n = moles of electrons transferred in the reaction and F is the **faraday** or Faraday constant. The faraday is the electrical charge per mole of electrons, and is equal to 96,500 C/mol e^-.

Note that ΔG and E have opposite signs: a spontaneous reaction is indicated by a negative free-energy change or a positive cell potential.

Standard cell potential, $E°$, is the cell potential when both reactants and products are in their standard states: solutes at 1 M concentration, gases at a partial pressure of 1 atm, solids and liquids in pure form, at 25 °C.

EXAMPLE:

If the standard cell potential for the following reaction is 2.37 V, what is the standard free-energy change? Write the shorthand notation for this reaction.

$$Al\,(s) + Fe^{3+}\,(aq) \rightarrow Al^{3+}\,(aq) + Fe\,(s)$$

SOLUTION: Use the equation $\Delta G° = -nFE°$

F is 96,500 C/mol, and n can be inferred from the balanced equation – three electrons are being transferred per mole of reactants, so $n = 3$.

$$\Delta G° = -3 \text{ x } 96{,}500 \frac{C}{mol} \text{ x } 2.37 \text{ V x } 1 \frac{J}{C \cdot V} = -686 \text{ kJ/mol}$$

The shorthand notation for this reaction is $Al\,(s) | Al^{3+}\,(aq) \,||\, Fe^{3+}\,(aq) \,|\, Fe\,(s)$

Workbook Problem 17.3

The reaction shown below in shorthand notation has a standard cell potential of 1.61 V. Write the equation for this reaction and calculate the standard free-energy change.

$$Mg\,(s)\,|\,Mg^{2+}\,(aq)\,||\,Zn^{2+}\,(aq)\,|\,Zn\,(s)$$

Strategy: Based on the shorthand notation above, write the equation. Using the formula $\Delta G° = -nFE°$, calculate the standard free-energy change.

Step 1: Write the balanced equation for the reaction:

Step 2: Identify n and use $\Delta G° = -nFE°$ to calculate $\Delta G°$.

17.4 Standard Reduction Potentials

The standard potential of a galvanic cell is the sum of the half-cell potentials:

$$E^\circ_{\text{cell}} = E^\circ_{\text{ox}} + E^\circ_{\text{red}}$$

Only a potential *difference* can be measured, so to determine half-reaction potentials, a standard electrode must be used against which all other reactions can be measured.

The standard hydrogen electrode (S.H.E.) is assigned an arbitrary potential of exactly 0 V, and all other voltages are relative to this reaction:

$$2\,H^+\,(aq,\,1\,M) + 2\,e^- \rightarrow H_2\,(g,\,1\,\text{atm}) \qquad E^\circ = 0.00\,V$$

The shorthand notation for S. H. E. is H^+ (1 M) | H_2 (1 atm) | Pt (*s*), where the Pt electrode is in contact with both H_2 gas and H^+ (*aq*).

Standard potentials for half–cells can be determined by constructing a galvanic cell in which the half–cell of interest is paired up with the standard hydrogen electrode: standard oxidation potential is the half–cell potential for an oxidation half–reaction, and standard reduction potential is the half–cell potential for a reduction half–reaction.

As with free-energy changes, when the direction of a half–reaction is reversed, the sign of E^0 must be reversed.

In a table of half-cell reaction, all the reactions are written as reductions. Oxidizing agents are on the reactant side, and reducing agents are on the product side. Reactions are listed in decreasing order of standard reduction potential, with the strongest oxidizing agents at the upper left and the strongest reducing agents at the lower right.

17.5 Using Standard Reduction Potentials

Tables of standard reduction potentials summarize an enormous quantity of chemical information in a small space. They make it possible to arrange oxidizing or reducing agents in order of strength, and predict the spontaneity of thousands of redox reactions.

To balance the equations in electrons, it may be necessary to place coefficients on the half reactions. What this means is that it may be necessary to multiply half–reactions by some factor to ensure that electrons cancel. In this case, the values of E° are NOT multiplied by the same factor: the voltage and free energy depend on the number of electrons transferred, not on the stoichiometry of the atoms.

An oxidizing agent can oxidize any reducing agent that lies below it in the table. For the reaction to be spontaneous, E° must be positive.

EXAMPLE:

Write the balanced net ionic equation, and calculate E° for the following galvanic cell:

Al (*s*) | Al^{3+} (*aq*) || Cl_2 (*g*) | Cl^- (*aq*)

SOLUTION: Al (*s*) is the anode and therefore undergoes oxidation, while Cl_2 (*s*) is the cathode and undergoes reduction. The half–reactions and their cell potentials are

$Al\ (s) \rightarrow Al^{3+}\ (aq) + 3\ e^-$ $E° = +1.66$ V

$Cl_2\ (g) + 2\ e^- \rightarrow Cl^-\ (aq)$ $E° = +1.36$ V

(Notice that the sign for $E°$ for the Al/Al^{3+} half-reactions has been reversed.) To write the balanced net ionic equation, we need to make sure that the electrons cancel out on both sides. Therefore, we need to multiply the top reaction by 2 and the bottom reaction by 3.

$2\ Al\ (s) \rightarrow 2\ Al^{3+}\ (aq) + 6\ e^-$ $E° = +1.66$ V

$3\ Cl_2\ (aq) + 6\ e^- \rightarrow 6\ Cl^-\ (aq)$ $E° = +1.36$ V

Remember that $E°$ values are independent of the amount of reaction and are not multiplied by stoichiometric coefficients.

$$E^{o}_{cell} = E^{o}_{Al \rightarrow Al^{3+}} + E^{o}_{Cl_2 \rightarrow Cl^-} = +1.66\text{ V} + 1.36\text{ V} = 3.02\text{ V}$$

Workbook Problem 17.4

Using the table of standard reduction potentials and without calculating the cell potential, determine if the following reactions are spontaneous.

$Cu\ (s) + 2\ Ag^+\ (aq) \rightarrow 2\ Ag\ (aq) + Cu^{2+}\ (aq)$

$2\ MnO_4^-\ (aq) + 16\ H^+\ (aq) + 10\ F^-\ (aq) \rightarrow 2\ Mn^{2+}\ (aq) + 8\ H_2O\ (l) + 5\ F_2\ (g)$

Strategy: Determine the position of the reducing agent relative to the position of the oxidizing agent, and based on those results, determine the spontaneity of the reactions.

17.6 Cell Potentials and Composition of the Reaction Mixture: The Nernst Equation

Like free-energy changes, cell potentials depend on temperature and on the composition of the reaction mixture. From $\Delta G = \Delta G° + RT \ln Q$, the following relationship can be derived:

$$-nFE = -nFE° + RT \ln Q$$

Dividing by $-nF$ yields the Nernst Equation:

$$E = E° - \frac{0.0592}{n} \log Q \text{ (in volts at 25 °C)}$$

This equation makes it possible to calculate cell potentials under nonstandard–state conditions.

EXAMPLE:

Calculate E_{cell} for the following reaction:

$2\ Fe^{3+}\ (aq) + Pb\ (s) \rightarrow 2\ Fe^+\ (aq) + Pb^{2+}\ (aq)$

$[Pb^{2+}] = 0.15$ M $[Fe^{3+}] = 0.50$ M $[Fe^+] = 0.10$ M

SOLUTION: The half–reactions for this equation are

$Pb\ (s) \rightarrow Pb^{2+} + 2\ e^-$ $E° = +0.13\ V$

$Fe^{3+}\ (aq) + 2\ e^- \rightarrow Fe^+\ (aq)$ $E° = 0.77\ V$

$E°_{cell} = 0.13\ V + 0.77\ V = 0.90\ V$

The Nernst equation for this reaction is:

$$E_{cell} = E^{o}_{cell} - \frac{0.0592}{2}\log\frac{[Pb^{2+}][Fe^{+}]^2}{[Fe^{3+}]^2}$$

Substituting the information given and solving for E_{cell} gives:

$$E_{cell} = +0.90 - \frac{0.0592}{2}\log\frac{(0.15)(0.10)^2}{(0.50)^2} = 0.97\ V$$

Workbook Problem 17.5

If $E_{cell} = 0.17$ V for the following reaction, what is the concentration of $[Sn^{4+}]$ ions? $[Cu^{2+}] = 0.63$ M, $[Sn^{2+}] = 0.72$ M.

$$Sn^{2+}\ (aq) + Cu^{2+}\ (aq) \rightarrow Sn^{4+}\ (aq) + Cu\ (s)$$

Strategy: Calculate the E°_{cell} for the cell reaction, substitute the information given into the Nernst equation, and solve for $[Sn^{4+}]$.

Step 1: Calculate $E°_{cell}$:

Step 2: Using $E°_{cell}$ and E, solve for $[Sn^{4+}]$:

17.7 Electrochemical Determination of pH

Using the Nernst equation, it is possible to derive an equation for the determination of pH.

When the following reaction is considered with a hydrogen electrode in a solution of unknown pH:

$$Pt\ (s)\ |\ H_2\ (1\ atm)\ |\ H^+\ (?\ M)\ ||\ \text{reference cathode}$$

The difference between E_{cell} and E_{ref} can be directly related to pH, by the following derived equation:

$$pH = \frac{E_{cell} - E_{ref}}{0.0592}$$

As a result of this equation, the pH of a solution can be measured by measuring E_{cell}. The awkwardness of the hydrogen electrode has resulted in its substitution with a glass electrode consisting of a silver wire coated with silver chloride immersed in a reference solution of dilute hydrochloric acid. The reference electrode is known as a *calomel* electrode: Hg_2Cl_2 in contact with liquid mercury and aqueous KCl.

EXAMPLE:

The following cell has a potential of 0.65 V. Calculate the pH of the solution in the anode compartment.

$$\text{Pt}(s) \mid \text{H}_2(g)\,(1\text{ atm}) \mid \text{H}^+\,(\text{pH} = ?) \,||\, \text{Cl}^-(aq)\,(1\text{ M}) \mid \text{Hg}_2\text{Cl}_2(s) \mid \text{Hg}(l)$$

SOLUTION: The cell reaction is

$$\text{Hg}_2\text{Cl}_2(s) + \text{H}_2(g) \rightarrow 2\text{ Hg}(l) + 2\text{ Cl}^-(aq) + 2\text{ H}^+(aq)$$

$E°$ for the calomel electrode is 0.28:

$$E^\circ = E^\circ_{\text{H}_2\rightarrow\text{H}^+} + E^\circ_{\text{Hg}_2\text{Cl}_2\rightarrow\text{Hg,Cl}^-} = 0.00\text{ V} + 0.28\text{ V} = 0.28\text{ V}$$

The pH can be calculated using the equation

$$\text{pH} = \frac{E_{\text{cell}} - E_{\text{ref}}}{0.0592}$$

$$\text{pH} = \frac{0.65\text{ V} - 0.28\text{ V}}{0.0592} = 6.25$$

Workbook Problem 17.6

If $E_{cell} = 0.62$ V for the following reaction, what is the pH in the anode compartment?

$$\text{Cu}^{2+}(aq) + \text{H}_2(g) \rightarrow 2\text{ H}^+(aq) + \text{Cu}(s)$$

Strategy: Determine the half-reactions occurring and calculate E_{ref}. Substitute into the modified Nernst equation to determine pH of the solution ($Cu^{2+}(aq) + 2e^- \rightarrow Cu(s)$ $E° = 0.34$).

Step 1: Determine E_{ref}:

Step 2: Calculate the pH of the anode solution:

17.8 Standard Cell Potentials and Equilibrium Constants

The standard cell potential is related to the standard free-energy change for the reaction. These are described by the following reactions:

$$\Delta G^\circ = -nFE^\circ$$

and

$$\Delta G^\circ = -RT\ln K$$

These equations can be set equal to each other and simplified:

$$E^\circ = \frac{RT}{nF}\ln K = \frac{2.303RT}{nF}\log K$$

$$E^\circ = \frac{0.0592}{n}\log K$$

or

$$\log K = E^\circ \frac{n}{0.0592}$$

This makes it possible to calculate equilibrium constants from cell potentials. This is convenient, as potentials are easily measured, but small concentrations are not as easily measured.

Compared with the equilibrium constants for acid-base reactions, those for redox reactions tend to be either very large or very small. In other words, redox reactions either go to completion or do not proceed at all. A positive value of $E°$ corresponds to $K > 1$, while a negative value of $E°$ corresponds to $K < 1$.

With the addition of this method, there are three different ways to determine the value of an equilibrium constant K.

K from concentration data: $K = \frac{[C]^c[D]^d}{[A]^a[B]^b}$.

K from thermochemical data: $\ln K = \frac{-\Delta G^\circ}{RT}$.

K from electrochemical data: $\ln K = \frac{nFE^\circ}{RT}$.

EXAMPLE:

Calculate the equilibrium constant for the following reaction at 25 °C.

$2\ Fe^{3+}\ (aq) + 3\ Pb\ (s) \rightarrow 2\ Fe\ (s) + 3\ Pb^{2+}\ (aq)$

SOLUTION: The half–reactions for this reaction are:

$Fe^{3+}\ (aq) + 2\ e^- \rightarrow Fe\ (s)$ $\quad E° = 0.32$ V

$Pb\ (s) \rightarrow Pb^{2+}\ (aq) + 2\ e^-$ $\quad E° = 0.13$ V

$E^\circ_{cell} = 0.32 + 0.13 = +0.45$ V

The value of n for this reaction is 6. We can now solve for K.

$\log K = \frac{(6)(0.45)}{(0.0592)} = 45.6$ $\quad K = 10^{45.6} = 3.98 \times 10^{45}$

Workbook Problem 17.7

Find K for the reaction:

$$H_2O_2\,(l) + 2\,H^+\,(aq) + 2\,Br^-\,(aq) \rightarrow 2\,H_2O\,(l) + Br_2\,(l)$$

Strategy: Determine the half-reactions occurring and calculate $E°$ from the standard potentials of the half-cell reactions. Calculate K from the equation $\log K = E°\dfrac{n}{0.0592}$.

Step 1: Determine $E°$:

Step 2: Calculate K from $E°$:

17.9 Batteries

Galvanic cells can be linked in series, with the final voltage the sum of all the individual voltages. Battery features depend on the application, but in general, batteries should be rugged, compact, lightweight, inexpensive, and provide stable power for a relatively long time.

Lead storage batteries have been generating power for automobiles for nearly a century. Six 2 V cells are connected to generate 12 V of power. Inside the battery are a series of grids packed with spongy lead, while the cathode is a series of grids packed with lead dioxide. These are dipped in a 38% solution of sulfuric acid.

Anode: $Pb\,(s) + HSO_4^-\,(aq) \rightarrow PbSO_4\,(s) + H^+\,(aq) + 2\,e^-$ $\quad E° = 0.296$ V

Cathode: $PbO_2\,(s) + 3\,H^+\,(aq) + HSO_4^-\,(aq) + 2\,e^- \rightarrow PbSO_4\,(s) + 2\,H_2O\,(l)$ $\quad E° = 1.628$ V

Overall: $Pb\,(s) + PbO_2\,(s) + 2\,H^+\,(aq) + 2\,HSO_4^-\,(aq) \rightarrow 2\,PbSO_4\,(s) + 2\,H_2O\,(l)$ $\quad E° = 1.924$ V

As the reaction proceeds, $PbSO_4$ adheres to the surface of the electrodes. By running power back through the battery, the lead sulfate is driven off the surfaces. Eventually, this is no longer possible, and the battery must be recycled.

Dry–cell batteries are common household batteries. They originated with Leclanché cells, which were patented in 1866. Leclanché cells consist of a zinc metal can (the anode), an inert graphite rod surrounded by a paste of solid MnO_2 and carbon black (cathode) and a paste of NH_4Cl and $ZnCl_2$ in starch (electrolyte).

Leclanché cells have been displaced by alkaline dry cells, in which the acidic NH_4Cl solution (which corrodes the zinc) is replaced by NaOH or KOH. This slows the corrosion, which increases battery life, power, and stability.

Anode: $Zn\,(s) + 2\,OH^-\,(aq) \rightarrow Zn(OH)_2\,(aq) + 2\,e^-$

Cathode: $2\,MnO_2\,(s) + H_2O\,(l) + 2\,e^- \rightarrow Mn_2O_3\,(s) + 2\,OH^-\,(aq)$

Nickel–Cadmium batteries or "ni-cad" batteries are rechargeable: like lead storage batteries, the solid products of the reactions adhere to the surface of the electrodes. The anode is cadmium, and cathode is NiO(OH).

Anode: $Cd\ (s) + 2\ OH^-\ (aq) \rightarrow Cd(OH)_2\ (s) + 2\ e^-$

Cathode: $NiO(OH)\ (s) + H_2O\ (l) + e^- \rightarrow Ni(OH)_2\ (s) + OH^-\ (aq)$

Nickel-metal hydride batteries or "NiMH" batteries have replaced ni-cad batteries in many applications because cadmium is expensive and toxic. These are more environmentally friendly, have the same voltage as ni-cad batteries, but have twice the energy density: half as much battery mass is required for the same voltage. The cathode is the same as in ni-cad batteries, but the anode is a special metal alloy that can absorb and release large amounts of H_2 at ordinary temperature:

Anode: $MH_{ab}\ (s) + OH^-\ (aq) \rightarrow M\ (s) + H_2O\ (l) + e^-$

The overall reaction transfers hydrogen from the anode to the cathode.

These batteries are used in consumer products and also in hybrid automobiles, where they supply energy to the electric motor, then are recharged when the brakes are applied, using some of the kinetic energy of the vehicle.

Lithium and lithium-ion batteries are lightweight, high–voltage, and rechargeable. They are used in cell phones, laptops, cameras, tools, and Tesla electric cars. With an atomic weight of 6.97 g/mol, less than 7 g of lithium can provide 1 mol of electrons. In addition, it has the highest standard oxidation potential of any metal.

The anode is lithium metal, the cathode is typically MnO_2, and the electrolyte a lithium salt such as $LiClO_4$ in an organic solvent.

Anode: $x\ Li\ (s) \rightarrow x\ Li^+\ (soln) + x\ e^-$

Cathode: $MnO_2\ (s) + Li^+\ (soln) + x\ e^- \rightarrow Li_xMnO_2\ (s)$

Lithium ion batteries contain a lithiated graphite anode. Abbreviated Li_xC_6, this is graphite with lithium atoms inserted between the layers. The cathode is CoO_2, which is also able to incorporate lithium ions into its structure. The electrolyte is a solution of lithium salt in an organic solvent or a solid-state polymer that can transport lithium ions.

Anode: $Li_xC_6\ (s) \rightarrow x\ Li^+\ (soln) + 6\ C\ (s) + x\ e^-$

Cathode: $Li_{1-x}CoO_2\ (s) + x\ Li^+\ (soln) + x\ e^- \rightarrow LiCoO_2\ (s)$

There are safety concerns about lithium ion batteries that are being addressed by the development of alternative electrode materials.

17.10 Fuel cells

Fuel cells are galvanic cells in which one of the reactants is a fuel such as hydrogen or methanol. Hydrocarbon fuels must first be converted to hydrogen, and the reactants are supplied from an external reservoir.

Hydrogen-oxygen fuel cells are used in space vehicles as a source of electric power. They contain porous carbon electrodes impregnated with metallic catalysts in a hot, aqueous KOH solution. The oxygen and hydrogen flow into separate compartments: the H_2 is oxidized at the anode, while the O_2 is reduced at cathode. The overall cell reaction is the conversion of hydrogen and oxygen to water.

Proton-exchange membrane (PEM) fuel cells are used to power "green" electric vehicles. The electrolyte is a plastic membrane that conducts protons but not electrons. Protons pass through the membrane from the anode to the cathode, while electrons move through the external circuit from the anode to the cathode. The only reaction product is water.

As of 2011, commercialization of fuel-cell vehicles will require improvements in performance and reduction in cost for the membranes, development of less expensive catalysts for the electrodes, safer methods of onboard hydrogen storage, and hydrogen-delivery infrastructure.

Direct methanol fuel cells (DMFC) are similar to PEM fuel cells, but use the more available and safer aqueous methanol as their fuel. They may find use in small consumer electronics, or in power generators. They are lighter than conventional batteries, but are not as environmentally friendly as they produce CO_2:

Anode: $2\ CH_3OH\ (aq) + 2\ H_2O\ (l) \rightarrow 2\ CO_2\ (g) + 12\ H^+\ (aq) + 12\ e^-$

Cathode: $3\ O_2\ (g) + 12\ H^+\ (aq) + 12\ e^- \rightarrow 6\ H_2O\ (l)$

Overall: $2\ CH_3OH\ (aq) + 3\ O_2\ (g) \rightarrow 2\ CO_2\ (g) + 4\ H_2O\ (l)$

17.11 Corrosion

Corrosion is defined as the oxidative deterioration of a metal. The conversion of iron to rust provides a well-known example. It requires both oxygen and water, and involves pitting of the metal surface and deposition of hydrated iron(III) oxide at a location separate from the pit.

A possible mechanism involves an electrochemical process in which iron is oxidized in one region of the surface while oxygen is reduced in another region. Step 1 is:

Anode region: $Fe\ (s) \rightarrow Fe^{2+}\ (aq) + 2\ e^-$ $\quad E° = 0.45$ V
Cathode region: $O_2\ (g) + 4\ H^+\ (aq) + 4\ e^- \rightarrow 2\ H_2O\ (l)$ $\quad E° = 1.23$ V

Soluble Fe^{2+} ions migrate through the water droplets and encounter dissolved oxygen, leading to the further oxidation of Fe^{2+} to Fe^{3+}. Fe^{3+} then reacts with water in the final step, producing $Fe_2O_3 \cdot H_2O$ - rust.

This mechanism also explains the increased corrosion of cars in the presence of road salt – salt in the water greatly increases its conductivity, allowing the reaction to proceed more quickly.

O_2 is able to oxidize all metals except a few, as can be seen by the fact that the O_2/H_2O half-reaction lies above the M^{n+}/M half–reaction for all but a very few, such as gold. In the case of many metals, including aluminum, magnesium, chromium, titanium, and zinc the oxidation forms a hard coating that prevents further oxidation.

Corrosion can be prevented by protecting the surface from oxygen and water. This can be done with paint, but is more effective with another metal. Coating steel with molten zinc is known as **galvanizing**, and provides protection even if the zinc layer is scratched, because the zinc is preferentially oxidized, and will reduce any oxidized iron.

Cathodic protection, a similar method uses a *sacrificial anode*: a metal that is more easily oxidized need only be connected to the steel or iron by an electrical connection. This is the method used to protect ships and large buildings from corrosion.

17.12 Electrolysis and Electrolytic Cells

In a galvanic cell, a spontaneous reaction generates a current. In an electrolytic cell, a current is used to drive a nonspontaneous reaction. The process of using an electric current to bring about chemical change is called **electrolysis**.

An electrolytic cell has two electrodes connected to a source of current and dipped into an electrolyte. The battery acts as an electron pump, removing electrons from the anode, moving electrons through the electrolyte, and pulling electrons into the cathode.

Electrolysis of molted NaCl: the negative electrode (cathode) attracts Na^+, which is reduced to Na (*l*). The positive electrode (anode) attracts Cl^-, which is oxidized to Cl_2 (*g*).

Anode: $2\,Cl^-(l) \rightarrow Cl_2(g) + e^-$

Cathode: $Na^+(l) + e^- \rightarrow Na(l)$

Electrolysis of aqueous NaCl: when an aqueous solution is used, the reactions may differ as water may be involved.

Cathode: $Na^+(aq) + e^- \rightarrow Na(s)$ $E° = -2.71$ V *or*
$2\,H_2O + 2\,e^- \rightarrow H_2(g) + 2\,OH^-(aq)$ $E° = -0.83$ V

Anode: $2\,Cl^- \rightarrow Cl_2(g) + e^-$ $E° = -1.36$ V *or*
$2\,H_2O \rightarrow O_2(g) + 4\,H^+(aq) + 4\,e^-$ $E° = -1.23$ V

At the cathode, water is split as that reaction is far less negative than the reduction of sodium, and hydrogen bubbles are seen. At the anode, however, chlorine gas is formed. This is due to a phenomenon known as *overvoltage.* A basic solution of NaOH is formed.

Overvoltage is the amount of voltage needed above the calculated standard reduction (or oxidation) potential for electrolysis to occur. It is needed when the rate of electron transfer is very slow. Only a small overvoltage is needed for the solution or deposition of metals – they are good conductors of electrons – but a large overvoltage is required for the formation of gases, especially O_2 or H_2. Current theories do not allow this to be predicted easily, so experimental evidence is needed if cell potentials are similar.

Electrolysis of water: if the electrolyte in solution is less easily oxidized and reduced than water, water will react at both electrodes:

Anode: $2\,H_2O(l) \rightarrow O_2(g) + 4\,H^+(aq) + 4\,e^-$

Cathode: $4\,H_2O(l) + 4\,e^- \rightarrow 2\,H_2(g) + 4\,OH^-(aq)$

Equal quantities of H^+ (*aq*) and OH^- (*aq*) are formed, so the solution will remain neutral.

Overall cell reaction: $2\ H_2O\ (l) \rightarrow O_2\ (g) + 2\ H_2\ (g)$

EXAMPLE:

Predict the half-cell reactions when a solution of aqueous KCl is electrolyzed. Write the overall cell reaction:

SOLUTION: The possible half–reactions are:

Cathode:

$K^+\ (aq) + e^- \rightarrow K\ (s)$ $\quad E° = -2.93$ V *or*

$2\ H_2O\ (l) + 2\ e^- \rightarrow H_2\ (g) + 2\ OH^-\ (aq)$ $\quad E° = -0.83$ V

Anode:

$2\ Cl^-\ (aq) \rightarrow Cl_2\ (g) + 2\ e^-$ $\quad E° = -1.36$ V *or*

$2\ H_2O\ (l) \rightarrow O_2\ (g) + 4\ H^+\ (aq) + 4\ e^-$ $\quad E° = -1.23$ V

At the cathode, water will be reduced, generating hydrogen gas. At the anode, remember that the overvoltage needed to generate chlorine gas is less than that needed to generate oxygen gas, so chlorine gas will be produced, despite its higher oxidation potential.

Overall reaction: $2\ H_2O\ (l) + 2\ Cl^-\ (aq) \rightarrow H_2\ (g) + Cl_2\ (g) + 2\ OH^-\ (aq)$

Workbook Problem 17.8:

Predict the half-cell and overall reaction for the electrolysis of an aqueous Li_2SO_4 solution.

Strategy: Determine the possible half-reactions and their standard voltages to predict the reactions at the cathode and anode, then determine the overall reaction.

Step 1: Determine the possible reactions at the cathode, and their standard voltages:

Step 2: Determine the possible reactions at the anode, and their standard voltages:

Step 3: Predict which reactions will occur and calculate the overall reaction:

17.13 Commercial Applications of Electrolysis

Sodium is produced commercially in a *Downs cell* by electrolysis of a molten mixture of NaCl and $CaCl_2$. The addition of the $CaCl_2$ lowers the melting point of the mixture by more than 200 °C. The liquid Na produced at the cathode is less dense than the mixture, and can be drawn off from the top. Requires extremely high currents, so manufacture is near hydroelectric power plants.

Chlorine and sodium hydroxide are produced by the electrolysis of aqueous sodium chloride in what is known as the *chlor–alkali industry*. The anode reaction, which produces Cl_2 gas is separated from the cathode compartment by a membrane that permits the passage of sodium ions. When water is hydrolyzed at the cathode, a sodium hydroxide solution is produced.

Aluminum is produced in the **Hall–Heroult process**, which involves the electrolysis of a molten mixture of Al_2O_3 and cryolite (Na_3AlF_6) at 1000 ^{o}C in a cell with graphite electrodes. At the cathode, molten aluminum is produced and sinks to the bottom of the cell. At the anode, O_2 gas is formed. This reacts with the electrode to generate CO_2 gas, so frequent replacement of the anode is necessary. The reactions at the electrodes are not well understood, and a great deal of current is required for the process, making aluminum production the largest single consumer of electricity in the United States.

Electrorefining is the purification of a metal by means of electrolysis. As an example, copper can be purified by using an impure copper anode and a pure copper cathode. Copper ions move away from the impure electrode and deposit on the pure electrode, leaving other metals either dissolved in the electrode solution or on the bottom of the container as *anode mud.*

Electroplating is a similar process in which one metal is coated on the surface of another using electrolysis. The cathode is the carefully-cleaned object to be plated, while the solution in the electrolytic cell contains ions of the metal to be deposited.

17.14 Quantitative Aspects of Electrolysis

The amount of substance produced at an electrode by electrolysis depends on the quantity of electrons passed through the cell. This is directly related to the stoichiometry of the reaction: one mole of electrons are needed to reduce 1 mol of sodium ions to sodium metal.

Moles of electrons passed through a cell can be calculated from electric current and time by the following formulas:

$$\text{Charge (C)} = \text{Current (A)} \text{ x } \text{Time (s)}$$

$$\text{Moles of e}^- = \text{charge(C)} \times \frac{1\text{ mol e}^-}{96{,}500\text{ C}}$$

Sequence of conversion used to calculate the mass or volume of product produced by passing a known current for a fixed period of time (see Figure 17.19, page 723).

Current and time → Charge → Moles of e^- → Moles of product → Grams or liters of product

EXAMPLE:

How many grams of Cl_2 gas would be produced in the electrolysis of molten NaCl by a current of 5.75 A for 50.0 min? How many grams of sodium will be produced in the same time?

SOLUTION: (Remember that a coulomb is an A·s or that an ampere is C/s.)

$2\ Cl^- \rightarrow Cl_2 + 2\ e^-$

Moles of electrons = 2

$$5.75\ \frac{\text{C}}{\text{s}} \times 50.0\ \text{min} \times \frac{60\ \text{s}}{1\ \text{min}} \times \frac{1\ \text{mol e}^-}{96,500\ \text{C}} \times \frac{1\ \text{mol Cl}_2}{2\ \text{mol e}^-} \times \frac{70.9\ \text{g Cl}_2}{1\ \text{mol Cl}^-} = 6.34\ \text{g Cl}_2$$

$$5.75\ \frac{\text{C}}{\text{s}} \times 50.0\ \text{min} \times \frac{60\ \text{s}}{1\ \text{min}} \times \frac{1\ \text{mol e}^-}{96,500\ \text{C}} \times \frac{1\ \text{mol Na}}{1\ \text{mol e}^-} \times \frac{23.0\ \text{g Na}}{1\ \text{mol}} = 4.11\ \text{g}$$

Workbook Problem 17.9

The Dow process isolates Mg metal from seawater. The final step in this process involves the electrolysis of molten $MgCl_2$ to metal. How long would it take to produce 12 kg of Mg (*s*) at a current of 15 A? How many liters of Cl_2 gas would be produced (at STP) in the same amount of time?

Strategy: Write the electrolysis reaction and consider the conversion process to calculate time.

Step 1: Treat the electrons as reactants in the chemical equation, and solve for the time required to produce 12 kg of magnesium. Remember, the conversion factor of 96,500 C (or A·s) per mole of electrons.

Step 2: Use the information calculated above to determine the volume of Cl_2 gas.

Putting It Together

Determine $E°$ for the following reaction:

$$O_2\ (g) + 4\ H^+\ (aq) + 4\ Br^-\ (aq) \rightarrow 2\ H_2O\ (l) + 2\ Br_2\ (l)$$

Is this reaction spontaneous? A buffer containing 0.35 M sodium formate ($NaCHO_2$) and 0.20 M formic acid is added to adjust the pH of the reaction. Assume [Br^- = 1.0 M]. Determine E after the addition of the buffer. Now is the reaction spontaneous?

Self–Test

This section is intended to test your knowledge of the material covered in this chapter. Think through these problems, and make certain you understand what they are asking. Make sure your answers make sense. Successful completion of these problems indicates that you have mastered the material in this chapter. You will receive the greatest benefit from this section if you use it as a mock exam, as this will allow you to determine which topics you need to study in more detail.

True–False

1. In galvanic or voltaic cells, a spontaneous reaction occurs.

2. A salt bridge allows electrons to flow between the two halves of a galvanic cell.

3. If the cell potential of a reaction is positive, the free-energy change of the reaction is also positive.

4. When balancing electrochemical equations, $E°$ is never multiplied by the stoichiometric coefficients.

5. E for a reaction at nonstandard conditions can only be determined by experiment.

6. Cell potential and pH have a linear relationship.

7. The voltage of galvanic cells is added when they are connected in series to form a battery.

8. Direct methanol fuel cells have been proposed for use in small consumer electronics.

9. Corrosion of any metal can be prevented by contact with another metal above it in the list of standard reduction potentials.

10. The results of the electrolysis of an aqueous salt solution can be predicted using only standard half-cell voltages.

Multiple Choice

1. For the following galvanic cell $Ni\ (s)\ |\ Ni^{2+}\ (aq)\ ||\ Br^-\ (aq)\ |\ Br_2\ (l)\ |\ Pt\ (s)$
   a. the cathode is Ni (*s*)
   b. the electrons flow from the Ni (*s*) electrode to the Pt (*s*) electrode
   c. the Ni^{2+} ions flow to the anode
   d. electrons flow from the Pt (*s*) electrode to the Ni (*s*) electrode

2. The reaction carried out in a galvanic cell
   a. is spontaneous, and therefore has a positive $E°$ value
   b. is nonspontaneous, and therefore has a positive $E°$ value
   c. is spontaneous, and therefore has a negative $E°$ value
   d. is nonspontaneous, and therefore has a negative $E°$ value

3. For the net reaction $Cr\ (s) + Fe^{3+}\ (aq) \rightarrow Cr^{3+}\ (aq) + Fe\ (s)$
   a. Cr (*s*) is the reducing agent
   b. Fe (*s*) is the anode
   c. Fe^{3+} (*aq*) is the reducing agent
   d. Cr (*s*) undergoes reduction

4. Given the following reduction half–cells:
   $PbO_2\ (s) + 3\ H^+\ (aq) + HSO_4^-\ (aq) + 2\ e^- \rightarrow PbSO_4\ (s) + 2\ H_2O\ (l)$ $E° = 1.628$ V
   $Cr_2O_7^{2-}\ (aq) + 14\ H^+\ (aq) + 6\ e^- \rightarrow 2\ Cr^{3+}\ (aq) + 7\ H_2O\ (l)$ $E° = 1.33$ V
   $SO_4^{2-}\ (aq) + 4\ H^+\ (aq) + 2\ e^- \rightarrow H_2SO_3\ (aq) + H_2O\ (l)$ $E° = 0.17$ V
   $2\ CO_2\ (g) + 2\ H^+\ (aq) + 2\ e^- \rightarrow H_2C_2O_4\ (aq)$ $E° = -0.49$ V

   a. the strongest oxidizing agent is $PbSO_4$ (*s*)
   b. $PbSO_4$ (*s*) will spontaneously react with CO_2 (*g*)
   c. the strongest oxidizing agent is PbO_2 (*s*)
   d. $H_2C_2O_4$ (*aq*) will not spontaneously react with PbO_2 (*s*)

5. In the electrolysis of molten BaI_2
   a. the Ba^{2+} ions migrate toward the anode
   b. the I^- ions migrate toward the anode

c. water undergoes oxidation at the anode
d. the system becomes increasingly basic

6. In the table of standard reduction potentials
   a. the strongest reducing agents are located in the bottom left of the table
   b. the strongest reducing agents are located in the top left of the table
   c. the strongest reducing agents are located in the lower right of the table
   d. the strongest oxidizing agents are located in the bottom left of the table

7. In an electrolytic cell
   a. the anode has a positive sign
   b. reduction occurs at the anode
   c. the anode has a negative sign
   d. oxidation occurs at the cathode

8. In an electrolytic cell
   a. anions migrate towards the cathode
   b. cations migrate towards the anode
   c. ions migrate through a salt bridge
   d. a current forces nonspontaneous reactions to occur

9. Alkaline dry cells
   a. contain an electrolyte of a moist paste of NaOH and $ZnCl_2$
   b. have a cathode in which steel is in contact with HgO in an alkaline medium
   c. contain an electrolyte of a moist paste of NH_4Cl and $ZnCl_2$
   d. are rechargeable nickel–cadmium batteries

10. The amount of substance produced at an electrode by electrolysis depends
    a. on the quantity of reactant present
    b. on the quantity of charge passed through the cell
    c. on the spontaneity of the reaction
    d. on the size of the electrolytic cell

Fill–in–the–Blank

1. Car batteries are ____________________ batteries, while common household batteries are ___________________.

2. The oxidative deterioration of a metal is called ___________________. In the case of steel, this can be prevented by coating with zinc, also called ___________________ for small objects, or with the use of another metal as a ____________________________ for buildings or ships.

3. The refining of _______________ and ____________ metals require very large quantities of _________________. As a result, these processes are generally located near _________________.

4. Standard half–cell potentials are written as ______________________, and are listed in __________ tendency to occur as written, and ____________ tendency to occur in reverse. The "zero" on the scale is the ______________________________________.

5. The ________________________ of a reaction can be calculated from the standard ____________________, but these tend to be either very _________ or very ____________, indicating that these reactions either go to _______________, or not at all.

Matching

Anode	a. The process of using an electric current to bring about a chemical change
Cathode	b. using current to deposit a thin layer of one metal on another
Corrosion	c. electrode at which reduction occurs
Electrolysis	d. covering one metal with another to prevent corrosion
Electroplating	e. the voltage above the standard potential needed to speed a reaction.
Electrorefining	f. electrode at which oxidation occurs
Galvanizing	g. mechanism for allowing ion flow in a galvanic cell.
Overvoltage	h. oxidative deterioration of a metal
Salt bridge	i. Using current to purify impure metals

Problems

1. Describe the galvanic cell that uses the following reaction:

$$Ni^{2+}\ (aq) + Mg\ (s) \rightarrow Ni\ (s) + Mg^{2+}\ (aq)$$

Write the half-reactions, identify the anode and cathode, and provide the shorthand notation.

2. Determine standard reaction potentials for the following unbalanced reactions and indicate whether they are spontaneous in the forward or reverse direction:

a. $2\ Al\ (s) + 3\ Pb^{2+}\ (aq) \rightarrow 2\ Al^{3+}\ (aq) + 3\ Pb\ (s)$

b. $Br_2\ (l) + 2\ Cl^-\ (aq) \rightarrow 2\ Br^-\ (aq) + Cl_2\ (g)$

c. $Ni\ (s) + Pb^{2+}\ (aq) \rightarrow Ni^{2+}\ (aq) + Pb\ (s)$

d. $Fe\ (s) + HNO_3 \rightarrow Fe^{3+}\ (aq) + NO_3^-(aq) + H_2\ (g)$

3. Calculate E for the following cell:

$$Mg\ (s)\ |\ Mg^{2+}\ (aq)\ ||\ Fe^{3+}\ (aq)\ |\ Fe\ (s)$$

Under the following conditions:

a. standard conditions

b. $[Mg^{2+}] = 0.1$ M, $[Fe^{3+}] = 2.0$ M

c. $[Mg^{2+}] = 2.0$ M, $[Fe^{3+}] = 0.1$ M

4. Given the following reaction:

$$Cu^{2+}\,(aq) + 2\,Ag\,(s) + 2\,Br^{-}\,(aq) \rightarrow Cu\,(s) + 2\,AgBr\,(s) \qquad E° = 0.27\text{ V}$$

Calculate $\Delta G°$ and K.

5. $E°$ for a galvanic cell in which Zn^{2+} is reduced to Zn (s) is +1.61 V. What is the potential at the anode? What metal is oxidized at the anode? Write the half–reaction that occurs at the anode.

6. If the following cell has a potential of 0.14 V, what is the pH in the anode compartment?

$$Pt\,(s)|\,H_2\,(1\text{ atm})\,|\,H^{+}\,(aq)||\,Ni^{2+}\,(1\text{ M})|\,Ni\,(s)$$

7. Which of these metals, if coated onto iron, would prevent corrosion? Explain your answers.

   a. Sn

   b. Mn

   c. Cu

8. Calculate the K_{sp} for AgCl (s) using the table of standard reduction potentials.

9. Write the reactions that occur at the anode and cathode in the electrolysis of molten NaBr. How much current would be required to produce sodium metal at the rate of 3 g/hour?

10. If 3 A of current are applied to a copper object immersed in a solution of $AgNO_3$ and copper, how much time will be required to deposit 15 g of silver?

11. An acidic solution of $NaMnO_4$ can be used to oxidize zinc metal. If 3.27 g of zinc are to be oxidized in 250 mL of solution, what is the minimum number of grams of $NaMnO_4$ that need to be dissolved?

12. How many liters of O_2 are produced at 356 mm Hg when 2.97×10^3 C are passed through water at 25 °C?

Challenge Problem

The K_{sp} for lead(II) sulfate is 1.8×10^{-8}. The standard reduction potential for Pb^{2+}/Pb is $E° = -0.126$ V. Use this information to determine $E°$ for the reaction:

$$PbSO_4 + 2\,e^{-} \rightarrow Pb\,(s) + SO_4^{2-}\,(aq)$$

CHAPTER 18

HYDROGEN, OXYGEN, AND WATER

Chapter Learning Goals

A. Properties of Hydrogen

1. Describe the properties of hydrogen, including appearance, structure, occurrence in nature, synthesis, and industrial use.
2. Describe the isotopes of hydrogen, and compare and contrast the properties of the isotopes and compounds containing the isotopes.

B. Reactions of Hydrogen

1. Apply the gas laws to problems involving hydrogen.
2. Assign oxidation numbers and identify the oxidizing agent and reducing agent for redox reactions of hydrogen. Balance equations for these reactions, using either the oxidation–number method or the half–reaction method.
3. Classify binary hydrides as ionic, covalent, or metallic.
4. Discuss properties and reactions of hydrides, including nonstoichiometric interstitial hydrides.

C. Properties of Oxygen

1. Describe the properties of oxygen, including appearance, structure, occurrence in nature, synthesis, and industrial use.
2. Classify oxides as basic, acidic, or amphoteric.
3. Use periodic properties to predict the properties of oxides.
4. Identify compounds as oxides, peroxides, or superoxides.
5. Use structure and bonding concepts to explain the physical and chemical properties of elemental oxygen and ozone.

D. Reactions of Oxygen

1. Apply the gas laws to problems involving oxygen.
2. Assign oxidation numbers and identify the oxidizing agent and reducing agent for redox reactions of oxygen. Balance equations for these reactions.

E. Properties and Reactions of Water

1. Discuss properties of water and methods for purification and treatment.
2. Identify and describe the general properties of a hydrate.
3. Determine the empirical formula of a hydrate.

Chapter in Brief

This chapter examines more closely the production and reactions of hydrogen and oxygen, as well as the compounds they form: water and hydrogen peroxide. Hydrides – ionic, covalent, and metallic – are examined, as well as the many types of compounds that can be formed with oxygen as it varies its oxidation state to form oxides, peroxides, and superoxides. Hydrogen peroxide is an unstable compound that can act as both an oxidizing agent and a reducing agent, and can also disproportionate – both oxidize and reduce itself – as it decomposes to water and oxygen. Water is the most abundant compound on Earth – methods for purification and treatment are examined, as are hydrates, ionic compounds that contain water molecules bound within their crystal structures.

18.1 Hydrogen

Hydrogen was first isolated by Henry Cavendish, an English chemist who showed that acid acting on metals generated a low-density, flammable gas.

The name "hydrogen," which means "water former" was given by French chemist Antoine Lavoisier who noted that it combines with oxygen to form water.

Hydrogen is a colorless, odorless, and tasteless gas that primarily exists as a nonpolar, diatomic molecule. Its lack of polarity and small size lead to very weak intermolecular forces, and low melting and boiling points, but it has the highest dissociation energy of any diatomic element.

Hydrogen is thought to be approximately 75% of the mass of the universe, but is very rare in the Earth's atmosphere because the Earth's gravity is not strong enough to hold it. Combined with other elements, it is the ninth most abundant element in the Earth's crust and oceans.

18.2 Isotopes of Hydrogen

Protium is the lightest and most common form of hydrogen. ${}^{1}_{1}H$ comprises 99.985% of atoms in naturally occurring hydrogen.

Deuterium, ${}^{2}_{1}H$ or D, is also known as heavy hydrogen, and is present in small amounts (0.015% atom %), while *tritium*, ${}^{3}_{1}H$ or T is radioactive, and is present only in trace amounts.

All three isotopes have the same electron configuration, and thus the same chemical behavior. However, their very different masses cause quantitative differences in properties known as **isotope effects**. These are greater for hydrogen than for any other element, as the mass differences are so great.

The heavier the isotope, the higher the melting point, and boiling point. The same is true of the isotopes when bound into water. When bonded in water, the D-O and T-O bonds are stronger than H-O bonds, which makes it possible to separate the isotopes.

The effect of isotopic mass on reaction rates is called a *kinetic-isotope effect.* D_2O can be separated from H_2O because its stronger bonds separate more slowly. As water is electrolyzed, the remaining molecules are enriched in D_2O. Reducing water from 2400 L to 83 mL yields 99% pure D_2O. 150 metric tons of D_2O are manufactured in the U.S. per year for use as a coolant and moderator in nuclear reactors.

Workbook Problem 18.1

Benzene has the formula C_6H_6. What is the molecular mass of C_6D_6? What is the percent increase in the molecular mass when benzene is made with deuterium instead of protium? Make the same calculations for H_2O and D_2O, then qualitatively describe the changes expected for the melting and boiling points of C_6D_6 compared to C_6H_6. Would more or less energy be released from burning C_6D_6 compared to C_6H_6?

Strategy: Calculate the molecular masses of the four molecules, calculate the percent differences, and describe the kinetic isotopic effects.

Step 1: Calculate the molecular weights for the four molecules:

Step 2: Calculate the percent difference in the two pairs:

Step 3: Predict the expected properties:

Workbook Problem 18.2

What volume of D_2O can be purified from 25 L of water?

Strategy: Set up a ratio to calculate the volume, choose an appropriate unit for the answer.

18.3 Preparation and Uses of Hydrogen

Electrolysis of water produces hydrogen of 99.95% purity, but is impractical for large-scale production due to the large amount of energy required – 286 kJ/mol of energy for 1 mol – 2g – of hydrogen.

Small-scale hydrogen production can be accomplished in the lab by reacting dilute acid with an active metal such as zinc.

Large–scale industrial preparation methods use inexpensive reducing agents to extract the oxygen from steam. The steam–hydrocarbon re–forming process is a three-step process.

In the first step, a mixture of steam and methane are reacted at high temperature in the presence of a nickel catalyst to produce *synthesis gas*, so called as it can also be used to synthesize liquid fuels:

$$H_2O\ (g)\ +\ CH_4\ (g)\ \rightarrow\ CO\ (g)\ +\ 3\ H_2\ (g)$$

In the second step, the CO from the synthesis gas is mixed with more steam and passed over a metal oxide catalyst at 400 °C. This **water-gas shift reaction** removes the toxic carbon monoxide and produces more hydrogen gas:

$$CO\ (g)\ +\ H_2O\ (g)\ \rightarrow\ CO_2\ (g)\ +\ H_2\ (g)$$

In the third step, the gas mixture is passed through a basic aqueous solution, which dissolves the carbon dioxide into carbonate ion:

$$CO_2\ (g)\ +\ 2\ OH^-\ (aq)\ \rightarrow\ CO_3^{2-}\ (aq)\ +\ H_2O\ (l)$$

95% of H_2 is produced and consumed in the same place. The largest single consumer is the Haber process for synthesizing ammonia:

$$N_2\ (g)\ +\ 3\ H_2\ (g)\ \leftrightarrows\ 2\ NH_3\ (g)$$

Large quantities are also used in the synthesis of methanol:

$$CO\ (g)\ +\ H_2\ (g)\ \rightarrow\ CH_3OH\ (l)$$

This is an industrial solvent and is used in making formaldehyde, a precursor to plastics.

Workbook Problem 18.3

Write a balanced equation for the production of synthesis gas from ethane – C_2H_6. If each step in the process has a 60% yield, how many L of hydrogen gas at 25 °C and 1 atm can be produced from 15 kg of ethane?

Strategy: Write and balance the equation, determine the quantities of reactants and products at each step. Use the ideal gas law to calculate the volume of hydrogen gas.

Step 1: Write and balance the equation for the formation of synthesis gas:

Step 2: Determine the moles of reactants entering the first step, and the moles of products produced.

Step 3: Multiply the moles of products by 60%, and calculate the products of the second step.

Step 4: Multiply the moles of products by 60%, and calculate the products of the third step.

Step 5: Multiply the moles of products by 60%, and determine the volume of the H_2 gas.

18.4 Reactivity of Hydrogen

With its one valence electron, hydrogen has properties similar to both the alkali metals and the halogens. Like the alkali metals, it can ionize to form H^+ (E_i =1312 kJ/mol). Like the halogens, it can share its electron to form covalent compounds.

Complete ionization of hydrogen is only possible in the gas phase. In liquids or solids, a bare proton is too reactive to exist by itself, so it will attach to a molecule with a lone pair of electrons.

Also like the halogens, hydrogen will accept an electron (E_{ea} = -73 kJ/mol) from an active metal to make an ionic hydride.

Gain electron — H^-.

E_{ea} = –73 kJ/mol

will accept an electron from an active metal to give an ionic hydride

Because of the strength of the H-H bond, hydrogen is relatively unreactive. But in the presence of oxygen, an explosive reaction can occur that requires as little as 4% hydrogen in a mix of gases. This is the potentially dangerous and very exothermic reaction:

$$2\ H_2\ (g) + O_2\ (g) \rightarrow 2\ H_2O\ (l) \qquad \Delta H = -572\ \text{kJ}$$

Workbook Problem 18.4

What is the change in volume involved when 15 g of hydrogen and 100 g of oxygen react at STP? What volume of water is produced? What volume of gas is left over?

Strategy: Determine the initial volume, moles of each gas, and how much is left over (assume that the volume of the water produced is negligible). How many kJ are released in this reaction?

Step 1: Determine the moles of each gas, and the limiting reagent.

Step 2: Determine the moles of excess reagent remaining after the reaction.

Step 3: Determine the volume change of the reaction.

Step 4: Determine the kJ released by the reaction.

18.5 Binary Hydrides

Binary hydrides contain hydrogen and just one other element. They can be ionic, covalent, or metallic.

Ionic hydrides are formed by the alkali metals and Ca, Sr, and Ba. They can be formed by direct reaction at 400 °C. They are salt-like, with the alkali metal hydrides forming a face-centered cubic cell like that of sodium chloride.

The hydride ion is an electron donor – a Brønsted-Lowry base – and a good reducing agent. When dissolved in water, the hydride ion reduces water to generate H_2 gas and OH^- ions.

Covalent hydrides are molecules in which hydrogen is covalently bonded to another nonmetal. These tend to be small molecules with relatively weak intermolecular forces, and so are gases or volatile liquids, at room temperature. Many are quite familiar – H_2O, NH_3, and CH_4 are all covalent hydrides.

Metallic hydrides will also form with large metal atoms, especially the lanthanides, actinides and certain *d*–block transition metals. When these large metal atoms pack together, the hydrogen atoms – whether ionized or not is unknown – pack into the interstitial spaces. These compounds can be stoichiometric – as UH_3 – or not, as in the case of $ZrH_{1.9}$.

The properties of these compounds depend on their composition, which is also dependent on the partial pressure of H_2 gas in the surroundings. This makes these compounds of potential interest as hydrogen storage devices.

Workbook Problem 18.5

Write a balanced equation for the reaction of sodium hydride with water. How many liters of hydrogen gas (at STP) will be evolved if 3.65 g of sodium hydride are allowed to react with excess water?

Strategy: Write and balance the equation, determine the moles of sodium hydride, use the balanced equation to determine the moles of gas, and the standard molar volume to determine liters.

Step 1: Write and balance the equation.

Step 2: Determine the moles of sodium hydride and the moles of hydrogen evolved.

Step 3: Determine the volume of hydrogen evolved.

18.6 Oxygen

The English chemist Joseph Priestley and the Swedish chemist Karl Wilhelm Scheele were the first to isolate and characterize oxygen, a colorless, tasteless gas that supports combustion better than air. Antoine Lavoisier realized that it was a unique element and named it oxygen – "acid former."

Oxygen is pale blue when liquid or solid, and is paramagnetic in all three phases. It exists as a double-bonded diatomic element, with a bond length of 121 pm and a bond dissociation energy of 498 kJ/mol. It has a melting point of -219 °C, and a boiling point of -183 °C.

Oxygen is the most abundant element on the Earth's surface: it is 23% of the atmosphere, primarily as O_2 gas; it is 46% of the Earth's crust (the lithosphere) as oxides, silicates, carbonates, and other compounds; and it is more than 85% of the hydrosphere in the form of H_2O.

Despite its use in respiration and combustion, the amount of oxygen in the atmosphere is relatively constant as it is replenished by photosynthesis:

$$6\,CO_2 + 6\,H_2O \xrightarrow{h\upsilon} 6\,O_2 + C_6H_{12}O_6$$

In this process, plants use the energy from the sun to form carbohydrates and oxygen from carbon dioxide and water. When the plant matter is metabolized or burned, the process is reversed.

This process makes it possible to link solar energy to metabolism.

Workbook Problem 18.6

If a can of soda contains 39 g of glucose, how many Calories of energy does this supply?

ΔH°_f for CO_2 = -393.5 kJ/mol
ΔH°_f for H_2O = -285.8 kJ/mol
ΔH°_f for $C_6H_{12}O_6$ = -1260 kJ/mol

4.184 J = 1 cal

Strategy: Write and balance the equation for combustion of glucose, determine the energy released in kilojoules, convert to calories, then kcal (dietary calories, C, recall, are kilocalories).

Step 1: Write and balance the equation.

Step 2: Use the heats of formation to calculate the heat evolved per mole.

Step 3: Convert the quantity given to moles, convert this to energy in kcal.

18.7 Preparation and Uses of Oxygen

Small amounts of oxygen can be prepared in the lab by the thermal decomposition of an oxoacid salt in the presence of a catalyst, with the gas being collected over water. This is rarely done, due to the ready availability of cylinders of oxygen gas.

The industrial synthesis of oxygen is performed by fractional distillation of liquefied air. As the liquid is slowly warmed, nitrogen (bp = -196 °C or 77 K), then argon (bp = -186 °C or 87 K) can be removed as gases, leaving the oxygen (bp = -183 °C or 90 K) behind. Oxygen ranks only behind sulfuric acid and nitrogen in industrial chemicals produced in the U.S.

More than two-thirds of the oxygen produced is used in steelmaking, where it can be used to oxidize impurities in iron. It is also used in sewage treatment, where it destroys malodorous compounds, and in bleaching paper. In the oxyacetylene torch, a highly exothermic reaction provides the heats over 3000 °C that are needed in cutting and welding metal.

Oxygen is an inexpensive and readily available oxidizing agent.

Workbook Problem 18.7

When heated in the presence of a catalyst, sodium chlorate decomposes to form oxygen gas. How many L of oxygen gas could be obtained from 15.0 g sodium chlorate at 250 °C? How many mL will this be at room temperature?

Strategy: Write and balance the equation for decomposition of sodium chlorate, determine the number of moles of salt present and the possible moles of oxygen generated. Use the ideal gas law to determine the number of L of oxygen.

Step 1: Write and balance the equation.

Step 2: Determine moles of salt present and moles of oxygen generated.

Step 3: Use the ideal gas law to calculate the volume of gas at 250 °C and 25 °C.

18.8 Reactivity of Oxygen

Oxygen is very electronegative, and needs only two electrons to obtain an octet. As a result, oxygen can accept two electrons from active metals to form ionic oxides (Na_2O, for example) or share two electrons with other nonmetals to form covalent bonds.

In covalent compounds, oxygen can form either two single bonds or one double bond. With smaller nonmetal atoms, such as C and N, there is good overlap of the π orbitals, which facilitates double-bonding. With larger atoms, this overlap is less efficient and less likely to occur.

Oxygen reacts directly with all the elements in the periodic table with the exception only of the noble gases, and a few inactive metals like platinum and gold. These reactions are slow at room temperature, but proceed rapidly at high temperatures.

18.9 Oxides

Oxides are classified based on oxidation state. A binary compound with oxidation state -2 is an oxide, oxidation state of -1 is a peroxide, and -1/2 is a superoxide. Oxides can be classified as basic, acidic, or amphoteric.

Basic oxides are ionic compounds formed by elements on the left side of the periodic table. Water-soluble oxides dissociate to form metal ions and OH^- ions, while water insoluble basic oxides will dissolve in acid, as the H^+ ions react with the oxygen to form water.

Acidic oxides (acid anhydrides) are covalent (N_2O_5, for example), and dissolve by reacting with water to form H^+ ions. Water insoluble acidic oxides such as SiO_2 will dissolve in strong bases.

Amphoteric oxides exhibit both acidic and basic properties, and are formed by elements with intermediate electronegativities. They are formed by elements that have intermediate electronegativities. Their bonds are strongly polar – intermediate between ionic and covalent.

The amphoteric character of Al_2O_3 is useful in purifying it during the **Bayer process** during the production of aluminum: impure Al_2O_3 is dissolved in base, forming $Al(OH)_4^-$ which is then precipitated with weak acid and heated to drive off water, leaving behind pure Al_2O_3. This is then electrolytically converted to metal.

The acid–base properties and ionic–covalent character of these oxides depend on the position of the other element in the periodic table. The more covalent the bonds, the more acidic they are, the more ionic, the more basic they are.

Acidic and covalent character also increases with increasing oxidation state: CO_2 is a good example: it is nonpolar covalent, and also acidic, and the oxidation state of C is +4.

As bonding changes from ionic to covalent, a corresponding structural change is seen from extended networks in more ionic oxides to discrete molecule in covalent oxides. Properties also follow these trends, with ionic substances having higher melting points due to lattice energy, and very covalent structures like SO_3 being liquid at room temperature.

The properties of these oxides determine their uses: MgO and Al_2O_3 are high-temperature insulators, while SiO_2 is used in optical fibers for communication. Acidic oxides are precursors to industrial acids.

Workbook Problem 18.8

Write and balance equations for the following reactions:

Dissolution of Rb_2O in water

Dissolution of BeO in acid and base

Dissolution of PbO in acid and base (to form $Pb(OH)_4^{2-}$)

Reaction of N_2O_3 with water

Strategy: Determine the nature of the oxide and how it will react with water or the given solution, then write and balance the equations.

Step 1: Determine the nature of each oxide:

Step 2: Write and balance the equations:

18.10 Peroxides and Superoxides

When heavier group 1A and 2A metals are heated in an excess of oxygen, they can form either peroxides, such as Na_2O_2 or BaO_2, or superoxides such as KO_2. The oxygen atoms in peroxides have an oxidation number of -1, while the oxygen atoms in superoxides have an oxidation number of -1/2.

All of these are ionic solids, and the product depends on the pressure of O_2, the sizes of the ions, the way they pack together, and the lattice energy of the resultant solid.

The trends in bond lengths and magnetic properties are explained by **molecular orbital theory**: the highest occupied orbitals in all of the diatomic O species have parallel-spin antibonding orbitals. As the charge on the ion increases, the number of antibonding electrons also increases, leading to a longer bond length.

The diamagnetic peroxide ion has an O–O single bond, and is a basic ion, tending to pick up H^+ ions. In the presence of a strong acid, it will form hydrogen peroxide, H_2O_2. Metal peroxides dissolve in water, and the O_2^{2-} ion pulls a proton off of water, forming HO_2^- and OH^- ions.

The paramagnetic superoxide ion, O_2^{2-}, has the longest O-O bond. When metal superoxides are dissolved in water, oxygen is evolved in a **disproportionation reaction**: the oxygen in the superoxide ion is oxidized from -1/2 to 0, in O_2, and reduced from -1/2 to -1 in HO_2^-.

$$\underset{\substack{\uparrow \\ -1/2}}{2\,KO_2\,(s)} + H_2O\,(l) \rightarrow \underset{\substack{\uparrow \\ 0}}{O_2\,(g)} + 2\,K^+\,(aq) + \underset{\substack{\uparrow \\ -1}}{HO_2^-\,(aq)} + OH^-\,(aq)$$

Workbook Problem 18.9

Identify the following as oxides, peroxides, or superoxides, and write their reaction with water.

MgO_2
KO_2

Li_2O

Strategy: Determine the oxidation state of the oxygen in the molecules: -2 is an oxide, -1 is a peroxide, and -1/2 is a superoxide.

Step 1: Determine the oxidation number of the oxygens in each oxide:

Step 2: Identify the type of oxide:

Step 3: Write the reaction with water of each molecule:

18.11 Hydrogen Peroxide

Because of its oxidizing properties, hydrogen peroxide (H_2O_2) is a mild antiseptic, a bleach for textiles, paper, and hair, and a starting material for other peroxides.

Pure H_2O_2 is a colorless, syrupy liquid that freezes at $-0.4\ ^{\circ}C$ and boils around $150\ ^{\circ}C$, but heating causes it to explode. The viscosity and high boiling point indicate strong **hydrogen bonding** in the pure liquid. It is a weak acid in aqueous solutions, dissociating partially to H_3O^+ and HO_2^-.

Hydrogen peroxide can act as both a strong oxidizing and a strong reducing agent. When acting as a reducing agent, O_2 gas is formed. As an oxidizing agent, the peroxide ion is reduced to an oxide ion.

Hydrogen peroxide can also oxidize and reduce itself by the following equation:

$$2\ H_2O_2\ (l) \rightarrow 2\ H_2O\ (l) + O_2\ (g) \qquad \Delta H^\circ = -196\ \text{kJ}$$

This reaction is slow at room temperature, but in the presence of heat or any of a broad range of catalysts, the decomposition is rapid, exothermic, and potentially explosive.

18.12 Ozone

Ozone – O_3 - is an unusual allotrope of oxygen naturally formed when an electric current is passed through oxygen, as in an electrical storm. It is a toxic, pale-blue gas with a characteristic sharp odor. At ground level, it is a pollutant, while in the upper atmosphere, it is a filter for harmful ultraviolet radiation.

There are two resonance structures for ozone, both bent due to resonance structures and lone pairs of electrons. The atoms have σ bonds and share a π bond over all three oxygen atoms for a net bond order of 1.5.

Ozone can be used to kill bacteria in drinking water and swimming pools.

18.13 Water

Water is the most familiar and abundant compound on Earth, with 97.3% of it contained in the oceans. The rest is distributed through polar ice caps, underground aquifers, and freshwater lakes and rivers. With as much water as occurs in lakes and rivers, these last are only 0.01% of the water on Earth.

Seawater cannot be used for drinking or agriculture because it contains 35 g of dissolved salts per kg. Sodium chloride is the most abundant salt, but over 60 elements are present in small amounts. Only NaCl, Mg, and Br_2 are commercially prepared from seawater.

Drinking water comes from freshwater lakes, rivers, and aquifers. It is filtered before being sedimented by the addition of lime (CaO) and aluminum sulfate. These react to form a loose precipitate of aluminum hydroxide that settles, carrying with it sediment and much of the bacteria in the water. It is then filtered through sand, sprayed into the air to allow organic impurities to evaporate, and sterilized with ozone.

Hard water contains appreciable concentrations of doubly charged cations such as Ca^{2+}, Mg^{2+}, and Fe^{2+}. These cations react to form soap scum and with carbonates to form boiler scale. Ion exchange with Na allows these compounds to regain their solubility.

18.14 Hydrates

Hydrates are solid compounds that contain loosely bound water molecules. A dot is used to indicate that the waters are not covalently bound, but part of the crystal structure of the compound as in $CuSO_4 \cdot 5H_2O$.

Because bonding interaction increases with increasing charge, hydrates are more likely to form with salts that contain +2 and +3 cations.

When the compound is heated, the water molecules can be driven off, leaving behind the anhydrous form of the compound.

Some anhydrous compounds have such a strong tendency to bind waters that they are **hygroscopic** compounds. They absorb water from the air and are therefore useful as drying agents.

Workbook Problem 18.10

3.76 grams of anhydrous copper(II) sulfate was left out on a benchtop over the weekend in the summer. On Monday morning, it was bluish and weighed 4.36 g. If hydrated copper sulfate has 5 associated waters per formula unit, has this sample become fully hydrated?

Strategy: Determine the moles of copper(II) sulfate, the moles of water, and the ratio between them.

Step 1: Determine the moles of copper(II) sulfate.

Step 2: Determine the moles of water.

Step 3: Determine the ratio of water to copper sulfate and compare to the ratio (5:1) for the fully hydrated compound.

Putting It Together

When sodium hydride reacts with liquid sulfur dioxide, solid sodium dithionate ($Na_2S_2O_4$) and hydrogen gas are produced. Sodium dithionate is used to bleach paper pulp, and these sulfur-containing compounds are responsible for the famously bad smell of paper mills. Write a balanced equation for this reaction. What volume of hydrogen gas is produced when 23.57 g of sodium hydride reacts with 100.0 mL of sulfur dioxide at a temperature of 25 °C and pressure of 1 atm? The density of sulfur dioxide is 1.434 g/mL.

Self–Test

This section is intended to test your knowledge of the material covered in this chapter. Think through these problems, and make certain you understand what they are asking. Make sure your answers make sense. Successful completion of these problems indicates that you have mastered the material in this chapter. You will receive the greatest benefit from this section if you use it as a mock exam, as this will allow you to determine which topics you need to study in more detail.

True–False

1. Hydrogen is thought to make up 75% of the mass of the universe, but is too light to be a major part of our atmosphere.

2. Due to isotope effects, D_2O is easier to electrolyze than H_2O.

3. Solitary H^+ ions can be produced in the gas phase, but are too reactive to be found elsewhere.

4. Interstitial metal hydrides may be useful as hydrogen-storage devices.

5. Oxygen levels in the atmosphere tend to fluctuate.

6. Almost no oxygen is found in the Earth's crust.

7. Oxygen is an inexpensive and readily available reducing agent.

8. Oxides can be acidic, basic, or amphoteric.

9. The properties of oxides can be predicted based on periodic trends.

10. When it is found in lower levels of the atmosphere ozone filters UV, but in the upper atmosphere it is a pollutant.

Multiple Choice

1. Which of the following ranks the isotopes of hydrogen from lightest to heaviest?
   a. tritium, protium, deuterium
   b. deuterium, protium, tritium
   c. protium, deuterium, tritium
   d. protium, tritium, deuterium

2. Hydrogen can be produced by
   a. reaction of metals with acid
   b. electrolysis of water
   c. the steam-hydrocarbon reforming process
   d. all of the above

3. Rank the following from most to least ionic: CH_4, H_2, AlH_3, KH
    a. AlH_3, H_2, KH, CH_4
    b. KH, AlH_3, CH_4, H_2
    c. H_2, KH, AlH_3, CH_4
    d. KH, AlH_3, H_2, CH_4

4. Which of these does not describe oxygen?
    a. rare
    b. paramagnetic
    c. double-bonded in elemental form
    d. reactive

5. Oxygen can be prepared by
    a. fractional distillation of liquefied air
    b. decomposition of H_2O_2 (*g*)
    c. heating oxoacid salts
    d. all of the above

6. Lithium oxide is
    a. a covalent oxide
    b. double-bonded
    c. a basic oxide
    d. an amphoteric oxide

7. BaO_2 is
    a. a peroxide
    b. a superoxide
    c. a hydride
    d. an oxide

8. Hydrogen peroxide is
    a. a reducing agent
    b. an oxidizing agent
    c. both
    d. neither

9. The bonding of ozone is
    a. a resonance hybrid
    b. one double bond and one single bond
    c. two single bonds
    d. two double bonds

10. The water molecule(s) in hydrates are
    a. covalent bonded
    b. loosely bound
    c. ionic bonded
    d. none of the above

Fill–in–the–Blank

1. Hydrogen is able to behave chemically as a (n) ________________ and as a(n) __________________________ depending on whether it is gaining or losing an __________________.

2. Binary hydrides can be classified as ______________________, ______________________, or ______________________.

3. Photosynthesis and metabolism of carbohydrates allow energy from the ________ to be converted into ______________ energy. Photosynthesis ____________ oxygen, and metabolism (combustion) ______________ oxygen.

4. Oxygen can form _______________ oxides with _________________and _______________ oxides with nonmetals.

5. With small nonmetals, oxygen is able to form ________________ bonds. This does not happen with _____________ atoms, as there is less efficient ________________ of π orbitals.

6. Binary oxygen compounds where the oxygen has an oxidation state of _____ are called "oxides". Compounds with oxygen in an oxidation state of _____ are called "peroxides", and compounds with an oxidation state of _____ are "superoxides".

7. The ____________________ and __________________ character of oxides increases from left to right across the periodic table. The __________________ and ________________ character of oxides increases from right to left.

8. Peroxides and superoxides can be formed between oxygen and the _________________________ __________________________ and are prepared by _________________________________.

9. Hard water contains high levels of _____________ cations. These combine with soaps and carbonates to form undesirable precipitates, which can be resolubilized using sodium in a process called _________________.

10. Hygroscopic compounds are __________________ compounds that have a tendency to form _____________ and are useful as ____________________________ since they absorb _______________ from the air.

Matching

Isotope effect	a. a reaction in which a substance is both oxidized and reduced
Hygroscopic	b. a compound that contains hydrogen and just one other element

Binary hydride	c. an ionic compound that contains loosely bound water molecules within its structure
Hydrate	d. attracting water to the extent that it can pull it out of the surrounding air
Nonstoichiometric	e. difference in properties due to differences in the mass of isotopes
Disproportionation	f. having a composition that can't be expressed as a ratio of small, whole numbers

Problems

1. Write and balance equations for the generation of hydrogen from:
   a. Fe and HCl (*aq*)
   b. Mg and H_2SO_4 (*aq*)
   c. Al and HNO_3 (*aq*)

2. To generate 100 mL of hydrogen gas at 35 °C and atmospheric pressure, how many grams of zinc should be added to hydrochloric acid?

3. Write formulas for hydrides of the following elements and identify them as ionic or covalent:
   a. Li
   b. P
   c. Cl
   d. Ca

4. If 1400 g of calcium hydride are reacted with water, how many liters of hydrogen will be produced if inflating a weather balloon on a cold morning at 2.00 °C?

5. Write equations for the dissolution of the following oxides in the most suitable solvent – water, acid, or base. In the case of amphoteric oxides, write equations for dissolution in both acid and base:
   a. K_2O (soluble in water)
   b. Al_2O_3 (insoluble in water, can form soluble $Al(OH)_4^-$)
   c. Cl_2O_7 (soluble in water)

6. Identify the following compounds as oxides, peroxides, or superoxides:
   a. RbO_2
   b. MgO_2
   c. H_2O
   d. CaO

7. By identifying the changes in oxidation state for the various reactants, identify which reaction uses hydrogen peroxide as an oxidant, and which as a reductant:

   a. $2\ Fe^{2+}(aq) + H_2O_2(l) + 2\ H^+(aq) \rightarrow 2\ Fe^{3+}(aq) + 2H_2O(l)$

   b. $2\ KMnO_4(s) + 3\ H_2O_2(l) \rightarrow 2\ MnO_2(aq) + 2\ KOH(aq) + 2\ H_2O(l) + 3\ O_2(g)$

8. If seawater contains 3.5 % salts by mass, how many kg of salt are present in one km^3 of seawater if the density of seawater is 1.025 g/cm^3?

9. 5.00 g of hydrated cobalt(II) chloride are weighted and gently heated until no more water can be driven off. Upon reweighing, the weight is found to have a mass of 2.72 g. How many waters are associated with cobalt(II) chloride in its hydrated state?

10. The reaction of 3.7 g of an alkaline earth metal hydride with dilute hydrochloric acid generates 1.30 L of hydrogen gas at room temperature and atmospheric pressure. What is the alkaline earth metal?

Challenge Problem

Household bleach is 6% NaOCl (sodium hypochlorite) by mass. Household hydrogen peroxide is 3% H_2O_2 by mass. When the two are mixed, they react to form water, sodium chloride, and oxygen gas. If 500 mL of household bleach and 500 mL of household hydrogen peroxide are mixed at room temperature (25 °C) and atmospheric pressure, what volume of oxygen gas is evolved, and what is the molarity of the resultant solution (assume 1 L total volume, and a density of 1 for each solution)? Will the final solution be acidic or basic?

CHAPTER 19

THE MAIN–GROUP ELEMENTS

Chapter Learning Goals

A. Periodic Properties

1. Determine which of two main–group elements has more metallic character, higher ionization energy, larger atomic radius, higher electronegativity, the more acidic oxide, the more ionic hydride, and/or the more ionic oxide.
2. Contrast the chemical and physical properties of the second–row main–group elements with the properties of the heavier members in the same groups.
3. Compare the properties of elements within a group and between groups. Include valence electron configurations, common oxidation states, and trends in atomic radii, first ionization energies, and electronegativities.

B. Group 3A and Boranes

1. Discuss the uses and origins of the group 3A elements, and the unique properties of boron and its electron-poor bonding.
2. Describe the unique bonding found in boranes. Draw the structure of diborane and explain its bonding.

C. Group 4A

1. Describe how each of the 4A elements is found in nature, and give methods of preparation and commercial uses.
2. Describe the carbon allotropes.
3. Briefly describe the chemistry of carbon oxides, carbonates, cyanides, and carbides, including commercial uses of these compounds.
4. Given the formula of a silicate–containing mineral, determine the charge, the number of shared oxygens, and the structure of the silicate.

D. Group 5A

1. Describe how each of the 5A elements is found in nature, and give one commercial use.
2. Give an example of a nitrogen–containing compound for each common oxidation state exhibited by nitrogen. For each compound, sketch its electron–dot structure and describe its geometry.
3. Briefly describe the chemistry of ammonia, hydrazine, and the nitrogen oxides.
4. Give an example of a phosphorus–containing compound for each common oxidation state exhibited by phosphorus. For each compound, sketch its electron–dot structure and describe its geometry.

E. Group 6A

1. Describe the properties of the group 6A elements.
2. Give an example of a sulfur–containing compound for each common oxidation state exhibited by sulfur. For each compound, write a balanced chemical equation for its preparation, sketch its Lewis electron–dot structure, and describe its geometry.
3. Describe the phase changes in sulfur and their basis.

F. Halogen Oxoacids

1. Give an example of a halogen oxoacid, HXO_n, for $n = 1, 2, 3$, and 4. Name each acid, sketch its Lewis electron–dot structure, and describe its geometry.

Chapter in Brief

The main-group elements demonstrate strong periodic trends in ionization energy, electronegativity, atomic radius, and metallic character. The second-row elements have some unique properties as a result of their small size and high electronegativities. The families will be examined as groups, with particular attention given to the unique properties of boron, carbon, silicon, nitrogen, phosphorus, and sulfur.

19.1 A Review of General Properties and Periodic Trends

Overall, the main group elements are divided into metals on the left of the periodic table, non-metals on the right of the periodic table, and semimetals with intermediate properties along a stairstep line dividing them.

From left to right across the periodic table, the effective nuclear charge (Z_{eff}) increases, as the additional electrons are being added into a shell that does not completely shield the additional nuclear charge.

As a result of the increased Z_{eff}, electrons are more strongly attracted to the nucleus. Ionization energy and electronegativity increase to the right, atomic radius and metallic character decrease.

From top to bottom down the periodic table, atomic radius increases as additional electron shells are occupied. As the valence electrons are farther from the nucleus, ionization energy and electronegativity decrease, and metallic character increases.

The more metallic an element, the more likely it is to form ionic compounds with nonmetals. Binary hydrides can be metallic, in which case they are ionic solids, or nonmetallic, in which case they are gases, liquids, or low-melting-point solids. Similar trends are seen with oxides: metallic oxides are high-melting-point solids, while nonmetal oxides are gases or volatile liquids at room temperature.

Workbook Problem 19.1

Predict which of these pairs of elements has more metallic character:

C or Si   Be or Li   Se or Br   As or I

Strategy: Examine the position of the elements on the periodic table.

Step 1: Predict the metallic character of the elements based on their position on the periodic table.

19.2 Distinctive Properties of the Second–Row Elements

The properties of second–row elements differ markedly from those below them in the same periodic group, as they have especially small sizes and high electronegativities. This causes the elements in this row to be more nonmetallic: they form mainly covalent molecular compounds, with a maximum of four covalent bonds, as they do not have access to *d* orbitals.

Hydrogen bonding is limited to N, O, and F.

The small size of second-row elements allows for the formation of multiple bonds when $2p$ orbitals overlap to form π bonds. $3p$ orbitals are more diffuse, leading to longer bond distances and poor π overlap. For example, O_2 is double-bonded, while S_8 consists of single-bonded, crown-shaped rings.

19.3 The Group 3A Elements

The group 3 elements are boron, aluminum, gallium, indium, and thallium. These are the first of the *p* block elements and have the valence electron configuration ns^2np^1 and oxidation state of +3. As expected, metallic character increases down the group: boron (addressed below) is the only semimetal: the others behave as metals. Aluminum (also addressed below) is the most commercially important member of the group.

Gallium is liquid from 29.2 °C to 2204 °C. It is used to make gallium arsenide, a semiconductor used in making diode lasers for laser printers, CD players, and fiber optic devices.

Indium is also used in making semiconductor devices such as transistors and thermistors – electrical resistance thermometers.

Thallium is extremely toxic and has no commercial uses.

19.4 Boron

Boron is relatively rare, but occurs in concentrated deposits of borate minerals.

Pure crystalline boron is obtained by the following reaction:

$$2\ BBr_3\,(g) + 3\ H_2\,(g) \rightarrow 2\ B\,(s) + 6\ HBr\,(g)$$

It is a strong, high-melting-point substance that is chemically inert at room temperature, making it a desireable component in high–strength composite materials.

Boron halides are highly reactive, volatile covalent compounds that consist of trigonal planar BX_3 molecules. They have a vacant $2p$ orbital that allows them to behave as Lewis acids: they will form adducts with Lewis bases by accepting a share of a nonbonded pair of electrons.

Boron will also react with metal fluorides to form the tetrahedral BF_4^- anion, and is a catalyst in industrially important organic reactions.

Boron hydrides (boranes) are volatile, molecular compounds with formulas B_nH_m, the simplest of which is dibrorane: B_2H_6. This is of interest because of its unusual structure: rather than having a structure similar to ethane (C_2H_6), it has an *electron-deficient* structure in which two BH_2 groups are connected by bridging hydrogens: there are only 12 valence electrons. The unusual B-H-B bonds are much longer than the terminal B-H bonds.

19.5 Aluminium

The most abundant metal in the Earth's crust, aluminum is very difficult to refine: until an economical method for its manufacture was developed in 1886, it was considered a precious metal.

Ruby and sapphire are impure forms of Al_2O_3, the red color of rubies coming from chromium impurities and the blue of sapphires from tin impurities.

Aluminum metal is purified from bauxite - $Al_2O_3 \bullet x\ H_2O$. The Al_2O_3 is removed by the Bayer process, which takes advantage of the fact that this is an amphoteric oxide, and the aluminum is then electrolyzed out of a mixture of Al_2O_3 and cryolite by the Haber process.

Aluminum is a reducing agent that loses 3 electrons to form Al^{3+} ions. It reacts vigorously with halogens to form halides with formula AlX_3. AlF_3 and $AlCl_3$ have extended crystal structures, but $AlBr_3$ and AlI_3 both form dimers with bridging halogens similar to diborane.

When exposed to air, aluminum oxidizes immediately to form a thin, hard oxide coating that protects the underlying metal from oxygen or water. However, aluminum will still react with acid or base to form hydrogen gas and either Al^{3+} ions in acid solutions, or aluminate ions $(Al(OH)_4^-$ in basic solutions.

19.6 The Group 4A Elements

Carbon, silicon, germanium, tin, and lead are important both in industry and in living organism, and exemplify the increase in metallic character down the periodic table.

Carbon is a nonmetal that is present in all plants and animals, and an essential part of biological molecules.

Silicon is a semimetal that makes up numerous silicate minerals, is the second-most abundant element in the Earth's crust, and with germanium is used to make solid-state electronic devices.

Germanium is a high-melting-point brittle semiconductor with the same structure as diamond and silicon.

Tin and lead are both soft, malleable, low-melting-point metals that have been known since ancient times. Tin has two allotropes: malleable, silvery metallic *white tin* and brittle, semiconducting *gray tin*. White tin is the most stable allotrope at room temperature, but when kept below 13 °C for long periods, it crumbles into grey tin, in what is known as tin disease. Lead is found only in metallic form.

This group has the valence electron configuration ns^2np^2, giving +4 as the most common oxidation state. The chlorides of the group 4A elements are all sp^3 covalently-bonded, volatile molecular liquids. An oxidation state of +2 is also possible for Sn and Pb in solution, and generally leads to ionic bonding. There are no +4 aqueous ions: these species instead exist as complex ions such as $Sn(OH)_6^{2-}$.

19.7 Carbon

Elemental carbon forms a number of allotropes, discussed below:

Diamond has an sp^3-bonded covalent network in which each C atom forms a tetrahedral array of σ bonds. This forms the hardest known substance, as well as the element with the highest melting point: approximately 9000 K at a pressure of 6-10 million atm. The tight localization of the electrons makes diamond an electrical insulator.

Graphite has a two-dimensional, sheet-like structure in which sp^2 hybrid orbitals form trigonal planar σ bonds to three neighboring C atoms, while the remaining *p* orbital forms a delocalized π bond with its neighbors. These delocalized electrons make graphite useful as an electrode: conductivity is 10^{20} greater than that of diamond.

Gases can get in between the sheets, which are held together only by dispersion forces. The sheets can slide over one another, making graphite useful as a lubricant.

Graphene is a two-dimensional array of hexagonal –arranged carbon atoms just one atom thick. It is very strong, flexible, and an excellent conductor, making it a possible material for electronics and composite materials.

Fullerene is a nearlyspherical allotrope of C_{60} that can be prepared by vaporizing graphite in a helium atmosphere. As a molecular substance, it is soluble in nonpolar solvents, unlike graphite and diamond.

Derivatives of fullerene include compounds in which other atoms are attached to the C_{60} cage, and compounds in which metal atoms are trapped in the C_{60} cage.

Similar compounds include the egg-shaped C_{70} molecule and *carbon nanotubes*. Carbon nanotubes can conduct electricity 10 times better than copper and will probably replace silicon in electronic devices.

Carbon forms more than 40 million compounds, most of which are organic. Some of the inorganic compounds of carbon are discussed below:

Oxides of carbon include carbon monoxide and carbon dioxide:

Carbon monoxide is a colorless, odorless, toxic gas that forms when carbon or hydrocarbs are burned in limited oxygen. The toxicity of CO arises from the fact that it binds 200 times more tightly to hemoglobin than oxygen. Thus, even low levels reduce the blood's ability to deliver oxygen to the tissues.

Carbon dioxide is a colorless, odorless, nonpoisonous gas. It is produced when fuels burn in an excess of O_2, is an end-product of metabolism, is a by-product of the yeast–catalyzed fermentation of sugar in the manufacture of alcoholic beverages, and is produced when metal carbonates are heated or reacted with acids.

Carbon dioxide has numerous uses, including providing the "fizz" in beverages, and in fighting fires. CO_2 is more dense than air, allowing it to settle over a small fire like a blanket, smothering it. Solid CO_2 is known as dry ice. It sublimes at -78 °C, and is used primarily as a refrigerant.

Supercritical carbon dioxide, abbreviated $scCO_2$ is a form that exists above 31 °C and 73 atmospheres. It is neither true liquid nor true gas, and is nontoxic, nonflammable, non-polar, and easily recovered, making it useful as a replacement for organic solvents. It is already used to decaffeinate tea and coffee and in "green" dry cleaning, with other uses being investigated in coating drug particles and manufacturing computer chips.

Carbonates also come in two forms: carbonates with the CO_3^{2-} ion, and hydrogen carbonates or bicarbonates, that contain the HCO_3^- ion.

Carbonates include Na_2CO_3, also known as soda ash. It is used in glassmaking, while $Na_2CO_3 \cdot 10\ H_2O$ or washing soda is used in laundering textiles: the carbonate ions precipitate cations from hard water, as well as producing OH^- ions which remove grease.

Sodium hydrogen carbonate, or sodium bicarbonate - $NaHCO_3$ – is also called baking soda. It reacts with acidic substances in foods to produce bubbles of CO_2 that cause dough to rise.

Hydrogen cyanide, HCN is a highly toxic volatile substance that is weakly acidic in solution, dissociating to form CN^- ions. These are known as *pseudohalides* because they behave like Cl^-, precipitating out Ag ions. In forming complex ions, cyanide behaves as a Lewis base, bonding through the lone pair of electrons on carbon.

The toxicity of HCN and cyanides is due to the strong and irreversible binding of CN^- to the iron (III) atom contained in cytochrome oxidase, an important enzyme in cellular metabolism. With this enzyme unable to function, cellular energy production ceases and death occurs rapidly.

CN^- ions are able to react with ores containing Au and Ag. When the cyanide salts are reduced with zinc, the precious metal can be recovered.

Carbides are binary compounds of carbon in which the carbon atom has a negative oxidation state. They can be ionic carbides of active metals (CaC_2), interstitial carbides of transition metals (Fe_3C) or covalent network carbides (SiC).

Workbook Problem 19.2

Determine the oxidation state of carbon in the following compounds:

CO CO_2 C_2O_3 CO_3^{2-} HCN CaC_2

Strategy: Use the rules for assigning oxidation numbers to determine the oxidation state of carbon.

Step 1: Determine the oxidation state of carbon.

19.8 Silicon

Silicon is a hard, gray, semiconducting solid that melts at 1414 °C. It has a diamondlike structure, but does not have a graphite-like allotrope as the silicon atoms are too large for effective overlap of π orbitals.

In nature, silicon is generally found in silica and silicate minerals. To generate elemental silicon, silica sand is reduced by carbon to form silicon and carbon monoxide. To make the ultrapure silicon needed for electronics, silicon is converted to $SiCl_4$, purified by fractional distillation, and reduced with hydrogen gas to form pure silicon and HCl. This can then be further purified by zone refining, in which a rod of silicon is slowly melted from top to bottom. Impurities concentrate in the molten zone, and are dragged to the bottom of the rod.

Silicates are ionic compounds that contain silicon oxoanions and metal cations. The basic building block is the SiO_4 tetrahedron. Alone, it is the orthosilicate ion SiO_4^{4-}, but more commonly, there is an oxygen shared between units, leading to a large number of possible mineral structures, including cyclic anions, chains, layers, and extended three-dimensional structures. This bonding flexibility means that silicates include such varied mineral types as fibrous asbestos, sheets of mica, and emeralds.

Aluminosilicates occur when partial substitution of the Si^{4+} ion with Al^{3+} occurs. Among these, *feldspar*s are the most common of all minerals. In *zeolites*, the tetrahedral are joined together in an open structure with a three-dimensional network of cavities that can only be entered by small molecules, making them useful as molecular sieves to separate small molecules from larger ones, and as catalysts in industrial processes including the manufacture of gasoline.

Workbook Problem 19.3

The formula unit of garnet has three SiO_4^{4-} units balanced by Ca^{2+} and Al^{3+} cations. What is the formula unit of garnet?

Strategy: From the information provided, determine the overall silicate structure, remembering that oxygens are shared in the tetrahedral units. Balance the charge on the silicate using the ions in a correct ratio.

Step 1: Determine the silicate structure.

Step 2: Determine how many of each cation is needed to balance the charge on the silicate unit.

Step 3: Write the formula unit.

19.9 The Group 5A Elements

Nitrogen, phosphorus, arsenic, antimony, and bismuth exhibit the expected trends down the periodic group — increasing atomic size, decreasing ionization energy, and decreasing electronegativity and acid-base properties of oxides. N and P oxides are acidic, As and Sb oxides are amphoteric, and bismuth oxides are basic.

The valence configuration of the group is ns^2np^3, which allows for a variety of oxidation states. Nitrogen and phosphorus will exhibit all oxidation states between -3 and +5, while As and Sb will exhibit both +3 and +5 oxidation states, but the +5 oxidation state is less stable as atomic size increases.

Sb^{3+} and Bi^{3+} are found in salts, but there are no simple cations of N or P.

As, Sb, and Bi are found in sulfide ores and are used in making various metal alloys. Arsenic is used in pesticides and semiconductors, and bismuth is present in some pharmaceuticals, such as Pepto-Bismol.

19.10 Nitrogen

Nitrogen is a colorless, odorless, tasteless gas that makes up 78% of Earth's atmosphere. It can be separated from liquid air by fractional distillation.

Nitrogen gas can be used as a protective inert atmosphere in manufacturing processes, while liquid nitrogen is commonly used as a refrigerant.

A large amount of energy is required to break the N≡N bond, so reactions involving N_2 typically have a high activation energy and/or an unfavorable equilibrium constant. As temperatures increase, equilibrium shifts to the right. The many compounds formed by nitrogen are conveniently classified by oxidation state.

Ammonia (NH_3) is the starting material for industrial synthesis of other nitrogen compounds. It is synthesized by the Haber process, which requires high temperatures, high pressures, and catalaysis.

Colorless, strong-smelling, and gaseous at room temperature, NH_3 molecules have a lone pair of electrons, making them polar, trigonal pyramidal, water soluble, and easily condensed, as it is able to hydrogen bond. This ability also makes it an excellent solvent for ionic compounds. It is a Brønsted–Lowry base, and reacts with acids to yield ammonium salts.

Hydrazine (NH_2NH_2) is an ammonia derivative in which one H is replaced by an *amino* (NH_2) group. It is prepared by reacting ammonia with OCl^-.

A poisonous, colorless liquid that smells like ammonia, hydrazine is explosive in the presence of oxidizing agents including air, and is used as a rocket fuel: when it reacts with N_2O_4 (*l*), the two liquids generate large volumes of nitrogen gas and water vapor. It is safely handled in aqueous solution, where it is a weak base and versatile reducing agent.

Oxides of nitrogen: many exist, three will be discussed:

N_2O, or nitrous oxide, is a colorless, sweet–smelling gas used as a propellant and dental anesthetic as it is mildly intoxicating in small doses, leading to its common name of "laughing gas." Produced by gently heating molten ammonium nitrate.

NO, or nitric oxide, is a colorless gas that can be produced in small quantities by reacting copper metal with dilute nitric acid. It is important in biological processes where it transmits nerve impulses, kills harmful bacteria, and dilates blood vessels to increase blood flow.

NO_2 , or nitrogen dioxide, is a toxic, reddish-brown gas that can be produced by reacting copper with concentrated nitric acid. It is paramagnetic, and tends to dimerize, forming a N-N bond between two units.

HNO_2, or nitrous acid, is produced when NO_2 reacts with water. This is a disproportionation reaction in which nitrogen goes from +4 in NO_2 to +3 in nitrous acid, and also +5 in HNO_3 – nitric acid. It tends to again disproportionate into nitric oxide and nitric acid.

HNO_3, or nitric acid, is a strong acid primarily used to manufacture ammonium nitrate for fertilizers, as well as explosives, plastics, and dyes. It is produced by the multistep Ostwald process in which ammonia is oxidized to nitric oxide, nitric oxide is oxidized to nitrogen dioxide, and nitrogen dioxide is disproportionated in water. Further removal of water generates concentrated nitric acid, which is 15 M, and 68.5% HNO_3 by mass. It often has a slight yellow color due to the presence of small amounts of NO_2 produced by decomposition.

Nitric acid is a stronger oxidizing agent than H^+ alone, and will oxidize relatively inactive metals, but the product depends upon the nature of the reducing agent and the reaction conditions.

Aqua regia is a 3:1 volume ratio of concentrated HCl and concentrated HNO_3. It is an even stronger oxidizing agent, and will oxidize even inactive metals such as gold, which do not react with either component separately.

Workbook Problem 19.4

Upon heating, ammonium nitrate decomposes to produce nitrous oxide (N_2O) and water. What volume of nitrous oxide, collected over water, at a total pressure of 705 mmHg and

22°C, can be produced from 5.3 g ammonium nitrate? (The vapor pressure of water at 22°C is 24 mmHg.)

Strategy: After writing the balanced chemical reaction, determine the volume of N_2O from the pressure of N_2O using the ideal gas law. You must determine the pressure of N_2O using Dalton's law.

Step 1: Write the balanced chemical equation.

Step 2: Calculate the pressure of N_2O, using the total pressure and the vapor pressure of water.

Step 3: Calculate the number of moles of N_2O produced from 3.5 g NH_4NO_3.

Step 4: Calculate the volume of N_2O using the ideal gas law.

19.11 Phosphorus

Found in phosphate rock as $Ca_3(PO_4)_2$, and in fluorapatite, $Ca_5(PO_4)_3F$. Apatites are minerals with the general formula 3 $Ca_3(PO_4)_2 \cdot CaX_2$ ($X^- = F^-$ or OH^-).

Phosphorus is also important in living systems, and is the sixth most abundant element in the human body. Tooth enamel is almost pure hydroxyapatite - $Ca_5(PO_4)_3OH$ – while bones are hydroxyapatite and collagen. Phosphate is also an essential part of the backbone of nucleic acids, and the phospholipids that are the major components of cell membranes.

Industrial production of phosphorus involves heating phosphate rock, coke, and silica sand at 1500 °C. The gaseous P_4 tetrahedra produced are condensed by passing it through water.

Two main allotropes of phosphorus exist:

White phosphorus is toxic, waxy, white solid with discrete P_4 tetrahedra. It is nonpolar, has a low melting point, and is highly reactive due to an unusual bonding structure that requires "bent" overlap of *p* orbitals. If exposed to air, white phosphorus will burst into flame.

Red phosphorus can be produced by heating white phosphorus without air. It is a polymeric form, which is nontoxic, less soluble, and less reactive.

Phosphorus will form compounds with all oxidation states from -3 to +5, but +3 and +5 are most common. It is less electronegative than nitrogen, so is more likely to be found in a positive oxidation state.

Phosphine (PH_3) is a colorless, extremely poisonous gas. It is neutral in solution, and easily oxidized, burning in air to form phosphoric acid – H_3PO_4.

Halides will form when phosphorus reacts with any halogen. When halogen is limited, the halides have the formula PX_3. When halogen is in excess, PX_5 is formed. These compounds are gases, volatile liquids, or low–melting solids.

Oxides of phosphorus follow a similar pattern. When oxygen is limited, P_4O_6 forms, but when oxygen is in excess, P_4O_{10} is formed. Both are acidic oxides that react with water to form acids. P_4O_{10} has such a high affinity for water that it is used as a drying agent for gases and organic solvents.

Phosphorous acid, H_3PO_3, is a weak diprotic acid in which only two of the three hydrogens are attached to oxygen.

Phosphoric acid, H_3PO_4, is a low-melting, colorless crystalline solid. In the laboratory, it is a syrupy, aqueous solution that is 82% H_3PO_4 by mass. For use as a food additive, it is manufactured by burning pure liquid phosphorus in the presence of steam. For fertilizer, an impure version is made by treating phosphorus rock ($Ca_3(PO_4)_2$) with sulfuric acid. When this is reacted again with phosphorus rock, the result is called triple superphosphate.

19.12 The Group 6A Elements

Oxygen, sulfur, selenium, tellurium and polonium exhibit the expected trends down the periodic group. O and S are typical nonmetals. Se and Te are largely nonmetallic and are both semiconductors. Polonium is a radioactive metal found in trace amounts in uranium ores.

The valence configuration of the group is ns^2np^4, which makes the -2 oxidation state the most common. As metallic character increases, this oxidation state is increasingly unstable, and S, Se, and Te are commonly found in positive oxidation states, especially +4 and +6.

Commercial uses of Se, Te, and Po are limited. Se is used in making red glass and in photocopiers, Te is used in alloys to improve machinability, and Po is a heat source for space equipment and a source of alpha particles.

19.13 Sulfur

Sulfur accounts for 0.016% of the Earth's crust by mass, has numerous allotropes, and is found in large underground deposits and as a component of numerous minerals. In biology, it is a component of proteins.

Rhombic sulfur is the most stable form. It is a yellow crystalline solid that contains crown-shaped S_8 rings. Above 95 °C, monoclinic sulfur is more stable. In this allotrope, the S_8 rings have a different packing arrangement than the crystals of rhombic sulfur.

Molten sulfur is a fluid, straw–colored liquid at just above its melting point, but between 160 °C and 195 °C, the S_8 rings open and form long chains that can contain more than 200,000 S atoms. In this state, the sulfur becomes viscous and dark, reddish-brown. Above 195 °C the chains break, and viscosity declines. If molten sulfur is dropped in water, it freezes in a disordered state known as plastic sulfur.

Sulfur forms numerous compounds:

Hydrogen sulfide (H_2S) is a colorless gas with a strong odor of rotten eggs. It is very toxic, and dulls the sense of smell, making exposure all the more dangerous. It can be produced in the lab by treating iron(II) sulfide with dilute sulfuric acid, or by hydrolysis of thioacetaminde for use in qualitative analysis. It is a weak diprotic acid and a mild reducing agent.

Sulfur dioxide and sulfur trioxide (SO_2 and SO_3) are the most important oxides of sulfur. Sulfur dioxide is produced by burning sulfur, and is a colorless, toxic gas with a pungent odor. It is toxic to microorganisms, so is used to sterilize wine and dried fruit. In the atmosphere, sulfur dioxide is oxidized to sulfur trioxide, which dissolves in rainwater to form sulfuric acid. Both species are acidic, but little sulfurous acid (H_2SO_3) is found in solution: SO_2 will dissolve without reacting.

Sulfuric acid is the world's most important industrial chemical. It is manufactured by the **contact process**: S is burned to SO_2, SO_2 is oxidized to SO_3 with a catalyst, and the SO_3 is dissolved in sulfuric acid to make more sulfuric acid, as SO_3 dissolves slowly in water. Concentrated sulfuric acid is 18 M, and 98% H_2SO_4 by mass.

Sulfuric acid is a strong acid at the first proton, and dissociates at the second proton with a K_{a2} of 1.2×10^{-2}. It forms two series of salts: hydrogen sulfates, with a hydrogen as one of the cations ($NaHSO_4$), and sulfates (Na_2SO_4). In dilute solutions at room temerature, it behaves like HCl, oxidizing metals above H in the activity series. Hot, concentrated H_2SO_4 is a stronger oxidant, and will come apart to reform SO_2. Sulfuric acid is used to manufacture fertilizers, and in many other industrial processes.

19.14 The Halogens: Oxoacids and Oxoacid Salts

Halogens have a valence configuration of ns^2np^5. They gain an electron to ionize, and share one electron in molecular compounds in which they have an oxidation state of -1.

Numerous positive oxidation states are also available for the larger halogens – Cl, Br, and I – allowing them to share valence electrons with oxygen. The oxidation states available are +1, +3, +5, or +7, and the general formula for an oxoacid is HXO_n, where the value of n depends on the oxidation state of the halogen.

The higher the oxidation state of the halogen, the stronger the acid. Acidic protons are bonded to the oxygen (despite how the formula is typically written), and these are all strong oxidizing agents.

Hypohalous acids are formed when halogens dissolve in water, disproportionating to form an aqueous -1 ion and a +1 oxidation state in HOX:

$$X_2\,(g,\,l,\text{ or }s) + H_2O\,(l) \leftrightharpoons HOX\,(aq) + H^+\,(aq) + X^-\,(aq)$$

Equilibrium lies to the left in this reaction, but shifts toward products in basic solution, generating OX^- ions. The most important salt of these is NaOCl – sodium hypochlorite – which is a strong oxidizing agent sold in 5% solution as chlorine bleach.

Further disproportionation of OCl^- to ClO_3^- is slow, but can be produced by the reaction of $Cl_2$with hot NaOH. Chlorate salts are used as weed killers and oxidizing agents: $KClO_3$ is used in matches, fireworks, and explosives, and reacts with organic matter.

Sodium perchlorate can be converted to perchloric acid by reaction with HCl: anhydrous perchloric acid is a colorless, shock–sensitive liquid that decomposes explosively on heating. It is a powerful and dangerous oxidizing agent that will oxidize even silver and gold: even the salts are powerful oxidants: ammonium perchlorate is used as a rocket fuel. Perchlorate ions are toxic, as they prevent uptake of iodide by the thyroid, but the source of environmental exposures is unclear.

Iodine forms multiple perhalic acids as a result of its large size:

Paraperiodic acid, H_5IO_6 is a weak, polyprotic acid obtained from evaporating HIO_4 solutions.

Metaperiodic acid, HIO_4, is a strong monoprotic acid produced when H_5IO_6 loses water.

Putting It Together

When sulfur dioxide reacts with aqueous hydrogen sulfide, elemental sulfur is produced. Write a balanced equation for the reaction. If a metric ton of coal, containing 3.3% sulfur by mass is burned, how much hydrogen sulfide is needed to remove the sulfur dioxide? How many kilograms of elemental sulfur are produced? What volume of sulfur dioxide gas is produced at 150 °C and 740 mm Hg?

Self–Test

This section is intended to test your knowledge of the material covered in this chapter. Think through these problems, and make certain you understand what they are asking. Make sure your answers make sense. Successful completion of these problems indicates that you have mastered the material in this chapter. You will receive the greatest benefit from this section if you use it as a mock exam, as this will allow you to determine which topics you need to study in more detail.

True–False

1. The periodic table is divided into metals on the left, nonmetals on the right, and semimetals in the center.
2. Metallic character decreases from top to bottom in a column.
3. Second-row elements have unique bonding properties because of their small sizes and high electronegativities.
4. It is common to find double (π) bonds in third- and fourth-row elements.
5. Boranes can be connected with bridging hydrogen atoms.
6. Carbon allotropes include nonconducting covalent solids and sheets with delocalized electrons that will conduct electricity.
7. Silicates are major components of the Earth's crust, and often consist of lengthy chains of double-bonded SiO_4^{4-} tetrahedra.
8. Nitrogen is obtained in pure form by fractional distillation of liquefied air.
9. White phosphorus is made up of tetrahedra of P atoms with strained, unstable bonds.
10. When it is liquefied, sulfur decreases steadily in viscosity to its boiling point.

Multiple Choice

1. Which of the following are characteristic of metals?
   a. brittle
   b. low ionization energy
   c. form covalent hydrides
   d. readily become anions

2. Because of their small size, second-row elements are able to form a maximum of ___ bonds.
   a. 2
   b. 3
   c. 4
   d. 5

3. Boron halides are able to act as Lewis acids because
   a. the empty *p* orbital can accept a lone pair from another atom
   b. they readily give up hydrogen ions in solution
   c. they are so reactive that they split water molecules
   d. all of the above

4. Nitrogen can exist in which of the following oxidation states?
   a. -5, -4,-3, 0, +1,+2
   b. -3, -2, -1, 0
   c. +1, +2,+3, +4, +5
   d. -3, -2, -1, 0, +1, +2, +3, +4, +5

5. Phosphorus is found in:
   a. bones and teeth
   b. cell membranes
   c. DNA and RNA
   d. all of the above

6. 6A elements have the electron configuration
   a. ns^2np^2
   b. ns^2np^3
   c. ns^2np^4
   d. ns^2np^5

7. The halogen that is capable of the most flexibility of bonding is
   a. F
   b. Cl
   c. Br
   d. I

8. The most electronegative elements in the periodic table are the
   a. halogens
   b. noble gases
   c. metals
   d. boranes

9. The only elements that will form triple bonds are
   a. N and P
   b. C and N
   c. Br and I

d. O and N

10. Which of the following correctly identifies the highest abundances
    a. crust: silicon, atmosphere: oxygen, ocean: phosphorus
    b. crust: silicon, atmosphere: nitrogen, ocean: oxygen
    c. crust: phosphorus, atmosphere: oxygen, ocean: nitrogen
    d. crust: calcium, atmosphere: oxygen, ocean: hydrogen

Fill–in–the–Blank

1. From left to right across the periodic table, effective nuclear charge__________________, ionization energy __________________, atomic radius __________________, and metallic character __________________.
2. Carbon has a number of allotropes, ranging from the tetrahedral covalent solid ______________, to the sheets of ________________, to the remarkable spheres of __________________.
3. When phosphorus is manufactured, an unstable allotrope known as ________________________ is produced. It is highly reactive because of the _____________ pi bonds in the P_4 tetrahedra. If it is heated in the absence of air, it is converted to the more stable, less toxic, polymeric form known as ________________________.
4. Silicates are typically found as chains of rings with a basic unit of _________.
5. The properties of sulfur are dramatically temperature dependent. At room temperature, sulfur is a _____________. When heated, it melts into a ___________, then becomes more viscous as the ___________ break and form ________________. When these begin to break down, the _____________ of the liquid begins to decrease with temperature.

Matching

Aluminosilicate	a. commercial process for making nitric acid from ammonia
Borane	b. purification technique in which a heater melts a narrow zone at the top of a rod of a material, then sweeps slowly down the rod, carrying impurities
Carbide	c. silicate mineral in which partial substitution of Al^{3+} for Si^{4+} has occurred
Contact process	d. any compound of boron and hydrogen
Ostwald process	e. an ionic compound containing silicon oxoanions along with other cations
Silicate	f. commercial process for making sulfuric acid from sulfur

Zone refining	g. binary compound in which carbon has a negative oxidation state

Problems

1. For the following pairs of elements, determine which has the more metallic character.
   a. Na or Rb b. In or Ga c. Cs or Sr

2. For the following pairs of elements, determine which has the more ionic hydride.
   a. K or Na b. Sn or In c. Sb or As

3. What are the general trends down a group in the periodic table for ionization energy, atomic radius, electronegativity, and basicity of oxides?

4. Describe the unique bonding in diborane.

5. Describe four of the allotropes of carbon in terms of their bonding, structure, and properties.

6. Give an example of a nitrogen–containing compound for each common oxidation state exhibited by nitrogen.

7. Give the name of the halogen oxoacid, HXO_n (X = Cl, Br, or I and n = 1, 2, 3, and 4).

8. Give the chemical equation for the production of NO_2 from nitric acid and copper.

Challenge Problem

Calcium carbide reacts with water to form acetylene (C_2H_2). If 15.0 g of calcium carbide react with 7.45 g of water, how many grams of acetylene are produced? If the resulting acetylene is burned to form carbon dioxide and water, how much energy is produced? (Assume 100% efficiency for all reactions.)

Possibly useful information:

ΔH_f C_2H_2 (*g*) = 227.4 kJ/mol
ΔH_f CO_2 (*g*) = -393.5 kJ/mol
ΔH_f H_2O (*g*) = -241.8 kJ/mol

CHAPTER 20

TRANSITION ELEMENTS AND COORDINATION CHEMISTRY

Chapter Learning Goals

A. d Block Elements: Properties

1. Write valence electron configurations for transition metal atoms and ions.
2. Compare the properties of the first–transition–series elements. Include appearance, valence electron configurations for atoms and ions, common oxidation states, and trends in melting points, atomic radii, densities, ionization energies, and standard oxidation potentials.
3. Balance equations for redox reactions of chromium, iron, and copper.

B. Coordination Compounds: Ligands

1. Write formulas of coordination complexes. Identify the ligands and their donor atoms. Determine the coordination number and the oxidation state of the metal and the charge on any complex ion.
2. Given the electron dot structure of a molecule or ion, determine whether it can serve as a chelate ligand.
3. Name coordination compounds.
4. Write formulas for coordination compounds given their names.

C. Coordination Compounds: Constitutional Isomers and Stereoisomers

1. Identify linkage isomers and ionization isomers.
2. Determine the number and structures of diastereoisomers possible for a given coordination complex.
3. Determine which coordination complexes are chiral, and draw the structures of the enantiomers.

D. Coordination Compounds: Valence Bond Theory

1. Give a valence-bond-theory description of a coordination complex, show the number of unpaired electrons and the hybrid orbitals used by the metal ion.

E. Coordination Compounds: Crystal Field Theory

1. Give a crystal-field-theory description of a coordination complex, and show the number of unpaired electrons.

Chapter in Brief

This chapter examines the properties and chemical behavior of transition-metal compounds, with particular emphasis on coordination compounds. Beginning with a review of the electronic configurations of the transition elements and how these configurations affect the properties and oxidation states of the ions formed, the chemistry of chromium, iron, and copper are examined in detail. Coordination chemistry is then explored, including structural properties, isomerization, color, and magnetism of the compounds. Two different theories, valence bond theory and crystal field theory, that account for these properties are then discussed.

20.1 Electron Configurations

Note: For convenience in this chapter, the orbitals will be listed by principal quantum number, as the $n + 1$ s electrons are lost first when transition metals ionize.

d orbitals begin at atomic number 21 – scandium – and continue through to zinc. The filling proceeds according to Hund's rule, with the $n + 1$ s shells filling first, followed by one electron in each of the d orbitals, thereafter doubling up the electrons.

There are two exceptions to this pattern: Cr and Cu are both able to gain stability by varying from this rule. Cr places one electron in each of the orbitals for a completely half-filled configuration: $4s^1 3d^5$. Copper pulls one electron from the 4s orbital to gain a fully filled d orbital and leaves a half-filled s orbital: $4s^1 3d^{10}$.

Exceptions from the expected orbital-filling pattern always result in completely half-filled subshells or completely filled shells. The nd and $n + 1s$ orbitals have very similar energies, allowing electrons to move between them if it is energetically favorable to do so.

Two half-filled subshells minimize electron-electron repulsions, as does a fully-filled d subshell with a half-filled s subshell.

When electrons are lost, the effective nuclear charge increases. The d orbitals experience a steeper drop in energy as the effective nuclear charge increases, making them quickly lower in energy than the s orbital. As a result, the s orbital empties first. Easy to predict electron configurations of transition metal cations.

The designation of transition metals as groups 1B – 8B was developed as these groups have analogous electron structures to the elements in groups 1A – 8A. Copper, in group 1B has Cu^+ as its most common ion, while Zn, in group 2B, will most commonly form a Zn^{2+} ion, like the main-group metals in group 2A. However, these designations are not as straightforward as those of the main-group elements, and should be used with care.

EXAMPLE:

Write the electron configuration of Cu^{3+}. Would this be expected to be a stable species?

SOLUTION: Writing the electron configuration for neutral copper gives:

$[Ar]3d^{10}4s^1$

Removing three electrons to make Cu^{3+} gives:

$[Ar]3d^8$, as the s orbital will empty first.

This will not be a stable species, as it has a high effective nuclear charge, and two half-filled d orbitals. The most stable ion for copper will be Cu^+, with electron configuration $[Ar]3d^{10}$.

Workbook Problem 20.1

Write the electron configurations for the following species:

Ti　　　Mn in MnO_4^-　　　Ni^{2+}

Strategy: Write the expected electron configuration for the neutral element, determine the charge, and remove electrons accordingly.

Step 1: Write the electron configuration of the neutral atom.

Step 2: Determine the oxidation state of the metal.

Step 3: Remove the necessary electrons.

20.2 Properties of Transition Elements

As metals, all these elements are malleable, ductile, lustrous, and good conductors of heat and electricity. The sharing of *d*, as well as *s*, electrons gives rise to stronger metallic bonding in these elements than is seen in the group 1A and 2A metals, which makes them harder and more dense, with higher melting and boiling points.

Melting points increase as the number of unpaired *d* electrons available for metallic bonding increases and then decrease as the *d* electrons pair up and become less available for bonding, meaning that the melting points reach a maximium in the middle of each series.

Atomic radii and densities demonstrate a similar pattern: as the unpaired *d* electrons increase to a half-filled orbital, the atomic radii decrease and densities increase. As the electrons begin to pair up, the electron repulsions increase and the effective nuclear charge decreases, so the atomic radii increases, and the density decreases.

The effects on atomic radii are more subtle in the transition metals than they are in the main-group elements. The similar atomic radii among the transition metals accounts for their ability to blend into alloys.

Across the *f*-block lanthanide elements, the increase in effective nuclear charge is almost exactly balanced by the increase in size expected from adding an entire quantum shell. As a result of the **lanthanide contraction**, the elements in the third transition series have almost the same radii as the elements in the second transition series. This means that the third series, which includes Os, the most dense metal, are unusually dense.

The ionization energies of transition metals increase from left to right across a series, due to an increase in the effective nuclear charge and the corresponding decrease in atomic radius. The $E°$ for the **oxidation** potential of the first transition series is positive for every metal except Cu, meaning that these metals are oxidized more readily than H_2 gas is oxidized to H^+.

These metals can be oxidized by simple acids that lack an oxidizing anion, but the oxidation of Cu (*s*) requires a stronger oxidizing agent such as HNO_3.

The general trend for $E°$ is the same as that for ionization energies. The easier a metal is to oxidize, the easier it is to ionize.

20.3 Oxidation States of Transition Elements

Transition metals exhibit a variety of oxidation states, which can be lower than their group number. All of the first series except scandium will form +2 cations, corresponding to the emptying of the 4*s* shell.

The first series can also lose 3*d* electrons to form ions with greater charges. The increased energy required to remove additional electrons is balanced by the larger $\Delta G°$ of hydration of the more highly charged cations. The most highly charged species are found in combination with the most electronegative elements, typically O.

For the group 3B–7B metals, the group number is the highest possible oxidation state, corresponding to the loss of all the valence *s* and *d* electrons. This is not true for the 8B elements, however, as the loss of all the valence electrons is energetically prohibitive.

Transition metal ions in a high oxidation state are good oxidizing agents, while early transition metal ions in a low oxidation state are good reducing agents, but divalent ions of the later metals are poor reducing agents because of the larger effective nuclear charge.

20.4 Chemistry of Selected Transition Metals

Chromium is obtained from chromite ore, ($FeO{\cdot}Cr_2O_3$ or $FeO{\cdot}Cr_2O_4$). If it is reduced with carbon, ferrochrome, or stainless steel is produced. To obtain pure chromium, Cr_2O_3 is reduced with aluminum. Solid chromium is used to electroplate metal objects with a protective coating: it takes a high sheen and is corrosion-resistant as it develops a hard invisible coating of Cr_2O_3 that prevents further corrosion.

In aqueous solution, chromium will exist in +2, +3, and +6 oxidation states, with +3 the most stable.

When chromium metal reacts with acid in the absence of air, +2 chromium ions form a bright blue complex with 6 water molecules. When atmospheric O_2 is present, these ions are quickly oxidized to the more stable +3 ion. Although the $Cr(H_2O)_6^{3+}$ complex is violet, Cr^{3+} solutions are typically green because other anions replace some of the water molecules.

In basic solution, chromium(III) precipitates as chromium(III) hydroxide, a pale green solid that will dissolve in both acid and base. Chromium(II) hydroxide is a typical hydroxide that dissolves in acid, but not in base, and chromium IV forms $CrO_2(OH)_2$ or chromic acid (also written H_2CrO_4). This is a strong acid. The higher the oxidation state of chromium, the more polar the O-H bond, and the greater the acidity.

In the +6 oxidation state, the two most important species are chromate - CrO_4^{2-} - in basic solution, and dichromate - $Cr_2O_7^{2-}$ - in acidic solution. Dichromate is a powerful oxidizing agent used as an oxidant in analytical chemistry.

Iron is the fourth-most abundant element in the Earth's crust and is immensely important in human civilization and in living systems. Alloys of iron with V, Cr, and Mn make steels that are less readily corroded and harder than iron alone.

The most important iron ores are hematite (Fe_2O_3) and magnetite (Fe_3O_4), which are reduced with coke in a blast furnace to give pure Fe.

The most important oxidation states of iron are +2 and +3.

Reaction with an acid that lacks an oxidizing anion (HCl, for example) in the absence of air will yield the iron (II) ion $Fe(H_2O)_6^{2+}$. The oxidation ceases at this point, as the standard potential for $Fe^{2+} \rightarrow Fe^{3+}$ is positive. In air, the iron ions slowly oxidize to Fe^{3+}.

Reaction with an acid that has an oxidizing anion directly yields Fe^{3+} ions.

The reaction of Fe^{3+} *(aq)* with a base yields the highly insoluble $Fe(OH)_3$. It is exceptionally insoluble, and forms if the pH rises above 2. It is the red-brown rust stain found in sinks and bathtubs.

Copper is relatively rare, but can be found in the elemental state. Its most important ores are sulfides, including chalcopyrite, $CuFeS_2$. Sulfides are concentrated, separated from the iron, and converted to copper(I) sulfide, which can then be oxidized to form copper and sulfur dioxide. The resultant metal is 99% pure, but can be further purified by electrolysis.

Copper has high conductivity and negative oxidation potential, making it valuable for electrical wiring and corrosion-resistant water pipes. It also can be used to form alloys such as brass (Cu and Zn) and bronze (Cu and Sn).

Prolonged environmental exposure will eventually lead to oxidation: moist air and CO_2 lead to the development of $Cu_2(OH)_2CO_3$. Subsequent reaction with highly dilute sulfuric acid in acid rain generates $Cu_2(OH)_2SO_4$, the green patina that develops on bronze.

Copper has two primary oxidation states: +1 (cuprous) and +2 (cupric). However, the $E°$ for $Cu^+ \rightarrow Cu^{2+}$ is less negative than that for the reaction of $Cu \rightarrow Cu^+$. As a result, anything strong enough to oxidize copper to Cu(I) ions is also strong enough to oxidize Cu(I) to Cu(II).

Cu^+ *(aq)* will disproportionate in solution:

$$2\ Cu^+(aq) \rightarrow Cu(s) + Cu^{2+}(aq) \qquad E° = +0.37\ V,\ K = 1.8 \times 10^6$$

As a result of the large equilibrium constant, Cu^+ is not an important species in aqueous solution, although it does exist in solid compounds: when Cl^- is present, the disproportionation is reversed, and the precipitation of CuCl shifts the reaction to the left.

The more common Cu^{2+} is found in the blue aqueous $Cu(H_2O)_6^{2+}$. If aqueous ammonia is added to this ion, it will generate a blue precipitate of copper(II) hydroxide. If more ammonia is added, it will redissolve and form the dark blue complex ion $Cu(NH_3)_4^{2+}$.

20.5 Coordination Compounds

A **coordination compound** is a compound in which a central metal ion is attached to a group of surrounding molecules or ions by coordinate covalent bonds. The molecules or ions surrounding the central metal ion are called **ligands**, while the atoms that are directly attached to the metal ion are the **ligand donor atoms**. Formation of these compounds is a Lewis acid-base interaction: the electron donors are the ligands, which always have lone pairs, and the electron acceptor –acid - is the central metal ion.

Some coordination compounds are salts containing a complex cation or anion along with enough ions of opposite charge to give a compound that is electrically neutral overall. The complex ion is enclosed in brackets when it is written to show that the other ions are simply balancing charges, while the complex ion is a discrete structural unit. The term **metal complex** refers to both neutral coordination compounds, and salts of ionic coordination compounds.

The number of ligand donor atoms surrounding the metal ion is the **coordination number**. The most common are 4 and 6, but others are also well-known. The coordination number depends on the size, charge, and electron configuration of the central atom, as well as the size and shape of the ligands.

The characteristic shape of a metal complex is determined by the metal ion's coordination number. A coordination number of two indicates a linear molecule, while a coordination number of four can be either tetrahedral or square planar. A coordination number of six will be octahedral.

The charge on a metal complex is equal to the oxidation state of the metal plus the sum of the charges on the ligands. If two are known, the third can be found.

EXAMPLE:

Silver forms a neutral complex with two ammonia molecules and a nitrate ion. What is the oxidation state of the metal, and what is the formula of the complex?

SOLUTION: The charge on the complex is the sum of all the charges. The ammonia molecules are neutral, and nitrate has a charge of -1. For the complex to be neutral, the oxidation state of silver must be +1.

The formula of the complex is $[Ag(NH_3)_2]NO_3$

Workbook Problem 20.2

Determine the oxidation state and coordination number of the metal in the following complexes:

$AgCl_2^-$ $[MnCl_4]^{-2}$ $[Fe(CN)_6]^{4-}$ $[Co(NH_3)_4Br_2]Br$

Strategy: With the charge and ligands, it is possible to determine the oxidation state of the metals. The coordination number is the number of ligand attachments.

Step 1: Determine the oxidation state of the metals.

Step 2: Determine the coordination numbers.

20.6 Ligands

Ligands can be classified as **monodentate** or **polydentate** depending on the number of electron pairs that bond to the metal atom. Water and ammonia are monodentate ligands, binding through the O and N atoms. Glycinate ($NH_2CH_2COO^-$) is a bidentate ligand, that binds through both the N and O.

Chelating agents are polydentate ligands. $EDTA^{4-}$ (Ethylenediamenetetraacetate) is a **hexadentate** ligand that will bind to metals. When the chelating agent is bound to a metal, it forms a **chelate ring**. A complex that contains one or more such rings is called a metal **chelate**.

EDTA forms particularly stable complexes, and can be used to hold metal ions in solution. It can be used in treating lead poisoning, preventing the precipitation of Pb in tissues, and is often added to foods to remove metals that could catalyze the oxidation of oils (rancidness).

Naturally occurring chelators are important components of many biomolecules. The heme group in hemoglobin is a planar, tetradentate ligand that binds an iron(II) ion. This will also reversibly bind an O_2 molecule to carry oxygen through the blood.

20.7 Naming Coordination Compounds

If the compound is a salt, name the cation first and then the anion.

In naming a complex ion or a neutral complex, name the ligands first, in alphabetical order (ignoring prefixes), and then the metal. Anionic ligands end in *–o*. The complex name is one word.

More than one ligand of a particular type, use Greek prefixes (*di-, tri-, tetra,* etc.) to indicate number.

If the name of a ligand itself contains a Greek prefix, put the ligand name in parentheses and use an alternate prefix (*bis-, tris-, tetrakis-).*

Use a Roman numeral in parentheses immediately following the name of the metal, to indicate the oxidation state of the metal.

In naming the metal, use the ending *–ate* if the metal is in an anionic complex.

Anionic Ligand	Ligand Name	Neutral Ligand	Ligand Name
Bromide, Br^-	Bromo	Ammonia, NH_3	Ammine
Carbonate, CO_3^{2-}	Carbonato	Water, H_2O	Aqua
Chloride, Cl^-	Chloro	Carbon monoxide, CO	Carbonyl
Cyanide, CN^-	Cyano	Ethylenediamine, en	Ethylenediamine
Fluoride, F^-	Fluoro		
Glycinate, gly^-	Glycinato		
Hydroxide, OH^-	Hydroxo		
Oxalate, $C_2O_4^{2-}$	Oxalato		
Thiocyanate, SCN^-	Thiocyanate (S ligand) Isothiocyanate (N ligand)		

EXAMPLE:

Name the following compounds:

$[Cr(H_2O)_5Cl](NO_3)_2$ $\qquad$ $Na_2[OsCl_4NO_3]^{3-}$ $\qquad$ $[Co(gly)NO_3]Cl$

SOLUTION: Using the rules for naming coordination complexes, we find

$[Cr(H_2O)_5Cl](NO_3)_2$	The ligands are aqua and chloro. With 5 H_2O, 1 Cl^-, and 2 charge-balancing NO_3^- ions, the charge on chromium is +3. This is pentaaquachlorochromium(III) nitrate.
$Na_2[OsCl_4NO_3]^{3-}$	The ligands in this compound are chloro and nitrato. With 2 Na^+, 4 Cl^-, and 1 NO_3^-, and an overall charge of -3, the charge on osmium must be +6. Since the complex is an anion, the ending on osmium is changed to *ate*. This is sodium tetrachloronitratoosmate(VI).
$[Co(gly)NO_3]Cl$	The ligands in this compound are glycinate and nitrato. With gly^-, NO_3^- and a balancing Cl^-, the charge on cobalt is +3. The compound is glycinonitratocobalt(III) chloride.

EXAMPLE:

Write formulas for the following compounds: hexaquanickel(II) chloride, ammonium diaquatribromochlorovanadate(III), sodium hexanitratochromium(III)

SOLUTION: Determine the formula for the compounds:

The aqua ligand is H_2O. The formula is $[Ni(H_2O)_6]Cl$.

The ammonium cation precedes the formula. $NH_3[V(H_2O)_2Br_3Cl]$

The nitrato cation is NO_3^-. $Na_3[Cr(NO_3)_6]$

20.8 Isomers

In coordination chemistry, it is possible for compounds to have the same formula, but a different arrangement of constituent atoms. This leads to different physical and chemical properties.

Constitutional isomers are isomers that have different connections among their constituent atoms.

Linkage isomers arise when a ligand can bond to a metal through either of two different donor atoms. NO_2^- can use N or O as a donor atom.

Ionization isomers differ in the anion that is bonded to the metal ion. For instance, $[AgCl_2]NO_3$ has different properties than $[AgClNO_3]Cl$, but the same type of atoms.

Stereoisomers have the same connections among atoms but have a different arrangement of the atoms in space.

Diastereoisomers or *geometric isomers* have different relative orientations of their metal–ligand bonds. In a *cis* isomer, identical ligands occupy adjacent corners of the square in a square planar complex. In a *trans* isomer, identical ligands are across from one another in a square planar complex.

These isomers have different properties: dipoles will differ and polarities will differ as a result of different geometries.

Square planar complexes of the type MA_2B_2 and MA_2BC (M = metal ion; A, B, C = ligands) can exist as *cis–trans* isomers. Tetrahedral complexes cannot have isomers, as all four corners are adjacent.

Octahedral complexes of the type MA_4B_2 or MA_4BC can also exist as diastereoisomers: two B ligands can be either on adjacent or on opposite corners of the octahedron.

EXAMPLE:

Of the following compounds, which have geometric isomers? (Assume square planar or octahedral geometry.)

$Co(H_2O)_2NO_3Cl$ $[Fe(CN)_6]^{4-}$ $Ni(CO)_2Cl_2$

SOLUTION: Determine the basic structure of the compounds and whether isomers are possible.

$Co(H_2O)_2NO_3Cl$ has the basic structure MA_2BC – it will have isomers

$[Fe(CN)_6]^{4-}$ has six identical ligands – it will not have isomers

$Ni(CO)_2Cl_2$ has the basic structure MA_2B_2 – it will have isomers

20.9 Enantiomers and Molecular Handedness

Enantiomers are molecules or ions that are nonidentical mirror images of one another. They are said to have different handedness or chirality. They are also called *optical isomers* because of their different effects on polarized light.

Chiral objects do not have a symmetry plane, **achiral** objects do have a symmetry plane.

Certain molecules and ions are chiral. If in order of size, the ligands spiral to the right (clockwise), the enantiomer is referred to as "right-handed." If the ligands spiral to the left (counterclockwise) the enantiomer is referred to as "left-handed."

Enantiomers have identical properties except for their reactions with other chiral substances and their effect on plane–polarized light: a solution of one enantiomer will rotate light to either the right or left, and the other enantiomer will rotate the light the same amount in the other direction.

A **racemic mixture** is a 50:50 mixture of the two enantiomers. This will have no effect on polarized light, as the two rotations cancel each other out.

20.10 Color of Transition Metal Complexes

The color of transition metal complexes depend on the identity of the metal and the ligands. The same metal with different ligands will have different colors, and different metals with the same ligand will also have different colors.

Metal complexes absorb light by undergoing an electronic transition from the lowest energy state (E_1) to a higher energy state (E_2). The wavelength absorbed depends on the energy difference between the two energy levels: $\Delta E = E_2 - E_1$.

$\Delta E = h\nu = hc/\lambda$ where h is Planck's constant, ν is the frequency, and λ is the wavelength.

Absorbance is the measure of the amount of light absorbed by a substance. The **absorption spectrum** is a plot of absorbance versus wavelength. The observed color is complementary to the color absorbed. For instance, if a complex strongly absorbs green light, it will appear orange.

EXAMPLE:

If a compound strongly absorbs at a frequency of $1.23 \times 10^{15}\ s^{-1}$, what is the difference between the ground and excited states of this compound in kJ/mol?

SOLUTION: Use the formula $\Delta E = h\nu$ to determine the energy of the absorbed photons, then convert to kJ/mol.

$\Delta E = h\nu = 6.626 \times 10^{-34}$ J•s x $1.23 \times 10^{15}\ s^{-1} = 8.15 \times 10^{-19}$ J/photon

8.15×10^{-19} J/photon x 6.022×10^{23} photons/mol = 491 kJ/mol

Workbook Problem 20.3

If a coordination compound strongly absorbs light with a wavelength of 550 nm, what is the difference between the ground and excited states of the complex in kJ/mol of complex?

Strategy: Using the formula $E = hc/\lambda$, determine the energy of the absorbed photons, then convert to kJ/mol.

Step 1: Determine the energy of the photons absorbed.

Step 2: Convert to kJ/mol.

20.11 Bonding in Complexes: Valence Bond Theory

In **valence bond theory**, bonding results when a filled ligand orbital containing a pair of electrons overlaps a vacant hybrid orbital on the metal ion to produce a coordinate covalent bond.

The geometry of these complexes is related to the hybrid orbitals used. Linear complexes have *sp* orbitals, tetrahedral complexes use sp^3 orbitals. Octahedral complexes have six equivalent hybrid orbitals that can be formed in either from d^2sp^3 or sp^3d^2. All are 90° from one another. Square planar complexes also use *d* orbitals in dsp^2 hybrids.

Transition metal complexes can be **paramagnetic**, containing unpaired electrons and attracted by magnetic fields, or **diamagnetic**, containing only paired electrons and weakly repelled by magnetic fields. The number of unpaired electrons can be measured by the force exerted on the complex by a magnetic field.

Metals can form complexes described as **high-spin complexes** – in which the *d* electrons are arranged to maximize unpaired electrons and minimize doubly-occupied orbitals – or **low-spin complexes** – in which *d* electrons are arranged to maximize filled orbitals and minimize unpaired electrons.

EXAMPLE:

$[Co(NH_3)_6]^{3+}$ is a low-spin complex. Draw an electron diagram that supports this.

SOLUTION: Determine which hybrid orbitals are being used to minimize spin.

The free Co^{3+} ion has the valence $[Ar]3d^6$. This is an octahedral complex, so will use either d^2sp^3 or sp^3d^2 hybrid orbitals.

Co^{3+} : ↑ ↑ ↑ ↑ ↑ (3d) ___ (4s) ___ ___ ___ (4p) d^6

$[Co(NH_3)_6]^{3+}$: ↑ ↑ ↑↓ ↑↓ ↑↓ (3d) ↑↓ (4s) ↑↓ ↑↓ ↑↓ (4p) d^2sp^3

or

$[Co(NH_3)_6]^{3+}$: ↑ ↑ ↑ ↑ ___ (3d) ↑↓ (4s) ↑↓ ↑↓ ↑↓ (4p) ↑↓ ↑↓ ___ ___ ___ (4d) sp^3d^2

The d^2sp^3 configuration has fewer unpaired electrons, and is also lower energy overall. Both support the choice of this configuration.

Workbook Problem 20.4

$[Co(Cl)_6]^{3-}$ is a high-spin complex. Draw an electron diagram that supports this.

Strategy: Draw the orbital diagram and determine how to maximize unpaired electrons.

Step 1: Determine the electron configuration of the metal alone.

Step 2: Identify the possible hybridization schemes.

Step 3: Draw electron diagrams with both hybridizations.

Step 4: Choose the hybridization that maximizes electron spin.

20.12 Crystal Field Theory

Crystal field theory views the bonding in complexes as arising from electrostatic interactions and considers the effect of the ligand charges on the energies of the metal ion *d* orbitals. It is able to account for the colors and magnetic properties of transition metal complexes, as well as the difference between the high- and low-spin complexes.

In crystal field theory, there are no covalent bonds, no sharing of electrons, and no hybrid orbitals. Interactions with neutral ligands are ion-dipole.

In octahedral complexes, metal is positively charged and the ligands are negatively charged. The anions adopt the configuration that allows them to get as far away from each other as possible. The metal ligand attraction is stronger than the ligand-ligand repulsions, so the complex is stable.

Ligand charges repel negatively charged *d* electrons, causing orbital energies to be higher in the complex: the lobes of the orbitals that are closest to the ligands are higher in energy than those that fall between ligands. This causes **crystal field splitting**, abbreviated Δ.

Crystal field splitting energy, Δ, corresponds to λ's in the visible region of the spectrum, providing an explanation for the colors of the complexes. The value of Δ can be calculated from the wavelength of the absorbed light.

The effect that ligands have on Δ allows ligands to be ranked in terms of the energy changes they cause. This is known as the **spectrochemical series. Weak-field ligands** produce relatively small values of Δ and low-spin complexes, while **strong-field ligands** produce high values of Δ and high-spin complexes.

If the value of Δ is greater than the value of P, the energy required to put two electrons in one orbital, as it is more likely to be with a strong-field ligand, the electrons will pair up in the lower-energy orbital. If the value of Δ is less than that of P, the electrons will spread out through the available orbitals.

This is only an issue for complexes in which there are 4-7 *d* electrons, known as d^4–d^7 complexes. For others, only one ground-state configuration is possible, and the compounds are colorless.

In tetrahedral complexes, the energy splitting for the *d* orbitals is different. In tetrahedral complexes, none of the ligands point directly at the *d* orbitals. As a result, P is almost always higher than Δ, and these tend to be high-spin.

In square planar complexes, the energy is similar to that of octahedral complexes, with the z–axis ligands missing. This leads to a large energy gap between the *d* orbitals that are pointing toward ligands, and those that are not, favoring low-spin complexes with the higher-energy orbitals vacant.

Putting It Together

The amount of iron in ore can be determined by dissolving a sample in a non-oxidizing acid, reducing all of the Fe^{3+} to Fe^{2+} and titrating with potassium dichromate. The reaction is the oxidation of the Fe^{2+} to Fe^{3+} and the reduction of the dichromate ion to Cr^{2+} in an acidic environment. Determine the mass percent of iron in a 1.513 g sample of ore if 33.28 mL of 0.090 M potassium dichromate is needed to reach the end point in a titration.

Self–Test

This section is intended to test your knowledge of the material covered in this chapter. Think through these problems, and make certain you understand what they are asking. Make sure your answers make sense. Successful completion of these problems indicates that you have mastered the material in this chapter. You will receive the greatest benefit from this section if you use it as a mock exam, as this will allow you to determine which topics you need to study in more detail.

True–False

1. Electron configurations are dependent both on orbital energies and electron-electron repulsions.

2. All of the transition elements are metals, and all are solids.

3. Densities of transition elements have a minimum in the center of the row.

4. Transition metals are rarely found in more than one oxidation state.

5. Complexes with a coordinate number of four are either tetrahedral or square planar.

6. Ligands connect to metal atoms by one or more ligand donor atoms.

7. The oxidation state of the transition metal is not given as part of the name of a coordination compound.

8. Optical activity is the result of chirality.

9. Complexes containing the SCN (thiocyanate) ion can have linkage isomers.

10. Crystal field theory assumes that all interactions between metal and ligand are covalent.

Multiple Choice

1. In a coordinate covalent bond,
   a. no electrons are shared
   b. a lone pair of electrons becomes a bonding pair
   c. metals bind to other metals
   d. ligands ionize

2. The multiple oxidation states of transition elements originate in
   a. the varying sizes of the elements
   b. increasing metallic character with ionization
   c. the closeness in energy of the *s* orbitals and the n-1 *d* orbitals.
   d. transition elements do not have multiple oxidation states

3. All of the following are properties of transition metals:
   a. malleable, ductile, good conductors of heat and electricity
   b. malleable, ductile, poor conductors of heat and electricity
   c. brittle, ductile, good conductors of heat and electricity
   d. malleable, dull, good conductors of heat and electricity

4. From top to bottom in the transition metals, the following property varies almost not at all:
   a. ionization energy
   b. electron configuration
   c. acidity
   d. atomic size

5. Chromium is used to electroplate other metals. It is useful for this because
   a. it does not oxidize
   b. it is dull and grey, discouraging thieves
   c. It forms a hard, invisible layer of oxidation that protects against further corrosion
   d. none of the above

6. Ligands in coordination compounds are always
   a. anions
   b. cations
   c. Lewis bases
   d. Lewis acids

7. Numerous spatial arrangements are possible for coordination compounds. The shape of these compounds is determined by
   a. the hybridization of the *d* orbitals
   b. the coordination number
   c. the polarity of the ligand
   d. the group number of the metal

8. Which of the following will have geometric isomers?
   a. MA_2B_4
   b. MA_6
   c. MA_4B_2
   d. a and c

9. The observed color of transition metal complexes is
   a. the maximum absorbed wavelength
   b. complementary to the absorbed wavelength
   c. the absorbed wavelength rotated 90°
   d. none of the above

10. In contrast to valence bond theory, crystal field theory is able to account for
    a. geometry of complexes
    b. color and spin
    c. paramagnetism
    d. all of the above

Fill–in–the–Blank

1. Transition elements exhibit a variety of ____________________ states. Ions in which this is high are good ________________ agents, while ions in which this is low are good _______________ agents.
2. _________________ compounds are compounds in which a central metal atom is connected to surrounding _______________ by ____________ _____________ bonds.
3. Many compounds exist as __________________ , which have the same formula, but different arrangements of the atoms. These can be ____________________ in which the connectivity is different, or ______________________ in which the connections are the same, but have a different spatial arrangement.

4. The ________________ of the light absorbed by a transition metal complex can be used to calculate the ______________ required for its most common transition, and can also be used to predict its observed _____________.

5. __________ __________ __________ describes metal complexes in terms of overlapping __________ orbitals. _______________ _____________ ______________ describes metal complexes as purely ___________________ in nature.

Matching

Absorption spectrum	a. a ligand that binds using two donor atoms
Achiral	b. having handedness
Bidentate ligand	c. the number of ligand donor atoms surrounding a metal ion
Chelate	d. isomer in which identical ligands or groups are opposite one another
Chiral	e. complex in which there are a maximum number of unpaired electrons
Cis isomer	f. a cyclic complex formed by a metal and a polydentate ligand
Coordination number	g. isomer in which identical ligands or groups are adjacent to one another
Crystal-field splitting	h. a plot of the amount of light absorbed versus wavelength
Enantiomer	i. atom attached directly to the metal
High-spin complex	j. optically inactive mixture of enantiomers in a 50:50 ratio
Ligand	k. the energy difference between two sets of *d* orbitals
Ligand donor atom	l. ligand that has strong crystal-field splitting
Low-spin complex	m. lacking handedness
Racemic mixture	n. stereoisomers that are mirror images of one another
Strong-field ligand	o. ligand that has weak crystal-field splitting
Trans isomer	p. complex in which there are a maximum number of paired electrons

Weak-field ligand — q. molecule or ion that bonds to a central atom in a complex

Problems

1. Write the valence electron configurations for the following atoms and ions: Mn^{5+}, Mo, Cr^{3+}, Fe^{2+}.

2. Identify the following neutral elements from the following electron configurations: [Ar] $4s^23d^5$, [Kr] $5s^24d^7$, [Xe]$6s^25d^1$.

3. Arrange the following from the strongest oxidizing agent to the strongest reducing agent:

   Co Cu Mn Zn

4. Arrange the following from the strongest oxidizing agent to the strongest reducing agent:

   Fe_2O_3 CoO MnO_4^- CrO_3^-

5. Identify the coordination number of each of the following compounds:

   $[HgCl_3]^-$ $[Mo(CN)_8]^{4-}$ $Fe(CO)_5$ $[CuCl_2]^-$

6. For the following compounds identify the ligands and their donor atoms. Determine the coordination number and the oxidation state of the metal and the charge on any complex ion. Determine if the compound can display geometric, linkage, or optical isomerism.

   a. $[Co(H_2O)_4(NH_3)_2]Cl_3$
   b. $[Co(NH_3)_4(H_2O)Br]NO_3$
   c. $[Cr(NH_3)_4(SCN)_2]$
   d. $(NH_4)_3[Fe(ox)_3]$
   e. $[Co(en)_3]I_3$

7. Name the compounds in question 6.

8. Write formulas for the following compounds:
   a. tetraamminediaquachromium(III) chloride
   b. ethylenediaminedithiocyanatocopper(II)
   c. pentacarbonylchloromanganate(I)
   d. potassium tetracyanochloroferrate(II) dihydrate
   e. tetraaquacopper(II) sulfate

9. Use valence bond theory to explain the bonding in $[Zn(NH_3)_4]^{2+}$ (tetrahedral complex) and $[Ni(CN)_4]^{2-}$ (square planar). What hybrid orbitals are used by the metal? Are the complexes low–spin or high–spin complexes?

10. Why is the crystal field splitting in a tetrahedral complex approximately half of the crystal field splitting in an octahedral complex?

Challenge Problem

Three different isomers with the formula $CrCl_3{\cdot}6\ H_2O$ exist. Two are green in color; one is violet. These isomers are labeled *A, B,* and *C.*

When a 0.500 g sample of isomer *A* reacts with a dehydrating agent, the resulting mass is 0.432 g. When 50 mL of a 0.125 M solution of isomer *A* is titrated with excess $AgNO_3$, 0.269 g of AgCl are produced.

When a 0.500 g sample of isomer *B* reacts with a dehydrating agent, the resulting mass is 0.466 g. Reaction of a 50 mL solution of 0.125 M isomer *B* with excess $AgNO_3$ produced 0.539 g of AgCl.

When a 0.500 g sample of isomer *C* reacts with a dehydrating agent, the resulting mass is 0.500 g. When 50 mL of a 0.125 M solution of isomer *C* is titrated with $AgNO_3$, 0.809 g of AgCl are produced.

Determine the formula of each isomer.

CHAPTER 21

METALS AND SOLID–STATE MATERIALS

Chapter Learning Goals

A. Minerals and Free Metals

1. Predict whether a given metal is likely to be found in nature as an oxide, a sulfide, a carbonate, a silicate, a chloride, or a free metal.
2. Describe the three steps in the metallurgical processes for isolating and purifying a metal from its ore.
3. Describe the Mond process.
4. Describe the processes by which iron ore is converted to steel.

B. Bonding Descriptions

1. Explain the properties of a metal according to the electron–sea and the band theory models.
2. Use band theory to explain transition metal melting–point trends.

C. Semiconductors

1. Classify doped semiconductors as *n*–type or *p*–type.

D. Ceramics and Composites

1. Define superconducting transition temperature (T_c) and the Meissner effect.
2. Define the terms ceramics, sintering, sol, and gel.
3. Describe the sol–gel method for the production of ceramic powders.
4. Classify composite materials as ceramic–ceramic, ceramic–metal, or ceramic–polymer.

Chapter in Brief

Most metals are found as minerals: the science and technology of extracting them from their natural sources, known as ores, is called metallurgy. There are two models for describing the bonding and properties of metals, the electron sea model and band theory. This second also accounts for the electrical conductivity of metals, as well as the behavior of semiconductors, both with and without doping materials. Superconductors, ceramics, and composite materials are also introduced.

21.1 Sources of the Metallic Elements

Most metals occur as minerals, the most common of which are silicates and aluminosilicates. As these are difficult to concentrate and reduce, they are not important sources of metals.

Oxides and sulfides such as hematite (Fe_2O_3), rutile (TiO_2), and cinnabar (HgS) are much more valuable. Substances such as these, from which metals can be produced economically are called **ores**.

The chemical composition of the most common ores correlates with location in the periodic table: early transition metals occur as oxides (group 3B as phosphates), while more electronegative electrons on the right on the periodic table occur as sulfides. Less electronegative metals bond with the more covalent character to less electronegative sulfur.

Only Au and the platinum-group metals (Ru, Os, Rh, Ir, Pd, and Pt) are unreactive enough to occur as free metals.

Because *s*-block oxides are strongly basic, and unable to exist around CO_2 and SiO_2, *s*-block metals occur as carbonates and silicates. Na and K exist as chlorides.

21.2 Metallurgy

Ores are complex mixtures of metal–containing minerals and economically worthless material called **gangue** that consists of sand, clay, and other impurities. **Metallurgy** is the science of extracting metals from their ores and making alloys.

The process of extracting metal from ores is a three-step process: concentration of the ore and chemical treatment prior to reduction if needed, reduction of the mineral to the free metal, and refining or purification of the metal

Concentration and chemical treatment of ores

Mineral is separated from the gangue by exploiting the different properties of the mineral and the gangue.

Density differences are used to separate dense gold from less dense silt. Magnetic properties can be used to separate magnetite iron ore from the nonmagnetic gangue.

Metal sulfides ores are concentrated by **flotation**, a process in which powdered ore is mixed with water, oil, and detergent. Ionic silicates sink into the water, while less-polar sulfides float in the oil and detergent at the top of the tank.

Chemical separation methods include the *Bayer process*, in which amphoteric aluminum oxides are dissolved out of impurities using hot NaOH, and **roasting**, in which sulfide minerals are converted to oxides by heating the sulfides in air. In modern mineral roasting facilities, the SO_2 produced in the process is converted to sulfuric acid.

Reduction

Concentrated ore is reduced to the free metal, either by chemical reduction or by electrolysis. The method used depends on the activity of the metal.

The most active metals – Au, and Pt – are found in nature in uncombined form.

Cu, Ag, and Hg are more active and are typically found in sulfide ores that are easily reduced by roasting: the sulfur is oxidized to SO_2, while the metal is reduced.

More active metals such as Cr, Zn, and W, can be reduced by reacting their oxides with C, H, or an even more active metal such as Na or Mg. C is the cheapest choice, but is unsuitable for metals that form stable carbides, such as tungsten, which is reduced using hydrogen.

The most active metals must be electrolytically reduced. Li, Na, and Mg are obtained by electrolysis of molten chlorides.

Refining

Purification methods for reduced ores include distillation, chemical purification, and electrorefining.

Zinc can be refined by distillation.

Nickel is purified by the **Mond process**, a chemical method in which CO is passed over impure nickel to form $Ni(CO)_4$, which is them decomposed at higher temperatures on pellets of pure nickel.

Zirconium is similarly purified by conversion into ZrI_4, which is decomposed on a tungsten or zirconium filament at high temperature.

Copper is purified by electrorefining, an electrolytic process in which Cu is oxidized to Cu^{2+} at an impure Cu anode, and Cu^{2+} from aqueous $CuSO_4$ is reduced to Cu at a pure Cu cathode.

Workbook Problem 21.1

Copper occurs in a mineral called covellite that has the formula CuS. Write balanced equations for the roasting of covellite, its reaction with sulfuric acid, and its electrolytic purification.

Strategy: Write and balance equations for purification of the metal from the mineral ore.

Step 1: Write the equation for roasting of covellite.

Step 2: Write the equation for the reaction of the resultant oxide with sulfuric acid.

Step 3: Write the equation for the electrolytic purification of the solution.

21.3 Iron and Steel

The metallurgy of iron is of special importance, as iron is the major constituent of steel, the most widely used metal.

Iron is produced by carbon monoxide reduction of Fe ore in a blast furnace. The overall reaction is:

$$Fe_2O_3\,(s) + 3\,CO\,(g) \rightarrow 2\,Fe(l) + 3\,CO_2\,(g)$$

Limestone is also added to the mix: at the high temperatures of the furnace, it decomposes to CaO, which reacts with SiO_2 and other acidic oxides to produce **slag,** which consists mainly of calcium silicate and floats on top of the molten iron.

The impure iron obtained is called *cast iron* or *pig iron*. This is brittle and contains about 4% elemental carbon and smaller amounts of other impurities formed in the reducing atmosphere of the furnace.

To further purify the cast iron and convert it to steel, the **basic oxygen process** is used. The molten iron is exposed to a jet of pure oxygen in a furnace lined with basic oxides. The acidic oxides that form react with the basic oxides to form a slag that can again be poured off. This produces steel with about 1% carbon and very small amounts of P and S.

The composition of liquid steel is monitored by chemical analysis, and the amounts of oxygen and impure steel continuously varied to achieve the desired concentrations of impurities.

Hardness, strength, and malleability of steel depend on chemical composition, rate of cooling, and subsequent heat treatment. For example, stainless steel is a corrosion-resistant alloy that contains up to 30% chromium as well as smaller amounts of nickel.

21.4 Bonding in Metals

Metals have some universal characteristics. They are malleable, ductile, lustrous, and good conductors of heat and electricity.

Two theoretical models are used to explain these properties: the *electron-sea model* and the *molecular orbital theory.*

Electron Sea Model of Metals

Metals do not have enough valence electrons to bond to adjacent molecules, therefore their crystals contain delocalized electrons that belong to the structure as a whole. This acts as an electrostatic glue that holds the metal cations together.

This is a simple qualitative explanation for the properties of metals: electrons are mobile, and therefore free to move from a negative electrode to a positive electrode when subjected to an electrical potential. Mobile electrons are also able to carry kinetic energy (heat) from one part of the crystal to another. Metals are malleable and ductile because no local bonds are broken when the solid is deformed. The more valence electrons there are, the harder the metal.

Molecular Orbital Theory for Metals

The number of molecular orbitals formed is the same as the number of atomic orbitals combined. As the number of atoms increases, the difference in energy between successive MOs decreases, allowing the orbitals to merge into an almost continuous band of energy levels.

For this reason, molecular orbital theory in metals is referred to as **band theory**. The bottom half of the band consists of filled bonding MOs, while the top half consists of empty antibonding MOs.

Electrons are equally likely to be traveling in one direction as another in the absence of a current. Therefore, energy levels occur in **degenerate** pairs – the same energy levels, but for different directions within the metal. If an electrical potential is applied, there is a reason for electrons to move into the higher energy antibonding orbitals.

If there are no vacant orbitals in the band, there are no orbitals available for electrons to pass through, making the material an electrical insulator.

Main-group metals have s and p subshells sufficiently close in energy that the s and p bands overlap, resulting in a partially filled composite band and electrical conductivity.

Transition metals have overlapping d and s bands, which allows six MOs per metal atom, three bonding and three antibonding. As a result, the metals with six valence electrons (bonding orbitals completely filled) are the hardest and have the highest melting points.

Workbook Problem 21.2

Rank Mo, Cd and Y in order of their expected melting points based on band theory.

Strategy: Determine the number of bonding and antibonding electrons, rank in order by maximum bonding electrons and minimum antibonding electrons.

Step 1: Determine the numbers of electrons of each type.

Step 2: Rank in order of expected bonding strength.

21.5 Semiconductors

A semiconductor has electrical conductivity intermediate between that of a metal and that of an insulator. In metals, there is no energy gap between the highest occupied orbitals and the lowest unoccupied orbital: electrons can move freely between these energy levels. In insulators, there is a large energy gap between the highest occupied orbitals and the lowest unoccupied orbital: electrons cannot move between these energy levels.

The bonding MOs are also referred to as the **valence band**. The higher-energy antibonding MOs are the **conduction band**. The difference in energy between the two is the **band gap**. Metals have no band gap, insulators have an insurmountable band gap, and semiconductors fall between the two.

In a semiconductor, a few electrons will have the energy to jump across the band gap. Conductivity increases with increasing temperature, as the number of high-energy electrons increases. This is the opposite of metals: as their temperature increases, conductivity drops, as the lattice of bonds is disrupted. This provides a convenient way to distinguish between the two types of material.

Doping is a process in which the conductivity of a semiconductor is increased by adding small amounts of impurities. Doping can increase the conductivity of a semiconductor by a factor of $\sim 10^7$.

If an impurity is introduced that has more electrons than the semiconductor, the extra electrons will reside in the conduction band. Because *negative* electrons are being added, these are called *n*–type semiconductors.

If an impurity is introduced that has fewer electrons than the semiconductor, there will be vacancies in the valence band that allow for increased conductivity. Because these can be thought of as adding *positive* holes in the valence band, these are *p*–type semiconductors.

Workbook Problem 21.3

Identify the following as *n*- or *p*-type semiconductors:

a. Silicon doped with gallium
b. Germanium doped with antimony
c. Silicon doped with arsenic
d. Germanium doped with phosphorus

Strategy: Determine the number of valence electrons in the semiconductor, and compare to the number of valence electrons in the dopant. As all of the elements are main-group, their group numbers can be used.

Step 1: Determine the numbers of valence electrons in each element.

Step 2: Determine whether the dopant has more or fewer valence electrons than the semiconductor.

Step 3: Identify the doped semiconductors as *n*- or *p*-type.

21.6 Semiconductor Applications

Doped semiconductors are essential components in modern solid-state electronic devices.

Diodes

These convert alternating current to direct current by permitting current to flow in one direction only. They consist of a *p*-type semiconductor in contact with a *n*-type semiconductor to give what is known as a *p-n* junction.

If the *n*-type side is connected to the negative terminal of a battery, current will flow through the junction: the electrons can find their way into the conductance band, and across the gap into the valence band. If the *p*-type side is connected to the negative terminal of a battery, the positive holes move toward the battery, and the electrons in the conductance band across the gap are repelled. In this case, no current will flow.

Light-Emitting Diodes (LEDs)

If there is a difference in energy levels between the conduction band and the valence band, electrons will be able to fall from one to the other, releasing energy as light. The energy of the light emitted is roughly equivalent to the size of the band gap. ($E_g = h\nu = hc/\lambda$)

LEDs are made of 3-5 semiconductors. These are 1:1 compounds of group 3A and 5A elements. They have the same basic structure as pure Si or Ge, as they have an average of four valence electrons.

Because these mixtures form a continuous series of solid solutions, the size of the band gap, and thus the color of the LED can be "tuned" by varying the composition of the mixture. The red light so commonly seen in LEDs is the result of a solid solution of $GaP_{0.4}As_{0.6}$ with a band gap of 181 kJ/mol.

LEDs are smaller, brighter, longer lived, more energy efficient, and faster than incandescent bulbs.

Diode Lasers

The word *laser* is an acronym for Light Amplification by Stimulated Emission of Radiation. These lasers produce light in the same way as LEDs, but the light is more intense, highly directional, and all of the same frequency and phase.

The essential features of a diode laser are very high forward bias and a laser cavity that allows emitted light to bounce back and forth, stimulating a cascade of electrons and holes that amplifies the amount of light produced.

These lasers are used in laser pointers, bar-code readers, CD players, and fiber optic data systems.

Photovoltaic (Solar) Cells

These work in the opposite direction of an LED– where the LED converts the energy difference of the band gap to light, a photovoltaic cell converts light to electricity.

When light shines on the *p-n* junctions, the photons from the light excite electrons from the valence band of the *p*-type semiconductor into the conduction band of the *n*-type semiconductor.

If the photovoltaic cell is part of an electrical circuit, light can flow in, charging the system, and the resultant energy can be used to power a device.

Current cells are only 20% efficient, so research is focusing on increasing efficiency and reducing cost.

Transistors

Transistors consist of *n-p-n* or *p-n-p* junctions that control or amplify electrical signals in modern integrated circuits. Vast numbers of these devices can be packed into tiny spaces, increasing the speed and decreasing the size of modern electronics.

Workbook Problem 21.4

If an LED has a band gap of 178 kJ/mol, what is the wavelength of the resulting light?

Strategy: Convert the band gap energy to photon energy, and then to wavelength.

Step 1: Determine the energy of an individual photon.

Step 2: Convert the photon energy to wavelength.

21.7 Superconductors

A **superconductor** is a material that loses all electrical resistance below a characteristic temperature called the **superconducting transition temperature** (T_c). Below that temperature, the material becomes a perfect conductor: once an electric current is started, it will flow indefinitely without loss of energy.

In 1986, $Ba_xLa_{2-x}CuO_4$ was demonstrated to have a $T_c = 35$ K, and soon after other copper-containing oxides were found to have even higher superconducting transition temperatures. The record as of this writing is 138 K.

This was unexpected, as most metal oxides, nonmetallic inorganic solids called *ceramics,* are insulators.

In this field, experimental data exceeds the ability of scientists to explain it: there is no generally accepted theory of superconductivity in ceramic superconductors.

If a superconductor is cooled to belowT_c and a magnet is lowered toward it, the magnet and superconductor repel each other, causing the magnet to levitate. Magnets induce a supercurrent in the superconductor that generated a magnetic field opposite to that of the magnet. This is called the *Meissner effect*, and it is already being used to operate a high-speed train in Shanghai, China.

Superconducting magnets are also used in MRI instruments and particle accelerators, but these uses currently require cooling with liquid helium, which is expensive and requires cryogenic equipment.

Once the T_c of a material goes above 77 K, it can be cooled using liquid nitrogen, an abundant refrigerant that is cheaper than milk.

Currently available high–temperature superconductors are brittle powders with high melting points, not easily made into wires and coils needed for electrical equipment. However, superconducting thin films are being used as microwave filters, and superconducting wires as long as 1 km are now available.

High–temperature superconductors are also being developed based on fullerene- C_{60} - molecules. K_3C_{60} is a metallic conductor at room temperature, and a superconductor at 18 K.

Flourine-doped lanthanum oxide iron arsenide also shows promise, superconducting at 26 K, with substitution of the La by other elements raising the temperature to 77 K.

21.8 Ceramics

Ceramics are inorganic, nonmetallic, nonmolecular solids, including both crystalline and amorphous materials, such as glasses. Traditional silicate ceramics are made by heating aluminosilicate clays to high temperatures, while **advanced ceramics**- materials that have high–tech engineering, electronic, and biomedical applications – include oxide ceramics such as alumina (Al_2O_3) and nonoxide ceramics such as silicon carbide and silicon nitride.

Oxide ceramics are named by adding an *–a* ending in place of the *–um* of the metal.

Compared to metals, ceramics have higher melting points, are stiffer, harder, and more resistant to wear and corrosion, maintain their strength at high temperatures, and are less dense than steel. This makes them attractive lightweight, high–temperature materials for replacing metal components in aircraft, space vehicles, and automobiles.

Non-oxide ceramics are covalent network solids with highly directional covalent bonds. This prevents planes of atoms from sliding over one another when the solid is subjected to the stress of a load or an impact. As the solid cannot deform as a metal would, the bonds instead give way. This brittleness is the primary drawback of these materials.

In oxide ceramics, the bonding is largely ionic, but the material is still brittle.

Ceramic processing, the series of steps that leads from raw material to the finished ceramic object, determines the strength and the resistance to fracture of the product.

Sintering is a process in which the particles of the powder are welded together without completely melting. It is done below the melting point of the material, and allows crystal grains to grow larger and the density of the material to increase as the void spaces between particles disappear. High–purity, fine powders that are tightly compacted prior to sintering are needed to prevent voids that will lead to cracking and material failure.

In the **sol–gel method** for preparing these high–purity fine powders, a metal oxide powder is synthesized from a metal alkoxide (a compound derived from a metal and an alcohol). This forms a colloidal dispersion called a *sol*, consisting of extremely fine particles that can then be dehydrated and linked with oxygen bridges. As this cross-linking occurs, the sol becomes a more rigid material called a *gel*.

Oxide ceramics have many uses. Alumina is used in spark plugs, dental crowns, and the heads of artificial hips, as well as being the substrate for circuit boards. Silica is used to make the heat-resistant tiles that protect the space shuttle on re-entry into the atmosphere. Military armor is now also ceramic rather than steel, reducing weight and increasing maneuverability.

21.9 Composites

To counteract the brittleness of ceramic, a ceramic powder can be mixed, prior to sintering, with a second ceramic material to combine the properties of both components. This second compound is often in the form of *whiskers* or fibers.

Fibers and whiskers increase the strength and fracture toughness of composite materials, as most of the chemical bonds are aligned along the fiber axis, giving the fibers great strength. The fibers can deflect cracks, preventing them from moving cleanly in one direction, or can bridge cracks to hold them together.

Composites can also be made of different materials. **Ceramic–metal composites** or **cermets**, are one type, while **ceramic-polymer composites** are another. These materials have a high strength-to-weight ratio, making them ideal for aerospace and other applications.

Ceramic fibers used in composites are usually made by high–temperature methods. For example, carbon fiber is made by heating polyacrylonitrile to 1400-2500 °C, converting it to graphite. Silicon carbide chains are made using a similar method.

Putting It Together

The presence of manganese in steel can be determined by the following titration procedure:

1. A steel sample is dissolved in an acidic solution that oxidizes the manganese to the permanganate ion.
2. The permanganate ion is reduced to Mn^{2+} by reaction with an excess of iron(II) sulfate.
3. The unreacted Fe^{2+} is oxidized to Fe^{3+} by reaction with potassium dichromate.

Determine the percent by mass of manganese in a 0.450 g sample of steel if 22.4 mL of 0.0100 M potassium dichromate is needed to react with the Fe^{2+} remaining after 50.0 mL of 0.0800 M iron (II) sulfate reacts with the steel sample.

Self–Test

This section is intended to test your knowledge of the material covered in this chapter. Think through these problems, and make certain you understand what they are asking. Make sure your answers make sense. Successful completion of these problems indicates that you have mastered the material in this chapter. You will receive the greatest benefit from this section if you use it as a mock exam, as this will allow you to determine which topics you need to study in more detail.

True–False

1. Most silicate minerals are valuable sources of metals.
2. Less electronegative metals form ionic oxides, while more electronegative metals form sulfides with more covalent character.
3. Roasting is used to convert sulfides to more easily reduced oxides.
4. Iron is produced by reduction with NO_2 in a blast furnace.
5. The electron-sea model of metal bonding accounts for the superconductivity of some materials.
6. Valence band theory accounts for the different conductivities of metals and semiconductors.
7. Doping of semiconductors increases their conductivity.
8. Heating a semiconductor reduces its conductivity.
9. Diodes can convert alternating current to direct current.
10. Ceramics are ideal for aerospace applications because they are strong, malleable, and light.

Multiple Choice

1. A commercially viable, naturally occuring source of refinable metal is referred to as a(n)
   a. mineral
   b. gangue
   c. ore
   d. flotation

2. Refining metals requires the following general steps in this order:
   a. concentration, oxidation, purification
   b. reduction, concentration and purification
   c. purification, oxidation, and concentration
   d. concentration, reduction, purification

3. Which of the following metals are found free in nature?
   a. aluminum
   b. gold
   c. copper
   d. tungsten

4. Carbon is an inexpensive reducing agent, but cannot be used with metals that form
   a. carbides
   b. alloys
   c. ceramics
   d. superconductors

5. Models of metallic bonding must account for metals'
   a. malleability and ductility
   b. dull appearance
   c. electrical conductivity
   d. a and c

6. Semiconducting metals conduct electricity but not efficiently. This is as a result of
   a. impurities
   b. a large band gap
   c. paramagnetism
   d. no available *p* orbitals

7. Doping silicon with which of the following would produce a *p*-type semiconductor?
   a. boron
   b. phosphorus
   c. nitrogen
   d. arsenic

8. Light-emitting diodes
   a. can be tuned to different wavelengths
   b. are energy efficient
   c. a and b
   d. none of the above

9. Superconductors
   a. are strongly magnetic
   b. operate best at high temperatures
   c. have no electrical resistance below T_c
   d. are easily made into wires

10. Most ceramics are
   a. malleable
   b. insulators
   c. shiny
   d. paramagnetic

Fill–in–the–Blank

1. The electron-sea model of metal bonding accounts for the ____________________ and __________________ of metals. These two abilities of metals to be shaped come from the fact that bonds and electrons are not ____________________. As a result, metals are not ____________: they will deform rather than shatter.

2. In the absence of an electrical potential, there is no net electric __________________ in a metal. If a potential is applied, electrons are able to access the _______________ band, the __________________ molecular orbitals that are _______________ in the absence of potential.
3. Superconductors are fascinating, poorly understood materials with some interesting properties. ___________ the T_c, the material loses all electrical _________________. The challenge for researchers is to find materials that can be cooled using liquid _________________, a common and inexpensive coolant that requires no specialized equipment.
4. Advanced ceramics have properties that make them superior to _______________ for many applications. They are lighter due to their lower _______________, electrical _______________, and resistant to ______________. Their durability is limited in one respect: they are not _________________, and will shatter under extreme stress.
5. Composite materials can consist of mixtures of any of the following: _______________, _________________, carbon or boron _______________, and _________________.

Matching

Alloy	a. molecular orbital theory for metals
Band gap	b. mineral deposit from which a metal can be economically extracted
Band theory	c. antibonding molecular orbitals in a semiconductor
Ceramic	d. material that has conductivity intermediate between metals and insulators
Conduction band	e. economically worthless material that accompanies an ore
Doping	f. crystalline inorganic constituent of rocks
Electron-sea model	g. energy difference between bonding MOs in the valence band and the non bonding MOs in the conductance band.
Flotation	h. heating a mineral in air
Gangue	i. process in which the particles of a powder are "welded" together without completely melting
Metallurgy	j. addition of a small amount of an impurity to increase the conductivity of a semiconductor
Mineral	k. science and technology of extracting metals from ore

Ore	l. an inorganic, nonmetallic, nonmolecular solid
P-type semiconductor	m. solid solution of two or more metals
Semiconductor	n. method of preparing ceramics involving synthesis of metal oxide powder from metal alkoxide
Roasting	o. by-product of iron production consisting mainly of calcium silicate
Sintering	p. model that visualizes metals as cations surrounded by delocalized electrons
Slag	q. temperature below which all electrical resistance is lost
Sol-gel method	r. semiconductor doped with a material that has fewer electrons than necessary for bonding
Superconducting transition temperature	s. bonding molecular orbitals in a semiconductor
Superconductor	t. process that exploits differences in the ability of water and oil to wet the surfaces of mineral and gangue.
Valence band	u. material that loses all electrical resistance below a certain temperature

Problems

1. The manganese ore rhodochrosite has the formula $MnCO_3$. It is refined by heating in the presence of oxygen to produce manganese(IV) oxide and drive off carbon dioxide, then reacted with aluminum metal to reduce it to elemental manganese. Write and balance the equations for these reactions. If all reactions occur with 100% efficiency, how many kg of elemental manganese can be obtained from 1500 kg of pure rhodochrosite? How many kg of aluminum metal will be needed for this reaction?

2. Cobalt, chromium, and copper are all period 4 transition metals. Rank them in order of melting point using molecular orbital theory.

3. Silicon and germanium are both semiconductors. Phosphorus, arsenic, and boron are all doping agents. Determine which type of semiconductor is formed by all of the combinations.

Challenge Problem

A concentration technique used to extract gold from low-grade ores is *cyanidation*. This technique involves reaction of the cyanide ion with the ore to produce $Au(CN)_2^-$ ions, which are then reduced with zinc to produce pure gold. Is it possible to use cyanidation to remove silver from the ore argentite, Ag_2S? $K_f\ [(Ag(CN)_2^-] = 1 \times 10^{21}$. $K_{sp}\ (Ag_2S) = 6 \times 10^{-51}$. Could cyanidation be used to remove silver from horn silver, AgCl? $K_{sp}\ (AgCl) = 1.8 \times 10^{-10}$.

CHAPTER 22

NUCLEAR CHEMISTRY

Chapter Learning Goals

A. Energy Changes

1. Calculate mass defects and binding energies for nuclides. Use values of binding energy per nucleon to compare the relative stabilities of two nuclides.
2. Calculate the energy released by a fission or fusion reaction.
3. Discuss how nuclear reactions relate to the law of conservation of mass and the law of conservation of energy.

B. Nuclear Reactions

1. Write balanced equations for nuclear reactions, identifying the types of radiation and nuclides involved.
2. Discuss how controlled fission can be used in the generation of electricity.
3. Identify the challenges involved in using fusion reactions.
2. Write balanced equations for nuclear transmutations, and identify products of these reactions.

C. Measurement of Radiation

1. Discuss the various units for radiation exposure and what they indicate.

D. Radionuclides

1. Show how radiocarbon dating is used to determine the age of an object.
2. Discuss how radiopharmaceuticals can be used to treat disease and image tissues.

Chapter in Brief

Nuclear chemistry is the study of the properties and reactions of atomic nuclei. The stability of an element is related to its proton/neutron ratio, and the binding energy of the nucleus can be calculated. Some elements can undergo fission, either spontaneously or by being induced, as in nuclear reactors. Other elements can fuse together under extreme conditions to form larger nuclei. All of the transuranium elements and many other isotopes have been made by nuclear transmutation, a process by which the nuclei of one element are bombarded with particles to increase their size. Radiation varies in its ability to damage tissues: a number of different units are used for radiation exposures depending on the reason for the measurement. There are numerous practical uses of nuclear chemistry, from radiocarbon dating of ancient objects to therapeutic and imaging uses in medicine.

Review

Radioactive decay is a first-order process whose rate is proportional to the number of radioactive nuclei times the rate constant (Rate = kN). Radioactive decay is characterized by a half-life ($t_{1/2}$), the time required for the number of radioactive nuclei to drop to ½ their original level. Half-life is related to the decay constant by the equation $t_{1/2} = 0.693/k$.

22.1 Energy Changes in Nuclear Reactions

The ratio of neutron to protons required for stability varies with the mass of the element. Lighter elements require a 1:1 ratio of neutrons to protons, but more neutrons are needed for heavier elements.

The more massive an atoms is, the more neutrons are needed to minimize the repulsions of the protons: neutrons serve as nuclear "glue."

The energy released when a nucleus is created cannot be measured, as the activation energy required to force the particles together is so high that temperatures rivaling those in the interior of the Sun are needed.

Although the energy changes cannot be measured, they can be calculated using Einstein's famous equation:

$$\Delta E = \Delta mc^2$$

Using the known masses of neutrons and protons, it is possible to calculate the **mass defect** of a nucleus. The mass of a nucleus is always less than the mass of the individual particles. This lost mass is a direct measure of the **binding energy** holding the nucleus together. The larger the binding energy, the more stable the nucleus.

The energy changes involved in nuclear reactions are millions of times greater than those involved in typical processes.

To make it easier to compare the stability of different nuclei, energies are usually expressed per **nucleon** in mega-electron volts – MeV. 1MeV = 1.6×10^{-13} J.

If binding energy is plotted against atomic number, a stability peak is found at Fe-56. Nuclear stability is increased when lighter nuclei fuse together, or heavier nuclei break apart.

The result of these calculations is the discovery that mass and energy are not independently conserved, but the combination of the two is conserved. For non-nuclear reactions, the effect is too small to be measured, and can thus be disregarded.

EXAMPLE:

Determine the mass defect and binding energy for Chlorine-35 in both J/mol and MeV/nucleon. The mass of ^{35}Cl is 34.968 85 amu.

SOLUTION:

First, calculate the total mass of the nucleons and the mass defect:

Mass of 17 protons = (17)(1.007 28 amu)	= 17.123 76 amu	
Mass of 18 neutrons = (18)(1.008 66 amu)	= 18.155 88 amu	
Mass of 17 protons and 18 neutrons		= 35.279 64 amu

Mass of chlorine-35 atom	= 34.968 85 amu
Mass of electrons = (17) (5.486×10^{-4})	= 0.009 33 amu
Mass of chlorine-35 nucleus	= 34.959 52 amu

Mass defect: 35.279 64 amu – 34.959 52 amu= 0.320 12 amu

Convert this mass to grams:

$$0.320\ 12 \text{ amu} \times \frac{1.6605 \times 10^{-24} \text{ g}}{\text{amu}} = 5.3156 \times 10^{-25} \text{ g}$$

Use the Einstein equation to convert the mass defect to binding energy:

$$\Delta E = 5.3155 \text{ x } 10^{-25} \text{ g x } \frac{1 \text{ kg}}{1000 \text{ g}} \text{ x } (3.00 \text{ x } 10^{8} \text{ m/s})^2 = 4.7840 \text{ x } 10^{-11} \text{ J}$$

Convert to J/mol:

$$4.7840 \text{ x } 10^{-11} \frac{\text{J}}{\text{nucleus}} \text{ x } \frac{6.022 \text{ x } 10^{23} \text{ nuclei}}{\text{mol}} = 2.881 \text{ x } 10^{13} \frac{\text{J}}{\text{mol}}$$

Convert to MeV/nucleon:

$$4.7840 \text{ x } 10^{-11} \frac{\text{J}}{\text{nucleus}} \text{ x } \frac{1 \text{ MeV}}{1.60 \text{ x } 10^{-13} \text{ J}} \text{ x } \frac{1 \text{ nucleus}}{35 \text{ nucleons}} = 8.54 \frac{\text{MeV}}{\text{nucleon}}$$

EXAMPLE:

Calculate the mass change in g/mol when N atoms combine to form N_2 molecules.

$$2 \text{ N} \rightarrow \text{N}_2 \qquad \Delta E = -945.4 \text{ kJ/mol}$$

SOLUTION:

With a known energy change, rearrange the Einstein equation to solve for Δm:

$$\Delta m = \Delta E/c^2$$

$$\Delta m = \frac{-945.4 \frac{\text{kJ}}{\text{mol}} \text{ x } \frac{1 \text{ kg} \cdot \text{m}^2/\text{s}^2}{\text{J}} \text{ x } \frac{1000 \text{ J}}{\text{kJ}} \text{ x } \frac{1000 \text{ g}}{\text{kg}}}{(3.00 \text{ x } 10^{8} \text{ m/s})^2} = 1.05 \text{ x } 10^{-8} \text{ g/mol}$$

This is an amount far too small to be measured.

Workbook Problem 22.1

Calculate the binding energy of a Xe-130 nucleus in Mev/nucleon. The mass of a ^{130}Xe nucleus is 129.903 amu.

Strategy: Calculate the mass defect and convert to energy using the Einstein equation.

Step 1: Determine the number of nucleons and their mass, subtract the mass of the electrons from the atomic mass, and calculate the mass defect in amu.

Step 2: Convert the mass defect to grams.

Step 3: Determine ΔE and convert to MeV/nucleon.

22.2 Nuclear Fission and Fusion

As the maximum stability is found in the middle of the periodic table, there is stability to be gained by the **fusion** of light nuclei and the **fission** of heavy nuclei.

Nuclear Fission

Certain large nuclei shatter into smaller pieces of roughly equal size when bombarded with neutrons. The fission of a given nucleus does not occur in the same way each time: more than 800 products have been identified from the fission of U-235.

When neutrons are released as part of a nuclear fission in addition to those used to begin the reaction, a **chain reaction** can occur in which the reaction continues even after the external supply of neutrons is cut off.

In a small sample, many of the neutrons will escape without initiating further fission, but if there is a sufficient amount of reaction – known as a **critical mass** – the chain reaction will become self-perpetuating.

EXAMPLE:

How much energy is released in both J and kJ/mol for the neutron-induced fission of U-235 (atomic mass = 235.043 92 amu) to form Te-140 (atomic mass = 139.938 85 amu) and Kr-93 (atomic mass = 92.931 27 amu)? The mass of a neutron is 1.008 66 amu.

SOLUTION:

To calculate the energy difference, the mass difference must be calculated and converted into energy.

Write and balance the equation: the overall masses need to balance

$$^{235}U + {}^{1}n \rightarrow {}^{140}Te + {}^{93}Kr + 2\ {}^{1}n$$

mass of reactants = 235.043 92 amu + 1.008 66 amu = 236.052 58 amu

mass of products = 139.938 85 amu + 92.931 27 amu + 2 (1.008 66 amu) = 234.887 44 amu

Calculate the mass difference:

Reactants – products = 1.165 14 amu

Convert mass to kg:

$$1.165\ 14\ \text{amu} \times \frac{1.6605 \times 10^{-24}\ \text{g}}{\text{amu}} \times \frac{1\ \text{kg}}{1000\ \text{g}} = 1.934\ 71 \times 10^{-27}\ \text{kg}$$

Use the Einstein equation to convert the mass difference to energy:

$$\Delta E = 1.934\ 71 \times 10^{-27}\ \text{kg} \times (3.00 \times 10^{8}\ \text{m/s})^{2} = 1.741\ 24 \times 10^{-10}\ \text{J}$$

Convert to J/mol:

$$1.741\ 24 \times 10^{-10}\ \frac{\text{J}}{\text{nucleus}} \times \frac{6.022 \times 10^{23}\ \text{nuclei}}{\text{mol}} = 1.049 \times 10^{14}\ \frac{\text{J}}{\text{mol}}$$

Nuclear Reactors

Controlled fission reactions can be used to generate electricity: uranium fuel is placed in a pressurized containment vessel surrounded by water and *control rods* made of boron and cadmium added. The water slows the escape of neutrons, and the control rods absorb neutrons to regulate the speed of the fission reaction: they are raised to speed the reaction – allowing more neutrons into the fuel– and lowered to slow the reaction. The energy generated is used to heat circulating coolant, which produces steam that drives a turbine to produce electricity.

Naturally occurring uranium is a mixture of 99.3% non-fissionable ^{238}U and 0.07% fissionable ^{235}U. Reactor fuel is compressed UO_2 pellets that have been enriched to 3% ^{235}U and encased in zirconium.

The amount and concentration of nuclear fuel in a reactor is not enough to sustain a nuclear explosion: the worst-case is that uncontrolled fission would melt the reactor and the containment vessel, releasing radioactivity into the environment. Newer reactors have *passive safety* designs that automatically slow runaway reactors.

Thirty-one countries now use nuclear energy to supply some of their electricity needs. The primary issue holding back further use of nuclear energy is safe disposal of the wastes: ^{90}Sr takes 600 years to decay to safe levels, and ^{239}Pu takes 20,000 years to decay to safe levels.

Nuclear Fusion

Very light nuclei also release enormous amounts of energy when they undergo *fusion*: it is hydrogen fusion reactions that power the Sun and other stars. Hydrogen is cheap and plentiful, and fusion products are non-radioactive and non-polluting, but with a temperature of approximately 40,000,000 K required to initiate the process, the technical barriers to making this a source of power are formidable.

Workbook Problem 22.2

In Oklo, Gabon, a natural nuclear reactor operated in uranium ore for hundreds of thousands of years. Part of the evidence for this is the presence of ^{136}Xe and ^{99}Ru in the surrounding rocks. Write and balance an equation for the fission of ^{235}U resulting in these isotopes, and calculate the energy generated by this reaction in J/mol. The mass of ^{235}U is 235.043 92 amu , the mass of ^{136}Xe is 135.9072195 amu, the mass of ^{99}Ru is 98.905 939 amu. One neutron is needed to induce the fission, and one neutron is produced by the fission.

Strategy: Balance the equation, calculate the mass defect, and convert to energy using the Einstein equation.

Step 1: Write and balance the equation.

Step 2: Calculate the mass defect in amu and convert to kilograms.

Step 3: Determine ΔE and convert to J/mol.

Workbook Problem 22.3
Hydrogen bombs utilize the reaction between deuterium (^{2}H – 2.014 10 amu) and tritium (^{3}H – 3.016 05 amu) that produces ^{4}He (4.002 60 amu) and a neutron (1.008 66 amu). Calculate the energy generated by the reaction per mol of ^{4}He produced.

Strategy: Calculate the mass defect and convert to energy using the Einstein equation.

Step 1: Determine the number of nucleons and their mass, and calculate the mass defect in amu.

Step 2: Convert the mass defect to grams.

Step 3: Determine ΔE and convert to J/mol.

22.3 Nuclear Transmutation

Of the 3600 known isotopes, only 300 occur naturally. The remainder have been made by **nuclear transmutation**. This is often accomplished by bombardment of an atom with a high energy particle such as a proton, a neutron, or an α-particle. In the collision, an unstable nucleus is momentarily generated, and a new element can be produced.

All of the transuranium elements-- those with atomic numbers above 92 – have been produced using transmutation.

The reactions and subsequent decays can be quite complex: as the elements can be produced in a variety of ways, they can also decay in a variety of ways.

EXAMPLE:
Ernest Rutherford was the first to accomplish nuclear transmutation, but the husband and wife team of Frederic and Irene Joliot-Curie were not far behind, and bombarded Al-27 with α-particles to produce P-30. Write and balance an equation for this reaction. What particle is also produced in this reaction?

SOLUTION:

According to the periodic table, Al has an atomic number of 13, and phosphorus an atomic number of 15. An α-particle is a helium nucleus: it has a mass of 4, and an atomic number of 2.

$$^{27}_{13}\text{Al} + ^{4}_{2}\text{He} \rightarrow ^{30}_{15}\text{P} + ?$$

The number of protons adds up, but the number of neutrons does not: a neutron must also be produced by the reaction:

$$^{27}_{13}\text{Al} + ^{4}_{2}\text{He} \rightarrow ^{30}_{15}\text{P} + ^{1}_{0}\text{n}$$

Workbook Problem 22.4
If U-238 is bombarded with C-12, the result is a new element and 6 neutrons. Write and balance the equation, and determine the identity of the element produced using the periodic table.

Strategy: Determine the overall mass, subtract out the neutrons, and determine the identity and mass of the element produced.

22.4 Detecting and Measuring Radioactivity

Radiation is invisible, but can be measured and detected by instruments, and the effect on biological systems can be observed.

Radiation intensity is measured in a number of different ways.

Decay events are measured as the *becquerel* (1 Bq = 1 disintegration/s) and the *curie* (1 Ci = 3.7×10^{10} disintegration/s). The curie is related to the decay rate of 1 g of radium.

Energy absorbed per kilogram of tissue is measured as the *gray* (1 Gy = 1 J/kg tissue) and the *rad*, which stands for *radiation absorbed dose*. A rad is one-hundredth of a Gy (1 rad = 0.01 Gy).

Tissue damage units take into account the type of radiation. The *sievert* (Sv) accounts for this. 1 Gy of α particles causes twenty times the damage of 1 Gy of γ rays, but 1 Sv of α particles causes the same damage as 1 Sv of γ rays. The *rem* (which stands for *roentgen equivalent for man*) is equivalent to one-hundredth of a sievert (1 rem = 0.01 Sv).

The seriousness of radiation exposures varies with the type of radiation, the energy of the radiation, the length of exposure, and whether the source is external or internal.

When external, X-rays and γ rays are the most harmful as they penetrate clothing and skin. They move at the speed of light, and have 1000 x the penetrating power of α particles. Several inches of lead are required to stop them.

When internal, α particles and β particles are more dangerous, as all their energy is absorbed by surrounding tissues. α particles move at one-tenth the speed of light, and can be stopped by the top layer of skin or a few pieces of paper. β particles are lighter and faster – they move at nine-tenths the speed of light, and require a block of wood or heavy protective clothing to stop them.

The biological effects of radiation exposure can be dire, but the usual exposure of an average person is only about 120 mrem. Transient health effects begin to be seen between 25 and 100 rem, where white counts begin to be temporarily decreased. The lethal dose of radiation is 600 rem – 5000 times higher than most people's annual exposure.

70% of yearly exposures are natural, from rocks and cosmic rays, with the remaining 30% coming from medical procedures such as X-rays.

22.5 Applications of Nuclear Chemistry

Dating with Radioisotopes

Radiocarbon dating depends on the slow production of C-14 in the upper atmosphere by neutron bombardment of nitrogen. From this reaction, $^{14}CO_2$ diffuses into the lower atmosphere, where it is taken up by plants and animals. While the organisms are living, the ratio of ^{14}C to ^{12}C is constant, and is the same as that in the atmosphere. When an organism dies, the ^{14}C decays and is not replenished. By measuring the $^{14}C/^{12}C$ ratio, the approximate age of the remains can be determined using the half-life of ^{14}C – 5715 years.

Variations on this technique use the fact that uranium-238 decays to lead-206 with a half-life of 4.47×10^9 years, making the dating of uranium-containing rocks possible. Potassium-40 decays with a half-life of 1.25×10^9 years to yield argon-40. The age of a rock can be estimated by crushing it, and comparing the amount of ^{40}Ar and the amount of ^{40}K.

EXAMPLE:

A skull hoped to be ancient was found to have a ^{14}C decay rate of 4.7 disintegrations/min. What age is implied by this if living organisms have a decay rate of 15.3 disintegrations/min, and the half-life of ^{14}C is 5715 years?

SOLUTION:

Radioactive decay is a first-order process, so the time since the beginning of the reaction can be determined by calculating using the integrated rate law:

$$\ln\frac{N_t}{N_0} = -0.693\left(\frac{t}{t_{1/2}}\right)$$

Substituting:

$$\ln\frac{4.7}{15.3} = -0.693\left(\frac{t}{5715 \text{ years}}\right)$$

$$\frac{\ln 0.307 \times 5715 \text{ years}}{-0.693} = 9734 \text{ years}$$

The skull is approximately 10,000 years old.

Medical Uses of Radioactivity

***In vivo* procedures** are those that take place inside the body to assess the function of a particular organ or body system. A radiopharmaceutical is administered, and its path in the body analyzed.

Therapeutic procedures are those in which radiation is used to kill diseased tissue. External radiation can be directed at tumors, often using γ rays from a Co-60 source. Although the radiation is directed as carefully as possible, patients typically develop some symptoms of radiation sickness. Internal radiation treatment is much more selective, and can target tissues very specifically, as when I-131, which is selectively taken up by the thyroid, is used in the treatment of thyroid disease.

Imaging procedures provide diagnostic information by analyzing the distribution of radioisotopes introduced into the body. Depending on the use, a diseased area might concentrate the isotope, showing up as a "hot" spot, or fail to take up the isotope, thus showing up as a "cold" spot.

Putting It Together

47.0 % of the chlorine atoms in a 78.3 mg sample of sodium perchlorate are ^{36}Cl, which is a radioactive isotope of chlorine. The half–life of ^{36}Cl is 3.0×10^5 yr. How many disintegrations/second are produced by this sample?

Self–Test

This section is intended to test your knowledge of the material covered in this chapter. Think through these problems, and make certain you understand what they are asking. Make sure your answers make sense. Successful completion of these problems indicates that you have mastered the material in this chapter. You will receive the greatest benefit from this section if you use it as a mock exam, as this will allow you to determine which topics you need to study in more detail

True–False

1. Radioactive decay is a second-order process.
2. Lighter elements require more neutrons for stability, but heavier elements have a 1:1 ratio of protons and neutrons.
3. When nucleons come together to form a nucleus, a small amount of mass is gained.
4. When elements undergo neutron-induced fission, they tend to break into pieces of roughly equivalent size.
5. The control rods in a reactor absorb excess neutrons to slow the reaction.
6. Nuclear fusion has no toxic by-products, but is unlikely to be a practical source of power in the near future.
7. Given their positions in the periodic table, a nuclear transmutation reaction is more likely to turn gold into lead than lead into gold.
8. Alpha particles are most damaging from an internal source.
9. The curie measures only disintegrations per second, and says nothing about tissue damage.
10. Radioisotopes can be used for disease treatment, imaging, and *in vivo* studies.

Multiple Choice

1. All of the following are nuclear processes *except*
   a. alpha emission
   b. electron capture
   c. beta-modification
   d. transmutation

2. Nuclei are most stable when they
   a. have a 1:1 ratio of neutrons to protons, if heavy
   b. have fewer neutrons than protons
   c. are ionized
   d. are ^{56}Fe

3. The mass defect in nuclei is the result of
   a. inaccurate masses of subatomic particles
   b. the conversion of mass to energy to hold the nucleus together
   c. the law of conservation of mass
   d. nuclear transmutation

4. The process of nuclear fission
   a. can self-perpetuate as a chain reaction
   b. generates very little energy
   c. generates only ^{4}He as a product
   d. b and c

5. In a reactor, nuclear fission is used to generate power by
   a. generating electrons.
   b. converting water to steam, which turns a turbine.
   c. driving a piston with repeated small explosions.
   d. none of the above.

6. The control rods in a reactor
   a. give off electrons
   b. generate neutrons
   c. absorb neutrons
   d. emit alpha particles

7. Nuclear transmutation is a process that can be used to
   a. generate new elements and isotopes
   b. generate electricity
   c. diagnose disease
   d. date ancient objects

8. Which of the following provides information about tissue damage?
   a. becquerel
   b. rem
   c. sievert
   d. b and c

9. Which of the following correctly ranks these types of emissions by speed, fastest to slowest?
   a. alpha particle, gamma ray, beta particle
   b. beta particle, alpha particle, X-ray

c. gamma ray, beta particle, alpha particle
d. X-ray, alpha particle, gamma ray

10. Radiocarbon dating can only be used to determine the ages of
    a. meteorites
    b. formerly living objects
    c. uranium-containing rocks
    d. all of the above

Fill–in–the–Blank

1. Neutrons appear to function as a nuclear ________ that overcomes the ___________ to ___________ repulsions that might otherwise cause a _______________ to fly apart. The energy to hold the nucleus together cannot be directly measured, but can be calculated using the _________ _____________ and the equation ________________.
2. Binding energy per nucleon reaches a ________________ in the middle of the periodic table. Therefore, small nuclei can increase their stability through ______________, and large nuclei can increase their stability through ________________ or radioactive _____________.
3. In nuclear _____________________, an atom is bombarded with any of a number of high-energy_______________. In the collision, an unstable _______________ is momentarily formed, and a new ________________ is produced.
4. The effects of radiation on a biological system vary according to a large number of variables. The most damage comes from large, slow particles such as _______________ and ________________ particles as an ________________ source. Also damaging are the fastest moving radiation, ________________ rays and _________________ rays, which do the most damage as ________________ sources. Large, slow particles are ________________ by skin. The fast-moving rays require several inches of _____________ to stop them.
5. Radiopharmaceuticals are used in a number of ways. They can be used in ______________ procedures in which they are administered and monitored, in therapeutic procedures in which they are used to target __________________ tissue (_______________ radiation therapy is always more specific and targeted than _________________ radiation therapy) , and in ____________, in which the isotope's location or absence is used to assist in diagnosis.

Matching

Binding energy	a. a reaction in which the nucleus of an element splits into two nearly equal pieces
Chain reaction	b. the loss in mass when nucleons come together to form a nucleus
Critical mass	c. the joining of nuclei in a reaction

Fission — d. the amount of material necessary for a nuclear reaction to be self-sustaining

Fusion — e. the energy that holds nucleons together

Mass defect — f. change of one element into another

Nuclear transmutation — g. a self-sustaining reaction whose product initiates further reaction

Problems

1. Calculate the binding energy of europium-130 (mass = 129.963 57 amu), europium-153 (mass = 152.921 23 amu) and europium-167 (mass = 166.953 21amu) in MeV/nucleon. From these data, which is the most stable isotope? Which is the least stable?

2. Calculate the change in mass for the following reaction in g/mol:

$$2\,O \rightarrow O_2 \qquad \Delta E = -498.4\ \text{kJ/mol}$$

3. How much energy in J/mol is released in the following fission reaction?

$$^{235}U + {}^{1}n \rightarrow {}^{92}Kr + {}^{141}Ba + 3\,{}^{1}n$$

The mass of ^{235}U is 235.0439 amu, the mass of ^{92}Kr is 91.9262 amu, the mass of ^{141}Ba is 140.9144 amu, and the mass of a neutron is 1.008 66 amu.

4. How much energy is released in the following fusion reaction?

$$^{3}H + {}^{3}H \rightarrow {}^{4}He + 2\,{}^{1}n$$

The mass of tritium is 3.016 049 amu, the mass of ^{4}He is 4.002 60 amu, and the mass of a neutron is 1.008 66 amu.

5. Write chemical reactions for the following nuclear transmutations:

   a. Be-9 is bombarded with hydrogen to form Li-6 and an alpha particle.

   b. Pu-239 reacts with an alpha particle to produce an unknown element and a neutron.

   c. Pb-208 is collided with another element to form Hs-265 and a neutron.

6. The half–life for zirconium–96 is 6.3×10^{26} s. What is the decay constant for this radioactive element?

7. What is the age of a bone fragment that shows an average of 3.5 disintegrations per minute per gram of carbon? The carbon in living organisms undergoes an average of 15.3 disintegrations per minute per gram, and the half–life of ^{14}C is 5715 y.

8. The decay of U-235 to Pb-207 takes place with a half-life of 704 million years. If an asteroid that originally contained no lead is found to contain 56 g of Pb-207 and 27 g of U-235, how old is it?

Challenge Problem

In an experimental form of cancer treatment known as boron neutron capture therapy (BNCT), boron-10 is injected into a cancer patient, where it binds preferentially to tumor cells. A beam of neutrons is then directed at the tumor, and the boron-10 disintegrates into lithium. Only one type of radiation is produced by the reaction, and only one neutron is absorbed in the reaction. Write and balance the equation for this reaction, and explain the rationale behind this therapeutic method.

CHAPTER 23

ORGANIC AND BIOLOGICAL CHEMISTRY

Chapter Learning Goals

A. Saturated Hydrocarbons

1. Draw structures for isomers of simple alkanes.
2. Write condensed structures of organic molecules.
3. Determine which structures represent different molecules and which are merely different conformations of the same molecule.
4. Given the structure of a straight or branched alkane, determine its IUPAC name and vice versa.
5. Given the structure of a cycloalkane, determine its IUPAC name and vice versa.
6. Identify functional groups in molecules. Draw structures of molecules containing functional groups.

B. Unsaturated Hydrocarbons

1. Given the structure of an alkene or alkyne, determine its IUPAC name and vice versa.
2. Given the structure of a substituted cycloalkene or cycloalkyne, determine its IUPAC name and vice versa.
3. Understand and apply concepts of *cis-trans* isomerization.
4. Predict the products and write balanced chemical equations for alkene addition reactions.

C. Aromatic Compounds

1. Given the structure of an aromatic compound, determine its IUPAC name and vice versa.
2. Predict the products and write balanced chemical equations for aromatic substitution reactions.
3. Explain the unique stability of aromatic rings as a result of hybridization and resonance.

D. Alcohol, Ethers, and Amines

1. Classify molecules as alcohols, ethers, or amines. State the general properties of these classes of molecules.

E. Carbonyl Containing Compounds

1. Classify molecules as ketones or aldehydes.
2. Classify molecules as carboxylic acids, esters, or amides. State the general properties of these classes of molecules.
3. Give systematic names of carboxylic acids, esters, and amides.
4. Predict the products and write balanced chemical equations for carbonyl substitution reaction and reactions of carboxylic acids, esters, and amides with water.

F. Biological Chemistry

1. Give a summary of the basic processes of metabolism.
2. Distinguish between amino acids, peptides, and proteins.
3. Classify amino acid side chains as polar, nonpolar, acidic, and basic.
4. Classify a carbohydrate as simple or complex. Classify a monosaccharide as a ketose or aldose.
5. Identify the characteristic properties of a lipid.
6. Distinguish between fats and oils on the basis of their side chains.
7. Draw the structures of simple nucleic acids.
8. Utilize the complementarity of base pairs in DNA and RNA to generate sequences.

9. Understand the flow of information from DNA to RNA to protein.

Chapter in Brief

Organic chemistry is the study of carbon compounds. This chapter provides an introduction to the vast array of compounds and reactions in organic chemistry. Naming of compounds, varying structures with multiple bonds, cyclic structures and functional groups, and the most basic reactions of these molecules will be introduced. The chapter then provides a basic introduction to biochemical energetics and the main classes of biomolecules. Proteins, carbohydrates, lipids, and nucleic acids are discussed in terms of their basic building blocks, functions, and properties.

23.1 Organic Molecules and Their Structures: Alkanes

There are a multitude of single-bonded compounds that contain only carbon and hydrogen. They are known as **hydrocarbons** and belong to a family of single-bonded organic molecules called **alkanes**.

Each carbon can form four bonds, either to hydrogen or to another carbon. There are only three possible structures for alkanes with one to three carbons, but at four carbons, **branched-chain** hydrocarbons become possible. This greatly increases the number of possible compounds. At six carbons, for example, there is one **straight-chain alkane** with no branches, and four branched-chain alkanes

Compounds such as these, with the same formula but different arrangements of atoms are called **isomers**. Despite their identical numbers and types of atoms, their differing structures give them different properties, both physical and chemical.

To reduce the complexity of drawing these molecules, **condensed structures** are used: in these structures, most single bonds are omitted, with only vertical bonds to branches shown:

```
     H   H                     H
     |   |                  H  |  H
  H—C—C—H                   H \C/ H
     |   |                   |  |  |
     H   H                H—C—C—C—H
                             |  |  |
                             H  H  H
```

become

CH_3CH_3

CH_3 | CH_3CHCH_3

The condensed structure provides information about connectivity only. As molecules get larger, they are more and more able to adopt different conformations, but their connectivity remains the same. As a result, all of the following are different representations of the same molecule:

```
CH3CH3              CH3CHCHCH2CH3       CH3CH2CHCHCH3          CH3 CH3
  \ |                   / |                  / \                  \ /
CH3CHCHCH2CH3         CH3CH3               CH3 CH3          CH3CH2CHCHCH3

          CH3                  CH3
          |                    |
   CH3CH2CHCHCH3        CH3CHCHCH2CH3
            |                  |
            CH3                CH3
```

These are all 2,3-dimethylpentane.

Single bonds are free to rotate, so there are an infinite number of possible *conformations* for organic molecules. Most alkanes will have an extended, zig-zag shape due to the tetrahedral arrangement of sp^3 orbitals.

EXAMPLE:
Draw the straight-chain alkane C_8H_{18} as a condensed structure.

SOLUTION:
Remembering that all carbons in an alkane must have four bonds, the structure will be:

$$CH_3CH_2CH_2CH_2CH_2CH_2CH_2CH_3$$

23.2 Families of Organic Compounds: Functional Groups

Organic compounds can be classified into families according to structural features, and members of a given family also have similar reactivity. Alkanes, alkenes, alkynes, and the aromatic rings known as arenes contain only carbon-carbon bonds. Others contain bonds to nitrogen or oxygen: some contain bonds to halogens as well.

The following is a condensed version of the much more extensive chart on page 913 of your book: for additional information, refer to that chart as well.

Family Name	Functional Group	Name Ending
Alkanes	single bonds	*-ane*
Alkenes	one or more C=C bonds	*-ene*
Alkynes	one or more C≡C bonds	*-yne*
Arenes	C_6 ring, alternating C-C and C=C bonds	none
Alcohol	-C-OH	*-ol*
Ether	-C-O-C-	*ether*
Amine	-C-N-	*-amine*
Aldehyde	-C-C(=O)-H	*-al*
Ketone	-C-C(=O)-C-	*-one*
Carboxylic acid	-C-C(=O)-OH	*-oic acid*
Ester	-C-C(=O)-O-C-	*-oate*
Amide	-C-C(=O)-N-	*-amide*

Workbook Problem 23.1
Identify the functional groups in the following compounds:

a.
$$CH_3\text{-}\overset{\overset{\displaystyle O}{\|}}{C}\text{-}OH$$

b.
$$CH_3\text{-}\overset{\overset{\displaystyle O}{\|}}{C}\text{-}CH_3$$

c.
$$CH_3\text{-}\overset{\overset{\displaystyle O}{\|}}{C}\text{-}H$$

d. $CH_3\text{-}O\text{-}CH_3$
e. $CH_3CH_2CHCH_2$

Strategy: Look for non-carbon atoms and multiple bonds. Compare with the chart above or the chart on page 913 of the textbook to name the functional groups.

Step 1: Name the functional groups.

23.3 Naming Organic Compounds

Because of the large numbers of organic molecules, and their complexity, the International Union of Pure and Applied Chemistry (IUPAC) has come up with a standard method for naming organic molecules.

In this system, a molecule is named with a prefix, which indicated substitutents, a parent name that indicates the number of carbons in the longest chain, and a suffix that indicates the chemical family of the molecule (these are the name endings in the chart above).

Straight-chain alkanes are names using Greek prefixes for those with chains of five carbons or more, thus, *pent*ane, *hex*ane, etc. For historical reasons, chains with one to four carbons are named using unique prefixes:

1 carbon = *meth-*
2 carbons = *eth-*
3 carbons = *prop-*
4 carbons = *but-*

Branched-chain alkanes are named using the following four steps:

Step 1: Name the parent chain. This requires finding the longest continuous carbon chain in the molecule, which may require taking a non-straight path.

Step 2: Number the carbon atoms in the main chain, beginning at the end *nearest* the first branch point. The first branch point should be at the lowest possible number.

Step 3: Identify and number the branching substituent. Sometimes, there will be more than one substituent on a carbon. In this case, assign the same number twice. Hydrocarbon substituents are called **alkyl groups**, and are named using the same prefixes as the main chain, but with the suffix *–yl*. In other words, a four-carbon substituent is a *butyl* group.

Step 4: Write the name as a single word. Use hyphens to separate prefixes, and commas to separate numbers if there are multiple substituents. If there are multiple identical substituents, name them as *di-* or *–tri*: *trimethyl-*, for instance, indicates that there are three methyl group substituents. If there are multiple non-identical substituents, name them in alphabetical order, ignoring any prefixes: *ethyl* before *dimethyl.*

EXAMPLE:

Name the following alkane:

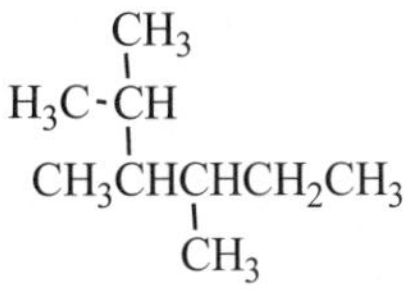

```
    CH3
    |
H3C-CH
    |
  CH3CHCHCH2CH3
        |
        CH3
```

SOLUTION:

Find the longest chain:

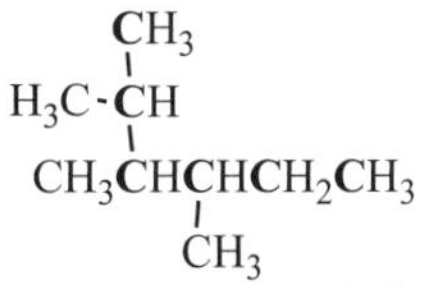

```
    CH3
    |
H3C-CH
    |
  CH3CHCHCH2CH3
        |
        CH3
```

This is a substituted *hexane*.

Number the chain so as to give the smallest numbers to the substituents:

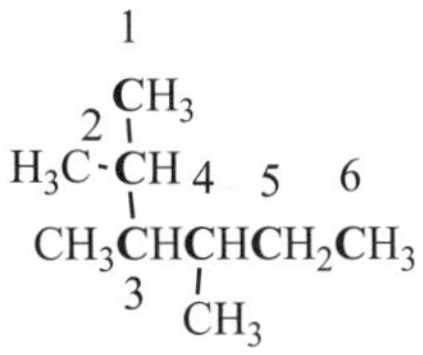

```
     1
    CH3
   2|
H3C-CH 4  5  6
    |
  CH3CHCHCH2CH3
     3  |
        CH3
```

There are *methyl* groups attached to carbons 2, 3, and 4. (Notice that if the chain had been numbered backward, these would be carbons 3, 4, and 5 – lower numbers)

Name the molecule: 2, 3, 4-trimethylhexane.

More About Alkyl Groups

Methyl and ethyl groups have only one possible attachment point, so where they attach is irrelevant. But with a three-carbon propyl group, the attachment can be at either end or in the middle, which gives an *isopropyl* group. Four-carbon butyl groups can be attached at either end, or at carbon-2 to form *sec-butyl.* But with as few as four carbons, the substituent itself can be branched to form *isobutyl*, in which the substituent attaches at the end of the chain, or *tert-butyl*, in which the substituent attaches in the middle of the chain. The prefixes *sec-* and *tert-* indicate the number of carbons attached: *sec* indicates that there are two carbons attached to the branching carbon, and *tert-* indicates that there are three.

Workbook Problem 23.2

Name the following molecules:

a.

$$\begin{array}{c} \qquad\quad CH_3 \\ \qquad\quad | \\ CH_3CH_2CHCCH_3 \\ \quad / \;\; \backslash \\ \quad CH_3\,CH_3 \end{array}$$

b.

$$\begin{array}{c} CH_2CH_3 \\ | \\ CH_3CH_2CH_2CCH_2CH_2CH_3 \\ | \\ CH_2CH_2CH_3 \end{array}$$

c.

$$\begin{array}{c} CH_2CH_3 \\ | \\ CH_3CHCH_2CH_2CH_3 \end{array}$$

Strategy: Use the naming strategy detailed above to identify the molecules.

Step 1: Identify and name the parent chain

Step 2: Number the chain, beginning at the end closest to the first branch point

Step 3: Identify and number the substituents

Step 4: Write the name of the molecule as a single word

Workbook Problem 23.3

Draw the following molecules:

a. 4-isopropylheptane
b. 2-methyl-3-ethylhexane
c. 2, 2-dimethylbutane
d. 2,3-dimethyl-4-propyloctane

Strategy: Draw the carbon backbones, and add substituents in the proper places.

Step 1: Draw the carbon backbones.

Step 2: Add the substituents, subtracting hydrogens as necessary.

23.4 Unsaturated Organic Compounds: Alkenes and Alkynes

Alkenes are molecules that contain one or more double bonds, while **alkynes** contain one or more triple bonds. These are known as **unsaturated** hydrocarbons, while **saturated** hydrocarbons are those that contain as many hydrogens as possible.

Alkenes are named by numbering the chain beginning closest to the double bond. If the double bond is in the center of the chain, numbering is begun closest to the substituent. The smallest alkenes are often named as *–ylene* – ethylene, for example, instead of ethene.

Isomers of alkenes can occur not only because of the position of the bond within the chain, but also as a result of the position of substituents around the bond. Double bonds cannot rotate the way single bonds can because of their pi-bonds, which leads to ***cis-trans*** **isomers**. In *cis* isomers, identical substituents are on the same side of the double bond, while in *trans* isomers, identical substituents are on opposite sides of the double bond.

EXAMPLE:

Name the following compound:

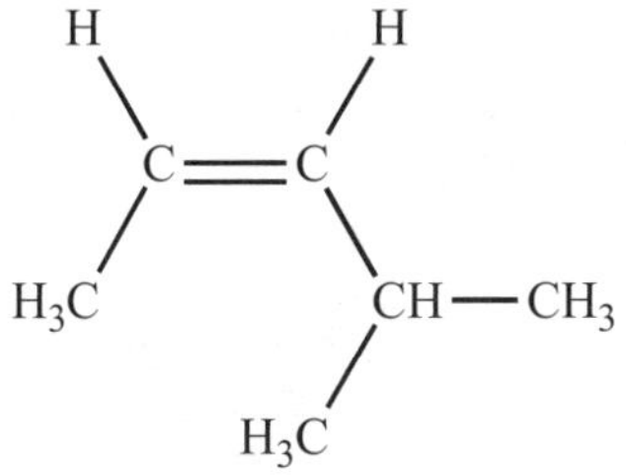

H H

C=C

H_3C $CH-CH_3$

H_3C

SOLUTION:

The longest chain has five carbons, and numbering begins at the right to minimize the number of the double bond. The substituents are both on the same side of the bond, and there is a methyl group on carbon 4- this is *cis*-4-methyl-2-pentene.

Alkynes are similar in many respects to alkenes, and are named using the suffix *–yne*. The smallest alkyne, *ethyne*, is often called *acetylene*. The geometry of triple bonds is linear, so *cis-trans* isomerization does not occur.

The most common reactions of alkenes and alkynes are **addition reactions** in which a reagent adds to the multiple bond. Alkenes can be converted into alkanes by the addition of hydrogen, into dihalides by adding halogens, or into alcohols by adding water:

Alkene + H_2 → Alkane

Alkene + X_2 → 1,2-Dihaloalkane

Alkene + H_2O → Alcohol

Workbook Problem 23.4

Draw 4-ethyl-6-methyl-2-heptyne.

Strategy: Draw the carbon backbone, and add substituents in the proper places.

Step 1: Draw the carbon backbone.

Step 2: Add the substituents, subtracting hydrogens as necessary.

Workbook Problem 23.5

What are the reaction products from 3-hexene and

a. water in the presence of an acid
b. chlorine
c. hydrogen in the presence of a metal catalyst

Strategy: Draw the carbon backbone, and determine the possible products of the addition reactions.

Step 1: Draw the carbon backbone.

Step 2: Add the substituents.

23.5 Cyclic Organic Compounds

Cyclic compounds are carbon rings from 3 carbons to 30 carbons and more. The simplest are the **cycloalkanes**. Even condensed structures are difficult to draw for these molecules, so polygonal **line-bond structures** are used to represent them in which every junction is a C, and the hydrogens are understood to be in the amount needed to give each carbon four total bonds.

Cyclobutane and cyclopropane have distorted bond angles, which make them weaker than other alkanes. Cyclopentane and higher are able to pucker into shapes that allow near-ideal bond angles.

Substituted cycloalkanes are names using alphabetical priority: The first substituent is named, after which the substituents are given the lowest possible combination of numbers. Cycloalkenes are numbered starting at the double bond, and then giving the other substituents the lowest possible numbers. The double bond itself is not given a number.

EXAMPLE: Name the following compound:

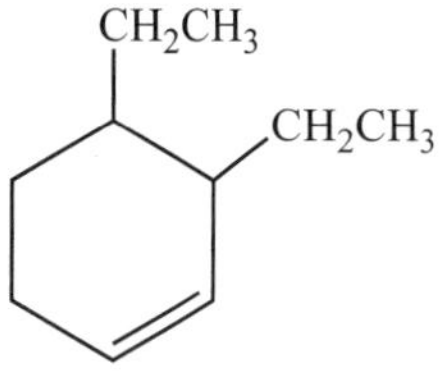

CH_2CH_3

CH_2CH_3

SOLUTION:

There are six junctions, indicating a six-carbon ring with one double bond: this is cyclohexene. To give the two substituents the lowest possible numbers, begin numbering clockwise from the double bond. The double bond therefore incorporates carbons 1 and 2, with the substituents on carbons 3 and 4. The substituents are both ethyl groups, so this is 3,4-diethylcyclohexene.

Workbook Problem 23.6

Draw 3,4,5-trimethylcyclopentene.

Strategy: Draw the carbon backbone, and add substituents in the proper places.

Step 1: Draw the cyclic backbone.

Step 2: Add the substituents.

23.6 Aromatic Compounds

Aromatic molecules are given that name because the original members of this group were fragrant substances found in natural sources. However, it was soon realized that these behaved chemically differently than other organic molecules.

Aromatic now refers to molecules containing six-membered rings represented as having alternating single and double bonds. The reality of these molecules is that the carbon molecules are sp^2-hybridized, with a delocalized pi-bond that spreads over the entire molecule generating a **resonance structure**.

Benzene (C_6H_6) is the most common aromatic molecule:

Substituted aromatic molecules are named using the suffix *–benzene*. Ethylbenzene, for example, has a single ethyl group. (It doesn't need a number, as all ring positions are equivalent). If the benzene ring is itself a substituent of a larger molecule, it is given the name *phenyl.*

Disubstituted benzenes are named based on the relative positions of the substituents. *Ortho-* substituted benzenes have substitutents directly next to one another on the ring, *meta-* substituted molecules have substituents separated by one carbon, and *para-* substituted benzenes have groups directly across the ring from one another.

Aromatic compounds typically undergo substitution reactions, as the benzene structure is so stable: one of the hydrogens on the ring is substituted by another group.

When benzene reacts with nitric acid in the presence of sulfuric acid, a *nitro* group is added to the ring. Nitrobenzene is a starting point for preparing dyes as well as explosives.

When halogens are added to a benzene ring in the presence of an iron-based catalyst (FeX_3 where X is the halogen being added), it yields a single-substituted benzene and HX.

Workbook Problem 23.7

Draw the following molecules:

o-dichlorobenzene *p*-methylnitrobenzene *m*-ethylmethylbenzene

Strategy: Draw the benzene ring, and add substituents in the proper orientations.

Workbook Problem 23.8

Provide the possible products from the following reactions:

a. *o*-dimethylbenzene with nitric acid in the presence of sulfuric acid (two possible products)
b. *m*-bromochlorobenzene with chlorine in the presence of $FeCl_3$ (four possible products)

Strategy: Draw the reactant, and add the new substituent in all places that give a unique configuration.

23.7 Alcohols, Ethers, and Amines

Alcohols have a hydroxyl (-OH) group in place of a hydrogen. Alcohols can form hydrogen bonds just like water. This makes small alcohols soluble in water, and gives them higher boiling points than alkanes.

Alcohols are named by the point of attachment, and are numbered beginning at the end of the chain closest to the alcohol group.

Methanol (CH_3OH) is also called wood alcohol. It is toxic, causing blindness in low doses and death in larger amounts. It is an important starting material for producing formaldehyde and acetic acid.

Ethanol is produced by fermentation of grains and sugars, and is sometimes called grain alcohol. It is the alcohol present in alcoholic beverages. Enzymes in yeast will break down sugars to make ethanol and CO_2.

Other alcohols: 2-propanol, also called isopropy alcohol or rubbing alcohol is a disinfectant; 1,2-ethanediol, also called ethylene glycol, is used in automobile antifreeze; 1, 2, 3-propanetriol is also called glycerol and is used as a moisturizing agent; and phenol is used in the manufacture of nylon, epoxy, and resins.

Ethers can be described as water derivatives in which both hydrogens have been replaced by organic substituents. They are relatively inert chemically, so are often used as reaction solvents. Diethyl ether was used as an anesthetic agent for many years but has now been replaced by other less flammable alternatives.

Amines are organic derivatives of ammonia. They are names using the suffix *–amine* to describe the substituents on the nitrogen. Like ammonia, amines are weak bases that can use the lone pair of electrons on the nitrogen to accept a proton and form ammonium salts. These salts are more soluble than the neutral compounds.

Many drugs are amines: they become more soluble in body tissues if they are converted to their ammonium salts,

23.8 Carbonyl Compounds

Carbon-oxygen double bonds are known generally as **carbonyl groups**. All biomolecules, most pharmaceuticals, and many synthetic polymers contain these groups. They are polar, as a result of the electronegative oxygen atom, but can be even more polar if there is an additional substituent on the carbonyl carbon.

Aldehydes and ketones contain a carbonyl group at the end of a carbon chain for aldehydes, and in the middle of a carbon chain for ketones. They are present in many biologically important compounds, as well as in small molecules such as formaldehyde, or, more properly *methanal*, which is used as a biological preservative and sterilizing agent, as well as a starting material for plastics manufacture. A common ketone is acetone, or *propanone*, a widely used organic solvent.

Carboxylic acids contain a $-\overset{\displaystyle O}{\overset{\|}{C}}-OH$ group and are found throughout the plant and animal kingdoms. Vinegar is primarily acetic or ethanoic acid, and long-chain carboxylic acids are constituents of all fats and oils. Systematic names replace the ending of the alkane parent chain with *–oic acid.*

These compounds are very weak acids, but will react with alcohols to produce an ester and water: the –OH of the acid and the H of the alcohol group combining to form water, and leaving a bond through the oxygen of the alcohol.

Esters contain a $-\overset{\displaystyle O}{\overset{\|}{C}}-O-C$ functional group. Esters are common in pharmaceuticals, in polymers such as *polyester*, and in fragrance molecules.

The most common reaction of esters is conversion into carboxylic acids. They undergo a *hydrolysis* reaction in which a water molecule is added into the ester to form a carboxylic acid and water. Note: this is the reverse of the reaction described above. This reaction is catalyzed by both acid and base.

When this hydrolysis is base-catalyzed, it is also called *saponification*: soap is made by hydrolysis of long-chain esters in animal fat, and is a mixture of the sodium salts of carboxylic acids.

Esters are named by identifying first the alcohol derivative and then the acid derivative, ending in *–ate.*

Amides contain a $-\overset{\displaystyle O}{\overset{\|}{C}}-N$ functional group. Amide bonds are the links that hold proteins together, as well as the links that hold some polymers, particularly nylon, together. They are also found in some pharmaceuticals.

Amides are neutral because of the strong influence of the adjacent oxygen. This pulls the lone pair of electrons tightly to the nitrogen, preventing them from reacting.

Acid or base-catalyzed hydrolysis of an amide generates a carboxylic acid and an amine. This acid-catalyzed hydrolyis is an important part of the digestion of protein.

23.9 An Overview of Biological Chemistry

All living organisms do work, both physical and chemical. In animals, the energy for these processes comes from their food, which is broken down in interconnected reactions collectively called **metabolism**. Food molecules are oxidized to carbon dioxide, water, and energy. The reactions that break down larger molecules into smaller ones release energy and are known as **catabolism**. The reactions that build larger molecules from smaller ones absorb energy and are known as **anabolism**.

The four stages of catabolism:

Stage one is digestion: large molecules are broken down into smaller molecules such as simple sugars, *fatty acids*, and amino acids.

Stage two is the conversion of these small molecules into two-carbon $CH_3C{=}O$ acetyl groups which are connected to the carrier molecule *coenzyme A* through the carbonyl carbon. The resultant molecule is *acetyl coenzyme A* or *acetyl CoA*. This is an intermediate in the breakdown of all types of food molecules.

Stage three is the oxidation of acetyl group in the *citric acid cycle* to yield carbon dioxide and water and energy. This energy is used in stage four.

Stage four is the *electron transport chain*. This takes the energy released by the citric acid cycle and uses it to make adenosine triphosphate (ATP), the "energy currency of the cell."

Catabolic reactions make ATP, anabolic reactions "spend" it.

23.10 Amino Acids, Peptides, and Proteins

Proteins are major components of biological systems. Fifty percent of the dry weight of a human body is protein, and the reactions that occur in the body are catalyzed by proteins. There are more than 150,000 proteins in the human body.

Proteins can be structural, they can carry messages as hormones, and they can catalyze reactions as **enzymes**.

Proteins are polymers made up of individual units called *amino acids*. These small molecules contain an amide group, a central carbon, and a carboxylic acid group. They are linked together by amide bonds which, in proteins, are referred to as **peptide bonds**.

Two amino acids liked together are known as a *dipeptide*, three as a *tripeptide*. Up to 100 amino acids linked together is known as a **peptide**, with the term *protein* reserved for chains that are even longer.

Twenty amino acids occur naturally, all of which have the basic structure below, with a variety of –R groups attached to the central carbon, which is known as the α-carbon. These are subsequently also referred to as α-amino acids. Each is known by its name, and a three-letter abbreviation (Serine = Ser, for example). There is a chart of all 20 on page 935 of the textbook.

$$H_2N-\underset{\large R}{\overset{\large H}{C}}-\overset{\large O}{\overset{\|}{C}}-OH$$

The R group is also known as a *side chain*. Of the 20 amino acids, humans can synthesize only 11, with the remaining 9 *essential amino acids* required from dietary sources. Fifteen of the 20 have neutral side chains, two are acidic, three are basic. Of the neutral amino acids, these can again be divided into polar and nonpolar or *hydrophobic* (nonpolar, literally, "water hating") and *hydrophilic* (polar or "water-loving").

The number of possible combinations increases rapidly with the number of amino acids: with only three amino acids, six different tripeptides can be formed. All combinations will have an *N-terminal amino acid* with a free NH_2 group, and a *C-terminal amino acid* with a free CO_2H group. By convention, sequences are written N-terminus on the left to C-terminus on the right.

23.11 Carbohydrates

Carbohydrates occur in all living organisms. The name originated with carbon, that has the empirical formula $C(H_2O)$, which led to the idea that it was a hydrate of carbon. Although the idea was discarded, the name persisted, and is now used for the large class of hydroxyl-containing aldehydes and ketones that we commonly call *sugars*

Monosaccharides, or *simple sugars* are carbohydrates that cannot be broken into smaller molecules by acid hydrolysis. An *aldose* contains an aldehyde carbonyl group, while a *ketose* contains a ketone carbonyl group. The *–ose* suffix designates a sugar.

Monosaccharides are typically found in a ring form on which a hydroxyl group near one end of the chain combines with the carbonyl at or near the other end of the chain. When this happens, more than one form is possible depending on whether the following hydroxyl is above (β) or below (α) the ring. Crystalline forms of monosaccharides are generally in the α-form, but in solution, all three forms, α, β, and linear are found.

EXAMPLE: Classify the following monosaccharides:

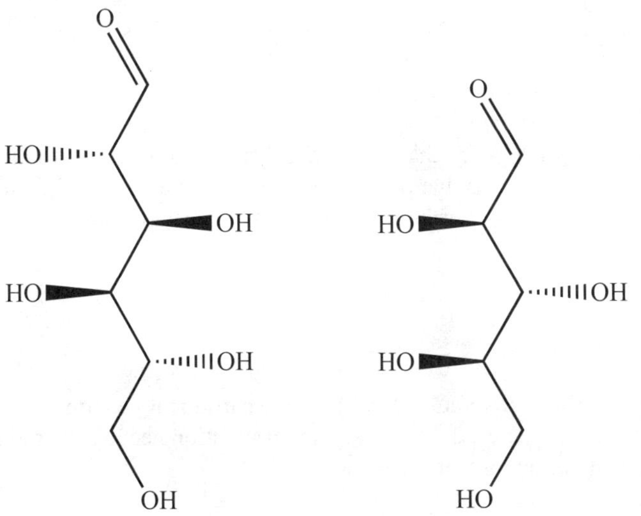

SOLUTION:

The molecule on the left has six carbons and an aldehyde group. This is an aldohexose.

The molecule on the right has five carbons and a keto group. This is a ketopentose.

Polysaccharides are formed when monosaccharides join together. Sucrose is a *disaccharide* formed of one glucose and one fructose. Cellulose, the structural material in plants, is made up of thousands of β-glucose molecules joined together in an immense chain. Starch is made up of thousands of α-glucose units. This small difference in configuration makes starches – found in rice, beans, and potatoes – digestible, while cellulose cannot be digested by humans.

23.12 Lipids

A **lipid** is an organic molecule that dissolves in nonpolar solvent. The fact that lipids are defined by a physical property means that they exist in a variety of forms, but all contain large hydrocarbon portions.

Animal fats and vegetable oils are the most abundant lipids. Fats and oils are both **triacylglycerols** or *triglycerides*. These are esters of glycerol (1,2,3-propanetriol) with three long-chain carboxylic acids called **fatty acids**.

As shown below, fatty acids can be saturated or unsaturated, and can also be *polyunsaturated:*

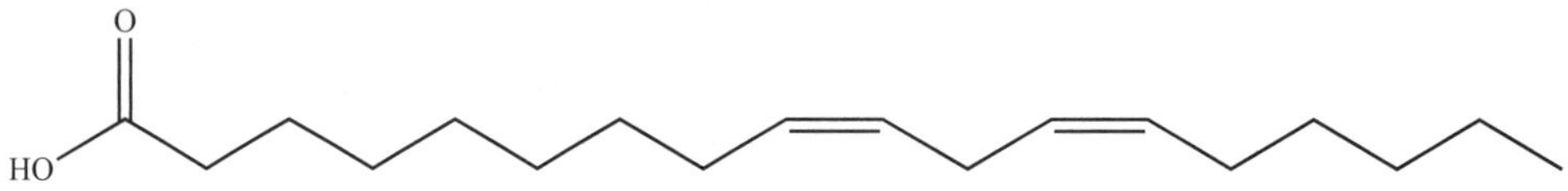

O

HO

linoleic acid

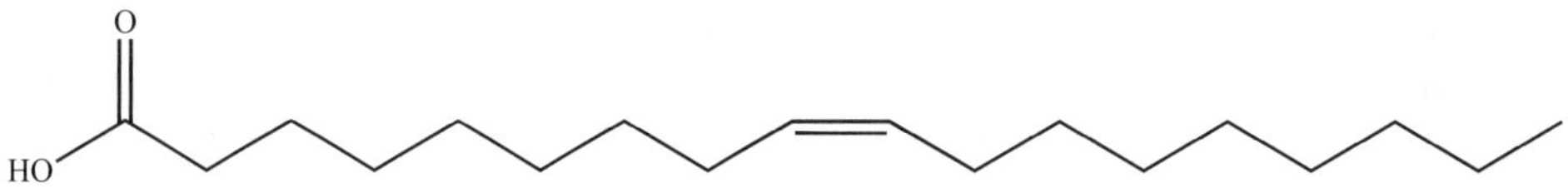

O

HO

oleic acid

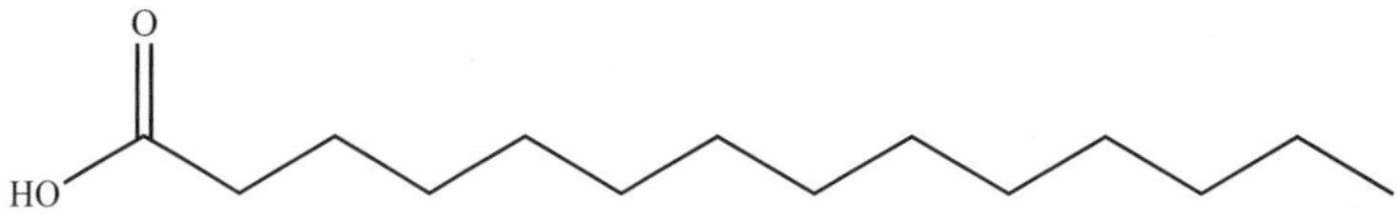

O

HO

myristic acid

It is the different properties of these fatty acids that lead to the different properties of fats and oils. Fats have primarily saturated fatty acids that pack together more efficiently and therefore have higher melting points. Oils contain primarily unsaturated fatty acids, but can be hydrogenated to yield saturated fats.

23.13 Nucleic Acids

Deoxyribonucleic acid (DNA) and **ribonucleic acid (RNA)** are the carriers of genetic information. Coded in an organism's nucleic acids are all the information needed to produce the many thousands of proteins required by that organism.

Nucleic acids are polymers built of **nucleotide** units. Nucleotides are built of **nucleosides**, which are amine bases linked to aldopentose sugars (ribose for RNA and deoxyribose for DNA), and phosphoric acid.

DNA and RNA share three amine bases:

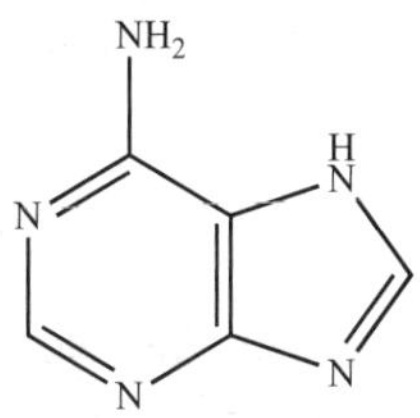

adenine 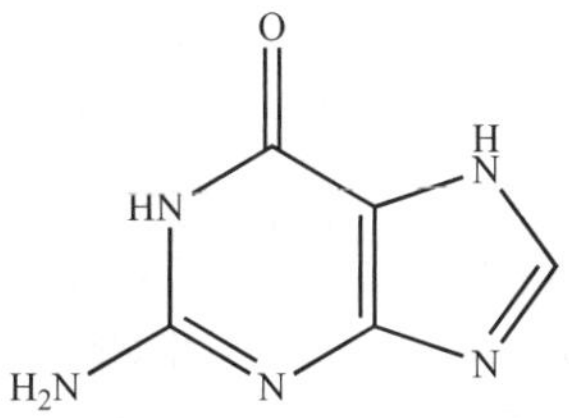 guanine 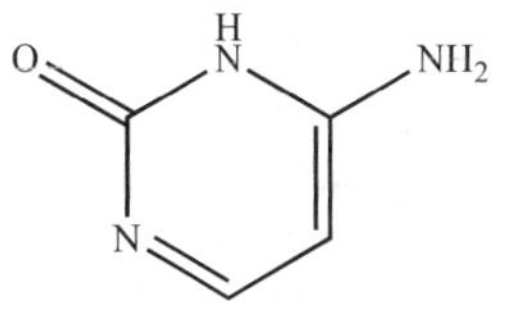 cytosine

Only DNA contains thymine:

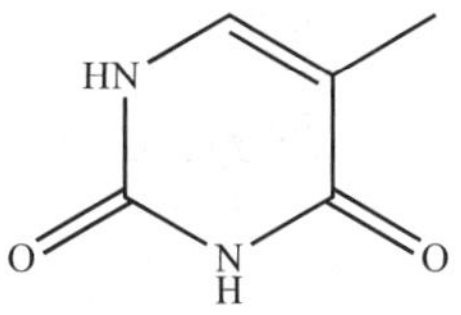

thymine

Only RNA contains uracil:

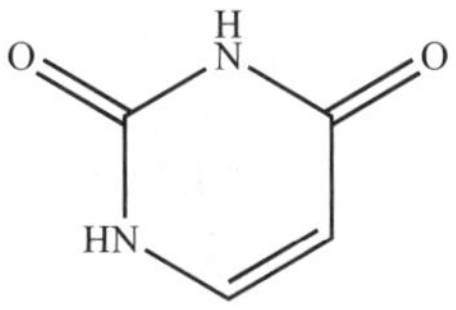

uracil

In both DNA and RNA, the amine base is bonded to the C1' of the sugar, and the phosphoric acid is bonded to the C5' carbon of the sugar. The "prime" indicates a sugar carbon. Other numbered carbons are in the amine base. When the nucleotides are joined to make polymers, the phosphates join to the C3' of the next nucleoside.

DNA sequences are listed beginning at the 5' end, using the first letters of the nucleotides: TACG. DNA sequences from different tissues of the same species have the same proportions of each nucleotide, but sequences from different species may have very different proportions.

There are always equal amounts of A and T, G and C, regardless of species. In the **Watson-Crick model** of DNA structure, two complementary strands hydrogen bond to one another, with A forming two hydrogen bonds with T, and G forming three hydrogen bonds with C. The two strands then coil around one another to form a double helix.

DNA is primarily found in the nuclei of cells coated with proteins and wound into *chromosomes*. Each chromosome contains several thousand *genes*, where a gene is a segment of DNA coding for a specific protein. DNA is the storage mechanism for genetic information, while RNA is used to make proteins.

Replication is the process by which additional copies of the DNA are made. The helix unwinds, and the bases are exposed. A new complementary strand is made for each half of the DNA, leading to two identical copies.

Transcription is the process by which the instructions in the DNA are copied for use. A single-stranded RNA copy of one of the two DNA strands is made, with uracil taking the place of thymine. Once completed, the RNA separates from the template, and the DNA helix reforms.

Translation is the conversion of the information contained in the nucleic acid template to protein. *Messenger RNA* or *mRNA* travel to ribosomes where the RNA is read and the protein made. Each three-nucleotide sequence codes for a different amino acid. For instance, the sequence C-U-G codes for the amino acid leucine. A *transfer RNA* or *tRNA* has a complementary sequence, and brings the next amino acid in to be joined to the chain. At the end of the mRNA there is a "stop" sequence that brings translation to a close.

Workbook Problem 23.9

Determine the mRNA sequence that would be produced from the following DNA sequence:

AGCTTGACACGTGTCACTAGA

How many amino acids would be coded for by this mRNA?

Strategy: Determine the complementary bases, determine the number of codons.

Putting It Together

Butyl butyrate is responsible for the distinctive scent of bananas. The percent composition of this ester is 67% carbon, 11% hydrogen, and 22% oxygen. Reaction with water yields butanol and an acid. The molar mass of the acid is 144 g/mol. What is the structure of butyl butyrate?

Self–Test

This section is intended to test your knowledge of the material covered in this chapter. Think through these problems, and make certain you understand what they are asking. Make sure your answers make sense. Successful completion of these problems indicates that you have mastered the material in this chapter. You will receive the greatest benefit from this section if you use it as a mock exam, as this will allow you to determine which topics you need to study in more detail

True–False

1. Different isomers will have different physical and chemical properties.
2. A double or triple bond is not considered a functional group.
3. Whether an alkene is *cis* or *trans* has very little practical significance.
4. Alcohols are able to participate in hydrogen bonding.
5. Organic carboxylic acids are strong acids.
6. The links between amino acids are amide bonds, but when they are in peptides they are referred to as peptide bonds.
7. Lipids are defined by their physical properties, not by their structure.
8. Monosaccharides can be aldoses or ketoses.
9. All polysaccharides are readily digested to form glucose monomers.
10. Nucleic acids are the primary structural components of plant cell walls.

Multiple Choice

1. In naming alkanes, the parent chain is
   a. the longest straight chain
   b. the longest continuous chain
   c. numbered to give the maximum numbers for substitutuents
   d. b and c

2. When alkenes react with halides
   a. one halogen adds to the side of the double bond
   b. the double bond is completely broken, and two molecules are formed
   c. one halogen adds to each carbon of the double bond
   d. a strong acid is needed as a catalyst

3. A three-carbon alkane is called
   a. propane
   b. triane
   c. propene
   d. butane

4. The smallest cycloalkane without stress in the bonds is
   a. propane
   b. butane
   c. pentane
   d. hexane

5. Aromatic rings
   a. contain six-membered carbon rings
   b. have delocalized *p* electrons
   c. are flat
   d. all of the above

6. Alcohols and ethers can both be considered water derivatives, but they differ in that
   a. ethers have double bonds to oxygen
   b. alcohols have one hydrogen of the water substituted, ethers have both
   c. ethers contain nitrogens
   d. alcohols have double bonds to oxygen

7. Which of the following are carbonyl compounds?
   a. aldehydes and ketones
   b. amides
   c. carboxylic acids
   d. all of the above

8. Which of the following places these in order from smallest to largest?
   a. protein, tripeptide, peptide
   b. amino acid, tripeptide, dipeptide
   c. amino acid, dipeptide, protein
   d. peptide, protein, amino acid

9. Complex carbohydrates are
   a. nonpolar
   b. long chains

   c. all digestible by humans
   d. a and b

10. Lipids
    a. are soluble in nonpolar solvents.
    b. are important components of cell membranes
    c. often contain fatty acids bonded to a glycerol backbone
    d. all of the above

Fill–in–the–Blank

1. Because of carbon's flexible ________________ there are more than 40 million organic compounds. These are organized into families according to the ________________ they contain.
2. ______________ contain only hydrogen and _________________. __________________ have only single bonds, __________________ have at least one double bond, and ______________ have at least one _______________ bond. When six-membered rings alternate single and ______________ bonds, the result is a(n) ________________ ring.
3. ________________________ are both structural components of plants and important sources of ______________ for animals.
4. Proteins are made up of long chains of _______________ ______________. When they catalyze reactions in biological systems, they are called _________________.
5. Nucelic acids are an essential part of biological systems. In _________________, copies of the molecules are made. In transcription, ________ molecules are made that are then _______________ into ______________.

Matching

Alcohol	a. protein monomer containing an $–NH_2$ group and a –COOH group
Aldehyde	b. metabolic processes that build larger molecules from smaller ones
Alkane	c. metabolic reactions that break larger molecules into smaller molecules, releasing energy
Alkene	d. carbonyl group with hydrogen on one side, and an alkyl group on the other
Alkyne	e. hydrocarbon containing only single bonds that contains a ring structure
Amino acid	f. C=O

Amide	g. molecule that contains two alkyl groups bonded to a single bonding oxygen
Amine	h. organic derivative of ammonia
Anabolism	i. hydrocarbon containing at least one double bond
Aromatic	j. carboxylic acid with a long hydrocarbon tail
Carbonyl group	k. organic molecule containing an –OH group
Catabolism	l. molecule containing two alkyl groups on either side of a carbonyl group
Cycloalkane	m. organic molecule containing an alkyl group and a nitrogen bonded to a carbonyl group
Enzyme	n. total of all the reactions that go on in a cell
Ester	o. carbohydrate that cannot be broken down by acid hydrolysis
Ether	p. organic molecule that contains a $-\overset{\overset{\displaystyle O}{\|}}{C}-O-R$
Fatty acid	r. containing a benzene ring
Hydrocarbon	s. hydrocarbon containing only single bonds
Ketone	t. protein that catalyzes biological reactions
Lipid	u. molecule that contains only hydrogen and carbon
Metabolism	v. making RNA molecules from DNA templates
Monosaccharide	w. hydrocarbon containing at least one triple bond
Peptide	x. having as many hydrogens as possible, containing only single bonds
Replication	y. naturally occurring molecule that is soluble in non-polar solvents
Saturated	z. mRNA to protein
Transcription	aa. not having as many hydrogens as possible: containing one or more multiple bonds
Translation	bb. process by which DNA molecules are exactly copied

Unsaturated — cc. chain of two to 100 amino acids

Problems

1. Draw the following compounds:
   a. *meta*-dibromobenzene
   b. 2,3,4-trimethyl-1-hexanol
   c. 2-hexanone
   d. 3-ethyl-2,4-dimethylpentanal
   e. 4-ethyl-6-methyl-1-heptene

2. Name the following compounds:

a. O, NH_2

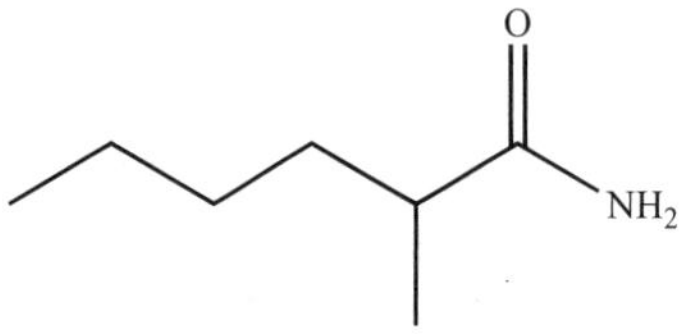

b. O, O

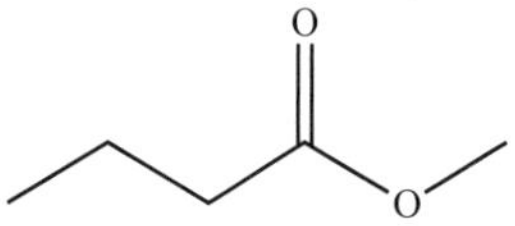

c. NH_2

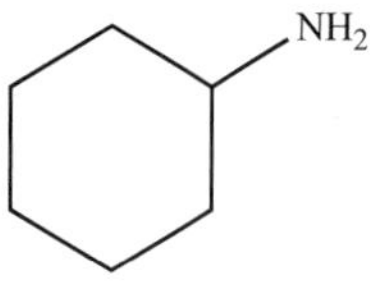

d. O

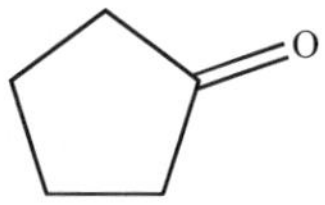

e.

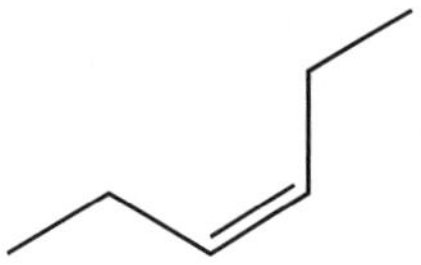

3. Identify the functional groups in the following molecules:

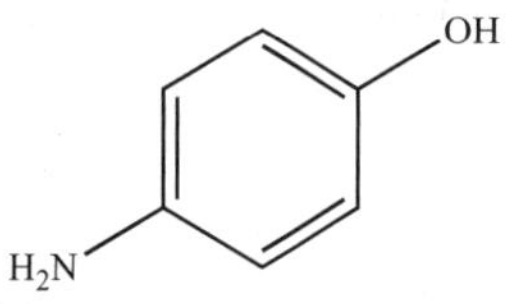

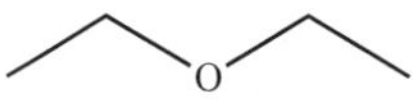

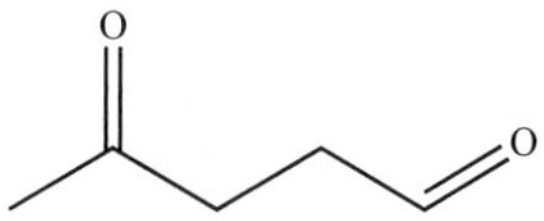

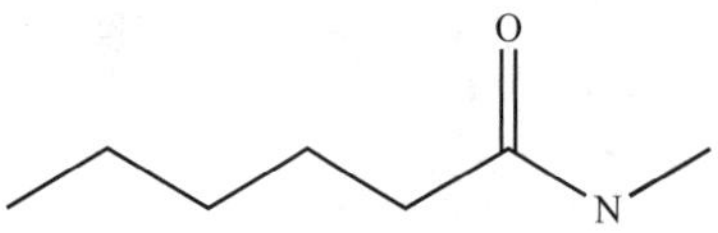

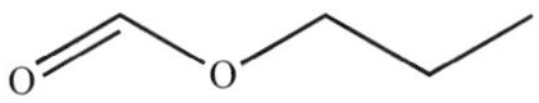

OH, H_2N — O — O, O — O, N — O, O

4. Predict the results of the following reactions giving all possible products:

    a. benzene plus nitric acid in the presence of sulfuric acid

    b. ethylbenzene plus chlorine in the presence of $FeCl_3$

    c. 3-hexene plus water

    d. 2-butene plus hydrogen

    e. ethene and chlorine

    f. dehydration of ethanoic acid and ethanol

    g. *N*-methylacetamide plus acid

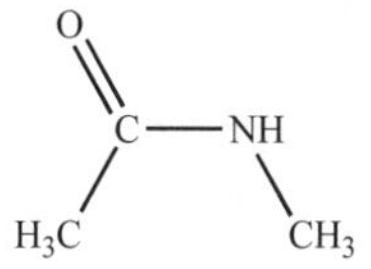

O
C—NH
H_3C CH_3

N-methylacetamide

5. Briefly describe the four stages of catabolism.

6. Draw the two possible dipeptides that can be formed by alanine and serine.

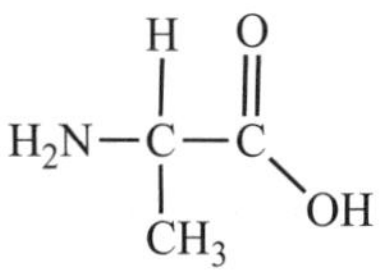

H O
H_2N—C—C
OH
CH_3

alanine

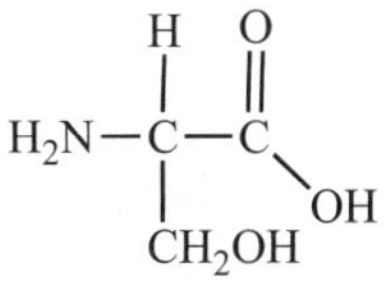

H O
H_2N—C—C
OH
CH_2OH

serine

7. List the six possible tripeptides that can be formed by alanine (ala), leucine (leu), and histidine (his).

8. Classify the following monosaccharides:

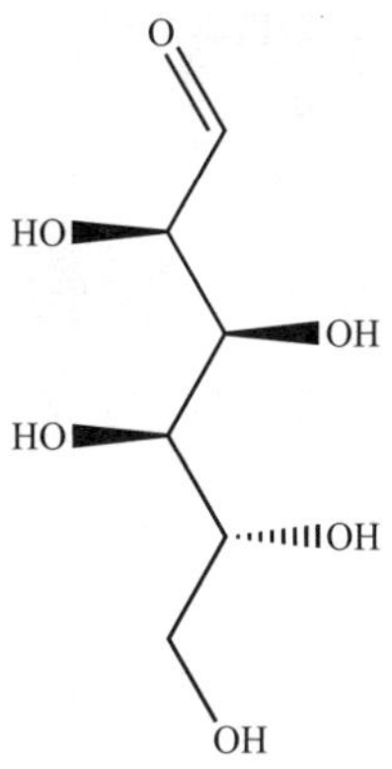

O
HO
OH
HO
OH
OH

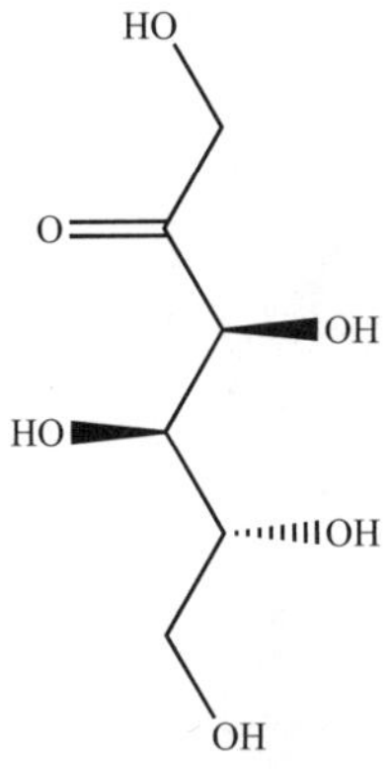

HO
O
OH
HO
OH
OH

9. Draw the basic structure of a saturated triglyceride.

10. From what DNA sequence was the following mRNA transcribed? For how many amino acids does it code?

UAGCACUGAUCAGUAAGC

Challenge Problem

Citric acid is an unsaturated, tricarboxylic acid. When 5.00 g of the acid undergoes combustion, 6.87 g of CO_2 and 1.87 g of H_2O are produced. When a 0.312 g sample of citric acid is titrated with 0.100 M sodium hydroxide, 48.8 mL of are needed for complete titration. Determine the molecular formula of the acid.

WORKBOOK PROBLEM SOLUTIONS

Chapter 1

WP 1.1

Strategy: Consider the differences in the degree size and zero point adjustments for the Fahrenheit and Celsius scales.

Step 1: A Fahrenheit degree is smaller than a Celsius degree. Just think about the melting (freezing) point of water. Water melts at 0° on the Celsius scale and at 32° on the Fahrenheit scale. Therefore, when converting from Fahrenheit to Celsius, the temperature should be lower.

Step 2: Apply the formula for the conversion of °F to °C. (Remember, for this conversion, you do a zero-point correction followed by a size correction.)

$$°C = \frac{5}{9} \times (102 - 32) = 38.9°C$$

Step 3: Does your answer in step 2 agree with your answer in step 1?
We predicted that the Celsius temperature should be lower, and it is.

Step 4: Apply the formula for the conversion of °C to K.

$$K = 38.9 + 273.15 = 312\ K$$

WP 1.2

Strategy: Think about the information you have been given. Set up a mathematical equation that allows you to solve for the mass of lead. Remember that the definition of density is mass per unit volume (or mass divided by volume).

$$\text{Density}\left(\frac{g}{mL}\right) = \frac{\text{Mass of sample (g)}}{\text{Volume of sample (mL)}}$$

Step 1: Substitute the information given in the problem into the mathematical equation you set up.

$$11.34 \frac{g}{mL} = \frac{\text{mass of lead (g)}}{53.43\ mL}$$

Step 2: Solve for the mass of lead.

$$11.34 \frac{g}{mL} \times 53.43\ mL = 605.9\ g$$

WP 1.3

Strategy: Following the example and workbook problem 1.2, set up an equation to determine the volume of the cylinder.

$$\text{Density}\left(\frac{g}{mL}\right) = \frac{\text{Mass of sample (g)}}{\text{Volume of sample (mL)}}$$

Step 1: Solve for the volume of the cylinder.

$$1.977\,\frac{\text{g}}{\text{L}} = \frac{4.84365\ \text{g}}{\text{volume of sample (L)}} \qquad \text{volume of sample (mL)} = \frac{4.84365\ \text{g}}{1.977\,\frac{\text{g}}{\text{L}}} = 2.451\ 365\ 706\ \text{L}$$

Step 2: Apply the rule for multiplication and division to determine the number of significant numbers in your answer.

When carrying out either multiplication or division, your answer cannot have more significant figures than either of the original numbers. 1.977 has four significant figures; 4.84365 has six significant figures. Your answer can't have more than four significant figures.

Step 3: If necessary, use the rules for rounding off numbers.

We must round off 2.451 365 706 L to four significant figures. The first digit we must remove is a 3, therefore the number can simply be truncated at four digits The final answer is 2.451 L.

WP 1.4

Strategy: Using the conversion table on the back cover of your book, determine the conversion factor for meters to miles and convert time to one unit.

1 mi = 1.6093 km

$$27\ \text{min} \times \frac{60\ \text{s}}{1\ \text{min}} = 1620\ \text{s}; \quad 1620\ \text{s} + 46\ \text{s} = 1666\ \text{s}$$

Step 1: Calculate the amount of time it will take for the athlete to run that distance, and convert back to minutes and seconds.

$$10\ \text{km} \times \frac{1\ \text{mi}}{1.6093\ \text{km}} \times \frac{1666\ \text{s}}{5\ \text{mi}} = 2070\ \text{s}$$

$$2070\ \text{s} \times \frac{1\ \text{min}}{60\ \text{s}} = 34.5\ \text{minutes or 34 min 30 seconds}$$

Chapter 2

WP 2.1

Step 1: Determine the carbon-to-hydrogen ratio for the first compound.

$$\text{First compound: C:H mass ratio} = \frac{3.43\ \text{g C}}{0.857\ \text{g H}} = 4.00$$

Step 2: Determine the carbon-to-hydrogen ratio for the second compound.

$$\text{Second compound: C:H mass ratio} = \frac{4.80\ \text{g C}}{0.400\ \text{g H}} = 12.0$$

Step 3: Divide the ratio found for the first compound by the ratio for the second compound.

$$\frac{\text{C : H mass ratio in 1st compound}}{\text{C : H mass ratio in 2nd compound}} = \frac{4.0}{12.0} = 0.33$$

Step 4: Can your answer be converted to a small whole-number ratio?

Yes. $0.33 = \frac{33}{100} \approx \frac{1}{3}$

WP 2.2

Step 1: First, it's necessary to know the chemical symbol for oxygen.

The chemical symbol for oxygen is O.

Step 2: Use the periodic table to determine the atomic number for oxygen.

From the periodic table, we find that the atomic number for O is 8.

Step 3: The number of neutrons can be determined from the definition for mass number.

A (mass number) = Z (atomic number) + number of neutrons.
number of neutrons = $A - Z$
For oxygen-16: number of neutrons = 16 – 8 = 8
For oxygen-17: number of neutrons = 17 – 8 = 9
For oxygen-18: number of neutrons = 18 – 8 = 10

Step 4: The standard symbol is written with the mass number as a superscript and the atomic number as a subscript, both to the left of the symbol.

$^{16}_{8}O$ $^{17}_{8}O$ $^{18}_{8}O$

WP 2.3

Strategy: Determine the atomic number from the information given.

We are told that element X has 51 protons; therefore, the atomic number of the element is 51.

Step 1: Knowing the atomic number, identify the element by using the periodic table.

From the periodic table, we find that the element with $Z = 51$ is Sb (antimony).

Step 2: The standard symbol is written with the mass number as a superscript and the atomic number as a subscript, both to the left of the symbol.

The mass number is determined by adding the number of protons and the number of neutrons.

$A = 51 + 70 = 121.$

The standard symbol is $^{121}_{51}Sb$.

WP 2.4

Strategy: Determine the number of grams of gold in the ring and find the conversion factors needed to convert from grams of sample to number of atoms,

If the ring is 14/24 gold, and weighs a total of 3.65 grams, there are 14/24 x 3.65 grams of gold present:

$\frac{14}{24} \text{ x } 3.65 = 2.13 \text{ g gold}$

We are beginning with grams and want to end up with atoms. The first conversion factor we will need is one that will allow us to cancel out the unit grams.

$$\frac{1\text{ amu}}{1.660\,54\times10^{-24}\text{ g}}$$

Using this conversion factor leaves us with the unit, amu. We now need a conversion factor that will cancel out the unit, amu.

$$\frac{1\text{ atom Au}}{196.7\text{ amu}}$$

(196.7 amu is the atomic mass for gold from the periodic table. One atom of gold weighs this much.) This conversion factor leaves us with the desired unit.

Step 1: Set up a mathematical equation such that all of the units except number of atoms cancel.

$$2.13\text{ g}\times\frac{1\text{ amu}}{1.660\;54\times10^{-24}\text{g}}\times\frac{1\text{ atom Au}}{196.7\text{ amu}}=6.52\times10^{21}\text{ atoms of Au}$$

WP 2.5

Strategy: Remember, the atomic mass of an element = Σ(mass of each isotope × the abundance of the isotope), and the abundances must add up to 100%.

Step 1: Use the information given to determine the abundance of ^{204}Pb.

The abundance of ^{204}Pb is not given, but the abundances of the other isotopes are, making it a straightforward process to find the missing abundance:

Abundance ^{204}Pb + 52.4% + 22.1% + 24.1% = 100%

Abundance ^{204}Pb = 1.4%

Step 2: Use the information given to determine the average atomic mass.

$$(0.524\times207.977\text{ amu})+(0.221\times206.976\text{ amu})+(0.241\times205.974\text{ amu})+(0.014\times203.973\text{ amu})=207.217\text{ am}$$

WP 2.6

Step 1: Determine whether the formula contains both metals and nonmetals or only nonmetals.

a) HI – two non-metals
b) NO_2 – two nonmetals
c) $Ba(OH)_2$ – metal and nonmetal
d) PCl_3 – two nonmetals

Step 2: Identify the compounds as either molecular or ionic, based on your conclusions in step 1.

a) , b), and d) – molecular compounds c) – ionic compound

Step 3: Determine if the substance is capable of providing either an H^+ or OH^- ion in water.

a) HI – produces H^+ when dissolved in water.
b) NO_2 – cannot produce either H^+ or OH^- when dissolved in water.
c) $Ba(OH)_2$ – produces OH^- when dissolved in water.

d) PCl_3 – cannot produce either H^+ or OH^- when dissolved in water.

Step 4: Identify the compounds as either acids or bases (if applicable) based on your conclusions in step 3.

a) HI – acid c) $Ba(OH)_2$ – base

WP 2.7

Step 1: When naming compounds, determine if the metal is a main group metal or a transition metal. If the metal is a transition metal, you must indicate the charge on the metal when naming the compound. The charge on the metal is determined from the number and charge on the anion. (Remember, you must maintain electrical neutrality.)

$MgCl_2$ Magnesium chloride. (Mg^{2+} is a group 2A metal and has only one charge.)

CaO Calcium oxide. (Ca^{2+} is a group 2A metal and has only one charge.)

TiI_4 Titanium is a transition metal; therefore, we need to determine and note the charge on the metal ion. Iodine is a group 7A non-metal, so carries a charge of -1. As there are four iodide ions, the name of the compound is titanium(IV) iodide.

CoF_3 Cobalt is a transition metal. We have three fluorides for a total charge of –3. To maintain electrical neutrality, the total charge on the cobalt ions must be +3. The name of the compound is cobalt(III) fluoride.

CuSe Copper (II) selenide. (Se is a group 6A metalloid and carries a charge of -2.)

Step 2: When writing formulas, use the periodic table or the name of the compound to determine the charge on the metal. Use the periodic table to determine the charge on the anion. (Remember to make sure that all charges add up to zero.

sodium chloride Sodium is a group 1A metal and forms only Sr^{2+}. Bromide has a –1 charge. For the charges to balance, one of each is needed: NaCl

chromium (III) oxide The name tells us that the charge on the chromium cation is +3. Using the periodic table, we find that the charge on the oxide anion is –2. The lowest common multiple of these charges is 6, which means that there will need to be six positive charges and six negative charges. These requirements are fulfilled by two chromium ions and three oxygen ions: Cr_2O_3.

bismuth (II) nitride Bismuth has a charge of +2, nitride has a charge of -3. Again, the lowest common multiple is 6, and six positive and six negative charges are needed: Bi_3N_2.

aluminum sulfide Aluminum is a group 3A metal and forms Al^{3+} cations. The sulfide anion has a –2 charge. Al_2S_3.

Step 3: Make sure the formula contains the smallest whole-number ratio of cation to anion.

The formulas above all have the smallest whole-number ratios of cation to anion.

WP 2.8

Step 1: When naming the molecules, remember that the element that is more anionlike uses the *–ide* suffix. Also, remember to use numerical prefixes to indicate the number of each atom.

PF_6 dinitrogen tetroxide selenium dioxide NBr_3 HF (*g*)

PF_6	phosphorus hexafluoride.
NBr_3	nitrogen tribromide
HF (*g*)	The formula was written to indicate that this binary hydrogen compound is a gas. Therefore, it is named as hydrogen fluoride gas.

Step 2: Use the correct numerical prefixes:

dinitrogen tetroxide	N_2O_4
selenium dioxide	SeO_2

WP 2.9

Step 1: When naming the compounds, refer to your textbook for the names of polyatomic ions, but begin memorizing them. Also, don't forget that you must indicate the charge on certain metal cations.

$Fe(NO_3)_3$	NO_3 is the nitrate ion and has a charge of –1. Since we have three nitrate ions, we have an overall charge of –3. Therefore, the charge on the iron must be +3. Iron is a transition metal, and we must include the charge on the metal when naming the compound. The name of the compound is iron(III) nitrate.
$Al_2(SO_4)_3$	SO_4 is the sulfate ion and has a charge of –2. Since we have three sulfate ions, we have an overall charge of –6. Therefore, the overall charge on the metal must be +6. There are two aluminum ions, which must each carry a charge of +3. Aluminum is a main-group metal, so a charge of +3 is expected, and need not be denoted with a Roman numeral. This is aluminum sulfate.

Step 2: When writing formulas, use the periodic table or the name of the compound to determine the charge on the metal. To determine the formula for a polyatomic ion, refer to your textbook. Remember, you must maintain electrical neutrality.

potassium chromate	The formula for chromate is CrO_4^{2-}. Potassium is a group 1A metal and has a charge of +1. To maintain electrical neutrality, we need to have two potassium cations for every one CrO_4^{2-} anion. The formula is K_2CrO_4.
magnesium phosphate	The formula for phosphate is PO_4^{3-} (this is the only polyatomic ion with a charge of -3 that you need to worry about). Magnesium is a group 2A metal and has a charge of +2. To balance the charges, three magnesium ions are needed for every two phosphate ions. The formula is $Mg_3(PO_4)_2$.
ammonium nitrate	The formula for ammonium is NH_4^+. The formula for nitrate is NO_3^-. There is no reason in the world that two polyatomic ions cannot bind

together to form an ionic compound, but ammonium is the only polyatomic ion you need to worry about that carries a positive charge. The formula for the compound is NH_4NO_3.

WP 2.10

Strategy: First, determine the type of compound. Is it a simple binary ionic compound, a molecule, an ionic compound containing a polyatomic ion, or an acid (either binary acid or oxoacid)? Once you have determined the type of compound, apply the appropriate nomenclature rules.

$MnSO_4$ — This compound contains a metal and a polyatomic ion. The charge on a sulfate ion is -2, making the charge on the manganese ion +2. Manganese is a transition metal, so its charge must be indicated: this is manganese(II) sulfate.

hydrochloric acid — From the name of the compound, we know that it is a binary acid containing hydrogen and chlorine. The chloride ion has a –1 charge. The formula for this compound is HCl (*aq*). It is necessary to indicate that this is in aqueous solution, as it would otherwise be a molecular gas.

H_2SO_3*(aq)* — This compound contains a nonmetal and a polyatomic ion. One of the nonmetals is hydrogen, and this is in aqueous solution, so it will be named as an acid. This is hydrogen sulfite, so is sulfurous acid.

H_2S(*g*) — This compound contains two nonmetals and is a molecule. Although it contains hydrogen, it is in the gas phase, so it is not an acid. This compound is hydrogen sulfide. (If it were dissolved in water, it would be hydrosulfuric acid.)

dinitrogen trioxide — This compound contains only nonmetals, so is a covalent molecule – N_2O_3

cobalt (II) sulfide — This compound contains a metal and a nonmetal. The charge of +2 on the cobalt will need to be balanced by the anion: from the periodic table, the ion of sulfur carries a charge of -2, so this is CoS.

MgF_2 — This compound contains a metal and a nonmetal, making it a simple binary ionic compound. Magnesium is a group 2A metal. Therefore, we do not need to indicate the charge on the metal cation. The name of the compound is magnesium fluoride.

sodium sulfite — This compound contains a metal and a polyatomic anion. It is an ionic compound. Sodium is a group 1A metal and has a +1 charge. The sulfite anion, SO_3^-, has a –2 charge. To balance the charges, two sodiums are needed. The formula will be Na_2SO_3.

All formulas for the ionic compounds have the smallest whole-number ratio of cation to anion.

Chapter 3

WP 3.1:

Step 1: Remember, the term *combustion* is used to indicate reaction with oxygen. When hydrocarbons (compounds containing primarily C and H) undergo a combustion reaction, carbon dioxide and water are the products.

$C_6H_{14} + O_2 \rightarrow CO_2 + H_2O$

Step 2: Use coefficients to balance the equation. (Remember, it helps to save oxygen for last.)

Begin with carbon. There are 6 carbons on the reactant side, but only 1 carbon on the product side. Place a 6 in front of CO_2.

$C_6H_{14} + O_2 \rightarrow 6\ CO_2 + H_2O$

There are 14 hydrogens on the reactant side, but only 2 hydrogens on the product side. Place a 7 in front of H_2O.

$C_6H_{14} + O_2 \rightarrow 6\ CO_2 + 7\ H_2O$

Now balance the oxygens. There are two oxygens on the reactant side and a total of 19 on the product side. This odd number of oxygens will require a doubling of everything we have done so far to make it even:

$2\ C_6H_{14} + O_2 \rightarrow 12\ CO_2 + 14\ H_2O$

Now there are 2 oxygens on the reactant side, and 38 oxygens on the product side. Place a 19 in front of the oxygen on the reactant side:

$2\ C_6H_{14} + 19\ O_2 \rightarrow 12\ CO_2 + 14\ H_2O$

Step 3: Reduce the coefficients to their smallest whole-number ratio.

The ratio is 2:19:12:14. This cannot be reduced, as 19 is a prime number.

Step 4: Check your answer.

Reactant side	Product Side
12 C	12 C
28 H	28 H
38 O	38 O

WP 3.2

Step 1: Write an unbalanced chemical equation based on the information given in the problem.

$H_2O_2\ (aq) \rightarrow H_2O\ (l) + O_2\ (g)$

Step 2: Use coefficients to balance the equation.

$2\ H_2O_2\ (aq) \rightarrow 2\ H_2O\ (l) + O_2\ (g)$

Step 3: Reduce the coefficients to their smallest whole-number ratio if necessary.

The ratio is 2:2:1 and cannot be reduced further.

Step 4: Check your answer.

Reactant side	Product side
4 H	4 H
4 O	4 O

Part b

Step 1: Determine the mole ratio for hydrogen peroxide and oxygen from your balanced chemical equation.

For every two moles of hydrogen peroxide, two moles of water will be produced: this is a 1:1 ratio.

For every two moles of hydrogen peroxide, 1 mole of oxygen gas will be produced: this is a 2:1 ratio.

Step 2: Use the mass and molecular weight of the hydrogen peroxide solution to determine the moles of hydrogen peroxide present.

$$60 \text{ g x} \frac{1 \text{ mol}}{34.02 \text{ g}} = 1.76 \text{ mol}$$

Step 3: Use the mole ratio from above as a conversion factor, and calculate the number of moles of oxygen generated by the decomposition reaction.

$$1.76 \text{ mol } H_2O_2 \text{ x} \frac{1 \text{ mol } O_2}{2 \text{ mol } H_2O_2} = 0.88 \text{ mol } O_2$$

Step 4: Use the mole ratio from above as a conversion factor, and calculate the number of moles of water generated by the decomposition reaction.

$$1.76 \text{ mol } H_2O_2 \text{ x} \frac{1 \text{ mol } H_2O}{1 \text{ mol } H_2O_2} = 1.76 \text{ mol } H_2O$$

Ballpark check: The number of moles of water generated should be two times the number of moles of oxygen generated based on the ratio found in the balanced equation.

WP 3.3

Step 1: Write an unbalanced chemical equation from the information given in the problem:

$Fe + O_2 \rightarrow Fe_2O_3$ (this provides a little nomenclature review!)

Step 2: Balance the equation.

$4Fe + 3O_2 \rightarrow 2Fe_2O_3$

Step 3: From the balanced chemical equation, determine the mole-to-mole ratio of iron to iron (III) oxide.

4 mol Fe : 2 mol Fe_2O_3 can be reduced (or not) to 2 mol Fe : 1 mol Fe_2O_3

Step 4: Find the atomic mass of iron.

From the periodic table, Fe = 55.85 g/mol

Step 5: Determine the molecular mass for iron (III) oxide.

(55.85 g/mol x 2) + (16.00 x 3) = 159.7 g/mol

Step 6: From the information in steps 3, 4, and 5, create conversion factors. Use these conversion factors, and set up a dimensional analysis problem, to convert from grams of iron to moles of iron to moles of iron oxide to grams of iron oxide.

$$14 \text{ g Fe x } \frac{1 \text{ mol Fe}}{55.85 \text{ g}} \text{x} \frac{2 \text{ mol Fe}_2\text{O}_3}{4 \text{ mol Fe}} \text{x} \frac{159.7 \text{ g Fe}_2\text{O}_3}{\text{mol Fe}_2\text{O}_3} = 20.0 \text{ g Fe}_2\text{O}_3$$

Step 7: Using the Law of Mass Conservation, how many grams of oxygen were added to the iron?

There are two ways to solve this: the easy way, as indicated in the question, and the hard way. The easy way to solve this is to see that on the process of rusting, the iron has gained 6.0 grams, and that must have come from the only other reagent, oxygen.

The hard way is to do another stoichiometry problem like above:

$$14 \text{ g Fe x } \frac{1 \text{ mol Fe}}{55.85 \text{ g}} \text{x} \frac{3 \text{ mol O}_2}{4 \text{ mol Fe}} \text{x} \frac{32.0 \text{ g O}_2}{\text{mol O}_2} = 6.0 \text{ g O}_2$$

As you continue through chemistry, you will often find situations like this, where there are numerous ways to get to the same answer. Always look for the easiest: 20 - 14= 6 beats the alternative any day.

WP 3.4

Step 1: Balance the equation.

$3H_2 + N_2 \rightarrow 2NH_3$

Step 2: Calculate the theoretical yield using dimensional analysis.

$$18.0 \text{ g N}_2\text{x} \frac{1 \text{ mol N}_2}{28.02 \text{ g}} \text{x} \frac{2 \text{ mol NH}_3}{1 \text{ mol N}_2} \text{x} \frac{17.01\text{g NH}_3}{\text{mol NH}_3} = 21.85 \text{ g}$$

Step 3: Calculate the percent yield using the actual yield stated in the problem and the theoretical yield just calculated.

$$\% \text{ yield } = \frac{\text{actual yield}}{\text{theoretical yield}} \text{x}100 = \frac{14.6 \text{ g}}{21.85 \text{ g}} \text{x}100 = 66.8\%$$

WP 3.5

Step 1: Using the information given, write an unbalanced chemical equation.

$Al + HCl \rightarrow AlCl_3 + H_2$

Step 2: Balance the chemical equation.

$2Al + 6HCl \rightarrow 2AlCl_3 + 3H_2$

Step 3: Determine the number of moles of each reactant present.

$$5.3 \text{ g Al x } \frac{1 \text{ mol Al}}{27.0 \text{ g Al}} = 0.196 \text{ mol Al}$$

$$9.2 \text{ g HCl x } \frac{1 \text{ mol HCl}}{36.5 \text{ g HCl}} = 0.252 \text{ mol HCl}$$

Step 4: Determine which reactant gives the smallest number of moles of product. This gives the limiting reagent and the theoretical yield.

$$0.196 \text{ mol Al x } \frac{2 \text{ mol AlCl}_3}{2 \text{ mol Al}} = 0.196 \text{ mol AlCl}_3$$

$$0.252 \text{ mol HCl x } \frac{2 \text{ mol AlCl}_3}{6 \text{ mol HCl}} = 0.084 \text{ mol AlCl}_3$$

The limiting reagent is HCl, and the theoretical yield is 0.084 mol $AlCl_3$.

Step 5: Determine how many moles of the excess reagent were used.

$$0.252 \text{ mol HCl x } \frac{2 \text{ mol Al}}{6 \text{ mol HCl}} = 0.084 \text{ mol Al used}$$

Step 6: Determine how many moles of the excess reagent remain.

$$0.196 \text{ mol Al - } 0.084 \text{ mol Al} = 0.112 \text{ mol Al remaining}$$

Step 7: Determine how many grams of the excess reagent are left over.

$$0.112 \text{ mol Al x } \frac{27.0 \text{ g Al}}{\text{mol Al}} = 3.02 \text{ g Al left over}$$

Step 8: Determine the percent yield of the reaction.

First, convert the theoretical yield to grams:

$$0.084 \text{ mol AlCl}_3 \text{ x } \frac{133.5 \text{ g AlCl}_3}{\text{mol AlCl}_3} = 11.2 \text{ g AlCl}_3$$

$$\% \text{ yield} = \frac{8.3 \text{ g AlCl}_3}{11.2 \text{ g AlCl}_3} \text{ x } 100 = 74\%$$

NOTE: The same answer could be obtained by converting the actual yield to moles:

$$8.3 \text{ g AlCl}_3 \text{ x } \frac{1 \text{ mol AlCl}_3}{133.5 \text{ g AlCl}_3} = 0.062 \text{ mol AlCl}_3$$

$$\% \text{ yield} = \frac{0.062 \text{ mol AlCl}_3}{0.084 \text{ mol AlCl}_3} \text{ x } 100 = 74\%$$

WP 3.6

Step 1: First, determine the number of moles found in 500 mL of a 0.30 M solution.

$$\text{moles} = \text{M x L} = 0.3 \frac{\text{moles}}{\text{L}} \text{ x } 0.500\text{L} = 0.150 \text{ moles}$$

Step 2: Determine the number of grams of sodium chloride from the number of moles.

$$0.150 \text{ mole NaCl x } \frac{58.5 \text{ g}}{\text{mol}} = 8.8 \text{ g NaCl}$$

WP 3.7

Step 1: Rearrange the equation $M_i \text{ x } V_i = M_f \text{ x } V_f$ to solve for the final molarity.

$$M_f = \frac{M_i \text{ x } V_i}{V_f} = \frac{50 \text{ mL x } 10 \text{ M}}{1000 \text{ mL}} = 0.5 \text{ M}$$

Note that here, the conversion of volumes *was* needed, so that they would match and cancel out.

WP 3.8

Step 1: Begin by writing a balanced chemical equation.

$$Zn\ (s) + 2\ HCl\ (aq) \rightarrow ZnCl_2\ (aq) + H_2\ (g)$$

Step 2: Determine the number of moles of zinc present.

$$\text{mol Zn} = 4.75 \text{ g Zn x } \frac{1 \text{ mol Zn}}{65.4 \text{ g}} = 0.073 \text{ mol Zn}$$

Step 3: Determine the number of moles of HCl needed.

$$0.073 \text{ mol Zn x } \frac{2 \text{ mol HCl}}{\text{mol Zn}} = 0.145 \text{ mol HCl needed}$$

Step 4: Calculate the volume of 5.0 M HCl needed for this reaction.

$$V = \frac{0.145 \text{ mol}}{5.0 \frac{\text{mol}}{\text{L}}} \text{ x } \frac{1000 \text{ mL}}{\text{L}} = 29 \text{ mL HCl}$$

Step 5: To calculate the volume of H_2 gas produced, we first need to calculate the moles, then grams of H_2 gas produced.

$$0.073 \text{ mol Zn x } \frac{1 \text{ mol H}_2}{\text{mol Zn}} \text{ x } \frac{2.016 \text{ g H}_2}{\text{mol H}_2} = 0.147 \text{ g H}_2$$

Step 6: Using the density of H_2 gas, determine the volume produced:

$$V = \frac{0.147 \text{ g } H_2}{0.0899 \ \frac{\text{g}}{\text{L}}} = 1.64 \text{ L } H_2$$

WP 3.9

Step 1: Write a balanced chemical equation for the reaction.

$$2NaOH + H_2SO_4 \rightarrow Na_2SO_4 + 2H_2O$$

Step 2: Calculate the number of moles of NaOH present.

$$0.153 \text{ g NaOH x } \frac{1 \text{ mol NaOH}}{40.0 \text{ g NaOH}} = 0.00383 \text{ mol}$$

Step 3: Determine the number of moles of sulfuric acid needed to react with that quantity of NaOH.

$$\text{moles } H_2SO_4 = 0.00383 \text{ mol NaOH x } \frac{1 \text{ mol } H_2SO_4}{2 \text{ mol NaOH}} = 0.00191 \text{ mol } H_2SO_4$$

Step 4: Calculate the molarity of the acid solution.

$$\text{Molarity} = \frac{0.00191 \text{ mol } H_2SO_4}{0.01637 \text{ L}} = 0.117 \text{ M } H_2SO_4$$

WP 3.10

Step 1: Convert the grams of each element into moles of each element.

$$18.9\% \text{ Li} = 18.9 \text{ g Li x } \frac{1 \text{ mol Li}}{6.94 \text{ g Li}} = 2.72 \text{ mol Li}$$

$$16.2\% \text{ C} = 16.2 \text{ g C x } \frac{1 \text{ mol C}}{12.01 \text{ g C}} = 1.35 \text{ mol C}$$

$$64.9\% \text{ O} = 64.9 \text{ g O x } \frac{1 \text{ mol O}}{16.0 \text{ g O}} = 4.06 \text{ mol O}$$

Step 2: Find the mole ratios.

There are no obvious ratios here – nothing is 1:1, so it makes sense to determine how many there are of each element relative to carbon – the element of which there is the least. Assume there is one carbon, and go from there:

$$\frac{2.72 \text{ mol Li}}{1.35 \text{ mol C}} = 2\frac{\text{Li}}{\text{C}}$$

$$\frac{4.06 \text{ mol O}}{1.35 \text{ mol C}} = 3\frac{\text{O}}{\text{C}}$$

Step 3: If necessary, convert the ratios to whole numbers, and write the empirical formula.

Li_2CO_3 - this is lithium carbonate.

WP 3.11

Step 1: Determine the formula of the compound.

NaOCl – this is common household bleach

Step 2: Determine the molecular weight, and the weights of each component

Na	=	23
O	=	16
Cl	=	35.5
Molecular Weight		74.5

Step 3: Determine the percent composition by dividing the mass of each element present by the total mass of the compound and multiplying by 100.

$$\%Na = \frac{23}{74.5} \text{ x } 100 = 30.9\%$$

$$\%O = \frac{16}{74.5} \text{ x } 100 = 21.5\%$$

$$\%Cl = \frac{35.5}{74.5} \text{ x } 100 = 47.6\%$$

WP 3.12

Step 1: Find the molar amounts of C and H in CO_2 and H_2O.

$$5.70 \text{ g } CO_2 \text{x} \frac{1 \text{ mol } CO_2}{44 \text{ g } CO_2} = 0.129 \text{ mol } CO_2 \text{x} \frac{1 \text{ mol C}}{\text{mol } CO_2} = 0.129 \text{ mol C}$$

$$2.31 \text{ g } H_2O \text{ x} \frac{1 \text{ mol } H_2O}{18 \text{ g } H_2O} = 0.128 \text{ mol } H_2O \text{ x} \frac{2 \text{ mol H}}{\text{mol } H_2O} = 0.257 \text{ mol H}$$

Step 2: Carry out mole-to-gram conversions to find the number of grams of C and H in the original sample.

$$0.129 \text{ mol C x} \frac{12 \text{ g C}}{\text{mol C}} = 1.55 \text{ g C}$$

$$0.257 \text{ mol H x} \frac{1.008 \text{ g H}}{\text{mol H}} = 0.259 \text{ g H}$$

Step 3: Subtract the masses of C and H from the mass of the starting sample to determine the mass of S.

$2.85 \text{ g} - 1.55 \text{ g} - 0.259 \text{ g} = 1.04 \text{ g}$

Step 4: Convert the mass of S to moles of S.

$$1.04 \text{ g S} \times \frac{1 \text{ mol S}}{32 \text{ g S}} = 0.0325 \text{ mol S}$$

Step 5: Find the mole ratios.

$$\frac{0.129 \text{ mol C}}{0.0325 \text{ mol S}} = 4\frac{\text{C}}{\text{S}}$$

$$\frac{0.257 \text{ mol H}}{0.0325 \text{ mol S}} = 8\frac{\text{H}}{\text{S}}$$

Step 6: Use the ratios above to write the empirical formula of the compound.

C_4H_8S

Step 7: Compare the empirical formula weight to the molecular formula weight to see if a conversion is needed.

The empirical formula weight is 88 amu – no conversion is needed.

Step 8: Write the molecular formula.

The molecular formula is C_4H_8S. Phew!

Chapter 4

WP 4.1

Strategy: $AlCl_3$ is a strong electrolyte and completely dissociates in water.

Step 1: Determine the total number of moles of ions formed when $AlCl_3$ completely dissociates in water.

$AlCl_3 \rightarrow Al^{3+} + 3Cl^-$

Step 2: Create a conversion factor comparing the total number of moles of ions in solution to 1 mol of $AlCl_3$.

Four ions are formed in the dissociation of aluminum chloride, so the conversion factor would be

$$\frac{4 \text{ mol ions}}{\text{mol AlCl}_3}$$

Step 3: Use the conversion factor to calculate the molar concentration of ions in solution.

$$\frac{0.75 \text{ mol AlCl}_3}{\text{L}} \times \frac{4 \text{ mol ions}}{\text{mol AlCl}_3} = 3.0 \text{ M ions}$$

WP 4.2

Strategy: From the information given, write a molecular equation and determine if any of the reactants or products are strong electrolytes.

Step 1: Write and balance the molecular equation.

$HNO_3\ (aq) + KOH\ (aq) \rightarrow H_2O\ (l) + KNO_3\ (aq)$

Step 2: Write the strong electrolytes as free ions for the ionic equation.

Ionic Eqn: $H^+\ (aq) + NO_3^-\ (aq) + K^+\ (aq) + OH^-\ (aq) \rightarrow H_2O\ (l) + K^+\ (aq) + NO_3^-\ (aq)$

Step 3: Eliminate any spectator ions to write the net ionic equation.

Spectator Ions: K^+ and NO_3^-

Net Ionic Equation: $H^+\ (aq) + OH^-\ (aq) \rightarrow H_2O\ (l)$

This net ionic equation is characteristic of acid-base neutralization reactions.

WP 4.3

Strategy: Using the solubility guidelines and solution stoichiometry, determine suitable reactants to use in the preparation of lead chloride and the amount of reactants needed.

Step 1: Determine reactants that are soluble and will produce the insoluble lead (II) chloride and another soluble product.

Using the guidelines in the study guide, an alkali metal will bring just about anything into solution, as will ammonium, so two easy ways to get the chloride ion into solution would be NH_4Cl or NaCl, both of which are highly soluble.

To bring a lead (II) ion into solution, nitrate will always work, but the textbook indicates that acetate and perchlorate can also be used to bring lead into solution, so some options for the lead (II) ion are $Pb(NO_3)_2$, $Pb(CH_3CO_2)_2$, and $Pb(ClO_4)_2$. Note that the ratios of ions are the same in all the compounds: the chloride ion is being brought into solution that has ratio MCl, and can be generically represented that way. The lead (II) cation will be brought into solution using a compound with two anions, which can be generically represented PbA_2. This will make it possible to devise a generic "recipe" for making lead (II) chloride.

Step 2: Write a balanced chemical equation.

$2MCl\ (aq) + PbA_2\ (aq) \rightarrow PbCl_2\ (s) + 2MA\ (aq)$

Step 3: Use solution stoichiometry to determine the volume of each reactant needed.

First it is necessary to determine how many moles of lead (II) chloride need to be made.

$$1.3\ g\ PbCl_2 x \frac{1\ mol\ PbCl_2}{278\ g\ PbCl_2} = 4.67x10^{-3}\ mol\ or\ 4.67mmol\ PbCl_2$$

From this, determine the moles of each component that are needed:

$$4.67\text{x}10^{-3}\text{ mol PbCl}_2\text{x}\frac{1\text{ mol PbA}_2}{\text{mol PbCl}_2} = 4.67\text{x}10^{-3}\text{ mol PbA}_2$$

$$4.67\text{x}10^{-3}\text{ mol PbCl}_2\text{x}\frac{2\text{ mol MCl}}{\text{mol PbCl}_2} = 9.35\text{x}10^{-3}\text{ mol MCl}$$

From this, the volume of solution can be determined using the definition of molarity: M = mol/V, so V = mol/M

$$\text{V PbA}_2 = \frac{4.67\text{x}10^{-3}\text{ mol PbA}_2}{1\text{M PbA}_2} = 4.67\text{x}10^{-3}\text{ L PbA}_2\text{ or }4.67\text{ mL PbA}_2$$

$$\text{V MCl} = \frac{9.35\text{x}10^{-3}\text{ mol MCl}}{1\text{M MCl}} = 9.35\text{x}10^{-3}\text{ L MCl or }9.35\text{ mL MCl}$$

Regardless of the exact composition of the solutions, as long as they follow the generic ratios, 4.67 mL of the lead solution combined with 9.35 mL of chloride-containing solution will have a theoretical yield of 1.3 g lead (II) chloride. If you really need 1.3 g of lead chloride, you'll want to add a little more, remembering that you always need twice as much of the chloride solution as the lead solution.

WP 4.4

Determine the reaction type by examining the reactants – are they acids or bases or neutral salts? – then write and balance the equations for the reactions.

Step 1: Write the balanced molecular equation for each reaction.

$$2LiOH + H_2SO_4 \rightarrow Li_2SO_4 + 2H_2O$$

$$NH_4Cl + AgNO_3 \rightarrow NH_4NO_3 + AgCl$$

Step 2: Determine the presence of any strong electrolytes. Write the ionic equation, showing the strong electrolytes in terms of their free ions.

$$2Li^+ + 2OH^- + 2H^+ + SO_4^{2-} \rightarrow 2Li^+ + SO_4^{2-} + 2H_2O$$

$$NH_4^+ + Cl^- + Ag^+ + NO_3^- \rightarrow NH_4^+ + NO_3^- + AgCl$$

Step 3: Write the net ionic equation by removing spectator ions.

$2OH^- + 2H^+ \rightarrow 2H_2O$ – This is an acid-base reaction.

$Cl^- + Ag^+ \rightarrow AgCl$ – This is a precipitation reaction.

WP 4.5

Strategy: Follow the rules for determining oxidation states found above:

Step 1: Identify the elements that have a fixed oxidation state.

VO_4^{3-} – oxygen carries an oxidation state of -2

$NaVO_3$ – oxygen carries an oxidation state of -2, Na carries an oxidation state of +1

Step 2: Determine the oxidation number of the remaining elements, keeping in mind that the sum of all oxidation numbers must be equal to the ionic charge or to zero for a neutral molecule.

VO_4^{3-} – oxygen carries an oxidation state of -2
The overall charge is -3, so (4 x -2) + V = -3, vanadium has an oxidation state of +5

$NaVO_3$ – oxygen carries an oxidation state of -2, Na carries an oxidation state of +1
This is a neutral molecule, so +1 + (3 x -2) + V = 0, vanadium is again +5.

WP 4.6

Strategy: Determine the oxidation number of all species present.

$$\overset{0}{Cu}\,(s) + 2\overset{+1}{Ag^+}\,(aq) + 2\overset{-1}{NO_3^-}\,(aq) \rightarrow 2\overset{0}{Ag}\,(s) + \overset{+2}{Cu^{2+}}\,(aq) + 2\overset{-1}{NO_3^-}\,(aq)$$

Step1: Identify the species which have a change in oxidation number.

Copper goes from an oxidation number of 0 to an oxidation number of 2+
Silver goes from an oxidation number of 1+ to an oxidation number of 0

Step 2: Based on the change in oxidation number, identify the species oxidized, the species reduced, and the oxidizing and reducing agent.

Copper loses electrons – it is oxidized, and is the reducing agent in this reaction.
Silver gains electrons – it is reduced, and is the oxidizing agent in this reaction.

WP 4.7

Strategy: Remember that any element higher in the activity series will react with the ion of any element lower in the activity series. Also remember, metals above the H^+ ion in the activity series will displace the hydrogen ion from an acid to form H_2 gas.

Step 1: Predict the outcome of these reactions.

a. $Cu\,(s) + HCl\,(aq) \rightarrow$ no reaction: copper is below H^+ in the activity series: most metals will dissolve in acid, but mercury, platinum, and what are sometimes called the coinage metals – silver, copper, and gold – will not.

b. $Mn\,(s) + ZnCl_2\,(aq) \rightarrow MnCl_2\,(aq) + Zn\,(s)$

c. $3Mg\,(s) + 2Al(NO_3)_3\,(aq) \rightarrow 3Mg(NO_3)_2\,(aq) + 2Al\,(s)$

WP 4.8

Strategy: Follow the steps in the preceding worked example.

Step 1: Write two unbalanced half-reactions.

Oxidation:	$Na_2SO_3 \rightarrow Na_2SO_4$	(S goes from +4 to +6)
Reduction:	$KMnO_4 \rightarrow MnO_2$	(Mn goes from +7 to +4)

Step 2: Balance each half-reaction for atoms other than H and O.

$Na_2SO_3 \rightarrow Na_2SO_4$

$KMnO_4 \rightarrow MnO_2 + K^+$

Step 3: Add H_2O to balance in oxygen, and H^+ to balance in hydrogen.

$H_2O + Na_2SO_3 \rightarrow Na_2SO_4 + 2H^+$

$4H^+ + KMnO_4 \rightarrow MnO_2 + K^+ + 2H_2O$

Step 4: Balance each reaction for charge.

$H_2O + Na_2SO_3 \rightarrow Na_2SO_4 + 2H^+ + 2e^-$

$3e^- + 4H^+ + KMnO_4 \rightarrow MnO_2 + K^+ + 2H_2O$

Step 5: Make the electron count the same in both reactions.

$(H_2O + Na_2SO_3 \rightarrow Na_2SO_4 + 2H^+ + 2e^-) \text{ x } 3$

$(3e^- + 4H^+ + KMnO_4 \rightarrow MnO_2 + K^+ + 2H_2O) \text{ x } 2$

$3H_2O + 3Na_2SO_3 \rightarrow 3Na_2SO_4 + 6H^+ + 6e^-$

$6e^- + 8H^+ + 2KMnO_4 \rightarrow 2MnO_2 + 2K^+ + 4H_2O$

Step 6: Add the two half-reactions together, canceling anything that appears on both sides of the equation.

$3H_2O + 3Na_2SO_3 + 6e^- + 8H^+ + 2KMnO_4 \rightarrow 3Na_2SO_4 + 6H^+ + 6e^- + 2MnO_2 + 2K^+ + 4H_2O$

$3Na_2SO_3 + 2H^+ + 2KMnO_4 \rightarrow 3Na_2SO_4 + 2MnO_2 + 2K^+ + H_2O$

Step 7: Make the solution basic by adding 1 OH^- to each side for every H^+.

$2OH^- + 3Na_2SO_3 + 2H^+ + 2KMnO_4 \rightarrow 3Na_2SO_4 + 2MnO_2 + 2K^+ + H_2O + 2OH^-$

$2H_2O + 3Na_2SO_3 + 2KMnO_4 \rightarrow 3Na_2SO_4 + 2MnO_2 + 2K^+ + H_2O + 2OH^-$

Step 8: Cancel water molecules that appear on both sides of the equation.

$H_2O + 3Na_2SO_3 + 2KMnO_4 \rightarrow 3Na_2SO_4 + 2MnO_2 + 2K^+ + 2OH^-$

Step 9: Check your answer to make sure both atoms and charges are balanced.

Reactant side:	2H	6Na	3S	18O	2K	2Mn	net charge: 0
Product Side:	2H	6Na	3S	18O	2K	2Mn	net charge: 0

WP 4.9

Step 1: Determine the number of moles of thiosulfate ion that were required to react with the iodine solution to completely reduce the iodine.

Using the definition of molarity M = mol/V, so mol = V x M

$$\text{moles } S_2O_3^{2-} \text{ ion} = 0.15\frac{\text{mol}}{\text{L}} \text{ x } 0.0157 \text{ L} = 2.36\text{x}10^{-3} \text{ mol or } 2.36 \text{ mmol}$$

Step 2: Balance the equation for acid solution.

$I_2 + S_2O_3^{2-} \rightarrow I^- + S_4O_6^{2-}$

Reduction: $I_2 + 2e \rightarrow 2I^{-\,-}$

Oxidation: $2\ S_2O_3^{2-} \rightarrow S_4O_6^{2-} + 2e^-$

$I_2 + 2S_2O_3^{2-} \rightarrow S_4O_6^{2-} + 2I^-$

Step 3: Use mole ratios to determine the amount of iodine present in the reaction mixture.

$$2.36 \text{ x } 10^{-3} \text{ mol } S_2O_3^{2-} \text{ ion x } \frac{1 \text{ mol } I_2}{2 \text{ mol } S_2O_3^{2-} \text{ ion}} = 1.18\text{x}10^{-3} \text{ mol } I_2$$

Step 4: Determine the original concentration of I_2 in solution.

$$\frac{1.18\text{x}10^{-3} \text{ mol } I_2}{0.050 \text{ L solution}} = 0.024 \text{ M } I_2$$

Chapter 5

WP 5.1

Step 1: Choose the appropriate form of the equation, and solve.

$$\lambda = \frac{c}{\upsilon} = \frac{3.00 \text{ x } 10^8 \text{ m/s}}{88.5 \text{ x } 10^6\text{/s}} = 3.39 \text{ m}$$

As this demonstrates, wavelengths can be quite long – although visible light is in the nanometer range, it would not make sense to convert this wavelength to nanometers.

WP 5.2

Strategy: Determine the values of n that will make λ the longest and shortest. Remember that the value of λ is greatest when the value of n is smallest and the value λ is smallest when the value of n is greatest.

Step 1: Determine the values of n that will make λ the shortest and longest.

The longest wavelength line will be at the smallest possible value of n. If $m = 5$, and n must be greater than 5, the longest wavelength will occur when $n = 6$. The shortest wavelength will occur with the largest possible value of n. If $n = \infty$, $1/n = 0$.

Step 2: Use the Balmer-Rydberg equation with $m = 5$ and solve for the shortest wavelength.

$$\frac{1}{\lambda} = R\left[\frac{1}{m^2} - \frac{1}{n^2}\right] = 1.097x10^{-2}\,nm^{-1}\left[\frac{1}{25} - \frac{1}{\infty}\right] = 4.39x10^{-4}\,m$$

$$\frac{1}{\lambda} = \frac{1}{4.39x10^{-4}\,m} = 2280nm$$

Step 3: Solve for λ using the Balmer-Rydberg equation and the values of *m* that make λ the longest.

$$\frac{1}{\lambda} = R\left[\frac{1}{m^2} - \frac{1}{n^2}\right] = 1.097x10^{-2}\,nm^{-1}\left[\frac{1}{25} - \frac{1}{36}\right] = 1.34x10^{-4}\,nm^{-1}$$

$$\frac{1}{\lambda} = \frac{1}{1.34x10^{-4}\,nm^{-1}} = 7460nm$$

WP 5.3

Strategy: Use the equation for the energy of a photon, and multiply by Avogadro's number. Then convert frequency to wavelength.

Step 1: Determine the energy of one photon, then convert to one mole of photons.

$E = h\nu = 6.626 \text{ x } 10^{-34}\,J \bullet s \text{ x } 4.4x10^{14}\,/s = 2.91 \text{ x } 10^{-19}\,J$

$2.91 \text{ x } 10^{-19}\,J \text{ x } 6.022 \text{ x } 10^{23}\,/mol = 176{,}000\ J/mol = 176\ kJ/mol$

Step 2: Convert the frequency to wavelength.

$$c = \lambda\nu \therefore \lambda = \frac{c}{\nu} = \frac{3.00x10^8\,m/s}{4.4x10^{14}\,/s} = 6.82x10^{-7}\,m = 682nm$$

Step 3: Consult your textbook for the color of this wavelength of light.

Light of 682 nm will be reddish-orange.

WP 5.4

Strategy: Determine the subshell associated with a value of $l = 0$.

Step 1: Identify the subshell with the value of *n* and the letter designation for $l = 0$.

When $l = 0$, the subshell is s. This is a 3s subshell.

Step 2: Determine the number of orbitals in this subshell.

The number of orbitals in a subshell is determined by the magnetic quantum number, m_l. For $l = 0$, only one value of m_l is available, so there is only one orbital.

WP 5.5

Strategy: Identify *n* and *m*, solve for wavelength, then energy, and convert to kJ/mol.

Step 1: Identify n and m for this problem.

As the intention is to completely remove the electron and make a H^+ ion, $n = 3$, and $m = \infty$

Step 2: Solve the Balmer-Rydberg equation to get the wavelength of the energy needed.

$$\frac{1}{\lambda} = 1.097\text{x}10^{-2}\,/\text{nm}\left[\frac{1}{3^2}\right] = 0.00122\,/\text{nm}$$

$$\lambda = 820 \text{ nm}$$

Step 3: Solve for the energy of one photon of this energy.

$$E = \frac{hc}{\lambda} = \frac{6.626 \text{ x } 10^{-34}\,\text{J} \bullet \text{s x } 3.00 \text{ x } 10^{8}\,\text{m/s}}{820 \text{ x } 10^{-9}\,\text{m}} = 2.42 \text{ x } 10^{-19}\,\text{J}$$

Step 4: Convert to kJ/mol.

$$2.42 \text{ x } 10^{-19} \frac{\text{J}}{\text{photon}} \text{ x } 6.022 \text{ x } 10^{23} \frac{\text{photons}}{\text{mol}} \text{ X } \frac{1 \text{ kJ}}{1000 \text{ J}} = 146 \frac{\text{kJ}}{\text{mol}}$$

WP 5.6

Step 1: Determine the number of electrons in calcium.

Calcium has atomic number 20, so it has 20 electrons.

Step 2: Use the Aufbau principle to determine the ground-state electronic configuration.

Begin building up the electron configuration from $n = 1$

Ca: $1s^2\,2s^2 2p^6 3s^2 3p^6 4s^2$

Step 3: Determine the noble gas in the previous row, and remove the electrons associated with it to a shorthand notation. Specifically name the electrons in unfilled subshells.

The noble gas with fewer electrons than calcium is argon. With the argon atoms delineated using shorthand, the electron configuration becomes:

Ca: $[\text{Ar}]4s^2$

Step 4: Draw the orbital-filling diagram.

↑↓	↑↓	↑↓ ↑↓ ↑↓	↑↓	↑↓ ↑↓ ↑↓	↑↓
1*s*	2*s*	2*p*	3*s*	3*p*	4*s*

Chapter 6

WP 6.1

Step 1: The largest jump in ionization energies is between the fourth and fifth ionization energies. This shows that there are five valence electrons, so this must be a group 5A element. In fact, it is phosphorus. ($P \rightarrow P^{+} \rightarrow P^{2+} \rightarrow P^{3+} \rightarrow P^{4+} \downarrow \rightarrow P^{5+}$)

WP 6.2

Step 1: Identify the group number of the element and identify it as metal or nonmetal.

Rb:	Rubidium	Group 1A metal
Sc:	Scandium	Group 3B transition metal
Se:	Selenium	Group 6A nonmetal
As:	Arsenic	Group 5A nonmetal

Step 2: Find the nearest noble gas.

Rb:	Krypton (36 electrons)
Sc:	Argon (18 electrons)
Se:	Krypton (36 electrons)
As:	Krypton (36 electrons)

Step 3: Identify the number of electrons that must be gained/lost to achieve the electron configuration of that noble gas.

Rb:	37 electrons → Krypton (36 electrons)	One electron must be lost
Sc:	21 electrons → Argon (18 electrons)	Three electrons must be lost
Se:	34 electrons → Krypton (36 electrons)	Two electrons must be gained
As:	33 electrons →Krypton (36 electrons)	Three electrons will be gained

Step 4: Write the ionized form of the element.

Rb^+ Sc^{3+} Se^{2-} As^{3-}

WP 6.3

Strategy: Write and sum the energies of each of the steps in the reaction to calculate the overall energy change of the formation of rubidium chloride from its elements.

Step 1: Write each step of the reaction with its energy required or produced per mole.

1.	Sublimation of sodium	$Na\ (s) \rightarrow Na\ (g)$	+107.3 kJ/mol
2.	Vaporization of bromine	$\frac{1}{2}Br_2\ (l) \rightarrow \frac{1}{2}Br_2\ (g)$	+ 15.4 kJ/mol
3.	Splitting of bromine molecules	$\frac{1}{2}Br_2\ (g) \rightarrow Br\ (g)$	+112 kJ/mol
4.	Ionization of sodium	$Na\ (g) \rightarrow Na^+\ (g) + e^-$	+495.8 kJ/mol
5.	Ionization of bromine	$Br\ (g) + e^- \rightarrow Br^-\ (g)$	-325 kJ/mol
6.	Formation of ionic solid	$Na^+\ (g) + Br^-\ (g) \rightarrow NaBr\ (s)$	-747 kJ/mol

Step 2: Sum the energy terms to find the overall energy change of the reaction.

Total = -341.5 kJ/mol

Note: The values given for bond breaking and vaporizing bromine are divided in half. As given, they are for the following reactions: Br_2 (*l*) → Br_2 (*g*) and Br_2 (*g*) → 2Br (*g*).

WP 6.4
Strategy: Determine the relative sizes and charges for the ions and assign relative strengths based on charge and size.

Step 1: Identify the relative charges in the two sets of ions and how they change from one set to the other.

NaCl, KCl	same charges
AlF_3, $MgCl_2$	Al – 3+, Mg – 2+, aluminum will provide stronger lattice energy.
$CuBr_2$, CuCl	Cu (II) – 2+, Cu (I) – 1+, copper (II) will provide stronger lattice energy.

Step 2: If necessary, identify the relative ion sizes in the two sets of ions and how they change from one to the other.

NaCl, KCl same charges — Sodium is smaller than potassium, so will provide stronger lattice energy.

Step 3: Determine the relative lattice energies of the ion pairs.

NaCl > KBr $\quad$ AlF_3 > $MgCl_2$ $\quad$ $CuBr_2$ > CuCl

WP 6.5
Strategy: Based on the reactivity of the metals and the expected reactions:

Step 1: Determine what reaction, if any, will occur.

Sr (*s*) + O_2 (*g*) → strontium oxide will form spontaneously.

Ca (*s*) + H_2O (*l*) → calcium hydroxide will form and hydrogen gas will be generated. Non-vigorous.

Be (s) + H_2O (*l*) → no reaction expected. Beryllium is the least reactive of the alkaline earth metals.

Step 2: Write balanced equations for the reactions that occur.

2Sr (*s*) + O_2 (*g*) → 2SrO (*s*)

Ca (*s*) + $2H_2O$ (*l*) → $Ca(OH)_2$ (*aq*) + H_2 (*g*)

Chapter 7

WP 7.1
Step 1: Predict the polarity of the bonds based on their locations, then check with electronegativity values from Table 7.4 in the textbook.

H – F – nonmetals, far apart: polar covalent
F – Cl – nonmetals, close together: nonpolar
C – O – nonmetals, close together: nonpolar
K – Cl – metal and nonmetal: ionic
Ge – Br – semi-metal and nonmetal: polar covalent

Step 2: Put in the numbers to check.

H – F : 4.0 – 2.1 = 1.9, polar covalent
F – Cl : 4.0 – 3.0 = 1.0, polar covalent (estimate off – fluorine is really electronegative!)
C – O : 3.0 – 2.0 = 1.0, polar covalent
K – Cl : 3.0 – 0.8 = 2.2, ionic
Ge – Br : 2.8 – 1.8 = 1.0, polar covalent

Note: What this exercise shows is that the farther apart on the periodic table two elements are, the more polar/ionic their bond. Also, almost all nonmetal-nonmetal bonds are polar.

WP 7.2
Draw electron-dot structures for the following compounds:

HF, CH_2Br_2, Cl_2

Strategy: Determine valences of compounds, then combine them to form complete octets.

Step 1: Determine the number of electrons on each atom based on its position on the periodic table:

H – 1 valence electron
F – 7 valence electrons
C – 4 valence electrons
Br – 7 valence electrons
Cl – 7 valence electrons

Step 2: Combine the atoms to complete octets, substituting bond lines for paired electrons.

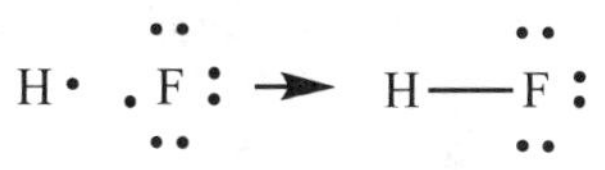

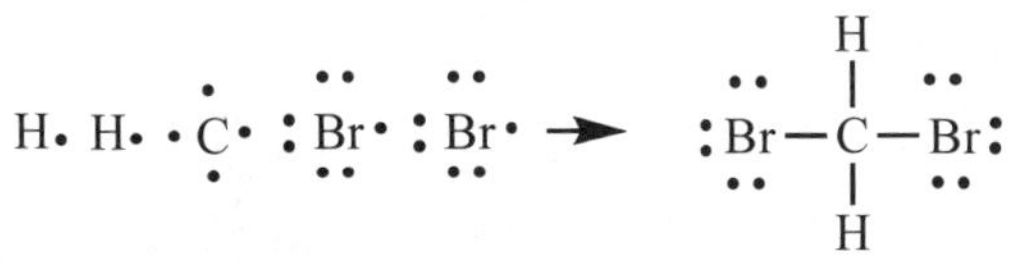

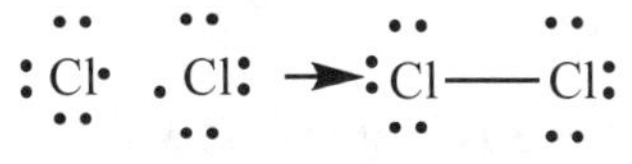

WP 7.3
Step 1: Determine the number of electrons on each atom based on its position on the periodic table.

H: one valence electron
C: four valence electrons
O: six valence electrons

Step 2: Combine the atoms to complete octets, substituting bond lines for paired electrons.

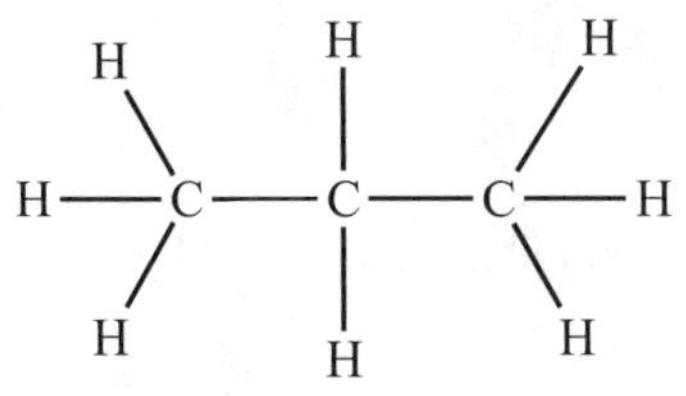

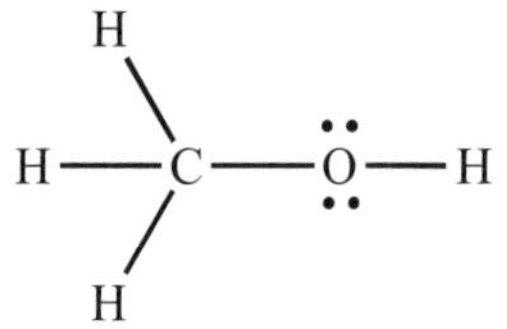

This compound does not have enough hydrogens to complete the octets on the carbon atoms, leaving two unpaired electrons. To complete these octets, the carbon atoms share these electrons as well, forming a double bond.

WP 7.4

Step 1: Determine the total number of valence-shell electrons.

5 (from P) + 35 (from 5 F) = 40

Step 2: Determine the connections.

P is the central atom. (This leads to the most symmetrical structure.)

Step 3: Draw the bonds, and subtract the number of electrons used from the total number of valence electrons available.

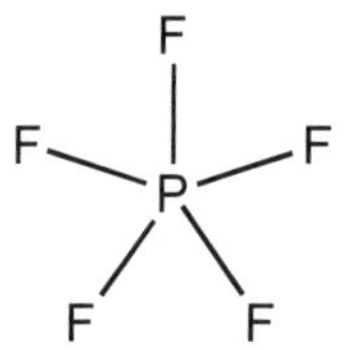

The five bonds use two electrons each for a total of 10. That leaves 30 electrons to be distributed among the fluorine atoms.

Step 4: Complete the octets of the outer atoms.

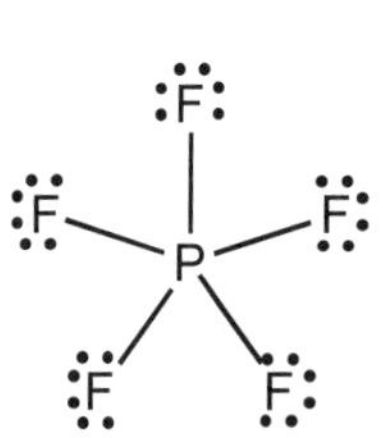

Step 5: Place any remaining electrons on the central atom.

There are no remaining electrons, but the central phosphorus atom has an expanded octet of 10.

WP 7.5

Step 1: Determine the total number of valence-shell electrons.

Total number of valence-shell electrons = 4 (from C) + 18 (from 3 O) + 2 (from the 2– charge) = 24

Step 2: Determine the connections.

Carbon is the central atom for symmetry's sake, as well as its being the most electronegative atom.

Step 3: Draw the bonds and subtract the number of electrons used from the total number of valence electrons available.

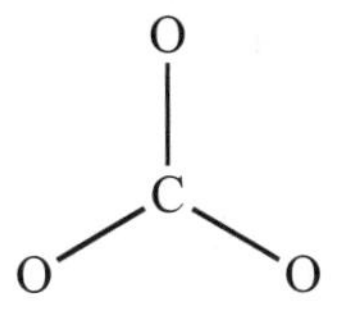

This uses 6 valence electrons, leaving 18.

Step 4: Complete the octets of the outer atoms.

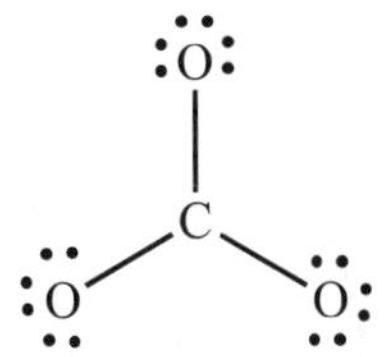

Step 5: Place any remaining electrons on the central atom, looking for multiple possibilities.

There are no electrons remaining, but carbon has an incomplete octet, and will need to share a pair of electrons from one of the oxygens. This leads to a number of resonance structures.

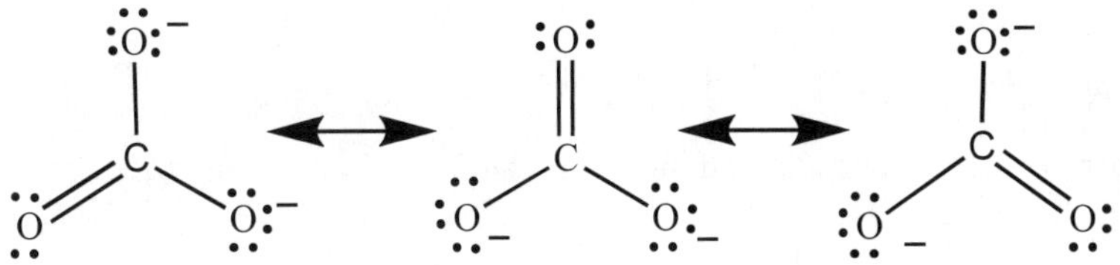

WP 7.6

Step 1: Determine the number of valence electrons in the ion

Total number of valence-shell electrons = 4 (from C) + 6 (from O) + 5 (from N) + 1 (from the 1– charge) = 16

Step 2: Determine three reasonable electron-dot structures with carbon central and this number of valence electrons.

$^{-}\ddot{N}=C=\ddot{O}$ $\quad$ $:N\equiv C-\ddot{O}:^{-}$ $\quad$ $:\ddot{N}^{-2}-C\equiv O:^{+}$

Step 3: Calculate the formal charge on each of the atoms in each of the structures.

Structure 1:	Structure 2:	Structure 3:
N = 5 – 2 - 4 = -1	N = 5 – 3 – 2 = 0	N = 5 – 1 – 6 = -2
C = 4 - 4 – 0 = 0	C = 4 – 4 – 0 = 0	C = 4 – 4 – 0 = 0
O = 6 – 2 – 4 = 0	O = 6 – 1 – 6 = -1	O = 6 – 3 – 2 = +1

Step 4: Choose the most stable structure.

Notice that all three have a net charge of -1, appropriate to this ion. Structure 3 (which is already violating rules about how many bonds oxygen likes to have) has formal charges on two atoms, and is therefore unstable. Structures 1 and 2 both have single negative charges on one atom. To choose between these two, choose the structure in which the negative charge is assigned to the most electronegative atom. This is structure 1.

WP 7.7

Strategy: Draw the electron-dot structure for each of the molecules and use the list above to determine the molecular shape.

Step 1: Draw the electron-dot structure for the molecules, using the number of valence electrons and the most symmetrical structure.

CS_2 has four valence electrons from carbon, and six from each of the sulfurs, for a total of 16.

$:\ddot{S}=C=\ddot{S}:$

Double bonds were needed to complete the octet around carbon. Although sulfur will form an expanded octet, here it does not need to.

SF_4 has 6 valence electrons from the sulfur, and 28 from the four fluorines, for a total of 34.

```
      :F:
       |
  :F — S: — F:
       |
      :F:
```

Step 2: Determine the number of charge clouds around the central atom. How many are used for bonding and how many are nonbonding?

CS_2 has two electron clouds around the central atom.

SF_4 has five electron clouds around the central atom, four bonding, and one nonbonding pair.

Step 3: Based on the chart, determine the shape.

CS_2 is linear.

SF_4 is seesaw shaped.

Workbook Problem 7.8
Given the following structure:

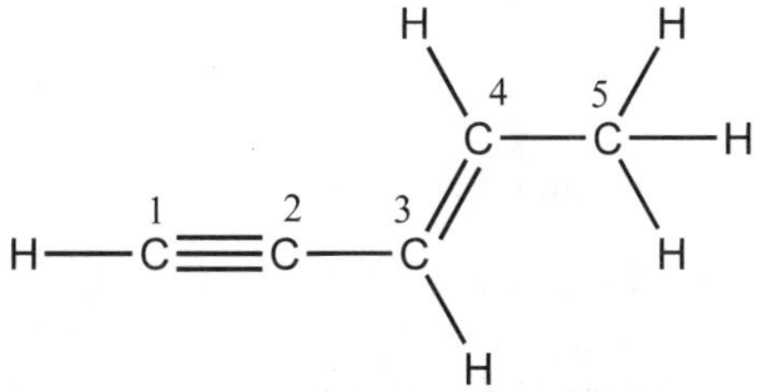

H—C≡C—C(H)=C(H)—C(H_3) (carbons numbered 1–5)

describe the hybridization and bonding of each of the carbon atoms.

Strategy: Determinc the number of electron clouds around each carbon to determine hybridization, then identify the number of σ and π bonds.

Step 1: Determine the number of electron clouds around each carbon and its hybridization.

Carbon 1: two electron clouds – *sp* hybridization
Carbon 2: two electron clouds – *sp* hybridization
Carbon 3: three electron clouds – sp^2 hybridization
Carbon 4: three electron clouds – sp^2 hybridization
Carbon 5: four electron clouds – sp^3 hybridization

Step 2: Determine the bond types present.

Carbon 1: σ bond to hydrogen, σ and 2π bonds to carbon 2
Carbon 2: σ and 2π bonds to carbon 1, σ bond to carbon 3
Carbon 3: σ bond to carbon 2, σ bond to hydrogen, σ and π to carbon 4
Carbon 4: σ and π to carbon 3, σ to hydrogen and σ to carbon 5
Carbon 5: σ bonds to carbon 4, and three hydrogens.

Chapter 8

WP 8.1

Step 1: Use $w = -(P \times \Delta V)$:

$w = -(13 \text{ atm x } [7.4 \text{ L} - (7.4 \text{ L} + 3.7 \text{ L})]) = 48.1 \text{ L} \bullet \text{atm}$

Step 2: Convert from L • atm to joules to kilojoules.

$48.1 \text{ L} \bullet \text{atm x } 101 \text{ J/L} \bullet \text{atm} = 4858 \text{ J} = 4.9 \text{ kJ}$

Step 3: What does the sign tell you? Was work done by or to the system?

The positive value of the work shows that the system gained: work was done on the system. The same thing can be determined by looking at the initial and final volume of the reactions: 11.1 L of reactants provided 7.4 L of products. The overall system contracted, so work was done on it.

WP 8.2

Step 1: Use $\Delta E = q - P\Delta V$

$\Delta E = -79.9 \text{ kJ} - 4.9 \text{ kJ} = -84.8 \text{ kJ}$

WP 8.3

Step 1: Determine the ΔH for the reaction given the amount of carbonic acid that dissociated.

0.5 mol x -20.7 kJ/mol = -10.4 kJ

Step 2: Determine $P\Delta V$ and convert from L • atm to kJ.

1 atm x 11.2 L = 11.2 L • atm x 101 J/L•atm = 1131 J = 1.13 kJ

Step 3: Determine ΔE from the equation: $\Delta H = \Delta E + P\Delta V$.

Rearrange the equation to $\Delta E = \Delta H - P\Delta V$.

$\Delta E = -10.4 - (-1.13) = -11.5$ kJ

Think it through to make sure this makes sense. Heat was released from the exothermic equation, leading to a negative ΔH, but work was done on the system when the volume decreased, which lessens the magnitude of the overall energy change.

WP 8.4

Step 1: Convert the grams of water to moles.

$$50\text{ g x}\frac{1\text{ mol}}{18\text{ g}} = 2.8\text{ mol}$$

Step 2: Multiply by the heat of vaporization.

$$2.8\text{ mol x }40.7\frac{\text{kJ}}{\text{mol}} = 113\text{ kJ}$$

Step 3: Make sure the sign of the heat transfer makes sense: the water is the system.

Vaporizing water is the same as boiling it on the stove. This is a way of adding heat, so it is expected that heat will need to be added to the system to vaporize the water. In fact, lots of heat will need to be added, as vaporizing water requires breaking all of the hydrogen bonds that hold it together as a liquid. The sign on the energy transfer is positive, so heat is being added to the system, as expected.

WP 8.5

Step 1: Determine the number of moles of calcium.

$$3.24\text{ g x}\frac{1\text{mol}}{40.08\text{ g}} = 0.0808\text{ mol}$$

Step 2: Using the stoichiometric ratio, determine the heat involved.

$$0.0808\text{ mol Ca x}\frac{2\text{ mol CaO}}{2\text{ mol Ca}}\text{x} - 634.9\frac{\text{kJ}}{\text{mol CaO}} = -51.3\text{ kJ}$$

Step 3: Based on the sign of the heat transfer, determine whether the reaction is exothermic or endothermic.

Based on the negative sign for the energy transfer, this is an exothermic reaction.

WP 8.6

Strategy: Determine the heat absorbed by the water, then reverse the sign to determine the number of joules evolved by the dissociation. Convert the grams of NaOH to moles, and divide the heat evolved by the number of moles of NaOH to determine the molar heat of dissociation.

Step 1: Determine the amount of heat gained by the water using the equation $q = mc\Delta T$.

$$q = 50 \text{ g} \times 4.184 \text{ J/g °C} \times (30.70 \text{ °C} - 22.73 \text{ °C}) = 1667\text{J}$$

Step 2: Determine the amount of heat released by the reaction.

As the dissociation released the energy absorbed by the water, it will have the same magnitude, but be opposite in sign: -1667 J.

Step 3: Determine the number of moles of NaOH involved.

$$1.5 \text{ g} \times \frac{1\text{mol}}{40 \text{ g}} = 0.0375 \text{ mol}$$

Step 4: Determine the heat of dissociation per mole of NaOH.

$$\frac{-1667 \text{ J}}{0.0375 \text{ mol}} = 44\,453 \frac{\text{J}}{\text{mol}} \times \frac{1 \text{ kJ}}{1000 \text{ J}} = -44.5 \frac{\text{kJ}}{\text{mol}}$$

WP 8.7

Strategy: Get the reactants on the right and the products on the left, flipping reactions around if needed, and multiplying them through if necessary to get the equation as written.

Step 1: Reactants on the left, multiplying through if necessary.

$Fe_2O_3\ (s) \rightarrow\ 2\ Fe\ (s) + 3/2\ O_2\ (g)\quad \Delta H° = +824.2$ kJ (turned around, and multiplied by ½)

$3\ CO\ (g) +\ 3/2O_2\ (g) \rightarrow 3\ CO_2\ (g)\quad \Delta H° = -848.1$ kJ (multiplied by 3/2)

Step 2: Products on the right, multiplying through if necessary.

Products are already on the right as a result of the manipulations above.

Step 3: Stack up the rewritten reactions, and add them together.

$$Fe_2O_3\ (s) \rightarrow 2\ Fe\ (s) + 3/2\ O_2\ (g) \qquad \Delta H° = +824.2 \text{ kJ}$$
$$\underline{3\ CO\ (g) + 3/2O_2\ (g) \rightarrow 3\ CO_2\ (g) \qquad \Delta H° = -848.1 \text{ kJ}}$$
$$Fe_2O_3\ (s) + 3\ CO\ (g) + 3/2O_2\ (g) \rightarrow 2\ Fe\ (s) + 3/2\ O_2\ (g) + 3\ CO_2\ (g) \quad \Delta H° = -23.9 \text{ kJ}$$

Step 4: Cancel terms that appear on both sides.

$$Fe_2O_3\ (s) + 3\ CO\ (g) \rightarrow 2\ Fe\ (s) + 3\ CO_2\ (g) \qquad \Delta H° = -23.9 \text{ kJ}$$

WP 8.8

Strategy: Write a balanced equation for the reaction, and use the standard heats of formation to determine the overall heat of reaction per mole of glucose, then convert grams to moles and determine the heat released by 10 grams of glucose.

Step 1: Write and balance the equation.

$C_6H_{12}O_6\,(s) + 6\,O_2\,(g) \rightarrow 6\,H_2O\,(l) + 6\,CO_2\,(g)$

Step 2: Determine the overall heat of reaction using $\Delta H^\circ = \sum \Delta H^\circ_{products} - \sum \Delta H^\circ_{reactants}$.

ΔH° = [(6 x -285.8 kJ/mol) + (6 x -393.5 kJ/mol)] – (-1273.3 kJ/mol) = -2800 kJ/mol

Step 3: Convert the grams of glucose to moles.

$$10 \text{ grams x } \frac{1 \text{ mol}}{180 \text{ g}} = 0.056 \text{ mol glucose}$$

Step 4: Determine the energy released by 10 g of glucose.

0.056 mol x -2800 kJ/mol = -157 kJ

Step 5: Using q = mcΔT, determine the temperature change of 500 g of water.

157,000 J = 500 g x 4.184 J/g□°C x ΔT

$$\Delta T = \frac{157{,}000 \text{ J}}{500\text{g x } 4.184 \text{ J/g}\cdot{}^\circ\text{C}} = 75\ {}^\circ\text{C}$$

WP 8.9

Predict the energy change for combustion of methane (CH_4) using bond dissociation energies.

Use the following information:

	Bond Energy
C-H	414 kJ/mol
O=O	498 kJ/mol
O-H	464 kJ/mol
C=O	799 kJ/mol

Strategy: Write a balanced equation for the reaction, and determine the bond energies for both products and reactants. Subtract products from reactants to determine the overall energy change of the reaction.

Step 1: Write and balance the equation.

$CH_4\,(g) + 2\,O_2\,(g) \rightarrow CO_2\,(g) + 2\,H_2O\,(l)$

Step 2: Determine the energy of the bonds on the reactant side.

(4 x 414 kJ/mol) + (2 x 498 kJ/mol) = 2652 kJ/mol

Step 3: Determine the energy of the bonds on the product side.

(2 x 799 kJ/mol) + (4 x 464 kJ/mol) = 3454 kJ/mol

Step 4: Determine the energy released by the reaction.

Reactants – products = 2652 kJ/mol – 3454 kJ/mol = -802 kJ/mol

WP 8.10

Given the following data, is the reaction of hydrogen and chlorine gas spontaneous at room temperature?

ΔH_f HCl (*g*) = -92.3 kJ/mol
$S°$ H_2 (*g*) = 130.6 J/K•mol
$S°$ Cl_2 (*g*) = 223 J/K•mol
$S°$ HCl (*g*) = 186.8 J/K•mol

Strategy: Write a balanced equation for the reaction, and determine ΔH and ΔS for the reaction. Using the Gibbs free-energy equation, determine ΔG for the reaction at 25 °C, not forgetting to convert to Kelvin first.

Step 1: Write and balance the equation.

$H_2(g) + Cl_2(g) \rightarrow 2HCl(g)$

Step 2: Determine ΔH.

ΔH = products – reactants = 2 x -92.3 kJ/mol = -184.6 kJ/mol

Step 3: Determine ΔS.

ΔS = 2 x 186.8 J/K•mol – (130.6 J/K•mol + 223 J/K•mol) = 20 J/ K•mol

Step 4: Determine the free-energy of the reaction at 25 °C, first converting the temperature to K.

ΔG = -184.6 kJ/mol – 298K(0.020 kJ/ K•mol) = -191 kJ/mol

Step 5: Determine the spontaneity of the reaction.

The negative sign on the free energy term indicates that this reaction is spontaneous at 25 °C. If the value here is written per mole of HCl formed, the value is -95.3 kJ/mol. This matches the value for the standard free energy of formation of HCl.

Chapter 9

WP 9.1

Strategy: Determine the pressure of the gas in mmHg, convert to atmospheres.

Step 1: Determine the pressure of the gas in mmHg.

The pressure is higher inside the manometer. Pressure = 760 mmHg + 73 mmHg = 833 mmHg

Step 2: Convert from mmHg to atmospheres.

$$833 \text{ mm Hg} \times \frac{1 \text{ atm}}{760 \text{ mm Hg}} = 1.10 \text{ atm}$$

WP 9.2

Strategy: Determine which gas law is involved, convert temperatures, solve for the final volume.

Step 1: Determine the gas law to use.

This problem involves only volume and temperature: Charles's Law will be used.

Step 2: Convert the temperatures to K.

37 °C + 273.15 = 310 K

350 °C + 273.15 = 623 K

Step 3: Solve for the final volume.

$$\frac{V_1}{T_1} = \frac{V_2}{T_2} \qquad \frac{27\ \text{L}}{310\ \text{K}} = \frac{V_2}{623\ \text{K}}$$

$$V_2 = \frac{27\ \text{L x } 623\text{K}}{310\ \text{K}} = 54.3\ \text{L}$$

WP 9.3

Strategy: Rearrange the ideal gas law to solve for pressure, convert grams to moles, and °C to K, solve.

Step 1: Rearrange the ideal gas law.

$$PV = nRT \text{ becomes } P = \frac{nRT}{V}$$

Step 2: Convert grams to moles and °C to K.

$$6.7\ \text{g}\ F_2 \text{ x } \frac{1\ \text{mol}\ F_2}{38\ \text{g}\ F_2} = 0.176\ \text{mol}\ F_2$$

25.0 °C + 273.15 = 298 K

Step 3: Solve for the final pressure.

$$P = \frac{nRT}{V} = \frac{0.176\ \text{mol x } 0.08206 \frac{\text{L} \bullet \text{atm}}{\text{mol} \bullet \text{K}} \text{ x } 298\ \text{K}}{3.0\ \text{L}} = 1.43\ \text{atm}$$

WP 9.4

If 12.3 g of neon were compressed from 1.32L at 1 atm and 25 °C to 18 atm and 750° C, what would the final pressure be in the container?

Strategy: Determine what form of the gas laws is needed, convert temperatures and moles if necessary, and solve for final pressure.

Step 1: Determine the form of gas laws needed.

Although this problem again requires solving for one variable, it is a very different type of problem. Whenever a change is indicated, this is an indication that the full ideal gas law is not needed. Instead, a version of the simple gas laws will be required.

Here, the temperature, pressure and volume are all changing, but the number of moles is not.

$$\frac{P_1V_1}{T_1} = \frac{P_2V_2}{T_2}$$

Step 2: Convert grams to moles and °C to K.

There is no need to convert to moles of fluorine.

25 °C + 273.15 = 298 K

750 °C + 237.15 = 1023 K

Step 3: Solve for the final pressure.

$$\frac{1 \text{ atm} \cdot 3.2 \text{ L}}{298 \text{ K}} = \frac{18 \text{ atm} \cdot V_2}{1023 \text{ K}}$$

$$V_2 = \frac{1 \text{ atm} \cdot 3.2 \text{ L} \cdot 1023 \text{ K}}{298 \text{ K} \cdot 18 \text{ atm}} = 0.61 \text{ L}$$

WP 9.5

Strategy: Determine which gas law is involved, solve for the volumes of oxygen and nitrogen dioxide.

Step 1: Determine the gas law to use.

This is Avogadro's Law, or at least a corollary of it. Since the volume of a molar amount of gas is constant, gases can be made to react in equivalent volumes. The volumes will be in stoichiometric ratios.

Step 2: Solve for the volume of oxygen.

$$7.45 \text{ L NO x } \frac{1 \text{ L } O_2}{2 \text{ L NO}} = 3.73 \text{ L } O_2$$

Step 3: Solve for the volume of nitrogen dioxide.

$$7.45 \text{ L NO x } \frac{2 \text{ L } NO_2}{2 \text{ L NO}} = 7.45 \text{ L } O_2$$

WP 9.6

Strategy: Convert temperatures and pressures to make it possible to solve for the volume of one mole under these conditions. Divide the mass of one mole of xenon by the volume to determine the density.

Step 1: Convert the temperature and pressure.

100 °C + 273.15 = 373.15 K

$$300 \text{ mm Hg} \times \frac{1 \text{ atm}}{760 \text{ mm Hg}} = 0.395 \text{ atm}$$

Step 2: Use the ideal gas law to calculate the volume of the gas.

$$V = \frac{nRT}{P} = \frac{1 \text{ mol} \times 0.08206 \frac{\text{L} \bullet \text{atm}}{\text{mol} \bullet \text{K}} \times 373\text{K}}{0.395 \text{ atm}} = 77.5 \text{ L}$$

Step 3: Determine the density of the gas.

mass of 1 mol of xenon = 131.3 g

$$\text{Density} = \frac{131.3 \text{ g}}{77.5 \text{ L}} = 1.69 \frac{\text{g}}{\text{L}}$$

WP 9.7

Strategy: Decide on a volume of gas, then solve for *n*. Using the density makes it possible to solve for the mass for that volume, and then for the molecular weight.

Step 1: Determine the number of moles in 1 L of gas.

$$n = \frac{PV}{RT} = \frac{3.75 \text{ atm} \times 1 \text{ L}}{0.08206 \frac{\text{L} \bullet \text{atm}}{\text{mol} \bullet \text{K}} \times (700°\text{C} + 273.15)\text{K}} = 0.0470 \text{ mol}$$

Step 2: Solve for the mass in 1 L of gas using the density.

Mass = density x volume = 0.798 g/L x 1 L = 0.798 g

Step 3: Determine the molecular weight by dividing g/moles.

M.W. = 0.798g/0.0470 mol = 17.0 g/mol. Ammonia is a gas that fits this molecular weight.

WP 9.8

Strategy: Determine the number of moles of each gas, the total number of moles, and the mole fraction of each component. Using these, calculate the total pressure and the partial pressures using the ideal gas law and Dalton's law.

Step 1: Determine the number of moles of each gas.

$$18.3 \text{ g } CO_2 \times \frac{1 \text{ mol}}{44.0 \text{ g}} = 0.416 \text{ mol } CO_2$$

$$23.1 \text{ g Ne} \times \frac{1 \text{ mol}}{20.2\text{g}} = 1.14 \text{ mol Ne}$$

$$5.26 \text{ g } H_2 \times \frac{1 \text{ mol}}{2.02 \text{ g}} = 2.61 \text{ mol } H_2$$

Step 2: Use the total number of moles to calculate the total pressure.

$$P = \frac{nRT}{V} = \frac{(0.416\text{ mol} + 1.14\text{ mol} + 2.61\text{ mol})\ 0.08206\frac{\text{L}\bullet\text{atm}}{\text{mol}\bullet\text{K}}(25°\text{C} + 273.15)\text{K}}{30\text{ L}}$$

$$= \frac{4.17\text{ mol x }0.08206\frac{\text{L}\bullet\text{atm}}{\text{mol}\bullet\text{K}}\text{ x }298\text{ K}}{30\text{ L}} = 3.40\text{ atm}$$

Step 3: Use the mole fraction (X) of each component to determine the partial pressures.

$$X_{CO_2} = \frac{0.416\text{ mol}}{4.17\text{ mol}} = 0.100;\ X_{Ne} = \frac{1.14\text{ mol}}{4.17\text{ mol}} = 0.273;\ X_{H_2} = \frac{2.61\text{ mol}}{4.17\text{ mol}} = 0.626$$

$$P_{CO_2} = 0.100\text{ x }3.40\text{ atm} = 0.340\text{ atm}$$

$$P_{Ne} = 0.273\text{ x }3.40\text{ atm} = 0.928\text{ atm}$$

$$P_{H_2} = 0.626\text{ x }3.40\text{ atm} = 2.13\text{ atm}$$

These problems are particularly easy to check – the mole fractions must add up to 1, and the partial pressures must add up to the total pressure, as they do here.

WP 9.9

Strategy: Determine the speed of H_2 molecules at 25 °C, then solve the equation again for temperature needed for the much larger I_2 molecules.

Step 1: Determine the speed of H_2 at 25 °C.

$$u = \sqrt{\frac{3RT}{M}} = \sqrt{\frac{3\text{ x }8.314\frac{\text{kg}\bullet\frac{\text{m}^2}{\text{s}^2}}{\text{K}\bullet\text{mol}}\text{x}(25+273)\text{K}}{0.00202\frac{\text{kg}}{\text{mol}}}} = \sqrt{3{,}680{,}000\frac{\text{m}^2}{\text{s}^2}} = 1918\frac{\text{m}}{\text{s}}$$

Step 2: Determine the temperature required for the iodine molecules to achieve the same speed.

$$1918\frac{\text{m}}{\text{s}} = \sqrt{\frac{3\text{ x }8.314\frac{\text{kg}\bullet\frac{\text{m}^2}{\text{s}^2}}{\text{K}\bullet\text{mol}}\text{x T}}{0.2538\frac{\text{kg}}{\text{mol}}}} = \sqrt{98.27\frac{\frac{\text{m}^2}{\text{s}^2}}{\text{K}}\text{x T}}$$

Square both sides to solve for T:

$$3{,}680{,}000\frac{\text{m}^2}{\text{s}^2} = 98.27\frac{\frac{\text{m}^2}{\text{s}^2}}{\text{K}}\text{x T}$$

$$T = 37{,}450\text{ K}$$

WP 9.10

Oxygen-16 has an atomic mass of 15.995, and Oxygen-18 an atomic mass of 17.999. What will be their relative rates of effusion?

Strategy: Solve using Graham's law.

Step 1: Solve using Graham's law.

$$\frac{\text{Rate of effusion of } ^{16}\text{O}}{\text{Rate of effusion of } ^{18}\text{O}} = \sqrt{\frac{35.998 \text{ g } ^{18}\text{O/mol}}{31.990 \text{ g } ^{16}\text{O/mol}}} = 1.061$$

WP 9.11

Strategy: Solve using the ideal gas law and the van der Waals equation.

Step 1: Solve using the ideal gas law.

$$P = \frac{nRT}{V} = \frac{1 \text{ mol x } 0.08206 \frac{\text{L} \bullet \text{atm}}{\text{mol} \bullet \text{K}} \text{ x } 300 \text{ K}}{0.100 \text{ L}} = 246 \text{ atm}$$

Step 2: Solve using the van der Waals equation.

$$P = \frac{nRT}{V - nb} - \frac{an^2}{V^2} = \frac{1 \text{ mol x } 0.08206 \frac{\text{L} \bullet \text{atm}}{\text{mol} \bullet \text{K}} \text{ x } 300 \text{ K}}{0.100 \text{ L } - (1 \text{ mol x } 0.0318 \text{ L/mol})} - \frac{(1\text{mol})^2 \text{ x } 1.38 \text{ (L}^2 \bullet \text{atm)/mol}^2}{(0.100 \text{ L})^2}$$

$$= \frac{24.6 \text{ L} \bullet \text{atm}}{0.0682 \text{ L}} - \frac{1.38 \text{ L}^2 \bullet \text{atm}}{0.0100 \text{ L}^2} = 361 \text{ atm} - 138 \text{ atm} = 223 \text{ atm}$$

Step 3: Compare the two values.

The ideal and real values are reasonably close, given the large pressures involved. Whether the real value would be acceptable would depend on the precision required in the work.

Chapter 10

WP 10.1

Strategy: Look for asymmetrical molecules with electronegativity differences.

Step 1: Determine which molecules are asymmetrical.

All of these molecules are asymmetrical, even trimethylamine, which has a lone pair of electrons on the nitrogen atom.

Step 2: Determine which element is most electronegative, and thus the area of highest electron concentration.

H_2CO – oxygen $N(CH_3)_3$ – nitrogen HBr – bromine

All of these molecules will have dipoles.

WP 10.2

Strategy: Calculate the dipole if it were purely ionic, then calculate the percent ionic character.

Step 1: Calculate ionic dipole.

$$\mu = Q \text{ x } r = 1.160 \text{ x } 10^{-19} \text{ x } 176 \text{ x } 10^{-12} \text{ m x } \frac{1 \text{ D}}{3.336 \text{ x } 10^{-30} \text{ C} \bullet \text{m}} = 6.12 \text{ D}$$

Step 2: Determine percent ionic character.

$$\% \text{ ionic character} = \frac{1.42 \text{ D}}{6.12 \text{ D}} \text{ x } 100 = 23.2\% \text{ ionic}$$

WP 10.3

Strategy: Examine the structures of the molecules and determine what interactions are present, then rank in order of strength.

Step 1: List the interactions experienced by each molecule.

CH_3OH : hydrogen bonding, dipole-dipole, and dispersion forces.

Xe : dispersion forces only

CH_3Cl : dipole-dipole and dispersion forces.

Step 2: Rank the molecules in order of increasing strength of interactions.

$$Xe < CH_3OH < CH_3Cl$$

WP 10.4

Strategy: Rearrange the equation $\Delta G = \Delta H - T\Delta S$ to solve for the entropy change.

Step 1: Rearrange to solve for ΔS.

$$0 = \Delta H - T\Delta S$$

$$\Delta S = \frac{\Delta H}{T}$$

Step 2: Solve for ΔS, being careful to maintain the correct sign.

$$\Delta S = \frac{\Delta H}{T} = \frac{-5970 \frac{J}{mol}}{(-107 + 273.15)K} = -35.9 \frac{J}{mol \bullet K}$$

It is important to remember that in freezing, heat is released and order increases. This means using $-\Delta H_{fus}$. It will be immediately apparent if this was not done correctly, as an increase in entropy will result, while freezing decreases entropy.

WP 10.5

Strategy: Use the Clausius-Clapeyron equation to solve for ΔH.

Step 1: Convert temperatures to K.

78.4 °C + 273.15 = 351.55 K
63.5 °C + 273.15 = 336.65 K

Step 2: Solve for ΔH.

$$\Delta H_{vap} = \frac{(\ln P_2 - \ln P_1)R}{\left(\frac{1}{T_1} - \frac{1}{T_2}\right)} = \frac{(\ln 760 - \ln 400)8.314\frac{J}{mol\bullet K}}{\left(\left[\frac{1}{336.65K}\right]-\left[\frac{1}{351.55K}\right]\right)} = \frac{(6.63-5.99)8.314\frac{J}{mol\bullet K}}{0.000126\frac{1}{K}} = 42.3 \text{ kJ/mol}$$

WP 10.6

Strategy: Use the Clausius-Clapeyron equation to solve for the vapor pressure at 25 °C.

Step 1: Use the Clausius-Clapeyron equation to solve for the vapor pressure.

$$\ln P_1 + \frac{42{,}300 \text{ J/mol}}{8.314\frac{J}{mol\bullet K}(25+273.15)K} = \ln 760 + \frac{42{,}300 \text{ J/mol}}{8.314\frac{J}{mol\bullet K}(351.55K)}$$

$$\ln P_1 + 17.06 = 21.10$$

$$\ln P_1 = 4.035$$

$$P_1 = e^{4.035} = 56.8 \text{ mm Hg}$$

WP 10.7

Polonium has a density of 9.3 g/cm^3 and a simple cubic.

Strategy: Use the number of atoms per unit cell and the molecular weight to determine the mass of a unit cell, then use the density to solve for the volume of a cell. With the volume, determine the length of one side, then determine the atomic radius.

Step 1: Determine the mass of a unit cell.

A simple cubic structure has a primitive cubic unit cell, which contains one atom.

$$1 \text{ atom x} \frac{1 \text{ mol}}{6.022\text{x}10^{23} \text{ atoms}} \text{x } 208.98\frac{g}{mol} = 3.470\text{x}10^{-22} \text{ g}$$

Step 2: Solve for the volume of a unit cell.

$$\frac{3.470 \text{ x } 10^{-22} g}{9.3\frac{g}{cm^3}} = 3.73\text{x}10^{-23} cm^3$$

Step 3: Determine the edge length of a unit cell.

$$\text{Edge length} = \sqrt[3]{3.731\text{x}10^{-23} cm^3} = 3.34\text{x}10^{-8} cm \text{ x} \frac{1 \text{ m}}{100 \text{ cm}} \text{ x } \frac{1\text{x}10^{12} \text{ pm}}{m} = 334 \text{ pm}$$

Step 4: Determine the atomic radius:

For a primitive cubic unit cell, the radius of an atom is ½ the edge length.

Atomic radius = 0.5 x 334 pm = 167 pm

Chapter 11

WP 11.1

Strategy: Determine which is the major component and which is the minor component.

In a 3% solution of hydrogen peroxide, hydrogen peroxide is the solute, and the water, which makes up the remaining 97%, is the solvent.

WP 11.2

Strategy: Charge and molecular size determine the hydration energy.

Smaller, more highly charged ions will have the greatest hydration energy.

$$Cl^- < Li^+ < Mg^{2+}$$

WP 11.3

Strategy: Calculate mass percent of solution, volume of solution, and moles of solute to provide all units.

Step 1: Determine mass %.

$$\text{mass } \% = \frac{15.43 \text{ g}}{315.43 \text{ g}} \text{x} 100 = 4.89\% \text{ glucose}$$

Step 2: Calculate the moles of solute, and the molality of the solution.

$$15.43 \text{ g x } \frac{1 \text{ mole}}{180 \text{ g}} = 0.0857 \text{ mol}$$

$$\text{m} = \frac{0.0857 \text{ mol}}{0.31543 \text{ kg}} = 0.272 \text{ m}$$

Step 3: Calculate the volume of the solution, and the molarity of the solution.

$$\frac{315.43 \text{ g}}{1.09 \frac{\text{g}}{\text{mL}}} = 289 \text{ mL}$$

$$\text{M} = \frac{0.0857 \text{ mol}}{0.289 \text{ L}} = 0.296 \text{ M}$$

WP 11.4

Strategy: Calculate the initial and final amounts of CO_2 in a saturated solution, determine the moles, and then the liters of CO_2.

Step 1: Calculate the moles of dissolved CO_2 under pressure.

$$\text{Solubility} = 0.0032 \frac{\text{mol}}{\text{L} \bullet \text{atm}} \text{ x } 2.3 \text{ atm} = 0.00736 \frac{\text{mol}}{\text{L}}$$

dissolved CO_2 = 0.00736 mol/L x 0.500 L = 0.00368 mol

Step 2: Calculate the moles of dissolved CO_2 after pressure is released and equilibrium with the atmosphere reached.

$$\text{Solubility} = 0.0032 \frac{\text{mol}}{\text{L} \cdot \text{atm}} \text{ x } 0.30 \text{ mm Hg x } \frac{1 \text{ atm}}{760 \text{ mm Hg}} = 0.00000126 \frac{\text{mol}}{\text{L}}$$

$$0.00000126 \frac{\text{mol}}{\text{L}} \text{ x } 0.5 \text{ L} = 0.000000632 \text{ mol}$$

Step 3: Calculate the change in dissolved CO_2 – this is the CO_2 that has been released.

0.00368 mol – 0.000000632 mol = 0.00368 mol. For all practical purposes, all of the CO_2 has bubbled away.

Step 4: Determine the volume of gas at 25 °C and 1 atm pressure.

$$V = \frac{nRT}{P} = \frac{0.00368 \text{ mol x } 0.08602 \frac{\text{L} \cdot \text{atm}}{\text{mol} \cdot \text{K}} \text{ x } (25 + 273.15)\text{K}}{1 \text{ atm}} = 0.0944\text{L} = 94.4\text{mL}$$

WP 11.5

Strategy: Determine the mole fraction of ions needed to lower the vapor pressure by this amount, then convert to grams of each ionic substance.

Step 1: Calculate the mole fraction of water in a solution with this vapor pressure lowering.

$$230.7 \text{ mm Hg} = 233.7 \text{ mm Hg} \times X_{H_2O}$$

$$0.987 = X_{H_2O}$$

Step 2: Calculate the moles of dissolved ions that will lead to this mole fraction.

500 g/18.0 g/mol = 27.78 mol H_2O

$$X_{H_2O} = \frac{27.78 \text{ mol}}{(27.78 \text{ mol} + \text{x mol})} = 0.987$$

$$27.78 \text{ mol} = 0.987\text{x}(27.78 \text{ mol} + \text{x mol})$$

$$27.78 - 27.42 = 0.987\text{x}$$

$$\text{x} = 0.3659 \text{ mol ions}$$

Step 3: Convert to moles, then grams of $CaCl_2$.

$$0.3659 \text{ mol ions x } \frac{1 \text{ CaCl}_2}{3 \text{ ions}} = 0.1220 \text{ mol CaCl}_2$$

0.1220 mol $CaCl_2$ x 111 g/mol = 13.6 g $CaCl_2$

Step 4: Convert to moles, then grams of LiCl.

$$0.3659 \text{ mol ions x } \frac{1 \text{ LiCl}}{2 \text{ ions}} = 0.1830 \text{ mol LiCl}$$

0.1830 mol LiCl x 42.5 g/mol = 7.77 g LiCl

WP 11.6

Strategy: Determine the mole fraction of each component, and calculate the partial pressure of each.

Step 1: Solve for the mass and moles of each component.

0.13 x 150 g = 19.5 g ethanol/46 g/mol = 0.424 mol ethanol

0.87 x 150 g = 130.5 g water/18.0 g/mol = 7.25 mol water

Step 2: Determine the mole fraction of each component.

$$X_{H_2O} = \frac{7.25 \text{ mol}}{7.25 \text{ mol} + 0.424 \text{ mol}} = 0.945$$

$$X_{EtOH} = \frac{0.424 \text{ mol}}{7.25 \text{ mol} + 0.424 \text{ mol}} = 0.0552$$

Step 3: Determine the partial pressure of each component and the total pressure:

$$P_{H_2O} = 0.945 \text{ x } 23.8 \text{ mm Hg} = 22.5 \text{ mm Hg}$$

$$P_{EtOH} = 0.0552 \text{ x } 61.2 \text{ mm Hg} = 3.38 \text{ mm Hg}$$

$$P_{total} = 22.5 \text{ mm Hg} + 3.38 \text{ mm Hg} = 25.9 \text{ mm Hg}$$

WP 11.7

Strategy: Using the equation for freezing-point depression, solve for the molality of ions needed, then convert to moles and grams of ammonium sulfate.

Step 1: Determine the molality of ions needed to generate this freezing-point depression.

15.0 °C = 1.86 °C/*m* x X *m*

X = 8.06 *m*

Step 2: Determine the moles of $(NH_4)_3SO_4$ needed to generate this quantity of ions.

$$8.06 \text{ m} = \frac{\text{x moles}}{0.300 \text{ kg}}$$

$$8.06 \frac{\text{moles}}{\text{kg}} \text{ x } 0.300 \text{ kg} = 2.42 \text{ mol ions}$$

$$2.42 \text{ mol ions x } \frac{1 \text{ mol salt}}{4 \text{mol ions}} = 0.605 \text{ mol salt}$$

Step 3: Convert to grams of ammonium sulfate needed.

$$0.605 \text{ mol x } \frac{150\text{g}}{\text{mol}} = 90.8 \text{ g}$$

WP 11.8

***Strategy*:** Using the equation for osmotic pressure, solve for the molarity of the solution.

$$7.34\,\text{atm} = x\ \text{M} \times 0.08206 \frac{\text{L} \cdot \text{atm}}{\text{mol} \cdot \text{K}} \times 298\text{K}$$

$x = 0.300$ M

WP 11.9

Strategy: Using the equation for osmotic pressure, calculate the molarity of the solution, and from that the molecular weight of the compound.

Step 1: Calculate the molarity of the solution.

$$\Pi = \text{M} \times \text{R} \times \text{T}$$

$$8.428\ \text{mm Hg} \times \frac{1\ \text{atm}}{760\ \text{mm Hg}} = \text{M} \times 0.08206 \frac{\text{L} \cdot \text{atm}}{\text{mol} \cdot \text{K}} \times 298\text{K}$$

$\text{M} = 0.000453\ \text{M}$

Step 2: Calculate the number of moles in the solution.

$$0.00453 \frac{\text{mol}}{\text{L}} \times 0.100\ \text{L} = 0.0000453\ \text{mol or } 45\ \mu\text{mol}$$

Step 3: Determine the molecular weight.

$$23 \times 10^{-3}\,\text{g} = 0.0000453\ \text{mol} \times \text{M.W.}$$

$$\text{M. W.} = 507.18\ \frac{\text{g}}{\text{mol}}$$

Chapter 12

WP 12.1

Step 1: Balance the equation.

$$2\,\text{HBr}\,(g) \rightarrow \text{H}_2\,(g) + \text{Br}_2\,(g)$$

Step 2: Write the relative reaction rates.

$$\text{General Rate} = -\frac{\Delta\text{HBr}}{2\Delta t} = \frac{\Delta\text{H}_2}{\Delta t} = \frac{\Delta\text{Br}_2}{\Delta t}$$

Step 3: Determine the rate of formation of bromine from the rate of disappearance of HBr.

$$-\frac{-3.26\text{x}10^{-3}\,\text{M/s}}{2} = 0.00163\ \text{M/s}$$

This is also the rate of formation of hydrogen.

WP 12.2

Step 1: Zeroth-order reaction:

Rate = k – units will be M/s.

Step 2: First-order reaction:

Rate = k[A], so M/s = k • M, $k = s^{-1}$

Step 3: Second-order reaction:

Rate = $k[A]^2$, so M/s = $k \bullet M^2$, $k = M^{-1}s^{-1}$

WP 12.3

Step 1: Determine the order of the reaction with respect to NO_2.

In reactions 1 and 2, the concentration of NO_2 is doubled, while the concentration of F_2 is held constant. Doubling [NO_2] causes the rate to double, so the reaction is first order in NO_2.

Step 2: Determine the order of the reaction with respect to H_2.

In reactions 2 and 3, the concentration of NO_2 is held constant while the concentration of F_2 is varied. Doubling [F_2] causes the rate of the reaction to quadruple, so the reaction is second-order in F_2.

Step 3: Write the equation for the rate law.

Rate = $k[NO_2][F_2]^2$

Step 4: Calculate the value of k.

We can calculate the value of k, using the data from any one of the experiments. Using the data in experiment 1 gives:

$$0.026\frac{M}{s} = k(0.10\ M)(0.10\ M)^2 \qquad k = \frac{0.026\frac{M}{s}}{1.0\times10^{-3}M^3} = 26.0\frac{1}{M^2\cdot s}$$

WP 12.4

Step 1: Solve for k:

$$\ln\frac{0.03\ M}{0.10\ M} = -k(73\,\text{min})$$

$$k = 0.0165\ \text{min}^{-1}$$

Step 2: Determine the concentration of the reactant at 80% completion.

0.10 M x 0.80 = 0.08 M

0.10 M – 0.08 M = 0.02 M

Step 3: Solve for t at 80% completion.

$$\ln\frac{0.02\text{ M}}{0.10\text{ M}} = -0.0165\text{ min}^{-1}\text{x t}$$

$$t = 97.5\text{ min}$$

WP 12.5

Strategy: Substitute into the half-life equation.

$$t_{1/2} = \frac{0.693}{0.0165\text{ min}^{-1}} = 42\text{ min}$$

WP 12.6

Step 1: Solve for *k*.

$$5730\text{ years} = \frac{\ln 2}{k}$$

$$k = 1.21\text{x}10^{-4}\text{ year}^{-1}$$

Step 2: Substitute into the integrated rate law and solve for *t* after 23% decay.

$$\ln(1.0\text{-}0.23) = -1.21\text{ x }10^{-4}/\text{year x } t$$

$$t = 2160\text{ years old}$$

WP 12.7

Step 1: Solve for *k.*

$$\frac{1}{[A]_t} = kt + \frac{1}{[A]_0}$$

$$\frac{1}{0.0495\text{ M}} = k\text{ x }400\text{ s} + \frac{1}{0.1000\text{ M}}$$

$$k = 0.0255\text{ M}^{-1}\text{s}^{-1}$$

Step 2: Use the equation for half-life to determine the time required for 50% completion.

$$t_{1/2} = \frac{1}{0.0255\text{ M}^{-1}\text{s}^{-1}\text{ x }0.100\text{ M}} = 392\text{ s}$$

Step 3: Determine the concentration of NO_2 at 95% completion (5% of NO_2 remaining).

$$0.100\text{ M x }0.05 = 0.00500\text{ M}$$

Step 4: Substitute the concentrations into the integrated rate law, and solve for *t*.

$$\frac{1}{0.00500\text{ M}} = 0.0255\text{ M}^{-1}\text{s}^{-1}\text{ x t} + \frac{1}{0.100\text{ M}}$$

$$t = 7450\text{ s x }\frac{1\text{ hour}}{3600\text{ s}} = 2.07\text{ hours, or 2 hours 4 minutes}$$

WP 12.8

Step 1: Solve for *k*.

$$\text{rate} = k = 0.026\ \text{M} \cdot \text{s}^{-1}$$

Step 2: Substitute into the integrated rate law and solve for [A] after 20 seconds.

$$[\text{A}] = -0.026\ \text{M} \cdot \text{s}^{-1}\ \text{x}\ 20\ \text{s} + 1.0\ \text{M} = 0.48\ \text{M}$$

WP 12.9

Step 1: Write the expected rate law.

$$\text{Rate} = k[\text{A}]^2$$

Step 2: Sum for the overall reaction.

$2\ A + A_2 + B \rightarrow A_2 + A_2B$, cancels to

$2\ A + B \rightarrow A_2B$

WP 12.10

If a reaction has a rate constant of 0.0123 s^{-1} at 700 K and 0.00342 s^{-1} at 250 K, what is the activation energy of the reaction, and what will the rate constant be at 500 K?

Strategy: Solve for E_a using the two-point Arrhenius equation, then use the same equation to solve for the rate constant at 500 K.

Step 1: Solve for E_a using the two-point Arrhenius equation.

$$\ln\left(\frac{0.00342\ \text{s}^{-1}}{0.0123\ \text{s}^{-1}}\right) = \left(\frac{-E_a}{8.314\ \text{J/mol} \cdot \text{K}}\right)\left(\frac{1}{250\ \text{K}} - \frac{1}{700\ \text{K}}\right)$$

$E_a = 4138$ J/mol

Step 2: Solve for *k* at 500 K:

$$\ln\left(\frac{k_2}{0.0123\ \text{s}^{-1}}\right) = \left(\frac{-4138\ \text{J/mol}}{8.314\ \text{J/mol} \cdot \text{K}}\right)\left(\frac{1}{500\ \text{K}} - \frac{1}{700\ \text{K}}\right)$$

$$\ln\left(\frac{k_2}{0.0123\ \text{s}^{-1}}\right) = -0.284$$

$$\left(\frac{k_2}{0.0123\ \text{s}^{-1}}\right) = e^{-0.284}$$

$$k_2 = 0.00926\ \text{s}^{-1}$$

Chapter 13

WP 13.1

Step 1: Write the equation for the equilibrium constant.

$$K_c = \frac{[H][A]}{[HA]}$$

Step 2: Determine the concentration of each species.

If the solution is 1.37% ionized, the concentrations of H and A will be 0.0137 x 0.250 M = 3.43×10^{-3} M.

The concentration of HA will be 0.247 M.

Step 3: Solve for K_c.

$$K_c = \frac{[3.43\text{x}10^{-3}][3.43\text{x}10^{-3}]}{[0.247]} = 4.77\text{x}10^{-5}$$

WP 13.2

Strategy: Write the equilibrium equation for K_c.

$$K_c = \frac{[PCl_3][Cl_2]}{[PCl_5]}$$

Step 1: Solve for K_c, using the molar concentrations given.

$$K_c = \frac{(1.56)(1.56)}{(2.35)} = 1.04$$

Step 2: Determine Δn.

$\Delta n = 2 - 1 = 1$

Step 3 Solve for K_p, using the equation that expresses the relationship between K_c and K_p.

$$K_p = (1.04)\left[\left(\frac{0.0821\ \text{L}\cdot\text{atm}}{\text{mol}\cdot\text{K}}\right)(250\ \text{K})\right]^1 = 21.3$$

WP 13.3

Step 1: Write the balanced equation for the reaction.

$C(s) + H_2O(g) \leftrightarrows CO(g) + H_2(g)$

Step 2: Make a table listing the initial concentration, the change in concentration, and the equilibrium concentration. (Let x = the amount of substance that reacts.)

Principal Reaction	C (*s*)	+	H_2O (*g*)	$\leftrightarrows$	CO (*g*)	+	H_2 (*g*)
Initial Concentration (M)			11.75 M		0		0
Change (M)			$-x$		$+x$		$+x$
Equilibrium Conc. (M)			$11.75 - x$		$+x$		$+x$

Step 3: From the balanced equation, write the equilibrium equation. Substitute the algebraic expressions for the equilibrium concentrations into the equilibrium equation, and solve for *x*. Use the quadratic equation if necessary.

$$K_c = 3.0 \times 10^{-2} = \frac{[CO][H_2]}{[H_2O]} = \frac{(x)(x)}{(11.75 - x)}$$

$$3.0 \times 10^{-2}(11.75 - x) = (x)^2$$

$$x^2 + (3.0 \times 10^{-2})x - 0.353 = 0$$

$$x = \frac{-(3.0 \times 10^{-2})^2 \pm \sqrt{(3.0 \times 10^{-2}) - 4(-0.353)}}{2}$$

$x = 0.579$ or $x = -0.609$. The meaningful solution is $x = 0.579$. Substituting the value of *x* back into the equilibrium concentrations, we have

$$[CO] = [H_2] = 0.579 \text{ M}$$

$$[H_2O] = 11.75 - 0.579 = 11.17 \text{ M}$$

WP 13.4

Step 1: Determine the direction of reaction needed to minimize each stress.

BrNO added:	reaction will shift away from BrNO to the right toward products
BrNO removed:	reaction will shift toward BrNO to the left toward reactants
NO added:	reaction will shift away from NO, to the left toward reactants
Br_2 removed:	reaction will shift toward Br_2, to the right toward products.

WP 13.5

Step 1: Determine the direction of reaction needed to minimize each stress.

pressure increased:	reaction will shift toward N_2O_4, to the left toward reactants.
pressure decreased:	reaction will shift away from N_2O_4, to the right toward products.
Ne added:	equilibrium will be unaffected.

WP 13.6

Step 1: Determine the direction of reaction needed to minimize each stress, and its effect on $[CO_2]$.

Increased pressure:	increased $[CO_2]$ (3 moles of gas on left, 2 moles on right)
Decreased pressure:	decreased $[CO_2]$
Removed catalyst:	decreased $[CO_2]$
Lowered temperature:	increased $[CO_2]$
Added O_2:	increased $[CO_2]$

Chapter 14

WP 14.1

Strategy: Write the reactions with water, in which the acid protonates the water, and the base deprotonates the water.

Step 1: Write the reactions in the forward direction.

$$HNO_2 + H_2O\ (l) \leftrightarrows NO_2^- + H_3O^+\ (aq)$$

$$SO_4^{-2}\ (aq) + H_2O\ (l) \leftrightarrows HSO_4^- + OH^-\ (aq)$$

Step 2: Identify the conjugate acid and conjugate base.

Reaction 1: acid = HNO_2, conjugate base = NO_2^-

Reaction 2: base = SO_4^{-2}, conjugate acid = HSO_4^-

WP 14.2

Strategy: Determine which species is the stronger acid and which is the stronger base, then transfer protons accordingly.

Step 1: Write the reactions, transferring protons from the stronger acid to the stronger base.

$$HCl\ (aq) + CO_3^{2-}\ (aq) \leftrightarrows Cl^-\ (aq) + HCO_3^-\ (aq)$$

$$CN^-\ (aq) + H_3PO_4\ (aq) \leftrightarrows HCN\ (aq) + H_2PO_4^-\ (aq)$$

WP 14.3

Strategy: Use the water dissociation constant to determine the concentrations. A basic solution will have $[OH^-] > [H_3O^+]$, and an acid solution will have $[H_3O^+] > [OH^-]$.

Step 1: Use K_w to calculate the ion concentrations.

$$1\times10^{-14} = [OH^-](2.35\times10^{-5})$$

$$[OH^-] = \frac{1\times10^{-14}}{2.35\times10^{-5}} = 4.26\times10^{-10}$$

This solution is acidic: the concentration of hydronium ions is much larger than the concentration of hydroxide ions.

$$1\times10^{-14} = [H_3O^+](4.26\times10^{-10})$$

$$[H_3O^+] = \frac{1\times10^{-14}}{4.26\times10^{-10}} = 2.35\times10^{-5}$$

This solution is also acidic.

WP 14.4

Strategy: Calculate the $[H_3O^+]$, determine pH.

Step 1: Calculate the $[H_3O^+]$, using the expression for the dissociation of water.

$$1 \times 10^{-14} = [H_3O^+](2.35 \times 10^{-6})$$

$$[H_3O^+] = \frac{1 \times 10^{-14}}{2.35 \times 10^{-6}} = 4.25 \times 10^{-9}$$

Step 2: Calculate the pH.

$$pH = -\log(4.25 \text{ x } 10^{-9}) = 8.37$$

Step 3: Determine if the solution is acidic, basic, or neutral.

The pH is higher than 7, so the solution is basic.

WP 14.5

Step 1: Determine the concentration of $[OH^-]$.

KOH is a strong base and exists in aqueous solution as K^+ and OH^-. Therefore, $[OH^-]$ = initial concentration of KOH.

$[OH^-] = 7.2 \times 10^{-3}$ M

Step 2: Calculate the $[H_3O^+]$.

The H_3O^+ concentration is calculated from the OH^- concentration.

$$[H_3O^+] = \frac{K_w}{[OH^-]} = \frac{1.0 \times 10^{-14}}{7.2 \times 10^{-3}} = 1.4 \times 10^{-12} \text{ M}$$

Step 3: Calculate the pH.

$$pH = -\log(1.4 \text{ x } 10^{-12}) = 11.85$$

WP 14.6

Step 1: Determine the species present initially.

$HC_3H_5O_3$ and H_2O

Step 2: Write the proton-transfer reaction.

$HC_3H_5O_3\,(aq) + H_2O\,(l) \leftrightarrows H_3O^+\,(aq) + C_3H_5O_3^-\,(aq)$ $\qquad K_a = 1.4 \times 10^{-4}$

Step 3: Make a table showing the principal reaction and the initial and equilibrium concentrations.

Principal Reaction	$HC_3H_5O_3\,(aq)$	$\leftrightarrows$	$H_3O^+\,(aq)$	+	$C_3H_5O_3^-\,(aq)$
Initial Concentration	1.63		0		0
Change	$-x$		$+x$		$+x$
Eq. Concentration	$1.63 - x$		$+x$		$+x$

Step 4: Substitute the equilibrium concentrations into the equilibrium equation, and solve for x. (Since $K_a < 1.0 \times 10^{-3}$, assume x is negligible and that $1.63 - x \cong 1.63$)

$$K_a = \frac{[H_3O^+][C_3H_5O_3^-]}{[HC_3H_5O_3]} = \frac{(x)(x)}{(1.63 - x)}$$

$$1.4 \times 10^{-4} = \frac{x^2}{1.63} \qquad x = 1.51 \times 10^{-2}$$

Step 5: Calculate the equilibrium concentrations.

$[H_3O^+]_e = [C_3H_5O_3^-]_e = 1.51 \times 10^{-2}$ M

$[HC_3H_5O_3]_e = 1.63 - (1.51 \times 10^{-2}) = 1.61$ M

Step 6: Calculate the pH.

$pH = -\log(1.51 \times 10^{-2}) = 1.82$

WP 14.7

Strategy: To determine the percent dissociation we need to first determine the concentration of the dissociated HA. This requires a determination of the equilibrium concentration of H_3O^+ and $C_6H_5COO^-$.

Step 1: Determine the species present initially, and write the proton-transfer reaction.

The species present initially are C_6H_5COOH and H_2O (acid or base).

The proton-transfer reaction is

$C_6H_5COOH\ (aq) + H_2O\ (l) \leftrightharpoons H_3O^+\ (aq) + C_6H_5COO^-\ (aq)$ $\qquad K_a = 6.5 \times 10^{-5}$

Step 2: Make a table showing the principal reaction, initial concentration, change in concentration. and the equilibrium concentration.

Principal Reaction	$C_6H_5COOH\ (aq)$	$\leftrightharpoons$	H_3O^+	+	$C_6H_5COO^-$
Initial Concentration	0.79		0		0
Change	$-x$		$+x$		$+x$
Equilibrium Concentration	$0.79 - x$		$+x$		$+x$

Step 3: Substitute the equilibrium concentrations into the equilibrium expression, and solve for x.

$$K_a = 6.5 \times 10^{-5} = \frac{[H_3O^+][C_6H_5COO^-]}{[C_6H_5COOH]} = \frac{(x)(x)}{(0.79 - x)}$$

Given the value of K_a, assume that x is negligible compared with the initial concentration of the acid; therefore $0.79 - x \approx 0.79$. Using this value in the denominator, solve for x.

$x^2 \approx (6.5 \times 10^{-5})(0.79)$

$x \approx 7.2 \times 10^{-3}$

Step 4: Calculate the equilibrium concentrations of the major species present.

The equilibrium concentrations of H_3O^+ and $C_6H_5COO^-$ are $[H_3O^+] = [C_6H_5COO^-] = 0.0072$ M. This concentration also represents the amount of C_6H_5COOH that dissociated.

Step 5: Calculate the percent dissociation.

The percent dissociation can now be calculated.

$$\frac{0.0072}{0.79} \times 100 = 0.91\%$$

WP 14.8

Step 1: Write the stepwise dissociation for arsenic acid.

The stepwise dissociation for arsenic acid is

$H_3AsO_4\,(aq) + H_2O\,(l) \leftrightarrows H_3O^+\,(aq) + H_2AsO_4^-\,(aq)$
$H_2AsO_4^-\,(aq) + H_2O\,(l) \leftrightarrows H_3O^+\,(aq) + HAsO_4^{2-}\,(aq)$
$HAsO_4^{2-}\,(aq) + H_2O\,(l) \leftrightarrows H_3O^+\,(aq) + AsO_4^{3-}\,(aq)$

Step 2: Determine the principal reaction.

Since $K_{a1} > K_{a2}$, K_{a3}, and K_w, we know that the principal reaction is the first dissociation step and that all of the H_3O^+ present is produced from this step. Therefore, we only need to consider the first dissociation step when calculating the pH of the solution.

Step 3: Make a table showing the principle reaction, initial concentration, change in concentration, and equilibrium concentration of the reactant and products.

Principal Reaction	$H_3AsO_4\,(aq)$	$\leftrightarrows$	H_3O^+	+	$H_2AsO_4^-$
Initial Concentration	1.69×10^{-3}		0		0
Change	$-x$		$+x$		$+x$
Equilibrium Concentration	$(1.69 \times 10^{-3}) - x$		$+x$		$+x$

Step 4: Substitute the equilibrium concentrations into the equilibrium expression.

Because K_a is so close to the concentration of the acid, it is unlikely that x will be negligible. If it is assumed to be negligible, the result is

$$5.62 \times 10^{-3} = \frac{x^2}{1.69 \times 10^{-3}} \qquad x = 3.08 \times 10^{-3}$$

The quadratic equation must be used to solve for the equilibrium concentration of H_3O^+. Using $(1.69 \times 10^{-4}) - x$ and rearranging the equilibrium expression gives

$$5.62 \times 10^{-3}[(1.69 \times 10^{-3}) - x] = x^2 \qquad x^2 + (5.62 \times 10^{-3})x - (9.50 \times 10^{-6})$$

$$x = \frac{-(5.62 \times 10^{-3}) \pm \sqrt{(5.62 \times 10^{-3})^2 - 4(-9.50 \times 10^{-6})}}{2}$$

$x = 2.72 \times 10^{-3}$

Step 5: Calculate the pH of the solution.

$$pH = -\log (2.72 \times 10^{-3}) = 2.56$$

WP 14.9

Strategy: Solve using a table of equilibrium values.

Step 1: Write the principal reaction.

$K_b > K_w$; therefore, the principal reaction is

$$(CH_3)_3N\ (aq)\ +\ H_2O\ (l) \leftrightarrows (CH_3)_3NH^+\ (aq)\ +\ OH^-\ (aq)$$

Step 2: Construct a table with concentrations of the reactant and products.

Principal Reaction	$(CH_3)_3N\ (aq)$	$\leftrightarrows$	$(CH_3)_3N^+$	+	OH^-
Initial Concentration	0.975		0		0
Change	$-x$		$+x$		$+x$
Equilibrium Concentration	$0.975 - x$		$+x$		$+x$

Step 3: Substitute the equilibrium concentrations into the equilibrium expression, and solve for x.

$$K_a = 6.5 \times 10^{-5} = \frac{[(CH_3)_3NH^+][OH^-]}{[(CH_3)_3N]} = \frac{(x)(x)}{0.975 - x}$$

Assume that x is negligible compared with the initial concentration of the base; therefore, $0.975 - x \approx x$. Using this value in the denominator, we can now solve for x. Remember: x represents the equilibrium concentration of the OH^- ion.

$$6.5 \times 10^{-5} = \frac{x^2}{0.975} \qquad x = 0.0080$$

Step 4: Determine the equilibrium concentrations of the species present.

The equilibrium concentrations are $[(CH_3)_3N] = 0.975 - 0.0080 = 0.967$ M; $[OH^-] = [(CH_3)_3NH^+] = 0.0080$ M.

Step 5: Calculate the pH of the solution.

Knowing the equilibrium concentration of OH^-, we can calculate the H_3O^+ concentration from the K_w expression.

$$[H_3O^+] = \frac{1.0 \times 10^{-14}}{0.0080} = 1.3 \times 10^{-12}$$

$$pH = -\log (1.3 \times 10^{-12}) = 11.90$$

WP 14.10

Strategy: Determine if KOCl produces an acidic, basic, or neutral aqueous solution, write the hydrolysis reaction for this salt, and determine pH.

K^+ is an inert cation and will not react. However, OCl^- is the conjugate base of a weak acid and will react with water.

Step 1: Write the hydrolysis reaction for this salt.

$$OCl^- (aq) + H_2O (l) \leftrightarrows HOCl (aq) + OH^- (aq)$$

Step 2: Write an equilibrium expression for this reaction.

$$K_b = \frac{[HOCl][OH^-]}{[OCl^-]} \qquad K_b = \frac{1.0\times10^{-14}}{K_a(HOCl)} = \frac{1.0\times10^{-14}}{3.5\times10^{-8}} = 2.9\times10^{-7}$$

Step 3: Construct a table.

Principal Reaction	OCl^- (*aq*)	$\leftrightarrows$	HOCl (*aq*)	+	OH^-
Initial Concentration	0.137		0		0
Change	$-x$		$+x$		$+x$
Equilibrium Concentration	$0.137 - x$		$+x$		$+x$

Step 4: Substitute the equilibrium concentrations into the equilibrium expression.

Assume that x is negligible; therefore, $0.137 - x \approx 0.137$.

$$2.9\times10^{-7} = \frac{(x)(x)}{0.137} \qquad x = 2.0\times10^{-4}$$

Step 5: Calculate the pH of the solution.

The equilibrium concentration of OH^- is 2.0×10^{-4}. Determine the H_3O^+ concentration to calculate pH.

$$[H_3O^+] = \frac{1.0\times10^{-14}}{2.0\times10^{-4}} = 5.0\times10^{-11}$$

$pH = -\log(5.0 \times 10^{-11}) = 10.30$

Chapter 15

WP 15.1

Strategy: Write the net reaction, determine K_n.

Step 1: Write the net neutralization reaction by writing individual reactions for the acid, base, and water.

$HNO_2 (aq) + H_2O (l) \leftrightarrows NO_2^- (aq) + H_3O^+ (aq)$ $\qquad K_a$=4.5 x 10^{-4}

$H_3O^+ (aq) + OH^- (aq) \rightarrow H_2O (aq)$ $\qquad 1/K_w$=1.0 x 10^{14}

$HNO_2 (aq) + OH^- (aq) \rightarrow NO_2^- (aq) + H_2O (aq)$

Step 2: Calculate K_n based on the equilibrium constants for the individual reactions in step 1.

$K_n = K_a \times 1/K_w = (4.5 \times 10^{-4}) \times (1.0 \times 10^{14}) = 4.5 \times 10^{10}$

Step 3: Determine the position of equilibrium from the value of K_n.

Based on the value of K_n, the equilibrium will lie far to the right. The reaction will go to completion, leaving the conjugate base of the weak acid in solution.

Step 4: Predict the pH.

The solution will be basic: the pH will be greater than 7.

WP 15.2

Strategy: Write the reaction, determine the concentration of H_3O^+, $H_2PO_3^-$.

Step 1: Write the overall reaction.

$H_3PO_3\ (aq) + H_2O\ (l) \leftrightarrows H_2PO_3^-\ (aq) + H_3O^+\ (aq)$

Step 2: Make a chart of initial and equilibrium concentrations.

Principal Reaction	H_3PO_3 ⇆	$H_3O^+\ (aq)$	$H_2PO_3^-$
Initial Conc.	0.250	0	0.175
Change	$-x$	$+x$	$+x$
Equilibrium Conc.	$0.250 - x$	$+x$	$0.175 + x$

Step 3: Solve for the equilibrium concentrations of H_3O^+, $H_2PO_2^-$.

$$1.0\text{x}10^{-2} = \frac{x(0.175 + x)}{0.250 - x}$$

$$x^2 + (0.175x + 0.010x) - 0.0025 = 0$$

$$x^2 + 0.185x - 0.0025 = 0$$

$$x = \frac{-0.185 \pm \sqrt{(0.185)^2 + 4(0.0025)}}{2}$$

$x = 0.0126$

$[H_3O^+] = 0.0126$ M

$[H_2PO_3^-] = 0.0126 + 0.175 = 0.188$ M

Step 4: Determine the pH of the solution.

$-\log(0.0126) = 1.90$

Step 5: Determine the % dissociation of the acid.

$$\% \text{ dissociation} = \frac{0.126}{0.250} \text{x} 100 = 50.4\%$$

WP 15.3

Step 1: Determine the principal reaction and equilibrium concentrations.

The principal reaction and equilibrium concentration for this solution are

Principal Reaction	$HCOOH\ (aq) + H_2O\ (l) \leftrightarrows$	$H_3O^+\ (aq)$ +	$HCO_2^-\ (aq)$
Initial Conc.	0.35	0	0.25
Change	$-x$	$+x$	$+x$
Equilibrium Conc.	$0.35 - x$	$+x$	$0.25 + x$

Step 2: Using the equilibrium equation, solve for $[H_3O^+]$, assuming that the change (x) is negligible.

$$K_a = \frac{[H_3O^+][HCO_2^-]}{[HCOOH]}$$

$$[H_3O^+] = K_a \frac{[HCOOH]}{[HCO_2^-]}$$

K_a for formic acid equals 1.8×10^{-4}. Substituting the given information into the equilibrium expression gives

$$[H_3O^+] = (1.8 \times 10^{-4})\frac{(0.35)}{(0.25)} = 2.5 \times 10^{-4}$$

Step 3: Solve for the pH.

$$\text{pH} = -\log (2.5 \text{ x } 10^{-4}) = 3.60$$

Step 4: Write the neutralization reaction that occurs on addition of HCl, and set up a table showing the number of moles present before and after the addition of HCl.

Neutralization Reaction	$HCO_2^-\ (aq)$ +	$H_3O^+\ (l)$	$\leftrightarrows$	$HCOOH\ (aq)$ +	$H_2O\ (l)$
Before Reaction (mol)	0.35	0.10		0.25	
Change (mol)	–0.10	–0.10		+0.10	
After Reaction (mol)	0.25	0		0.35	

Step 5: Determine the concentrations of the buffer components after neutralization occurs.

Assuming that the volume of the solution does not change, the concentrations of the buffer components after neutralization are

$$[HCO_2^-] = \frac{0.25 \text{ mol}}{1.0 \text{ L}} = 0.25 \text{ M}$$

$$[HCOOH] = \frac{0.35 \text{ mol}}{1.0 \text{ L}} = 0.35 \text{ M}$$

Step 6: Substitute the concentrations of the buffer components into the equilibrium expression and calculate the pH.

$$[H_3O^+] = (1.8 \times 10^{-4})\frac{0.35}{0.25} = 2.5 \times 10^{-4}$$

$$pH = -\log(2.5 \times 10^{-4}) = 3.60$$

WP 15.4

Strategy: Use the Henderson-Hasselbalch equation to calculate the [base]/[acid] ratio. This information provides the mole ratio of acetate to acetic acid.

Step 1: Calculate the pK_a of acetic acid.

$$-\log(1.8 \text{ x } 10^{-5}) = 4.74$$

Step 2: Use the Henderson-Hasselbalch equation to calculate the [base]/[acid] ratio.

$$\log\frac{[\text{base}]}{[\text{acid}]} = pH - pK_a$$

$$\log\frac{[\text{base}]}{[\text{acid}]} = 5.75 - 4.74 = 1.01$$

$$\frac{[\text{base}]}{[\text{acid}]} = 10^{1.01} = 10.2$$

Step 3: Determine the mass of sodium acetate needed.

10.2 x 0.01 mol = 0.102 mol sodium acetate

$$0.102 \text{ mol x } \frac{82.03 \text{ g}}{\text{mol}} = 8.37\text{g}$$

WP 15.5

Strategy: Using the Henderson-Hasselbalch equation, calculate the log of the [base]/[acid] ratio, then determine the pH at which this value is obtained.

$$\log\frac{25}{75} = pH - 8.8$$

$$-0.477 = pH - 8.8$$

$$pH = 8.3$$

WP 15.6

Strategy: Determine the number of moles of H_3O^+ and OH^- present in the mixture and determine the pH based on the excess.

Step 1: Calculate the moles of OH^- and the moles of H_3O^+.

Moles OH^- = 0.05 L x 0.1 mol/L = 0.005 mol OH^-

Moles H_3O^+ = 0.015 x 1 mol/L = 0.015 mol H_3O^+

Step 2: Determine the excess quantity and the resultant $[H_3O^+]$.

H_3O^+ is in excess: 0.015 – 0.005 = 0.010 mol H_3O^+

$[H_3O^+]$ = 0.010 mol/0.065 L = 0.154 M HCl

Step 3: Determine the pH of the solution.

pH = -log (0.154) = 0.813

WP 15.7

Strategy: Determine the volume of NaOH required to provide equimolar amounts of base and acid, determine the concentration of conjugate base and pH.

Step 1: Calculate the moles of phenol present and thus the moles of OH^- needed.

0.100 L x 0.050 mol/L = 0.0050 moles

Step 2: Determine the volume of sodium hydroxide solution required.

0.0050 moles/0.1 mol/L= 0.050 L = 50 mL

Step 3: Determine the total volume of the resultant solution.

100 mL original volume + 50 mL NaOH solution = 150 mL

Step 4: Solve for the concentration of salt.

0.0050 mol/0.150 L = 0.033 M

Step 6: Solve for K_b and [OH].

$$K_b = \frac{K_w}{K_a} = \frac{1.0\text{x}10^{-14}}{1.3\text{x}10^{-10}} = 7.7\text{x}10^{-5}$$

$$7.7\text{x}10^{-5} = \frac{x^2}{(0.033 - x)} \quad \text{assume } 0.033 - x \approx 0.033$$

x = 0.00159 M (assumption was correct)

Step 7: Solve for $[H_3O\ ]$ and pH.

$$[H_3O^+] = \frac{1\text{x}10^{-14}}{0.00159} = 6.3 \text{ x } 10^{-12}$$

$$\text{pH} = -\log 6.3 \text{ x } 10^{-12} = 11.20$$

WP 15.8

a) mL of acid required to reach the equivalence point:

Step 1: Write the neutralization reaction that occurs, and determine the number of mmoles of $(CH_3)_3N$ in the initial solution.

$$(CH_3)_3N\ (aq) + HCl\ (aq) \leftrightarrows (CH_3)_3NH^+\ (aq) + Cl^-\ (aq)$$

$$\text{mol}\ (CH_3)_3N = 0.040\ \text{L} \times 0.375 \frac{\text{mol}\ (CH_3)_3N}{\text{L}\ (CH_3)_3N} = 0.015\ \text{mol}\ (CH_3)_3N$$

Step 2: Determine the mol of HCl needed to react with the $(CH_3)_3N$ and the volume of 0.100 M HCl needed.

From the neutralization reaction, we know that the mol ratio of $(CH_3)_3N$:HCl is 1:1. Therefore,

mol HCl =0.015

$$\text{mL HCl} = \frac{0.015\ \text{mol HCl}}{0.100\ \text{mol HCl/mL HCl}} = 0.150\ \text{L} = 150\ \text{mL HCl}$$

b) Initial pH of the $(CH_3)_3N$ solution:

Step 1: Determine the initial pH of the solution.

Principal Reaction	$(CH_3)_3N\ (aq)$	+	$H_2O\ (l)$	$\leftrightarrows$	$(CH_3)_3NH^+\ (aq)$	+	$OH^-\ (aq)$
Initial Conc.	0.375 M				0		0
Change	$-x$				$+x$		$+x$
Equilibrium Conc.	$0.375\ \text{M} - x$				$+x$		$+x$

$$K_b = \frac{[(CH_3)_3NH^+][OH^-]}{[(CH_3)N]}$$

$$6.5 \times 10^{-5} = \frac{(x)(x)}{(0.375 - x)} \cong \frac{x^2}{0.375} \qquad x = [OH^-] = 4.93 \times 10^{-3}$$

$$[H_3O^+] = \frac{K_w}{[OH^-]} = \frac{1.0 \times 10^{-14}}{4.93 \times 10^{-3}} = 2.02 \times 10^{-12}; \qquad \text{pH} = -\log(2.02 \times 10^{-12}) = 11.7$$

c) pH of the solution after addition of 10.00 mL of acid:

Step 1: Calculate the concentration of $(CH_3)_3N$ and its conjugate acid after the addition of 10.00 mL of HCl.

mol $(CH_3)_3N$ = 0.015 mol – (0.100 M)(0.010 L) = 0.014 mmol

The total volume present = 40.00 mL $(CH_3)_3N$ + 10.00 mL HCl added.

$$[(CH_3)_3N] = \frac{0.014\ \text{mol}\ (CH_3)_3N}{0.050\ \text{L soln}} = 0.280\ \text{M}$$

The conjugate acid of $(CH_3)_3N$ is $(CH_3)_3NH^+$.

mol $(CH_3)_3NH^+$ = mol HCl used = (0.100 M)(0.010 L) = 0.0010 mol

$$[(CH_3)_3NH^+] = \frac{0.0010 \text{ mol } (CH_3)_3NH^+}{0.050 \text{ L}} = 0.020 \text{ M}$$

Step 3: Determine the K_a and pK_a of the conjugate acid of $(CH_3)_3N$.

$$K_a(CH_3)_3NH^+ = \frac{K_w}{K_b(CH_3)_3N} = \frac{1.0\times10^{-14}}{6.5\times10^{-5}} = 1.5\times10^{-10}$$

$pK_a = -\log(1.5 \times 10^{-10}) = 9.81$

Step 4: Using the Henderson-Hasselbalch equation, calculate the pH of the solution.

$$pH = pK_a + \log\frac{[\text{base}]}{[\text{acid}]} = 9.81 + \log\frac{0.280}{0.020} = 11.0$$

d) pH halfway to the equivalence point:

Step 1: Use the Henderson-Hasselbalch equation to calculate the pH. Halfway to the equivalence point, [base] = [acid].

$pH = pK_a = 9.81.$

e) pH at the equivalence point:

Step 1: Calculate the concentration of the conjugate acid, using the initial moles of base, and the total volume of solution.

At the equivalence point, mol of $(CH_3)_3NH^+$ = mol of $(CH_3)_3N$ reacted (= amount of $(CH_3)_3N$ present initially).

mol $(CH_3)_3NH^+$ = 0.015

Total volume = 40.00 mL $(CH_3)_3N$ + 150 mL HCl to reach eq point = 190 mL

$$[(CH_3)_3NH^+] = \frac{0.015 \text{ mol}}{0.190 \text{ L}} = 0.079 \text{ M}$$

Step 2: Determine the pH of the solution.

Principal Reaction	$(CH_3)_3NH^+$ (*aq*)	+	H_2O (*l*)	⇆	$(CH_3)_3N$ (*aq*)	+	H_3O^+ (*aq*)
Initial Conc.	0.079				0		0
Change	$-x$				$+x$		$+x$
Equilibrium Conc.	$0.079 - x$				$+x$		$+x$

$K_a(CH_3)_3NH^+ = 1.5 \times 10^{-10}$

$$K_a = \frac{[H_3O^+][(CH_3)_3N]}{[(CH_3)_3NH^+]} \qquad 1.5\times10^{-5} = \frac{(x)(x)}{0.079 - x} \cong \frac{x^2}{0.079}$$

$x = [H_3O^+] = 1.09 \times 10^{-3}$ $\qquad$ $pH = -\log(1.09 \times 10^{-3}) = 2.96$

f) pH after the addition of 200.00 mL of 0.100 M HCl:

Step 1: Determine the number of mmoles of excess HCl.

The first 150 mL titrated the base, the remaining 50 mL are excess.

Moles HCl in excess = 0.050 L x 0.100 mol/L = 0.005 mol

Step 2: Determine the concentration of excess HCl, and calculate the pH.

$$[HCl]_{xs} = \frac{0.005 \text{ mol}}{0.240 \text{ L}} = 0.021 \text{ M} \qquad pH = -\log(0.021) = 1.68$$

WP 15.9

Strategy: Determine how far this amount of sodium hydroxide takes the titration: use the Henderson-Hasselbalch equation to determine pH. Calculate isoelectric point from pK_{a1} and pK_{a2}.

Step 1: Determine how many moles of proline and how many moles of sodium hydroxide are present.

Moles proline = 0.100 L x 0.050 mol/L = 0.0050 mol proline

$$\text{Moles NaOH} = \frac{0.025 \text{ g}}{40 \text{ g/mol}} = 0.000625 \text{ mol}$$

Step 2: Use the Henderson-Hasselbalch equation to determine the pH.

$$pH = pK_a + \log\frac{[\text{base}]}{[\text{acid}]} = 1.952 + \log\frac{0.000625\text{mol}/0.100\text{L}}{(0.005 - 0.000625 \text{ mol})/0.100\text{L}} = 1.1$$

Step 3: Calculate the isoelectric point.

$$\text{isoelectric point} = \frac{pK_{a1} + pK_{a2}}{2} = 6.30$$

WP 15.10

Step 1: Write the solubility equilibrium expression for $CdCl_2$.

The solubility equilibrium for $CdCl_2$ is

$CdCl_2\ (s) \leftrightarrows Cd^{2+}\ (aq) + 2\ Cl^-\ (aq)$

and the K_{sp} expression for this reaction is

$$K_{sp} = [Cd^{2+}][Cl^-]^2$$

Step 2: Determine the concentration of Cd^{2+} and Cl^- and calculate the K_{sp}.

If $[Cd^{2+}] = 1.10 \times 10^{-5}$ then the $[Cl^-]$ is 2 times that or 2.20×10^{-5}. Substituting these values into the equilibrium expression gives

$$K_{sp} = (1.10 \times 10^{-5})(2.20 \times 10^{-5})^2 = 5.32 \times 10^{-15}$$

WP 15.11

Strategy: Write the solubility equilibrium for $Cu(OH)_2$ and the equilibrium expression. Using the method described in Chapter 13, calculate the molar solubility of $Fe(OH)_3$.

The solubility equilibrium for $Cu(OH)_2$ is

$Cu(OH)_2.\ (s) \leftrightarrows Cu^{2+}\ (aq) + 2\ OH^-\ (aq)$

The equilibrium expression is

$K_{sp} = [Cu^{2+}][OH^-]^2$

Step 1: Let x be the number of mol/L of $Cu(OH)_2$ that dissolves. The saturated solution then contains x mol/L of Cu^{2+} and $2x$ mol/L of OH^-. Solve for K_{sp}.

Substitute this into the equilibrium expression:

$$K_{sp} = 2.18 \times 10^{-20} = [Cu^{2+}]\,[OH^-]^2 = (x)(2x)^2$$

Solve for x:

$2.18 \times 10^{-20} = 4x^3; \quad x = 1.75 \times 10^{-7}$ M

WP 15.12

Step 1: Write the solubility equation and solubility expression.

$Sn(OH)_2\ (s) \leftrightarrows Sn^{2+}\ (aq) + 2\ OH^-\ (aq)$

The equilibrium expression is

$K_{sp} = [Sn^{2+}][OH^-]^2$

Step 2: Using the pH, determine the $[OH^-]$ concentration.

$[H_3O^+]$ = antilog (–9.45) = 3.55×10^{-10}

$$[OH^-] = \frac{1 \times 10^{-14}}{3.55 \times 10^{-10}} = 2.82 \times 10^{-5}$$

Step 3: Let x be the number of mol/L of $Sn(OH)_2$ that dissolves. Construct a table showing the equilibrium concentrations.

Solubility Equilibrium	$Sn(OH)_2\ (s)$	$\leftrightarrows$	$Sn^{2+}\ (aq)$	+	$OH^-\ (aq)$
Initial Concentration			0		2.82×10^{-5}
Equilibrium Concentration			$+x$		$2.82 \times 10^{-5} + x$

Step 4: Calculate the solubility of $Sn(OH)_2$.

Given the very small value of K_{sp}, we can assume that x is negligible. Substituting the equilibrium concentrations into the K_{sp} expression gives

$5.4 \times 10^{-27} = (x)(2.82 \times 10^{-5}) \quad x = 1.92 \times 10^{-22}$

WP 15.13

Strategy: Write a metathesis reaction for the reaction between $Zn(NO_3)_2$ and Na_2CO_3, and apply the solubility rules in Chapter 4 to determine if a precipitate will form.

$Zn(NO_3)_2\ (aq) + Na_2CO_3\ (aq) \rightarrow 2\ NaNO_3\ (aq) + ZnCO_3\ (s)$

Step 1: Write the solubility equation and IP expression for the precipitate.

$ZnCO_3\ (s) \rightarrow Zn^{2+}\ (aq) + CO_3^{2-}\ (aq) \qquad K_{sp} = 1.2 \times 10^{-10}$

The ion product expression is

$IP = [Zn^{2+}][CO_3^{2-}]$

Step 2: Calculate the IP and compare its value to the K_{sp}. Will a precipitate form?

Calculate the IP from the concentrations of Zn^{2+} and CO_3^{2-}, remembering that two solutions were mixed for a total volume of 350 mL.

$$[Zn^{2+}] = \frac{(0.75\ M)(100\ mL)}{350\ mL} = 0.21\ M$$

$$[CO_3^{2-}] = \frac{(1.50\ M)(250\ mL)}{350\ mL} = 1.07\ M$$

$IP = (0.21)(1.07) = 0.23$

$IP > K_{sp}$; Therefore, a precipitate will form.

Chapter 16

WP 16.1

Strategy: Determine the number of moles of hydrogen, calculate ΔS.

Step 1: Calculate the moles of hydrogen.

$$3\ g\ H_2 \times \frac{1\ mol}{2.0\ g} = 1.5\ mol$$

Step 2: Calculate ΔS.

$$\Delta S = nR \ln\frac{V_{final}}{V_{initial}} = 1.5\ mol \times 8.314\ J/K \ln\frac{20}{5} = 17\ J/K$$

WP 16.2

Strategy: Write the formula for the reaction, calculate ΔS.

Step 1: Write the formula.

$$2H_2\,(g) + O_2\,(g) \rightarrow 2H_2O\,(g)$$

Step 2: Calculate ΔS.

$$2(188.7) - [2(130.6) + 205.0] = -88.8 \text{ J/mol•K}$$

WP 16.3

Step 1: Write the net ionic equation for this reaction.

$$Ag^{+}\,(aq) + Br^{-}\,(aq) \rightarrow AgBr\,(s)$$

Step 2: Calculate ΔS_{system}.

$$\Delta S_{system} = [S°(AgBr] - [S°(Ag^{+}) + S°(Br^{-})]$$

$$\Delta S_{system} = [107] - [72.7 + 82.4] = -48.1 \text{ J/K}$$

Step 3: Calculate ΔH_{rxn}

$$\Delta H^{\circ}_{rxn} = [\Delta H^{\circ}_{f}(AgBr)] - [\Delta H^{\circ}_{f}(Ag^{+}) + \Delta H^{\circ}_{f}(Br^{-})]$$

$$\Delta H^{\circ}_{rxn} = [-100.4] - [105.6 + (-121.5)] = -84.5 \text{ kJ}$$

Step 4: Calculate ΔS_{total}.

$$\Delta S_{total} = -48.1 \text{ J/K} - \frac{-84{,}500 \text{ J}}{298 \text{ K}} = 235 \text{ J/K}$$

The reaction is spontaneous.

WP 16.4

Strategy: Solve for ΔS and ΔH, solve for the temperature at which $\Delta G = 0$.

Step 1: Calculate ΔS.

$$\Delta S = 238 - 127 = 111 \text{ J/mol•K}$$

Step 2: Calculate ΔH.

$$\Delta H = -201.2 - -238.7 = 37.5 \text{ kJ/mol}$$

Step 3: Calculate T.

$$T = \frac{\Delta H^{\circ}}{\Delta S^{\circ}} = \frac{37{,}500 \text{ J/mol}}{111 \text{ J/mol•K}} = 338 \text{ K} = 65 \text{ °C}$$

WP 16.5

Strategy: From the information given, you can determine $\Delta G°$ from the equation $\Delta G° = \Delta H° - T\Delta S°$.

Step 1: Determine ΔH°_{rxn}.

$$\Delta H^\circ_{rxn} = [(-127.1) + (-447.5)] - [(-407.3) + (-101.8)] = -65.5 \text{ kJ/mol}$$

Step 2: Determine ΔS°_{rxn}.

$$\Delta S^\circ_{rxn} = (96.2 + 205.4) - (219.1 + 115.5) = -33.0 \text{ J/mol} \bullet \text{K}$$

Step 3: Calculate ΔG°_{rxn}.

$$\Delta G° = \Delta H° - T\Delta S° = -65{,}500 \text{ J/mol} - 298\text{K}(-33 \text{ J/mol}\bullet\text{K}) = -55.7 \text{ kJ/mol}$$

WP 16.6

Strategy: Determine ΔG°_{rxn}, then calculate $\Delta H°$ and $\Delta S°$ to determine the temperature at which the reaction becomes spontaneous.

Step 1: Based on the value of ΔG°_{rxn}, determine if the reaction is spontaneous.

$$\Delta G^\circ_{rxn} = (-65.3 + 0) - [(-50.8) + (2\text{x}0)] = -14.5 \text{ kJ/mol}$$

Step 2: Determine the temperature at which the reaction will become spontaneous.

Calculate $\Delta H°$ and $\Delta S°$ for this reaction.

$$\Delta H° = [(-135.4) + (0)] - [(-74.8) + (2 \text{ x } 0)] = -60.6 \text{ kJ/mol}$$

$$\Delta S° = [(216.4 + (2 \text{ x } 130.6)] - [(186.2) + (2 \text{ x } 223)] = -154.6 \text{ J/mol} \bullet \text{K}$$

$$T = \frac{\Delta H°}{\Delta S°} = \frac{-60{,}600 \text{ kJ/mol}}{-154.6 \text{ J/mol} \bullet \text{K}} = 392 \text{ K} = 119 \text{ °C}$$

Step 3: Determine the spontaneity at the temperatures listed.

In step 1, it was determined that the reaction was spontaneous at 25 °C. In step 2, it was determined that the reaction became spontaneous at 119 °C. Therefore, the reaction is spontaneous below 119 °C, and non-spontaneous above 119 °C.

400 °C = non-spontaneous

45 °C = spontaneous

-20 °C = spontaneous

WP 16.7

Consider the following reaction:

$$2\ NO\ (g) + O_2\ (g) \leftrightarrows 2NO_2\ (g)$$

Determine the spontaneity under standard conditions, and again under the following conditions: $P_{NO} = 0.150$ atm, $P_{O_2} = 0.250$ atm, $P_{NO_2} = 0.001$ atm. Under which conditions is the reaction more spontaneous?

Strategy: Determine ΔG°_{rxn}, then calculate ΔG_{rxn} and determine which reaction is more spontaneous.

	ΔG° (kJ/mol)
$NO\ (g)$	86.6
$O_2\ (g)$	0
$NO_2\ (g)$	51.3

Step 1: Calculate ΔG°_{rxn}, determine if the reaction is spontaneous.

$$\Delta G^\circ_{rxn} = 2(51.3) - 2(86.6) = -70.6\ \text{kJ/mol}$$

Step 2: Calculate ΔG.

$$\Delta G = -70{,}600\frac{\text{J}}{\text{mol}} + 8.314\frac{\text{J}}{\text{mol}\cdot\text{K}}(298\ \text{K})\ln\frac{(0.001)^2}{(0.150)^2(0.250)} = -91{,}993\frac{\text{J}}{\text{mol}} = -92.0\frac{\text{kJ}}{\text{mol}}$$

The reaction is more spontaneous under the non-standard conditions.

WP 16.8

Consider the following reaction:

$$I_2\ (g) + Cl_2\ (g) \leftrightarrows 2ICl\ (g)$$

The equilibrium constant, K_p, for this reaction is 81.9 at 25 °C. What is the standard free-energy change for this reaction?

Strategy: Use the equation $\Delta G^\circ = -RT \ln K$

Step 1: Solve for the free energy.

$$\Delta G^\circ = -RT\ \ln K = -8.314\frac{\text{J}}{\text{mol}\bullet\text{K}} \times 298\ \text{K} \times \ln 81.9 = -10{,}900\ \text{J/mol} = -10.9\ \text{kJ/mol}$$

Chapter 17

WP 17.1

Describe the construction of a galvanic cell based on the following reaction:

$$Pb^{2+}\ (aq) + Cd\ (s) \rightarrow Pb\ (s) + Cd^{2+}\ (aq)$$

Strategy: Break the overall reaction into half-reactions and design a galvanic cell, identifying the cathode and the anode.

Step 1: Write the half-reactions, identifying the oxidation and reduction reactions.

Pb^{2+} (*aq*) + 2 e^- → Pb (*s*) reduction - cathode

Cd (*s*) → Cd^{2+} (*aq*) + 2 e^- oxidation - anode

Step 2: Identify solutions for each half-cell, and a suitable inert electrolyte for the salt bridge.

Cathode solution: $Pb(NO_3)_2$
Anode solution: $Cd(NO_3)_2$

Salt bridge electrolyte: $NaNO_3$

Any soluble salt containing the metal ions is suitable for the half-cell solutions, and a common ion in the salt bridge will prevent any side reactions.

WP 17.2

Step 1: Write the half-reaction occurring at the anode. Remember, the reactant in each half-cell is written first, followed by the product.

Oxidation (loss of electrons) occurs at the anode. The half-reaction is

Mg (*s*) → Mg^{2+} (*aq*) + 2 e^-

Step 2: Write the half-reaction occurring at the cathode.

Reduction (gain of electrons) occurs at the cathode. The half-reaction is

Cr^{3+} (*aq*) + 3 e^- → Cr (*s*)

Step 3: Combine the two half-reactions to obtain the overall cell reaction (recall that the equation must be balanced in electrons).

For the number of electrons lost to equal the number of electrons gained, we must multiply the anode half-reaction by three and the cathode half–reaction by two. The two half-reactions would then be

3 Mg (*s*) → 3 Mg^{2+} (*aq*) + 6 e^-

2 Cr^{3+} (*aq*) + 6 e^- → 2 Cr (*s*)

Adding the two half-reactions together gives the overall equation

3 Mg (*s*) + 2 Cr^{3+} (*aq*) → 3 Mg^{2+} (*aq*) + 2 Cr (*s*)

Step 5: Describe the cell indicated in the shorthand notation.

This cell would consist of a strip of magnesium as the anode dipping into an aqueous solution containing the Mg^{2+} ion, such as $Mg(NO_3)_2$. The cathode would consist of a strip of chromium dipping into an aqueous solution of Cr^{3+}, such as $Cr(NO_3)_3$.

WP 17.3

Step 1: Write the balanced equation for the reaction.

$Mg\ (s) + Zn^{2+}\ (aq) \rightarrow Mg^{2+}\ (aq) + Zn\ (s)$

Step 2: Identify n and use $\Delta G° = -nFE°$ to calculate $\Delta G°$.

In this reaction, 2 moles of electrons are transferred per mole of reactants.

$$\Delta G° = -2 \text{ x } 96{,}500\ \frac{C}{mol} \text{ x } 1.61\ V \text{ x } 1 \frac{J}{C \bullet V} = -311\ kJ/mol$$

WP 17.4

Strategy: Determine the position of the reducing agent relative to the position of the oxidizing agent, and based on your results, determine the spontaneity of the reactions.

In the first reaction, the reducing agent (the species undergoing oxidation) is Cu. Cu lies above the oxidizing agent in the table of standard reduction potentials. In the second reaction, the reducing agent is F^-, which lies above the oxidizing agent, MnO_4^-.

In a spontaneous reaction, the reducing agent must lie below the oxidizing agent in the table of standard reduction potentials. For the first reaction, the reducing agent, Cu, lies below the oxidizing agent, Ag^+. Therefore, the first reaction is spontaneous. For the second reaction, the reducing agent, F^-, lies above the oxidizing agent, MnO_4^-. Therefore, the second reaction is non-spontaneous.

WP 17.5

Strategy: Calculate the E°_{cell} for the cell reaction, substitute the information given into the Nernst equation, and solve for $[Sn^{4+}]$.

Step 1: Calculate $E°_{cell}$.

$$E^o_{cell} = E^o_{Sn^{2+} \rightarrow Sn^{4+}} + E^o_{Cu^{2+} \rightarrow Cu} = -0.15\ V + 0.34\ V = 0.19\ V$$

Step 2: Using $E°_{cell}$ and E, solve for $[Sn^{4+}]$.

$$E_{cell} = E° - \frac{0.0592}{n} \log \frac{[Sn^{4+}]}{[Sn^{2+}][Cu^{2+}]}$$

$$0.17 = 0.19 - \frac{0.0592}{2} \log \frac{[Sn^{4+}]}{[0.63][0.72]}$$

$$0.02 \text{ x } \frac{2}{0.0592} = \log \frac{[Sn^{4+}]}{[0.45]}$$

$$10^{0.676} = \frac{[Sn^{4+}]}{[0.45]}$$

$$[Sn^{4+}] = 2.13\ M$$

WP 17.6

Strategy: Determine the half-reactions occurring and calculate E_{ref}. Substitute into the modified Nernst equation to determine pH of the solution ($Cu^{2+}(aq) + 2e^- \rightarrow Cu(s)$ $E° = 0.34$).

Step 1: Determine E_{ref}.

$$E_{cell} = E_{H_2 \rightarrow 2H^+} + E_{Cu^{2+} \rightarrow Cu}$$

$$0.62\ V = E_{H_2 \rightarrow 2H^+} + 0.34\ V$$

$$E_{H_2 \rightarrow 2H^+} = 0.28\ V = E_{cell} - E_{ref}$$

Step 3: Calculate the pH of the anode solution.

$$pH = \frac{0.28}{0.0592} = 4.73$$

WP 17.7

Find *K* for the reaction:

$$H_2O_2\,(l) + 2\,H^+\,(aq) + 2\,Br^-\,(aq) \rightarrow 2\,H_2O\,(l) + Br_2\,(l)$$

Strategy: Determine the half-reactions occurring and calculate $E°$ from the standard potentials of the half-cell reactions. Calculate *K* from the equation $\log K = E° \frac{n}{0.0592}$.

Step 1: Determine $E°$.

$H_2O_2\,(l) + 2\,H^+\,(aq) + 2\,e^- \rightarrow 2\,H_2O\,(l)$ $\quad E° = 1.78\ V$

$2\,Br^-\,(aq) \rightarrow Br_2\,(l) + 2\,e^-$ $\quad E° = -1.09\ V$

$E° = -1.09\ V + 1.78\ V = 0.69\ V$

Step 2: Calculate *K* from $E°$.

$$\log K = E° \frac{n}{0.0592} = 0.69 \frac{2}{0.0592} = 23.3$$

$$K = 10^{23.3} = 1.99 \times 10^{23}$$

WP 17.8:

Strategy: Determine the possible half-reactions and their standard voltages to predict the reactions at the cathode and anode, then determine the overall reaction.

Step 1: Determine the possible reactions at the cathode, and their standard voltages.

$Li^+\,(aq) + e^- \rightarrow Li\,(s)$ $\quad E° = -3.04\ V$ *or*

$2\,H_2O\,(l) + 2\,e^- \rightarrow H_2\,(g) + 2\,OH^-\,(aq)$ $\quad E° = -0.83\ V$

Step 2: Determine the possible reactions at the anode, and their standard voltages.

$2\ SO_4^{2-}\ (aq) \rightarrow S_2O_8^{2-}\ (aq)\ +\ 2e^-$ $\qquad E° = -2.01\ V$ *or*

$2\ H_2O \rightarrow O_2\ (g)\ + 4\ H^+\ (aq)\ + 4\ e^-$ $\qquad E° = -1.23\ V$

Step 3: Predict which reactions will occur and calculate the overall reaction.

In both cases, the hydrolysis of water requires far less energy than the reduction or oxidation of the other species, so the overall reaction will be.

$2\ H_2O\ (l) \rightarrow O_2\ (g)\ +\ 2\ H_2\ (g)$

WP 17.9

Strategy: Write the electrolysis reaction and consider the conversion process to calculate time.

Step 1: Treat the electrons as reactants in the chemical equation, and solve for the time required to produce 12 kg of magnesium. Remember the conversion factor of 96,500 C (or A·s) per mole of electrons.

$Mg^{2+}\ +\ 2e^- \rightarrow Mg\ (s)$

$2\ Cl^- \rightarrow Cl_2\ (g) + 2e^-$

$$12\text{ kg Mg} \times \frac{1000\text{ g}}{1\text{ kg}} \times \frac{1\text{ mol Mg}}{24.3\text{ g Mg}} \times \frac{2\text{ mol e}^-}{1\text{ mol Mg}} \frac{96,500\text{A}\cdot\text{s}}{1\text{ mol e}^-} \times \frac{1}{15\text{ A}} \times \frac{1\text{ h}}{3600\text{ s}} \times \frac{1\text{ day}}{24\text{ h}} = 73.54\text{ days}$$

Step 2: Use the information calculated above to determine the amount of Cl_2 gas.

$$15\ \frac{\text{C}}{\text{s}} \times 73.54\text{ days} \times \frac{24\text{ hours}}{\text{day}} \text{X} \frac{60\text{ min}}{\text{hour}} \text{X} \frac{60\text{ s}}{1\text{ min}} \times \frac{1\text{ mol e}^-}{96,500\text{ C}} \times \frac{1\text{ mol Cl}_2}{2\text{ mol e}^-} \times \frac{22.4\text{ L Cl}_2}{\text{mol}} = 11,062\text{ L Cl}_2$$

Putting it Together:

Write the 2 half-reactions and solve for $E° = E_{anode} + E_{cathode}$.

$O_2\ (g)\ +\ 4\ H^+\ (aq)\ +\ 4\ e^- \rightarrow 2\ H_2O\ (l)$ $\qquad E° = 1.23\ V$

$2\ Br^-\ (aq) \rightarrow Br_2\ (l)\ +\ 2\ e^-$ $\qquad E° = -1.09\ V$

$E° = 1.23\ V\ +\ (-1.09\ V) = 0.14$

The reaction is spontaneous since $E°$ is positive.

To determine the value of E after addition of the buffer, determine $[H^+]$ and substitute this value into the Nernst equation.

$K_a\ (HCHO_2) = 1.7 \times 10^{-4}$; $\qquad pK_a = 3.74$

$$\text{pH} = \text{pK}_a + \log\frac{[\text{CHO}_2^-]}{[\text{HCHO}_2]} = 3.74 + \log\frac{0.35}{0.20} = 3.98$$

$$[\text{H}^+] = \text{antilog}(-3.98) = 1.05\times10^{-4}$$

Substitute into the Nernst equation.

$$E = E^\circ - \frac{0.0592}{n}\log\frac{1}{[\text{Br}^-][\text{H}^+]}$$

After balancing the number of electrons in the two half-reactions, $n = 4$.

$$\text{E} = 0.14\ \text{V} - \frac{0.0592}{4}\log\frac{1}{(1.05\times10^{-4})} = 0.0811\ \text{V}$$

The reaction is still spontaneous under these non-standard state conditions.

Chapter 18

WP 18.1

Strategy: Calculate the molecular masses of the four molecules, calculate the percent differences, and describe the kinetic isotopic effects.

Step 1: Calculate the molecular weights for the four molecules.

C_6H_6 = (6 x 12.01) + (6 x 1.01) = 78.12 g/mol

C_6D_6 = (6 x 12.01) + (6 x 2.02) = 84.18 g/mol

H_2O = 16.00 + (2 x 1.01) = 18.02

D_2O = 16.00 + (2 x 2.02) = 20.04

Step 2: Calculate the percent difference.

$$\text{benzene \% difference} = \frac{84.18\ -\ 78.12}{84.18}\text{x}100 = 7.20\%$$

$$\text{water \% difference} = \frac{20.04\ -\ 18.02}{20.04}\text{x}100 = 10.1\%$$

Step 3: Predict the expected properties.

The increase in melting point from H_2O to D_2O is less than 4 °C, and the increase in boiling point is even smaller. The increases in these properties for the deuterated benzene would be expected to be smaller still.

Given the increased strength of O-D bonds compared to O-H bonds, it can be expected that C-D bonds will also be stronger than C-H bonds, so more energy would be released by burning C_6D_6.

WP 18.2

Strategy: Set up a ratio to calculate the volume, choose an appropriate unit for the answer.

$$\frac{0.083 \text{ L } D_2O}{2400 \text{ L } H_2O} = \frac{x \text{ L } D_2O}{25 \text{ L}}$$

$2400x \text{ L} = 2.075 \text{ L}^2$

$x = 0.0008645 \text{ L} = 0.8645 \text{ mL or } 864.5 \ \mu\text{L}$

WP 18.3

Write a balanced equation for the production of synthesis gas from ethane – C_2H_6. If each step in the process has a 60% yield, how many L of hydrogen gas at 25 °C and 1 atm can be produced from 15 kg of ethane?

Strategy: Write and balance the equation, determine the quantities of reactants and products at each step. Use the ideal gas law to calculate the volume of hydrogen gas.

Step 1: Write and balance the equation for the formation of synthesis gas.

$$2\ H_2O\ (g) + C_2H_6\ (g) \rightarrow 2\ CO\ (g) + 5\ H_2\ (g)$$

Step 2: Determine the moles of reactants entering the first step, and the moles of products produced.

$$15{,}000 \text{ g ethane x } \frac{1 \text{ mol}}{30.08 \text{ g}} = 500 \text{ moles ethane}$$

$$500 \text{ mol ethane x } \frac{2 \text{ mol CO}}{\text{mol ethane}} = 1000 \text{ mol CO}$$

$$500 \text{ mol ethane x } \frac{5 \text{ mol } H_2}{\text{mol ethane}} = 2500 \text{ mol } H_2$$

Step 3: Multiply the moles of products by 60%, and calculate the products of the second step.

$1000 \text{ mol CO x } 0.60 = 600 \text{ mol CO}$

$2500 \text{ mol } H_2 \text{ x } 0.60 = 1500 \text{ mol } H_2$

Only the CO continues on by the following reaction:

$$CO\ (g) + H_2O\ (g) \rightarrow CO_2\ (g) + H_2\ (g)$$

$$600 \text{ mol CO x } \frac{1 \text{ mol } H_2}{\text{mol CO}} = 600 \text{ mol } H_2$$

Step 4: Multiply the moles of products by 60%, and calculate the products of the third step.

$600 \text{ mol } H_2 \text{ x } 0.60 = 360 \text{ mol}$

The third step simply removes the CO_2 into solution.

Step 5: Multiply the moles of products by 60%, and determine the volume of the H_2 gas.

360 mol H_2 x 0.60 = 216 mol

Total moles of gas = 1500 mol from step 1 + 216 mol from step 3 = 1716 mol H_2 gas.

$$V = \frac{nRT}{P} = \frac{1716 \text{ mol x } 0.08206 \frac{L \bullet atm}{mol \bullet K} x 298K}{1 \text{ atm}} = 42{,}000 \text{ L}$$

WP 18.4

Strategy: Determine the initial volume, moles of each gas, and how much is left over (assume that the volume of the water produced is negligible). How many kJ are released in this reaction?

Step 1: Determine the moles of each gas, and the limiting reagent.

$$15 \text{ g } H_2 \text{ x} \frac{1 \text{ mol}}{2.02 \text{ g}} = 7.43 \text{ mol } H_2$$

$$100 \text{ g } O_2 \text{ x} \frac{1 \text{ mol}}{32.0 \text{ g}} = 3.13 \text{ mol } O_2$$

Oxygen is the limiting reagent.

Step 2: Determine the moles of excess reagent remaining after the reaction.

2 mol of hydrogen react for every 1 mol of oxygen, so 6.25 mol of hydrogen will react.

This leaves 0.62 mol of hydrogen unreacted.

Step 3: Determine the volume change of the reaction.

Initial volume = 7.43 mol + 3.13 mol = 7.74 mol x 22.4 L/mol = 173 L

Final volume = 0.62 mol x 22.4 L/mol = 13.9 L

ΔV = final – initial = 13.9 L – 173 L = -159.1 L

Step 4: Determine the kJ released by the reaction.

1 mol O_2 reacting releases 572 kJ

3.13 mol O_2 reacting releases 1790 kJ

WP 18.5

Strategy: Write and balance the equation, determine the moles of sodium hydride, use the balanced equation to determine the moles of gas, and the standard molar volume to determine liters.

Step 1: Write and balance the equation.

$NaH\ (s) + H_2O\ (l) \rightarrow H_2\ (g) + NaOH^-\ (aq)$

Step 2: Determine the moles of sodium hydride and the moles of hydrogen evolved.

$$3.65\text{ g x }\frac{1\text{ mol NaH}}{24.01\text{ g}}\text{ x }\frac{1\text{ mol H}_2}{\text{mol NaH}} = 0.152\text{ mol H}_2$$

Step 3: Determine the volume of hydrogen evolved.

$0.152\text{ mol H}_2 \text{ x } 22.4\text{ L/mol} = 3.4\text{ L H}_2\ (g)$

WP 18.6

Strategy: Write and balance the equation for combustion of glucose, determine the energy released in kilojoules, convert to calories, then kcal (dietary calories, C, recall, are kilocalories).

Step 1: Write and balance the equation.

$$6\text{ O}_2 + \text{C}_6\text{H}_{12}\text{O}_6 \rightarrow 6\text{ CO}_2 + 6\text{ H}_2\text{O}$$

This is simply the equation for photosynthesis turned around.

Step 2: Use the heats of formation to calculate the heat evolved per mole.

$\Delta H°_{\text{rxn}} = \text{products} - \text{reactants}$

$\Delta H°_{\text{rxn}} = [(6 \text{ x } -393.5\text{ kJ/mol}) + (6 \text{ x } -285.8\text{ kJ/mol})] - [(6 \text{ x } 0) + (1 \text{ x } -1260\text{ kJ/mol})]$
$\Delta H°_{\text{rxn}} = -4076\text{ kJ/mol} - (-1260\text{ kJ/mol}) = -2816\text{ kJ/mol}$

Step 3: Convert the quantity given to moles, convert this to energy in kcal.

$$39\text{ g x }\frac{1\text{ mol}}{180\text{ g}}\text{ x }\frac{2816\text{ kJ}}{\text{mol}}\text{ x }\frac{1000\text{ J}}{\text{kJ}}\text{ x }\frac{1\text{ cal}}{4.184\text{ J}}\text{ x }\frac{1\text{ kcal}}{1000\text{ cal}} = 146\text{ kcal}$$

The negative sign on the heat per mol is dropped here, as it is understood that the energy is being transferred into a metabolic system. A can of soda contains about 150 Calories, meaning that essentially all of the energy provided comes from sugar.

WP 18.7

Strategy: Write and balance the equation for decomposition of sodium chlorate, determine the number of moles of salt present and the possible moles of oxygen generated. Use the ideal gas law to determine the number of L of oxygen.

Step 1: Write and balance the equation.

$$2\text{ NaClO}_3\ (s) \rightarrow 2\text{ NaCl}\ (s) + 3\text{ O}_2\ (g)$$

Step 2: Determine moles of salt present and moles of oxygen generated.

$$15\text{ g NaClO}_3\text{ x }\frac{1\text{ mol NaClO}_3}{106.5\text{ g}}\text{ x }\frac{3\text{ mol O}_2}{2\text{ mol NaClO}_3} = 0.211\text{ mol O}_2$$

Step 3: Use the ideal gas law to calculate the volume of gas at 250 °C and 25 °C.

$$V = \frac{nRT}{P} = \frac{0.211\text{ mol x }0.08206\ \frac{\text{L}\bullet\text{atm}}{\text{mol}\bullet\text{K}}\text{ x}523\text{K}}{1\text{ atm}} = 9.06\text{ L}$$

$$V = \frac{nRT}{P} = \frac{0.211 \text{ x } 0.08206 \frac{L \bullet atm}{mol \bullet K} \text{ x} 298K}{1 \text{ atm}} = 5.17 \text{ L}$$

Workbook Problem 18.8

Write and balance equations for the following reactions:

Dissolution of Rb_2O in water

Dissolution of PbO in acid and base (to form $Pb(OH)_4^{2-}$)

Dissolution of BeO in acid and base

Reaction of N_2O_3 with water

Strategy: Determine the nature of the oxide and how it will react with water or the given solution, then write and balance the equations.

Step 1: Determine the nature of each oxide.

Rb_2O – basic oxide. Alkali metal
BeO – amphoteric oxide. Topmost of the amphoteric oxides
PbO – amphoteric oxide. Center of periodic table

N_2O_3 – acidic oxide. Covalent, with high oxidation states.

Step 2: Write and balance the equations.

$Rb_2O\ (s) + H_2O\ (l) \rightarrow 2\ Rb^+\ (aq) + 2\ OH^-\ (aq)$

$BeO\ (s) + 2\ H^+\ (aq) \rightarrow Be^{2+}\ (aq) + H_2O\ (l)$

$2\ BeO\ (s) + 2\ OH^-\ (aq) + H_2O\ (l) \rightarrow 2\ [Be(OH)_4]^{2-}\ (aq)$

$PbO\ (s) + 2\ H^+\ (aq) \rightarrow Pb^{2+}\ (aq) + H_2O\ (l)$

$PbO\ (s) + 2\ OH^-\ (aq) + H_2O\ (l) \rightarrow Pb(OH)_4^{2-}(aq)$

$N_2O_3\ (g) + H_2O\ (l) \rightarrow 2HNO_2\ (aq)$

WP 18.9

Strategy: Determine the oxidation state of the oxygen in the molecules: -2 is an oxide, -1 is a peroxide, and -1/2 is a superoxide.

Step 1: Determine the oxidation number of the oxygens in each oxide.

MgO_2 - oxidation state is - 1
KO_2 - oxidation state is - 1/2
Li_2O - oxidation state is -2

Step 2: Identify the type of oxide.

MgO_2 - magnesium peroxide
KO_2 - potassium superoxide
Li_2O - lithium oxide

Step 3: Write the reaction with water of each molecule.

$$MgO_2 + H_2O\ (l) \rightarrow Mg^{2+}\ (aq) + HO_2^-\ (aq) + OH^-\ (aq)$$
$$2\ KO_2 + H_2O\ (l) \rightarrow O_2\ (g) + 2\ K^+\ (aq) + HO_2^-\ (aq) + OH^-\ (aq)$$
$$Li_2O + H_2O\ (l) \rightarrow 2\ Li^+(aq) + 2\ OH^-\ (aq)$$

WP 18.10
Strategy: Determine the moles of copper (II) sulfate, the moles of water, and the ratio between them.

Step 1: Determine the moles of copper (II) sulfate.

$$3.76\ g \times \frac{1\ mol}{160\ g} = 0.0235\ mol\ anhydrous\ CuSO_4$$

Step 2: Determine the moles of water.

The grams of absorbed water is the difference between the initial and final weights.

$$(4.36\ g\ - 3.76\ g) \times \frac{1\ mol}{18\ g} = 0.0333\ mol\ water$$

Step 3: Determine the ratio of water to copper sulfate and compare to the ratio (5:1) for the fully hydrated compound.

$$\frac{0.0333\ mol\ H_2O}{0.0235\ mol\ CuSO_4} = 1.42$$

As the fully hydrated compound has 5 waters per formula unit, this is only partially hydrated. If anhydrous compound is needed, it can be heated to drive off the excess water.

Chapter 19

WP 19.1
Strategy: Examine the position of the elements on the periodic table.

Step 1: Predict the metallic character of the elements based on their position on the periodic table.

C or **Si** : Silicon is below carbon in the periodic table, so has more metallic character

Be or **Li**: Lithium is to the left of beryllium, so has more metallic character

Se or Br: Selenium is to the left of bromine, so has more metallic character

As or I: Astatine is above iodine, but iodine is to the right of astatine. This ambiguity is resolved by the fact that iodine is a nonmetal, while astatine is a semi-metal: astatine is more metallic than iodine.

WP 19.2
Strategy: Use the rules for assigning oxidation numbers to determine the oxidation state of carbon.

Step 1: Determine the oxidation state of carbon.

CO	Oxidation state for O is -2, neutral molecule, so the oxidation state for C is +2
CO_2	Oxidation state for O is -2, neutral molecule, so the oxidation state for C is +4

C_2O_3	Oxidation state for O is -2, neutral molecule, so the oxidation state for C is +3
CO_3^{2-}	Oxidation state for O is -2, charge on molecule is -2, oxidation state for C is +4
CaC_2	Oxidation state for Ca is +2, neutral molecule, so the oxidation state for C is -1

WP 19.3

Strategy: From the information provided, determine the overall silicate structure, remembering that oxygens are shared in the tetrahedral units. Balance the charge on the silicate using the ions in a correct ratio.

Step 1: Determine the silicate structure.

As the SiO_4^{4-} subunits join together, they share oxygens.

$$SiO_4^{4-} + SiO_4^{4-} = Si_2O_7^{6-}$$

$$Si_2O_7^{6-} + SiO_4^{4-} = Si_3O_{10}^{8-}$$

Step 2: Determine how many of each cation is needed to balance the charge on the silicate unit.

To balance a charge of -8 with cations having charges of +3 and +2 will require 2 of the +3, and one of the +2. There is no other way to use both and get to +8

Step 3: Write the formula unit:

$CaAl_2\ Si_3O_{10}$

WP 19.4

Step 1: Write the balanced chemical equation.

$$NH_4NO_3\,(s) \rightarrow N_2O\,(g) + 2\,H_2O\,(l)$$

Step 2: Calculate the pressure of N_2O, using the total pressure and the vapor pressure of water.

$$P_T = P_{H_2O} + P_{N_2O}$$

$$P_{N_2O} = 705\text{ mm Hg} - 24\text{ mm Hg} = 681\text{ mm Hg}$$

Step 3: Calculate the number of moles of N_2O produced from 3.5 g NH_4NO_3.

$$3.5\text{ g }NH_4NO_3 \times \frac{1\text{ mol }NH_4NO_3}{80.0\text{ g }NH_4NO_3} \times \frac{1\text{ mol }N_2O}{1\text{ mol }NH_4NO_3} = 4.38\times10^{-2}\text{ mol }N_2O$$

Step 4: Calculate the volume of N_2O, using the ideal gas law.

$$V = \frac{nRT}{P} = \frac{\left(4.38\times10^{-2}\right)\left(0.0821\frac{L\cdot atm}{mol\cdot K}\right)(295.15\text{ K})}{681\text{ mm Hg}\times\frac{1\text{ atm}}{760\text{ mm Hg}}} = 1.18\text{ L}$$

Chapter 20

WP 20.1

Strategy: Write the expected electron configuration for the neutral element, determine the charge, and remove electrons accordingly.

Step 1: Write the electron configuration of the neutral atom.

Ti: $[Ar]3d^2 4s^2$

Mn: $[Ar]3d^5 4s^2$

Ni: $[Ar]3d^8 4s^2$

Step 2: Determine the oxidation state of the metal.

Ti: 0 – no electrons need be removed, answer complete

Mn: each oxygen is -2, one negative charge remains, so oxidation state is +7

Ni: 2+

Step 3: Remove the necessary electrons.

Ti: $[Ar]3d^2 4s^2$ remove 0 electrons, $[Ar]3d^2 4s^2$

Mn: $[Ar]3d^5 4s^2$ remove 7 electrons, [Ar]

Ni: $[Ar]3d^8 4s^2$ remove 2 electrons, $[Ar]3d^8$

WP 20.2

Determine the oxidation state and coordination number of the metal in the following complexes:

$AgCl_2^-$ $[MnCl_4]^{-2}$ $[Fe(CN)_6]^{4-}$ $[Co(NH_3)_4Br_2]Br$

Strategy: With the charge and ligands, it is possible to determine the oxidation state of the metals. The coordination number is the number of ligand attachments.

Step 1: Determine the oxidation state of the metals.

$Ag + 2(-1) = -1; \; Ag = +1$

$Mn + 4(-1) = -2; \; Mn = +2$

$Fe + 6(-1) = -4; \; Fe = +2$

$Co + 4(0) + 2(-1) = +1$ (the complex has a +1 charge, balanced by a Br^- ion); $Co = +3$

Step 2: Determine the coordination numbers.

$AgCl_2^-$ Coordination number is 2

$[MnCl_4]^{-2}$ Coordination number is 4

$[Fe(CN)_6]^{4-}$ Coordination number is 6

$[Co(NH_3)_4Br_2]Br$ Coordination number is 6

WP 20.3

Strategy: Using the formula $E = hc/\lambda$, determine the energy of the absorbed photons, then convert to kJ/mol.

Step 1: Determine the energy of the photons absorbed.

$$E = 6.626\text{x}10^{-34}\ \text{J}\cdot\text{s x}\ \frac{3.00\ \text{x}\ 10^{8}\ \text{m/s}}{550\ \text{x}\ 10^{-9}\ \text{m}} = 3.61\ \text{x}\ 10^{-19}\ \text{J}$$

Step 2: Convert to kJ/mol.

$$3.61\ \text{x}\ 10^{-19}\ \text{J x}\ \frac{1\ \text{kJ}}{1000\ \text{J}}\ \text{x}\ 6.022\ \text{x}\ 10^{23}\ \text{mol}^{-1} = 218\ \text{kJ/mol}$$

WP 20.4

$[Co(Cl)_6]^{3-}$ is a high-spin complex. Draw an electron diagram that supports this.

Strategy: Draw the orbital diagram and determine how to maximize unpaired electrons.

Step 1: Determine the electron configuration of the metal alone.

The free Co^{3+} ion has the valence $[Ar]3d^6$.

Co^{3+} : ⇅ ↑ ↑ ↑ ↑ (3*d*) ___ (4*s*) ___ ___ ___ (4*p*) $3d^6$

The d^2sp^3 configuration has fewer unpaired electrons, and is also lower energy overall. Both support the choice of this configuration.

Step 2: Identify the possible hybridization schemes.

This is an octahedral complex, so will use either d^2sp^3 or sp^3d^2 hybrid orbitals. There are 4 valence electrons from the cobalt ion, and 2 from each chloride ion.

Step 3: Draw electron diagrams with both hybridizations.

$[Co(Cl)_6]^{3-}$: ↑ ↑ ⇅ ⇅ ⇅ (3*d*) ⇅ (4*s*) ⇅ ⇅ ⇅ (4*p*) d^2sp^3

or

$[Co(Cl)_6]^{3-}$: ↑ ↑ ↑ ↑ ___ (3*d*) ⇅ (4*s*) ⇅ ⇅ ⇅ (4*p*) ⇅ ⇅ ___ ___ ___ (4*d*) sp^3d^2

Step 4: Choose the hybridization that maximizes electron spin.

sp^3d^2 has the greater number of unpaired electrons.

Chapter 21

WP 21.1
Strategy: Write and balance equations for purification of the metal from the mineral ore.

Step 1: Write the equation for roasting of covellite.
$2\ CuS\ (s) + 3\ O_2\ (g) \rightarrow 2\ CuO\ (s) + 2\ SO_2\ (g)$

Step 2: Write the equation for the reaction of the resultant oxide with sulfuric acid.

$CuO\ (s) + H_2SO_4\ (aq) \rightarrow H_2O\ (l) + CuSO_4\ (aq)$

Step 3: Write the equation for the electrolytic purification of the solution.

$Cu^{2+}\ (aq) + 2\ e^- \rightarrow Cu\ (s)$

WP 21.2
Strategy: Determine the number of bonding and antibonding electrons, rank in order by maximum bonding electrons and minimum antibonding electrons.

Step 1: Determine the numbers of electrons of each type.

Mo: $4d^4 5s^2$ - six electrons: bonding orbitals are filled, and antibonding orbitals are empty.

Cd: $4d^{10} 5s^2$ – twelve electrons: bonding and antibonding orbitals are filled.

Y: $4d^1 5s^2$ - three electrons: bonding orbitals half-filled, antibonding orbitals empty.

Step 2: Rank in order of expected bonding strength.

Cd < Y < Mo

The published melting points are 321 °C for Cd, 1523 °C for Y, and 2623 °C for Mo, which agrees with this determination.

WP 21.3
Strategy: Determine the number of valence electrons in the semiconductor, and compare to the number of valence electrons in the dopant. As all of the elements are main group, their group numbers can be used.

Step 1: Determine the numbers of valence electrons in each element.

Si: Group 4A – 4 valence electrons
Ge: Group 4A – 4 valence electrons
Ga: Group 3A – 3 valence electrons
Sb: Group 5A – 5 valence electrons
As: Group 5A – 5 valence electrons
P: Group 5A – 5 valence electrons

Step 2: Determine whether the dopant has more or fewer valence electrons than the semiconductor.

a. Si > Ga
b. Ge < Sb
c. Si < As
d. Ge < P

Step 3: Identify the doped semiconductors as *n*- or *p*-type.

a. *p*-type
b. *n*-type
c. *n*-type
d. *n*-type

WP 21.4
Strategy: Convert the band gap energy to photon energy, and then to wavelength.

Step 1: Determine the energy of an individual photon.

$$178\ \frac{\text{kJ}}{\text{mol}} \times \frac{1000\ \text{J}}{\text{kJ}} \times \frac{1\ \text{mol}}{6.022 \times 10^{23}\ \text{photons}} = 2.96 \times 10^{-19}\ \text{J/photon}$$

Step 2: Convert the photon energy to wavelength.

$$E = \frac{hc}{\lambda};\ \lambda = \frac{hc}{E} = \frac{6.626 \times 10^{-34}\ \text{J}\cdot\text{s} \times 3.00 \times 10^{8}\ \text{m/s}}{2.96 \times 10^{-19}\ \text{J}} = 673\ \text{nm}$$

This is red light.

Chapter 22

WP 22.1
Step 1: Determine the number of nucleons and their mass.

First, calculate the total mass of the nucleons and the mass defect.

Xe-130 has 54 protons, and 76 neutrons

Mass of 54 protons = (54)(1.007 28 amu)	= 54.393 12 amu
Mass of 76 neutrons = (76)(1.008 66 amu)	= 76.658 16 amu
Mass of 54 protons and 76 neutrons	=131.051 28 amu
Mass of xenon-130	=129.903 amu
Mass of electrons = (54)(5.486×10^{-4} amu)	= 0.030 amu
Mass of xenon-130 nucleus	=129.873 amu

Mass defect = 131.051 28 amu - 129.873 amu = 1.178

Step 2: Convert the mass defect to grams.

$$1.178\ \text{amu} \times \frac{1.6605 \times 10^{-24}\ \text{g}}{\text{amu}} = 1.956 \times 10^{-24}\ \text{g}$$

Step 3: Determine ΔE and convert to MeV/nucleon.

$$\Delta E = 1.957 \times 10^{-24}\ \text{g} \times \frac{1\ \text{kg}}{1000\ \text{g}} \times (3.00 \times 10^{8}\ \text{m/s})^2 = 1.761 \times 10^{-10}\ \text{J}$$

Convert to MeV/nucleon:

$$1.761 \times 10^{-10} \frac{\text{J}}{\text{nucleus}} \times \frac{1 \text{ MeV}}{1.60 \times 10^{-13} \text{ J}} \times \frac{1 \text{ nucleus}}{130 \text{ nucleons}} = 8.47 \frac{\text{MeV}}{\text{nucleon}}$$

WP 22.2

Strategy: Balance the equation, calculate the mass defect and convert to energy using the Einstein equation.

Step 1: Write and balance the equation.

$${}^{235}\text{U} + {}^{1}\text{n} \rightarrow {}^{136}\text{Xe} + {}^{99}\text{Ru} + {}^{1}\text{n}$$

Step 2: Calculate the mass defect in amu and convert to kilograms.

235.043 92 amu – (135.907 22 amu + 98.905 939 amu) = 0.230 76 amu

$$0.230\ 76 \text{ amu} \times \frac{1.6605 \times 10^{-27} \text{ kg}}{\text{amu}} = 3.8318 \times 10^{-28} \text{ kg}$$

Step 3: Determine ΔE and convert to J/mol.

$$\Delta E = 3.8318 \times 10^{-28} \text{ kg} \times (3.00 \times 10^{8} \text{ m/s})^2 = 3.35 \times 10^{-11} \text{ J}$$

$$3.35 \times 10^{-11} \frac{\text{J}}{\text{reaction}} \times 6.022 \times 10^{23} \frac{\text{reactions}}{\text{mol}} = 2.077 \times 10^{13} \text{ J/mol}$$

WP 22.3

Strategy: Calculate the mass defect and convert to energy using the Einstein equation.

Step 1: Write and balance the equation.

$${}^{2}\text{H} + {}^{3}\text{H} \rightarrow {}^{4}\text{He} + {}^{1}\text{n}$$

Step 2: Calculate the mass defect in amu and convert to kilograms.

(2.014 10 amu + 3.016 05 amu) – (4.002 60 amu +1.008 66 amu) = 0.018 89 amu

$$0.01889 \text{ amu} \times \frac{1.6605 \times 10^{-27} \text{ kg}}{\text{amu}} = 3.1367 \times 10^{-29} \text{ kg}$$

Step 3: Determine ΔE and convert to J/mol.

$$\Delta E = 3.136\ 7 \times 10^{-29} \text{ kg} \times (3.00 \times 10^{8} \text{ m/s})^2 = 2.82 \times 10^{-12} \text{ J}$$

$$2.82 \times 10^{-12} \frac{\text{J}}{\text{reaction}} \times 6.022 \times 10^{23} \frac{\text{reactions}}{\text{mol}} = 1.70 \times 10^{12} \text{ J/mol}$$

This is particularly impressive when you consider that 1 mol of ${}^{4}\text{He}$ has a mass of less than a nickel.

WP 22.4

Strategy: Determine the overall mass, subtract out the neutrons, and determine the identity and mass of the element produced.

The total number of protons in the new element is 98 – 92 from uranium plus 6 from carbon.

The mass of the new element will be the masses of the uranium and the carbon, minus the mass of the neutrons: 238 + 12 – 6 = 244.

The element with 98 protons is californium – Cf.
The final equation is: $^{238}_{92}U + ^{12}_{6}C \rightarrow ^{244}_{98}Cf + 6\ ^{1}_{0}n$

Chapter 23

WP 23.1

Strategy: Look for noncarbon atoms and multiple bonds. Compare with the chart above or the chart on page 913 of the textbook to name the functional groups.

Step 1: Name the functional groups.

a. A carbon double-bonded to an oxygen and single-bonded to carbons on each side is a *ketone*.

b. An oxygen that is double-bonded to a carbon that is also bonded to a hydrogen is an *aldehyde*.

c. An oxygen double-bonded to a carbon that is also bonded to an alcohol group is a *carboxylic acid.*

d. An oxygen that is single-bonded to two carbons is an *ether.*

e. There are no noncarbon atoms here, but there are also not enough hydrogens to satisfy the bonding requirements of the carbon atoms. This is an *alkene*.

WP 23.2

Strategy: Use the naming strategy detailed above to identify the molecules.

Step 1: Identify and name the parent chain.

a. pentane
b. heptane
c. hexane

Step 2: Number the chain, beginning at the end closest to the first branch point.

a. numbering begins on the right-hand side
b. numbering begins on the right-hand side on any three-carbon chain
c. numbering begins at the top of the structure.

*Step 3:*Identify and number the substituents.

a. 2-methyl, 2-methyl, 3-methyl
b. 4-propyl, 4-ethyl
c. 3-methyl

Step 4: Write the name of the molecule as a single word.

a. 2, 2, 3-trimethylpentane
b. 4-ethyl, 4-propylheptane
c. 3-methylhexane

WP 23.3

Strategy: Draw the carbon backbones, and add substituents in the proper places.

Step 1: Draw the carbon backbones.

a. $CH_3CH_2CH_2CH_2CH_2CH_2CH_3$ (heptane)
b. $CH_3CH_2CH_2CH_2CH_2CH_3$ (hexane)
c. $CH_3CH_2CH_2CH_3$ (butane)
d. $CH_3CH_2CH_2CH_2CH_2CH_2CH_2CH_3$ (octane)

Step 2: Add the substituents, removing hydrogens as necessary to maintain four bonds per carbon.

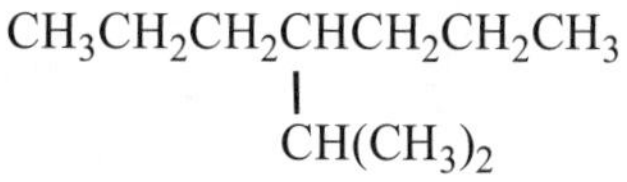

$$\begin{array}{l} \mathrm{CH_3CH_2CH_2\underset{\displaystyle |\atop \displaystyle CH(CH_3)_2}{CH}CH_2CH_2CH_3} \\ \text{a.} \end{array} \qquad \begin{array}{l} \mathrm{CH_3\overset{\displaystyle CH_3 \atop \displaystyle |}{C}H\underset{\displaystyle |\atop \displaystyle CH_2CH_3}{C}HCH_2CH_2CH_3} \\ \text{b.} \end{array} \qquad \begin{array}{l} \mathrm{CH_3\overset{\displaystyle CH_3 \atop \displaystyle |}{\underset{\displaystyle |\atop \displaystyle CH_3}{C}}CH_2CH_3} \\ \text{c.} \end{array}$$

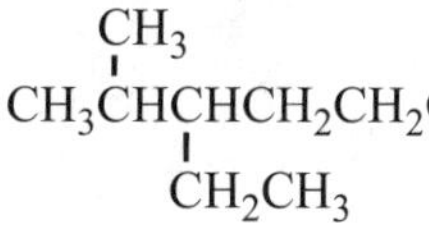

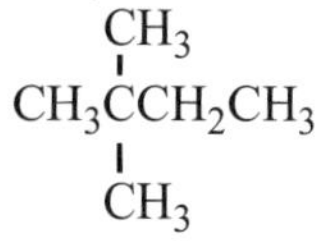

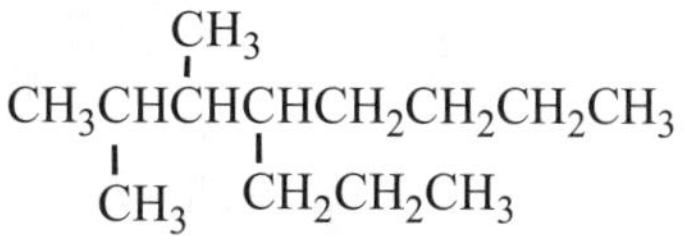

$$\begin{array}{c} \mathrm{CH_3\underset{\displaystyle |\atop \displaystyle CH_3}{C}H\overset{\displaystyle CH_3 \atop \displaystyle |}{C}H\underset{\displaystyle |\atop \displaystyle CH_2CH_2CH_3}{C}HCH_2CH_2CH_2CH_3} \\ \text{d.} \end{array}$$

WP 23.4

Strategy: Draw the carbon backbone, and add substituents in the proper places.

Step 1: Draw the carbon backbone.

$CH_3CH_2CH_2CH_2CH_2CH_2CH_3$

Step 2: Add the substituents, subtracting hydrogens as necessary.

$$\mathrm{CH_3C{=}\underset{\displaystyle |\atop \displaystyle CH_2CH_3}{C}CHCH_2\overset{\displaystyle CH_3 \atop \displaystyle |}{C}HCH_3}$$

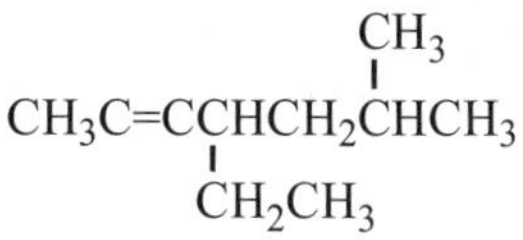

WP 23.5

Strategy: Draw the carbon backbone, and determine the possible products of the addition reactions.

Step 1: Draw the carbon backbone.

$CH_3CH_2CH{=}CHCH_2CH_3$

Step 2: Add the substituents.

a. 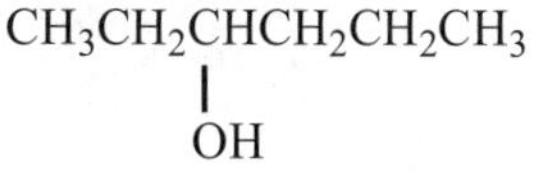 $$\mathrm{CH_3CH_2\underset{\displaystyle |\atop \displaystyle OH}{C}HCH_2CH_2CH_3}$$

b. 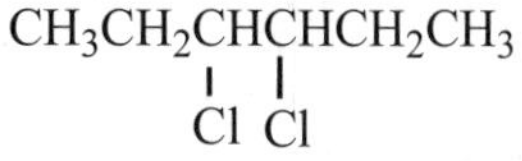 $$\mathrm{CH_3CH_2\underset{\displaystyle |\atop \displaystyle Cl}{C}H\underset{\displaystyle |\atop \displaystyle Cl}{C}HCH_2CH_3}$$

c. $CH_3CH_2CH_2CH_2CH_2CH_3$

WP 23.6
Strategy: Draw the carbon backbone, and add substituents in the proper places.

Step 1: Draw the cyclic backbone.

Step 2: Add the substituents.

WP 23.7
Strategy: Draw the benzene ring, and add substituents in the proper orientations.

o-dichlorobenzene *p*-methylnitrobenzene *m*-ethylmethylbenzene

WP 23.8
Strategy: Draw the reactant, and add the new substitutent in all places that give a unique configuration.

a. Although at first glance there appear to be four possibilities, two of them are identical.

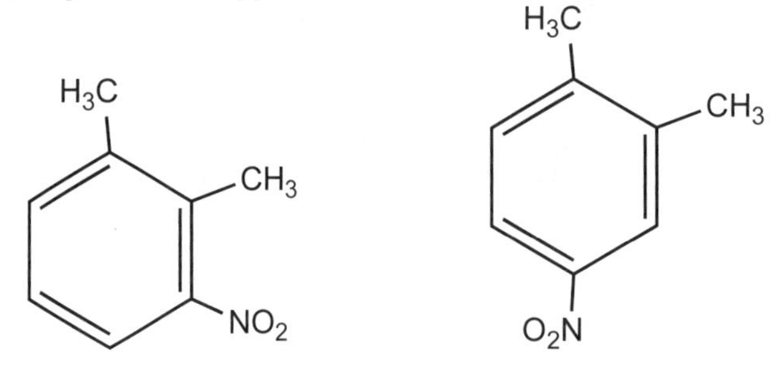

1-nitro-2,3-dimethylbenzene 1-nitro-3,4-dimethylbenzene

b. The two different original substituents greatly increase the possibilities for products in this reaction.

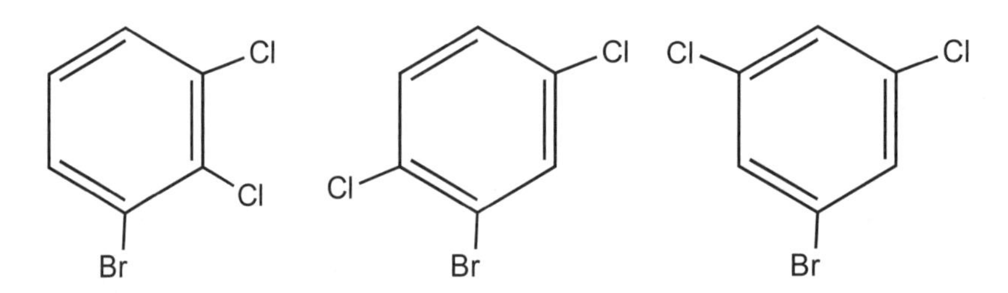

1-bromo-2,3-chlorobenzene 1-bromo-2,5-chlorobenzene 1-bromo-3,5-chlorobenzene

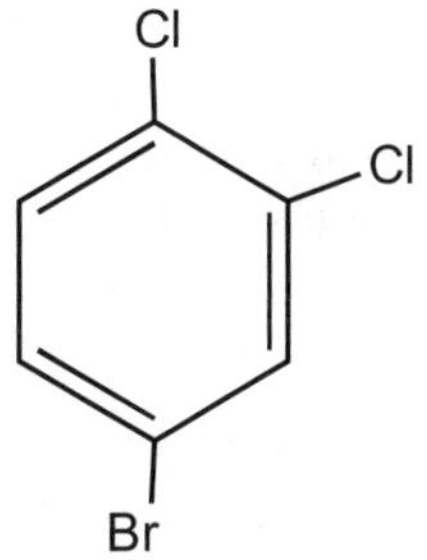

1-bromo-3,4-chlorobenzene

WP 23.9

Strategy: Determine the complementary bases, determine the number of codons.

A and T are usually complementary, but RNA substitutes U, G, and C are complementary in all nucleic acids. The complementary mRNA strand will be

UCGAACUGUGCACAGUGAUCU

Each codon is three nucleotides long. This sequence has 21 nucleotides, so will code for 7 amino acids.

PUTTING IT TOGETHER SOLUTIONS

Chapter Six

Strategy: Write and balance the reactions, do the stoichiometric calculations to determine the theoretical yield, and set up an algebraic equation to solve for the quantity of magnesium nitride in the first reaction mixture. Solve for the mass percent of magnesium nitride.

Step 1: Write and balance the equations.

$$2Mg + O_2 \rightarrow 2MgO$$

$$3Mg + N_2 \rightarrow Mg_3N_2$$

$$Mg_3N_2 + 3H_2O \rightarrow 3MgO + 2NH_3$$

Step 2: Determine the theoretical yield of MgO based on the amount of magnesium available.

$$0.237\text{g Mg} \times \frac{1\text{mol}}{24.3\text{g}} = 0.00975 \text{ mol Mg}$$

$$0.00975 \text{ mol Mg} \times \frac{2 \text{ mol MgO}}{2 \text{ mol Mg}} = 0.00975 \text{ mol MgO} \times \frac{40.3 \text{ g MgO}}{\text{mol}} = 0.393 \text{ g MgO}$$

Step 3: Set up a system of equations to determine the number of grams of each product in the intermediate step.

$x = \textit{moles of Mg reacted with } O_2$

$y = \textit{moles of Mg reacted with } N_2$

$$x + y = 0.00975$$

$$\left(x \bullet \frac{1\text{mol MgO}}{\text{mol Mg}} \times \frac{40.3\text{g MgO}}{\text{mol MgO}}\right) + \left(y \bullet \frac{1 \text{ mol } Mg_2N_3}{3 \text{ mol Mg}} \times \frac{100.93 \text{ g } Mg_2N_3}{\text{mol } Mg_2N_3}\right) = 0.277$$

Solve:

$$40.3x + 33.64y = 0.277$$

$$y = 0.00823 - 1.198x$$

Substitute for y:

$$40.3x + 25.5(0.00823 - 1.198x) = 0.277$$

$$40.3x - 30.5x + 0.210 = 0.277$$

$$9.8x = 0.067$$

$$x = 0.00684$$

$$y = 0.00291$$

Step 4: Solve for the mass percent of magnesium nitride.

$$0.00684 \text{ mol Mg x} \frac{1 \text{ mol } Mg_3N_2}{3 \text{ mol Mg}} \text{ x} \frac{100.93\text{g } Mg_3N_2}{\text{mol } Mg_3N_2} = 0.230\text{g } Mg_3N_2$$

$$\text{mass \% } Mg_3N_2 = \frac{0.230\text{g } Mg_3N_2}{0.277\text{g total}} \text{ x } 100 = 83.1\%$$

Given that nitrogen is over 70% of the atmosphere by mass, this is not a shocking result.

Chapter Seven

Strategy: Determine the empirical formula of the compound and the molecular formula. Determine the structure.

Step 1: Solve for the empirical formula.

26.68 g C = 2.221 mol C
71.09 g O = 4.443 mol O
2.224 g H = 2.222 mol H

Empirical formula = CO_2H

Step 2: Solve for the molecular formula.

Empirical formula weight = 45.01 g/mol

Molecular weight = 90.03 g/mol

Molecular formula = $C_2O_4H_2$

Step 3: Determine the structure of the molecule from the information given.

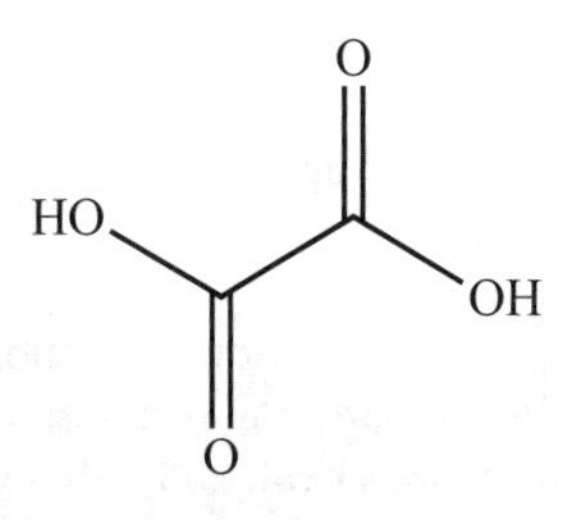

oxalic acid

Chapter 8

Balance the decomposition equation.

$N_2H_4\ (l) \rightarrow N_2\ (g) + 2H_2\ (g)$

Convert the grams of hydrazine to moles.

$$3\text{g x} \frac{1 \text{ mol}}{32 \text{ g}} = 0.0938 \text{ mol hydrazine}$$

Calculate ΔH = products – reactants = 0 – 95.4 kJ/mol = -95.4 kJ/mol

Calculate ΔS = products – reactants = [191.5 + 2(130.6)] – 121.2 = 331.5 J/mol • K

Calculate $\Delta G = \Delta H - T\,\Delta S$ = -95.4kJ/mol – 298K(0.3315 kJ/mol • K) = -194 kJ/mol

-194 kJ/mol x 0.0938 mol = -18.2 kJ

Chapter 9

Strategy: Determine the empirical formula, the number of moles present, and the molecular weight, then determine the molecular formula.

40.56 g Br = 0.5076 mol

18.02 g Cl = 0.5083 mol

28.93 g F = 1.522 mol

0.51 g H = 0.506 mol

12.19 g C = 1.015 mol

Empirical formula = C_2BrClF_3H, empirical formula weight = 197.4 g/mol

Solve for moles.

$$n = \frac{PV}{RT} = \frac{3.0 \text{ atm x } 13.67 \text{ L}}{0.08206 \frac{\text{L} \cdot \text{atm}}{\text{mol} \cdot \text{K}} \text{x} 500\text{K}} = 1.00 \text{ mol}$$

The empirical formula and the molecular formulas are the same.

Chapter 10

Strategy: Draw the electron-dot structure of SO_2 to determine the molecular shape and the intermolecular forces present, then use the information provided to calculate the heats of reaction.

Sulfur and oxygen both have 6 valence electrons, for a total of 18 valence electrons for the total molecule. These can be distributed to provide two valence structures, in which the double bond can occur on either side of the sulfur. The result of this is that, unlike carbon dioxide, sulfur dioxide is bent, and will have dipole/dipole forces:

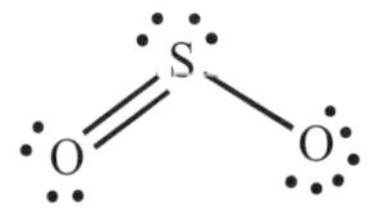

The equation for the reaction of sulfur dioxide with atmospheric oxygen to form sulfur trioxide is

$$SO_2\ (g) + \tfrac{1}{2}\ O_2 \rightarrow SO_3\ (g)$$

Using the provided heats of formation, the heat of reaction will be

$$\Delta H_{rxn} = (-395.7 \text{ kJ/mol}) - (-296.8 \text{ kJ/mol} + 0.0 \text{ kJ/mol}) = -98.9 \text{ kJ/mol}$$

When sulfur trioxide reacts with water to form sulfuric acid, the reaction is

$$SO_3\ (g) + H_2O\ (l) \rightarrow H_2SO_4\ (l)$$

Using the provided heats of formation, the heat of this reaction will be

$$\Delta H_{rxn} = (-814.0\ \text{kJ/mol}) - (-395.7\ \text{kJ/mol} + -285.8\ \text{kJ/mol}) = -132.5\ \text{kJ/mol}$$

Chapter 11

Calculate the empirical formula of the compound from the combustion analysis data.

$$2.026\ \text{g x } \frac{1\ \text{mol}}{44.0\ \text{g}} \text{x} \frac{1\ \text{C}}{CO_2} = 0.0460\ \text{mol C}$$

$$0.0460\ \text{mol C x } 12\frac{\text{g}}{\text{mol}} = 0.553\ \text{g C}$$

$$0.8288\ \text{g x } \frac{1\ \text{mol}}{18.0\ \text{g}} \text{x} \frac{2\ \text{H}}{H_2O} = 0.0921\ \text{mol H}$$

$$0.0921\ \text{mol H x } 1.008\frac{\text{g}}{\text{mol}} = 0.0928\ \text{g H}$$

Grams oxygen: 0.83 g – 0.0928 g H – 0.553 g C = 0.184 g O

Moles oxygen: $\dfrac{0.184\ \text{g O}}{16.0\dfrac{\text{g}}{\text{mol}}} = 0.0115\ \text{mol O}$

Empirical formula: oxygen has the smallest number of moles.

$$\frac{0.0115\ \text{mol O}}{0.0115\ \text{mol O}} = 1\ \text{oxygen}$$

$$\frac{0.0921\ \text{mol H}}{0.0115\ \text{mol O}} = 8\ \text{hydrogens per oxygen}$$

$$\frac{0.0460\ \text{mol C}}{0.0115\ \text{mol O}} = 4\ \text{carbons per oxygen}$$

Empirical formula = C_4H_8O

Determine the molarity of the solution formed.

$$1.02\ \text{atm} = x\ \text{M x } 0.08206\frac{\text{L} \bullet \text{atm}}{\text{mol} \bullet \text{K}} \text{x} 298\ \text{K}$$

$$x = 0.0417\ \text{M}$$

$$0.0416\frac{\text{mol}}{\text{L}} \text{x} 0.250\text{L} = 0.0104\ \text{mol}$$

$$\frac{1.50\ \text{g}}{0.0104\ \text{mol}} = 144.2\frac{\text{g}}{\text{mol}}$$

Determine the molecular formula.

Empirical formula weight = 72 g/mol

Molecular weight = 2 x empirical formula weight

Molecular formula = 2 x empirical formula: $C_8H_{16}O_2$

Chapter 12

a. The first step has a molecularity of 1. The second step is bimolecular.

b. The reaction is first order in $Ni(CO)_4$. As changing the concentration of reactants in the second step has no effect on the overall rate, the first step is rate-determining.

The rate law is.

$$\text{Rate} = k[Ni(CO)_4]$$

c. Substitute into the integrated first-order rate law.

$$[Ni(CO)_4]_0 = 0.30\text{ M},\ [Ni(CO)_4]_t = 0.30\text{ M x } 0.40 = 0.12\text{ M}.$$

$$\ln\frac{0.12\text{ M}}{0.30\text{ M}} = -9.3 \text{ x } 10^{-3}\text{s}^{-1} \text{ x t}$$

$$t = 98.5\text{ s}$$

d. Determine the overall reaction by summing the reaction steps.

$$Ni(CO)_4 + P(C_6H_5)_3 \rightarrow Ni(CO)_3[P(C_6H_5)_3] + CO$$

Determine the number of moles of nickel.

$$0.237\text{ g } Ni(CO)_4 \text{ x } \frac{1\text{ mol}}{170.69\text{ g}} = 1.39 \text{ x } 10^{-3}\text{ mol}$$

If the reaction goes to completion, the number of moles of triphenylphosphine will be equal to the starting moles of $Ni(CO)_4$, as the ratio is 1:1.

The amount of $Ni(CO)_4$ remaining after 2 minutes can be calculated as follows:

$$\ln\frac{\text{x mol}}{1.39 \text{ x } 10^{-3}\text{ mol}} = -9.3 \text{ x } 10^{-3}\text{s}^{-1} \text{ x } 120\text{ s}$$

$$\frac{\text{x mol}}{1.39 \text{ x } 10^{-3}\text{ mol}} = e^{-1.116}$$

$$x = 4.55 \text{ x } 10^{-4}\text{ mol}$$

The amount of $Ni(CO)_4$ consumed is equal to the amount of triphenylphosphine produced.

$$(1.39 \times 10^{-3} \text{ mol} - 4.55 \times 10^{-4} \text{ mol}) \times 433 \text{ g/mol} = 0.405 \text{ g}$$

The reaction is 32.7% complete.

Chapter 13

Dalton's law of partial pressures can be used to determine the pressure of all species present at equilibrium. As with any equilibrium problem, the first step is to set up a chart of initial values, changes, and final values, using the stoichiometry of the balanced equation.

First, convert to atmospheres.

$$NO = \frac{78.4 \text{ mm Hg}}{760 \frac{\text{mm Hg}}{\text{atm}}} = 0.103 \text{ atm;}$$

$$Br_2 = \frac{51.3 \text{ mm Hg}}{760 \frac{\text{mm Hg}}{\text{atm}}} = 0.0675 \text{ atm;}$$

$$\text{final pressure} = \frac{120.5 \text{ mm Hg}}{760 \frac{\text{mm Hg}}{\text{atm}}} = 0.159 \text{ atm}$$

	$2\,NO\,(g)$ +	$Br_2\,(g)$ ⇆	$2\,NOBr\,(g)$
Initial Pressure	0.103	0.0675	0
Change	$-2x$	$-x$	$+2x$
Eq Pressure	$0.103 - 2x$	$0.0675 - x$	$2x$

The total pressure at equilibrium is equal to the sum of the pressures of each species at equilibrium. The total pressure at equilibrium is 0.159 atm.

$$0.159 = (0.103 - 2x) + (0.0675 - x) + 2x = -x + 0.1705$$

$x = 0.0115$ atm

At equilibrium:

$P(NO) = 0.103 \text{ atm} - 2(0.0115 \text{ atm}) = 0.0800 \text{ atm}$

$P(Br_2) = 0.0675\text{atm} - 0.0115 \text{ atm} = 0.0560 \text{ atm}$

$P(NOBr) = 2(0.0115 \text{ atm}) = 0.0230 \text{ atm}$

$$K_p = \frac{(0.0230)^2}{(0.0800)^2(0.0560)} = 1.48$$

Chapter 14

To determine the amount of aniline used, determine the theoretical yield expected in this reaction.

$$\% \text{ yield} = \frac{\text{Experimental yield}}{\text{Theoretical yield}} \times 100$$

$$\text{Theoretical yield} = \frac{125 \text{ g}}{0.90} \times 100 = 139 \text{ g}$$

Calculate the number of moles of sulfanilic acid in 139 g.

$$139 \text{ g x} \frac{1 \text{ mol}}{173 \text{ g}} = 0.803 \text{ mol}$$

The mole ratio of aniline to sulfanilic acid is 1:1. Therefore, we have 0.803 mol of aniline. This needs to be converted to mass, then to volume.

$$0.803 \text{ mol x} \frac{93 \text{ g}}{\text{mol}} \text{ x} \frac{1 \text{ mL}}{1.02 \text{ g}} = 73.2 \text{ mL}$$

$Na(C_6H_4NH_2SO_3)$, sodium sulfanilate, contains the anion of a weak acid and the cation of a strong base. Therefore, the aqueous solution of this salt should be basic. The hydrolysis reaction for this salt is

$$C_6H_4NH_2SO_3^- (aq) + H_2O (l) \leftrightarrows HC_6H_4NH_2SO_3 (aq) + OH^- (aq)$$

$$K_b = \frac{K_w}{K_a} = \frac{1 \times 10^{-14}}{5.9 \times 10^{-4}} = 1.69 \times 10^{-11}$$

Determine the initial concentration of the sulfanilate ion.

$$\text{mol } C_6H_4NH_2SO_3^- = 5.73 \text{ g } NaC_6H_4NH_2SO_3H \times \frac{1 \text{ mol } NaC_6H_4NH_2SO_3H}{196 \text{ g } NaC_6H_4NH_2SO_3H} \times \frac{1 \text{ mol } C_6H_4NH_2SO_3^-}{1 \text{ mol } NaC_6H_4NH_2SO_3H}$$

$$= 2.92 \times 10^{-2} \text{ mol } C_6H_4NH_2SO_3^-$$

$$\text{M} = \frac{2.92 \times 10^{-2} \text{ mol } C_6H_4NH_2SO_3^-}{0.250 \text{ L}} = 0.117 \text{ M}$$

We can now set up our table.

Principal Reaction	$C_6H_4NH_2SO_3^- (aq)$ + $H_2O (l)$ ⇆	$HC_6H_4NH_2SO_3 (aq)$ +	$OH^- (aq)$
Initial concentration	0.117	0	0
Change	$-x$	$+x$	$+x$
Eq. concentration	$0.117 - x$	$+x$	$+x$

We can assume that x is negligible, given the value of K_b. Therefore, $0.117 - x \approx x$.

$$1.69 \times 10^{-11} = \frac{x^2}{0.117}$$

$$x^2 = 1.97 \times 10^{-12}$$

$x = [OH^-]_e = 1.41 \times 10^{-6}$ M

Calculate $[H_3O^+]$.

$K_W = [H_3O^+][OH^-]$

$$[H_3O^+] = \frac{1 \times 10^{-14}}{1.41 \times 10^{-6}} = 7.09 \times 10^{-9}$$

pH = $-\log(7.09 \times 10^{-9}) = 8.15$

Chapter 15

a. $AlCl_3\ (s) + H_3PO_4\ (aq) \rightarrow AlPO_4\ (s) + 3\ HCl\ (aq)$

b. Determine the limiting reactant: the mole ration of the reactants is 1:1.

$$\text{mol } AlCl_3 = 95 \text{ g } AlCl_3 \times \frac{1 \text{ mol } AlCl_3}{133.5 \text{ g } AlCl_3} = 0.711 \text{ mol } AlCl_3$$

$$\text{mol } H_3PO_4 = 1.50 \text{ L} \frac{0.75 \text{ mol } H_3PO_4}{\text{L}} = 1.125 \text{ mol } H_3PO_4$$

The limiting reactant is $AlCl_3$.

$$\text{g } AlPO_4 = 0.711 \text{ mol } AlCl_3 \times \frac{1 \text{ mol } AlPO_4}{1 \text{ mol } AlCl_3} \times \frac{122 \text{ g } AlPO_4}{1 \text{ mol } AlPO_4} = 86.74 \text{ g } AlPO_4$$

c. $AlPO_4\ (s) \leftrightharpoons Al^{3+}\ (aq) + PO_4^{3-}\ (aq)$ $\qquad K_{sp} = 1.3 \times 10^{-20}$

Let x represent the number of moles of $AlPO_4$ that will dissolve. Therefore, the number of moles of Al^{3+} and PO_4^{3-} formed is also x.

$K_{sp} = 1.3 \times 10^{-20} = (x)(x) = x^2$; $\qquad x = 1.1 \times 10^{-10}$ M

d. The solubility of $AlPO_4$ will increase on the addition of HCl: hydrogen ions can react with phosphate ions to produce HPO_4^{2-}, removing the phosphate ions from solution. Using LeChâtelier's principle, this will cause more $AlPO_4$ to dissolve.

$H^+\ (aq) + PO_4^{3-}\ (aq) \leftrightharpoons HPO_4^{2-}\ (aq)$

To determine if a precipitate will form, calculate Q and then compare this value to K_{sp}. (Remember that Q, the ion product, is calculated using the same formula for K_{sp}.)

$Q = [Al^{3+}][PO_4^{3-}]$

e. Calculate the concentration of Al^{3+} and PO_4^{3-} from the total volume of mixing the two solutions. Remember that when you dilute the solutions $M_f \times V_f = M_i \times V_i$. The total volume of the solution is 2.00 L.

$$[Al^{3+}] = \frac{0.75\ \text{L} \times (4.00 \times 10^{-3}\ \text{M})}{2.00\ \text{L}} = 1.50 \times 10^{-3}\ \text{M}$$

$$[PO_4^{3-}] = \frac{1.25\ \text{L}(0.700\ \text{M})}{2.00\ \text{L}} = 0.438\ \text{M}$$

$Q = (1.50 \times 10^{-3})(0.438) = 6.56 \times 10^{-4}$

$Q >> K_{sp}$; therefore, a precipitate will form. Given that the molar solubility of $AlPO_4$ is so small (1.1×10^{-10}), the mass of $AlPO_4$ can be calculated based on the number of moles of Al^{3+}, the limiting reactant.

mol $Al^{3+} = 3.00 \times 10^{-3}$; mol $PO_4^{3-} = 0.876$

$$\text{g AlPO}_4 = 3.00 \times 10^{-3} \times \frac{1\ \text{mol AlPO}_4}{1\ \text{mol Al}^{3+}} \times \frac{122\ \text{g AlPO}_4}{1\ \text{mol AlPO}_4} = 0.366\ \text{g AlPO}_4$$

Chapter 16

To solve for the percentage of hemoglobin molecules bound to carbon monoxide, first solve for the equilibrium constant, K, at 37 °C.

$\Delta G^{o} = -RT\ln K$; $T = (273.15 + 37) = 310.15$ K

$$-14{,}000\ \text{J} = -\left(8.314\frac{\text{J}}{\text{mol}\cdot\text{K}}\right)(310.15\ \text{K})\ln \text{K}$$

$\ln \text{K} = 5.43$ $\qquad$ $\text{K} = 228.1$

With the equilibrium constant and the proportions of carbon monoxide and oxygen, calculate the proportions of hemoglobin bound to each molecule.

$$\text{K} = \frac{[\text{Hb}\cdot\text{CO}]\,[\text{O}_2]}{[\text{Hb}\cdot\text{O}_2]\,[\text{CO}]} = \frac{[\text{Hb}\cdot\text{CO}]\,[0.90]}{[\text{Hb}\cdot\text{O}_2]\,[0.10]} = 228.1$$

$$\frac{[\text{Hb}\cdot\text{CO}]}{[\text{Hb}\cdot\text{O}_2]} = 25.3$$

There are 25.3 hemoglobin molecules bound to carbon monoxide for every one hemoglobin molecule bound to oxygen. This proportion, 25.3:1, can be converted to a percentage.

(25.3 Hb-CO/26.3 total Hb) x 100 = 96.4% bound to carbon monoxide.

Chapter 17

Write the 2 half-reactions and solve for $E° = E_{anode} + E_{cathode}$.

$O_2\,(g) + 4\,H^+\,(aq) + 4\,e^- \rightarrow 2\,H_2O\,(l)$ $\qquad E° = 1.23$ V

$2\,Br^-\,(aq) \rightarrow Br_2\,(l) + 2\,e^-$ $\qquad E° = -1.09$ V

$E° = 1.23\text{ V} + (-1.09\text{ V}) = 0.14$

The reaction is spontaneous since $E°$ is positive.

To determine the value of E after addition of the buffer, determine $[H^+]$ and substitute this value into the Nernst equation.

$K_a\,(HCHO_2) = 1.7 \times 10^{-4}$; $\qquad pK_a = 3.74$

$$\text{pH} = \text{pK}_a + \log\frac{[\text{CHO}_2^-]}{[\text{HCHO}_2]} = 3.74 + \log\frac{0.35}{0.20} = 3.98$$

$$[\text{H}^+] = \text{antilog}(\text{-}3.98) = 1.05 \times 10^{-4}$$

Substitute into the Nernst equation.

$$E = E^\circ - \frac{0.0592}{n}\log\frac{1}{[\text{Br}^-][\text{H}^+]}$$

After balancing the number of electrons in the two half-reactions, $n = 4$.

$$\text{E} = 0.14\text{ V} - \frac{0.0592}{4}\log\frac{1}{(1.05 \times 10^{-4})} = 0.0811\text{ V}$$

The reaction is still spontaneous under these non-standard state conditions.

Chapter 18

The balanced chemical equation is

$2\,NaH\,(s) + 2\,SO_2\,(l) \rightarrow Na_2S_2O_4\,(s) + H_2\,(g)$

Determine the limiting reactant.

$$\text{mol NaH} = 23.57\text{ g NaH} \times \frac{1\text{ mol NaH}}{24.0\text{ g NaH}} = 0.9821\text{ mol NaH}$$

$$\text{mass SO}_2 = 100.0\text{ mL} \times \frac{1.434\text{ g SO}_2}{1\text{ mL}} = 143.4\text{ g SO}_2$$

$$\text{mol SO}_2 = 143.4\text{ g SO}_2 \times \frac{1\text{ mol SO}_2}{64.0\text{ g SO}_2} = 2.241\text{ mol SO}_2$$

The limiting reactant is NaH. We can now calculate moles of H_2.

$$\text{mol } H_2 = 0.9821 \text{ mol NaH} \times \frac{1 \text{ mol } H_2}{2 \text{ mol NaH}} = 0.4911 \text{ mol } H_2$$

Using the ideal gas law, calculate the volume of hydrogen produced.

$$V = \frac{nRT}{P} = \frac{0.4911 \text{ mol } H_2 \left(0.08206 \frac{\text{L}\cdot\text{atm}}{\text{mol}\cdot\text{K}}\right)(298.15 \text{ K})}{1 \text{ atm}} = 12.01 \text{ L}$$

Chapter 19

Write and balance the equation.

$$SO_2\,(g) + 2\,H_2S\,(aq) \rightarrow 3\,S\,(s) + 2\,H_2O\,(l)$$

To determine the volume of H_2S needed, calculate the moles of SO_2 (*g*) generated in the burning of the coal. Assume 100% yield.

$$\text{mol } SO_2 = 1.0 \text{ tons coal x } 0.033 \frac{\text{S}}{\text{coal}} \times \frac{1000 \text{ kg}}{1 \text{ ton}} \times \frac{1000 \text{ g}}{1 \text{ kg}} \times \frac{1 \text{ mol S}}{32.0 \text{ g S}} \times \frac{1 \text{ mol } SO_2}{1 \text{ mol S}} = 1031 \text{ mol } SO_2$$

From the balanced equation, 2 moles of H_2S are needed for every mole of SO_2, so 2062 moles H_2S are needed.

Again from the balanced equation, 3 moles of S are produced for every mole of SO_2.

$$\text{kg S} = 1031 \text{ mol } SO_2 \times \frac{3 \text{ mol S}}{\text{mol } SO_2} \times \frac{32.0 \text{ g S}}{\text{mol S}} \times \frac{1 \text{ kg}}{1000 \text{ g}} = 99 \text{ kg S}$$

Now use the ideal gas law to calculate the volume of SO_2:

$$V = \frac{\left(0.08206 \frac{\text{L}\cdot\text{atm}}{\text{K}\cdot\text{mol}}\right)(1031 \text{ mol})(423 \text{ K})}{\left(740 \text{ mm Hg} \times \frac{1 \text{ atm}}{760 \text{ mm Hg}}\right)} = 3.68 \times 10^4 \text{ L}$$

Chapter 20

Write the balanced net ionic equation for the reaction, using the half-reaction method. The unbalanced half-reactions are

$$Cr_2O_7^{2-}\,(aq) \rightarrow Cr^{2+}\,(aq)$$

$$Fe^{2+}\,(aq) \rightarrow Fe^{3+}\,(aq)$$

The balanced half-reactions, after the addition of H^+ and H_2O, are

$$8\,e^- + 14\,H^+\,(aq) + Cr_2O_7^{2-}\,(aq) \rightarrow 2\,Cr^{2+}\,(aq) + 7\,H_2O\,(l)$$

$$Fe^{2+}\,(aq) \rightarrow Fe^{3+}\,(aq) + e^-$$

Balance half-reactions in electrons.

$$8\,e^- + 14\,H^+\,(aq) + Cr_2O_7^{2-}\,(aq) \rightarrow 2\,Cr^{2+}\,(aq) + 7\,H_2O\,(l)$$

$$8\,Fe^{2+}\,(aq) \rightarrow 8\,Fe^{3+}\,(aq) + 8\,e^-$$

Add the two half-reactions together for the net ionic equation.

$$14\,H^+\,(aq) + Cr_2O_7^{2-}\,(aq) + 8\,Fe^{2+}\,(aq) \rightarrow 2\,Cr^{2+}\,(aq) + 7\,H_2O\,(l) + 8\,Fe^{3+}\,(aq)$$

With this balanced reaction and the data provided, determine the mass of iron in the ore.

$$\text{g Fe} = 0.03328\text{ L} \times \frac{0.090\text{ mol } Cr_2O_7^{2-}}{1\text{ L}} \times \frac{8\text{ mol } Fe^{2+}}{1\text{ mol } Cr_2O_7^{2-}} \times \frac{1\text{ mol Fe}}{1\text{ mol } Fe^{2+}} \times \frac{55.8\text{ g Fe}}{1\text{ mol Fe}} = 1.34\text{ g Fe}$$

$$\%\text{ Fe} = \frac{1.34\text{ g Fe}}{1.513\text{ g sample}} \times 100 = 88.4\%$$

Chapter 21

Write and balance the equations for the reaction of Fe^{2+} with MnO_4^- and the reaction of Fe^{2+} with $Cr_2O_7^{2-}$.

The two half-reactions for the reaction of Fe^{2+} with MnO_4^- in acidic solution are

$$MnO_4^-\,(aq) \rightarrow Mn^{2+}\,(aq)$$

$$Fe^{2+}\,(aq) \rightarrow Fe^{3+}\,(aq)$$

For the first half-reaction, balance the oxygens by the addition of water and H^+.

$$5\,e^- + 8\,H^+\,(aq) + MnO_4^-\,(aq) \rightarrow Mn^{2+}\,(aq) + 4\,H_2O\,(l)$$

For the second half-reaction, balance the charge.

$$Fe^{2+}\,(aq) \rightarrow Fe^{3+}\,(aq) + e^-$$

Now multiply the second half-reaction by 5 so that the number of electrons lost equals the number of electrons gained.

$$5\,e^- + 8\,H^+\,(aq) + MnO_4^-\,(aq) \rightarrow Mn^{2+}\,(aq) + 4\,H_2O\,(l)$$

$$5\,Fe^{2+}\,(aq) \rightarrow 5\,Fe^{3+}\,(aq) + 5\,e^-$$

Adding the two half-reactions together gives

$$8\,H^+\,(aq) + MnO_4^-\,(aq) + 5\,Fe^{2+}\,(aq) \rightarrow Mn^{2+}\,(aq) + 4\,H_2O\,(l) + 5\,Fe^{3+}\,(aq)$$

The balanced, net ionic equation for the reaction of Fe^{2+} with $Cr_2O_7^{2-}$ is

$$14\,H^+\,(aq) + Cr_2O_7^{2-}\,(aq) + 8\,Fe^{2+}\,(aq) \rightarrow 2\,Cr^{2+}\,(aq) + 7\,H_2O\,(l) + 8\,Fe^{3+}\,(aq)$$

From the molarity and volume of $FeSO_4$, we know that we have 4.0×10^{-3} moles of Fe^{2+}. We also know that

Moles of Fe^{2+} consumed in the first reaction + Moles of Fe^{2+} consumed in the second reaction = Total moles of Fe^{2+} = 4.0×10^{-3} moles Fe^{2+}

By determining the number of moles of Fe^{2+} consumed in the second reaction, we can determine the moles of Fe^{2+} consumed in the first reaction, which will allow us to determine the moles of manganese present.

For the reaction

$$14\ H^+\ (aq) + Cr_2O_7^{2-}\ (aq) + 8\ Fe^{2+}\ (aq) \rightarrow 2\ Cr^{2+}\ (aq) + 7\ H_2O\ (l) + 8\ Fe^{3+}\ (aq)$$

$$\text{mol Fe}^{2+} = 0.0224\ \text{L} \times \frac{0.0100\ \text{mol Cr}_2\text{O}_7^{2-}}{1\ \text{L}} \times \frac{8\ \text{mol Fe}^{2+}}{1\ \text{mol Cr}_2\text{O}_7^{2-}} = 1.79 \times 10^{-3}\ \text{mol Fe}^{2+}$$

The moles of Fe^{2+} consumed in the reaction

$$8\ H^+\ (aq) + MnO_4^-\ (aq) + 5\ Fe^{2+}\ (aq) \rightarrow Mn^{2+}\ (aq) + 4\ H_2O\ (l) + 5\ Fe^{3+}\ (aq)$$

Mole Fe^{2+} = (4.00×10^{-3} mol Fe^{2+}) – (1.79×10^{-3} mol Fe^{2+}) = 2.21×10^{-3} Moles Fe^{2+} consumed in the first reaction.

We can now calculate the mass of manganese in the steel sample.

$$\text{g Mn} = \left(2.21 \times 10^{-3}\ \text{mol Fe}^{2+}\right) \times \frac{1\ \text{mol MnO}_4^-}{5\ \text{mol Fe}^{2+}} \times \frac{1\ \text{mol Mn}}{1\ \text{mol MnO}_4^-} \times \frac{54.9\ \text{g Mn}}{1\ \text{mol Mn}} = 0.0243\ \text{g Mn}$$

$$\%\ \text{Mn} = \frac{0.0243\ \text{g Mn}}{0.450\ \text{g steel}} \times 100 = 5.4\%$$

Chapter 22

The formula for sodium perchlorate is $NaClO_4$. The amount of chlorine in 78.3 mg in $NaClO_4$ is

$$0.0783\ \text{g NaClO}_4 \times \frac{1\ \text{mol NaClO}_4}{122.5\ \text{g NaClO}_4} \times \frac{1\ \text{mol Cl}}{1\ \text{mol NaClO}_4} \times \frac{35.5\ \text{g Cl}}{1\ \text{mol Cl}} = 0.0227\ \text{g Cl}$$

Of the 22.7 mg Cl present, 47.0% is ^{36}Cl. This is 10.7 mg.

Calculate *k*.

$$t_{1/2} = \left(3.0 \times 10^5\ \text{yr}\right) \times \left(\frac{365\ \text{d}}{1\ \text{yr}}\right) \times \left(\frac{24\ \text{h}}{1\ \text{d}}\right)\left(\frac{3600\ \text{s}}{1\ \text{h}}\right) = 9.46 \times 10^{12}\ \text{s}$$

$$k = \frac{0.693}{9.46 \times 10^{12}\ \text{sec}} = 7.32 \times 10^{-14}\ \text{s}^{-1}$$

The decay rate = k x N, where N is the number of radioactive nuclei. Calculate N.

$$N = \left(22.7 \times 10^{-2}\ \text{g}\ ^{36}\text{Cl}\right) \times \frac{1\ \text{mol}\ ^{36}\text{Cl}}{36.0\ \text{g}\ ^{36}\text{Cl}} \times \frac{6.022 \times 10^{23}\ \text{nuclei}\ ^{36}\text{Cl}}{1\ \text{mol}^{36}\text{Cl}} = 3.80 \times 10^{20}\ \text{nuclei}$$

Calculate the decay rate.

Rate = $(7.32 \times 10^{-14}\ s^{-1}) \times 3.80 \times 10^{20}$ nuclei = 2.78×10^{7} nuclei/s or 2.78×10^{7} disintegrations/s.

Chapter 23

Butyl butyrate is responsible for the distinctive scent of bananas. The percent composition of this ester is 67% carbon, 11% hydrogen, and 22% oxygen. Reaction with water yields butanol and an acid. The molar mass of the acid is 144 g/mol. What is the structure of butyl butyrate?

Assume we have 100 g of sample.

67 g C x 1 mol/12.01g = 5.6 mol C
11 g H x 1 mol/1.00 g = 11 mol H
22 g O x 1 mol/16.0 g = 1.4 mol O

Divide by the smallest number of moles present.

5.6 C ÷ 1.4 = 4.0 C
11 H ÷ 1.4 = 7.8 H
1.4 O ÷ 1.4 = 1.0 O

The empirical formula is C_4H_8O. This has a mass of 72. As a result, the formula mass must be $C_8H_{16}O_2$. When an ester undergoes hydrolysis, it separates into its constituent alcohol and carboxylic acid. Butanol has a molecular weight of 74, which includes a hydrogen from the added water. 73 g come from the original compound, leaving 71 for the carboxylic acid. Remembering that the carboxylic acid has had a hydrogen and an oxygen added to it from the water molecule brings the molecular weight of the acid to 88. COOH has a molecular weight of 45, leaving 43 for the alkene portion of the molecule. Removing 15 for the terminal methyl group leaves 28 for CH_2 groups, so there must be 2 of these, as they have a molecular weigh of 14 each. The carboxylic acid is $CH_3CH_2CH_2COOH$, and the structure of butyl butyrate is

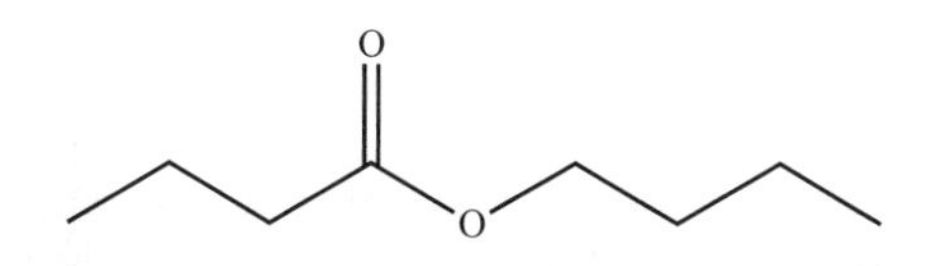

SELF TEST SOLUTIONS

Chapter 1

True–False

1. F. The symbol for ruthenium is Ru – also remember that the second letter in a chemical symbol is always lowercase.
2. T. Sodium belongs in the group referred to as the alkali metals.
3. T. Remember that leading zeroes only place the decimal – they are not significant, but internal zeroes are significant.
4. F. Aluminum is a metal; although it is below the zigzag line that separates metals from nonmetals, its behavior classifies it as a metal, not a semimetal, and definitely not a nonmetal.
5. F. Semimetals do have properties somewhere between those of metals and nonmetals, but they are poor conductors of electricity.
6. F. Mass is the amount of matter in an object, while weight is the gravitation pull on that object. They are often used interchangeably as it is rare to be making measurements in any gravitational field but Earth's, but it is important to understand the distinction.
7. F. The units most commonly used in the laboratory to measure volume are the liter (dm^{3}) and the milliliter (cm^{3}).
8. T
9. T
10. T, but be careful with numbers like this when it comes to significant figures.

Multiple Choice

1. d
2. b
3. a
4. c
5. d
6. a
7. d
8. b
9. c
10. d
11. d
12. b
13. a
14. c
15. a

Matching

Scientific method	e
Theory	h
Noble gases	f
Transition metals	j
Matter	i
Mass	a
Precision	d
Accuracy	c
Dimensional analysis	g
Chemical property	b

Fill–in–the–Blank

1. theory
2. elements
3. groups, periods
4. chlorine, Cl
5. intrinsic, physical
6. chemical
7. metals
8. kilogram, Kelvin, meter
9. Kelvin, Celsius, Fahrenheit
10. density
11. accurate, precise
12. scientific notation
13. round, significant figures
14. conversion factor
15. the number with the fewest digits to the right of the decimal
16. groups
17. metalloids, semimetals or semiconductors
18. giga, G
19. mass
20. volume

Problems

1. $246 \text{ cm} \times \frac{10 \text{ mm}}{1 \text{ cm}} = 2460 \text{ mm}$

$2460 \text{ mm} \times \frac{1 \text{ m}}{1000 \text{ mm}} = 2.46 \text{ m}$

$2.460 \text{ m} \times \frac{1 \text{ km}}{1000 \text{ m}} = 2.46 \text{ x } 10^{-3}\text{km}$

2. $\frac{8.3}{2.793} = 3.0$ (The least amount of significant figures, two, is found in the numerator.)

23.945 62 x 8.3421 = 199.76 (The least amount of significant figures, five, is found in 8.3421)

3.2 + 5.3875 = 8.6 (There is only one digit to the right of the decimal point in the number 3.2, so the answer must be rounded off to one digit after the decimal point.)

9434.21 – 4.199 = 9430.01 (There are only two digits to the right of the decimal point in 9434.21.)

3. $°\text{F} = \left(\frac{9}{5} \times 56\right) + 32 = 130 \text{ °F}$

4. The conversion factor needed to solve this problem is 1 mile = 1.6093 km

$$290{,}587 \text{ mi}^2 \times \frac{1.6093 \text{ km}}{1 \text{ mi}} \times \frac{1.6093 \text{ km}}{1 \text{ mi}} = 752{,}576 \text{ km}^2.$$

5. The conversion factor needed to solve this problem is 1 x 10^9 nm = 1 m.

$6.73 \times 10^{-7} \text{ m} \times \frac{1 \times 10^9 \text{ nm}}{1 \text{ m}} = 673 \text{ nm or } 6.73 \times 10^2 \text{nm}$

6. 78.5 °C + 273.15 = 351.6 K

$$ {}^{o}F = \left(\frac{9}{5} \times 78.5\right) + 32 = 173\ {}^{o}F $$

7. The conversion factors needed to solve these problems are 1 mile = 1.6093 km and 1 hour = 3600 s

Sprinter:

$$ \text{speed} = \frac{100\text{ m}}{9.58\text{s}} \times \frac{3600\text{ s}}{1\text{ h}} \times \frac{1\text{ km}}{1000\text{m}} \times \frac{1\text{mile}}{1.6093\text{ km}} = 23.4\text{ miles/hour} $$

Swimmer:

$$ \text{speed} = \frac{100\text{ m}}{47.85\text{s}} \times \frac{3600\text{ s}}{1\text{ h}} \times \frac{1\text{ km}}{1000\text{m}} \times \frac{1\text{mile}}{1.6093\text{ km}} = 4.68\text{ miles/hour} $$

8. density = mass/volume

$$ \text{density} = \frac{9.634\text{ g}}{(16.2\text{ mL - }11.5\text{ mL})} = 2.049787\text{ g/mL} = 2.05\text{ g/mL} $$

9. It is important to realize that the speed of light, 3.00 x 10^8 m/s can also be used as a conversion factor. 3.00 x 10^8 m = 1 s

$$ 13.6 \times 10^9\text{ km} \times \frac{1000\text{ m}}{1\text{ km}} \times \frac{1\text{ s}}{3.00 \times 10^8\text{ m}} \times \frac{1\text{ hour}}{3600\text{ s}} = 12.6\text{ hours} $$

10. The first problem is figuring out the volume of the pycnometer based on the mass of the water.

37.82 g – 23.75 g = 14.07 g water. The mass of this water is 14.07g/0.9982g/mL = 14.095 mL of water. This is the volume of the pycnometer.

When the beryllium is put in, we need to solve for the mass of the remaining water:

Total mass – mass of pycnometer – mass of beryllium = mass of remaining water

40.72 g – 23.75 g – 6.345 g = 10.625 g water.

The volume of this water will be 10.625 g/0.9982 g/mL = 10.644 mL

The difference between this volume and the total volume of the apparatus is the volume of the metal.

Beryllium volume = 14.095 mL – 10.644 mL = 3.451 mL beryllium.

Density = mass/volume = 6.345 g/3.451mL = 1.84 g/mL (At this point, the answer is rounded off to the correct number of significant figures.)

11. Volume = mass/density. 10.67g/4.55 g/mL = 2.35 mL

12. This problem requires a system of equations.

Let x = grams of oil, and y = grams of trichloroethane.

Assume 100 g total solution

$x + y = 100$ g

If there are 100 g total solution, and the liquids have been mixed so as to give the same density as water, it is possible also to determine the volume of the 100 g of solution.

100g/0.997 g/mL = 100.3mL

It is also possible to solve for the volumes of each of the components.

$x/0.918 + y/1.3492 = 100.3$ mL

Multiply through both sides by 1.3492.

$1.4697x + y = 135.32$

Set up a system of equations:

$$\begin{array}{rl} & 1.4697x + y = 135.32 \\ - & \underline{\quad\quad\quad x + y = 100} \\ & 0.4697x \quad = 35.32 \\ & x \quad = 75.2 \text{ g} \\ & y \quad = 24.8 \text{ g} \end{array}$$

13. Any of the following: color, smell, density, melting point, solubility, boiling point, conductivity. Note that these are all intrinsic properties: none of these properties are dependent on the quantity of the substance.

14. 0 °C is the freezing point of water, while 0 K is the coldest possible temperature, absolute zero.

15. The numbers in a calculation are not rounded or truncated until the final step. When a number is rounded because of significant figures, it can be changed slightly. If each number along the way is rounded, error will be introduced at each step. For example: if a box is measured as being 2.57 cm long, 3.634 cm wide, and 1 cm tall, the volume would properly be calculated 2.57 cm x 3.634 cm x 1 cm = 9.339 cm^3, which then is rounded to 9 cm^3. If an overzealous student were instead to round all the numbers to one significant figure before beginning the calculation, the answer would be 12 cm^3, a significant error.

Challenge Problem

On a mercury thermometer the liquid range of gallium is 2373.22 °C (the difference in temperature between the melting and boiling points). On the gallium thermometer the liquid range of gallium is 1000 °C. The size adjustments for these two scales are

$$1°C \times \frac{1000°Ga}{2373.22°C} = 0.421°Ga \qquad 1°Ga \times \frac{2373.22°C}{1000°Ga} = 2.37°C$$

Therefore, 1 °C is 0.421 times smaller than 1 °Ga, and 1 °Ga is 2.37 times larger than 1 °C.

The melting point of gallium is higher by 29.78° on the Celsius scale, so we would first do the size adjustment followed by the zero point adjustment when converting from °Ga to °C.

$$°C = (2.37 \times °Ga) + 29.78$$

Now convert the melting point of copper from °Ga to °C.

$$^{o}C = \left(2.37 \times 447^{o}Ga\right) + 29.78 = 1091^{o}C$$

We can determine the accuracy of the thermometer by comparing the difference between the converted and reported value to the reported value.

$$\frac{\left(\left|1085\ ^{o} - 1091\ ^{o}\right|\right)}{1085\ ^{o}} \times 100 = 0.52\ \%$$

Chapter 2

True–False

1. F. *Ions* are the same element carrying a different charge, *isotopes* are the same element with differing mass numbers.
2. T. This is the law of conservation of mass.
3. T. Broken atoms no longer retain their identity as elements.
4. F. Nonmetals have a strong tendency to form multiple compounds – think of carbon monoxide and carbon dioxide – and so must be named using prefixes.
5. F. Protons and neutrons are found in the nucleus, electrons in a cloud around the nucleus.
6. T. If the number of protons in a nucleus is changed, the identity of the atom also changes.
7. F. Atoms of the same element can have different masses.
8. T. This is what defines acids.
9. F. The ratio of elements in an ionic compound is identified by charge balance.
10. F. Transition metals can carry a variety of charges. Main group metals have invariant charges.

Multiple Choice

1. c
2. d
3. a
4. b
5. d
6. c
7. b
8. a
9. c
10. b

Fill–in–the–Blank

1. atoms, protons, neutron, and electrons
2. protons, atomic number
3. neutrons, isotope, electrons, ion
4. compounds, mixtures
5. covalent, shared
6. metal, nonmetal, ionic, opposite charges
7. polyatomic ion, nitrate, ammonium, sulfate
8. acid, base
9. di-, tri-, penta-
10. reactants, Law of Conservation of Mass

Matching

Ionic compound	g
Base	q

Molecules	j
Atomic number	b
Polyatomic ions	n
Binary compounds	a
Acid	c
Heterogeneous mixture	s
Structural formula	k
Electrons	d
Protons	m
Anion	p
Atoms	t
Neutrons	l
Ernest Rutherford	e
Isotopes	r
J. J. Thompson	o
Covalent bond	f
Cation	i
Atomic mass unit	h

Problems

1. mass of oxygen $3.2 \text{ L oxygen x } \frac{1.43 \text{ g}}{\text{L}} = 4.58 \text{g oxygen}$

   hydrogen + 4.58 g oxygen → 5.15 g water

   mass of hydrogen = 5.15 g – 4.58 g = 0.57 g hydrogen.

   volume of hydrogen $= \frac{0.57 \text{ g}}{0.0893 \text{ g/L}} = 6.4 \text{ L}$

2. Compound 1: $\text{O : P} = \frac{11.4 \text{ g O}}{4.3 \text{ g P}} = 2.65$

   Compound 2: $\text{O : P} = \frac{5.33 \text{ g O}}{3.34 \text{ g P}} = 1.59$

   $$\frac{\text{O : P ratio in compound 2}}{\text{O : P ratio in compound 1}} = \frac{1.59}{2.65} = 0.6 = \frac{6}{10} = \frac{3}{5}$$

3. a) p = 42, n = 55, e = 42
   b) p = 92, n = 143, e = 86
   c) p = 93, n = 142, e = 93
   d) p = 80, n = 122, e = 76

4. Abundances must all add up to 100%:
   Abundance of Neon-22 = 100% - 90.48% – 0.27% = 9.25%

   Solve for mass of Neon-22

   $= (19.992 \text{ amu} \times 0.9048) + (20.993 \text{ amu} \times 0.0027) + (x \text{ amu} \times 0.0925) = 20.18 \text{ amu}$

   mass = 21.99 amu

5. Average atomic mass

$= (0.7899 \times 23.985 \text{ amu}) + (0.1000 \times 24.986 \text{ amu}) + (0.1101 \times 25.983 \text{ amu}) = 24.305\text{amu}$

6. Naphthalene = C_8H_{10}

7. a) LiCl – lithium chloride b) Na_2S - sodium sulfide c) CaO – calcium oxide d) Mg_3N_2 – magnesium nitride

8. a) carbon monoxide - covalent b) strontium chloride – ionic
c) carbon tetrachloride – covalent d) platinum (IV) oxide - ionic

9. $9500 \text{ atoms} \times \frac{196.97 \text{ amu}}{1 \text{ atom}} \times \frac{1.6605 \times 10^{-24} \text{ g}}{1 \text{ amu}} = 3.11 \times 10^{-18} \text{ g}$

10. In dichloromethane, there are two chlorine atoms, two hydrogen atoms, and one carbon atom. From the periodic table, the masses of these atoms add up to

(2 x 35.5) + (2 x 1.00) + 12.01 = 85.01 amu

The percentage of chlorine will be (2 x 35.5)/85.01 x 100 = 83.5% Cl

The percentage of hydrogen will be (2 x 1.00)/85.01 x 100 = 2.35% H

The percentage of carbon will be 12.01/85.01 x 100 = 14.1% C

11. a. CaF_2 b. $Ca_3(PO_4)_2$ c. $Ca(NO_3)_2$ d. $CaCO_3$

12. a. tin (IV) chloride b. phosphoric acid c. magnesium acetate

d. hydrogen chloride e. rubidium sulfide f. carbon dioxide

g. chromium (III) phosphate h. difluorine monoxide i. phosphorous pentoxide

j. nickel (II) chloride k. iron (III) hydroxide l. hydrochloric acid

m. lead (II) iodide n. nitric acid o. barium bromide

p. aluminum nitrate

13. a. $Ca(NO_2)_2$ b. NaCl c. Cl_2O d. H_2O

e. NiH_2 f. BBr_3 g. $CuCl_2$ h. XeF_4

i. NO_2 j. Ba_3N_2 k. H_3PO_4(*aq*) l. HF (*aq*)

m. $LiMnO_4$ n. H_2CO_3 (*aq*) o. PCl_3 p. CBr_4

Challenge Problem

To calculate the volume of the sphere, it is necessary first to find the radius. It is also a good idea to convert the radius into centimeters, as the final desired unit is cm^3.

Diameter = 0.2 mm

Radius = 0.2 mm/2 x 1cm/10mm = 0.01cm

Volume of a sphere = $4/3\pi r^3 = 4.19 \times 10^{-6}$ cm^3

With the volume, and the densities of the two different metals, it is possible to calculate the masses of the spheres, then convert the masses to amu, and the amu to atoms:

$$\text{Osmium} = 22.61 \text{ g/cm}^3 \times 4.19 \times 10^{-6} \text{ cm}^3 \times \frac{1 \text{ amu}}{1.6605 \times 10^{-24} \text{ g}} \times \frac{1 \text{ atom}}{190.23 \text{ amu}} = 2.999 \times 10^{17} \text{ atoms}$$

$$\text{Aluminum} = 2.70 \text{ g/cm}^3 \times 4.19 \times 10^{-6} \text{ cm}^3 \times \frac{1 \text{ amu}}{1.6605 \times 10^{-24} \text{ g}} \times \frac{1 \text{ atom}}{26.98 \text{ amu}} = 2.525 \times 10^{17} \text{ atoms}$$

If these atoms were the size of sand grains 1 mm square, what volume would they take up?

Volume per “atom” = $(0.1 \text{ mm} \times 1 \text{ cm}/10 \text{ mm})^3 = (0.01 \text{ cm})^3 = 1 \times 10^{-6}$ cm^3

Osmium: 2.999×10^{17} atoms x 1×10^{-6} cm^3 = 2.999×10^{11} cm^3 = 299,900,000 liters

Aluminum: 2.525×10^{17} atoms x 1×10^{-6} cm^3 = 2.525×10^{11} cm^3 = 299,900,000 liters

To put this in perspective, an Olympic-sized swimming pool holds about 2,500,000 liters, so if the atoms in this miniscule sphere were expanded to the size of small grains of sand, they would be equivalent in volume to around 120 Olympic-sized swimming pools. Atoms are really, really tiny.

Chapter 3

True–False

1. F. The number of atoms of each type must be the same, but the number of molecules is often different.
2. T
3. T
4. F. The coefficients in a balanced chemical equation show the *mole* ratios of the components.
5. F. Theoretical yields of chemical reactions are determined through mathematical calculation, not experiment.
6. T
7. F. Molarity is moles of solute per liter of solution, not solvent.
8. F. Combustion analysis can only be used to determine the empirical formula of a compound.
9. T
10. T

Matching

Balanced	i
Yield	d
Mole	l
Coefficient	e
Stoichiometry	m
Dilution	k
Molecular mass	a
Empirical formula	b
Solute	c
Titration	f
Subscript	h
Combustion	g
Actual yield	j

Fill–in–the–Blank

1. balanced
2. coefficients, subscripts
3. molecular mass or formula mass
4. limiting reactant
5. excess reactant
6. theoretical yield, actual yield, percent yield
7. empirical formula, molecular formula, molecular mass
8. solvent, solution, molarity
9. carbon dioxide, water
10. Avogadro's number, 6.02×10^{23}

Problems

1. a. $2N_2 + 5O_2 \rightarrow 2N_2O_5$

   b. $C_3H_8 + 5O_2 \rightarrow 3CO_2 + 4H_2O$

   c. $H_3PO_4 + 3NaOH \rightarrow Na_3PO_4 + 3H_2O$

   d. $Ba(NO_3)_2 + 2NaOH \rightarrow Ba(OH)_2 + 2NaNO_3$

2. a. 45 g calcium chloride – formula is $CaCl_2$, molecular mass = 40 + (2 x 35.5) = 111 g/mol

   b. 23 g sucrose ($C_{12}H_{22}O_{11}$) – molecular mass = (12 x 12) + (22 x 1.008) + (11 x 16) = 342 g/mol

   c. 67 g phosphorus triflouride – formula is PF_3, molecular mass = 31 + (19 x 3) = 88 g/mol

   d. 5 grams ammonium carbonate – formula is $(NH_4)_2CO_3$, molecular mass is (18 x 2) + 60 = 96 g/mol

$$5\text{ g }(NH_4)_2CO_3 \text{ x } \frac{1\text{ mol }(NH_4)_2CO_3}{96\text{ g }(NH_4)_2CO_3} = 0.052\text{ mol }(NH_4)_2CO_3$$

3. To determine whether there is a limiting reactant, run the reaction mathematically using both reagents.

$$2N_2 + 5O_2 \rightarrow 2N_2O_5$$

$$4\text{ mol }N_2 \text{ x } \frac{2\text{ mol }N_2O_5}{2\text{ mol }N_2} = 4\text{ mol }N_2O_5$$

$$10\text{ mol }O_2 \text{ x } \frac{2\text{ mol }N_2O_5}{5\text{ mol }O_2} = 4\text{ mol }N_2O_5$$

The reactants are present in stoichiometric ratios, so the theoretical yield is 4 mol of N_2O_5. The molecular mass of dinitrogen pentoxide is 108 g/mol, so the theoretical yield in grams is

$$4\text{ mol }N_2O_5 \text{ x } 108\frac{\text{g}}{\text{mol}} = 432\text{ g}$$

If the actual yield is 376 grams, the percent yield is calculated

$$\frac{376\ g}{432\ g} x\ 100 = 87\%$$

4. Balance the equation.

$$2NaHCO_3 \rightarrow Na_2CO_3 + CO_2 + H_2O$$

Convert the 20 grams of sodium hydrogen carbonate to moles: the formula weight is 84.

$$20\ g\ NaHCO_3 x \frac{1\ mol\ NaHCO_3}{84\ g\ NaHCO_3} = 0.238\ mol\ NaHCO_3$$

Set up a stoichiometric ratio to determine the theoretical yield of water:

$$0.238\ mol\ NaHCO_3 x \frac{1\ mol\ H_2O}{2\ mol\ NaHCO_3} = 0.119\ mol\ H_2O$$

$$0.119\ mol\ H_2O\ x \frac{18\ g\ H_2O}{mol\ H_2O} = 2.14\ g\ H_2O$$

If the reaction has a typical percent yield of 87%, then only 87% of the theoretical yield can be expected.

$$2.14\ g\ H_2O\ x\ 0.87 = 1.86\ g\ H_2O$$

5. Write and balance the equation.

$$CH_4 + 2O_2 \rightarrow CO_2 + 2H_2O$$

Convert the quantities given to moles:

$$6.7g\ CH_4 x \frac{1\ mol\ CH_4}{16\ g\ CH_4} = 0.419\ mol\ CH_4$$

$$20\ g\ O_2 x \frac{1mol\ O_2}{32g\ O_2} = 0.625\ mol\ O_2$$

Run the reaction with each quantity to determine the limiting reagent and the theoretical yields.

$$0.419\ mol\ CH_4 x \frac{1\ mol\ CO_2}{mol\ CH_4} = 0.419\ mol\ CO_2 x \frac{44\ g\ CO_2}{mol\ CO_2} = 18.4\ g\ CO_2$$

$$0.419\ mol\ CH_4 x \frac{2\ mol\ H_2O}{mol\ CH_4} = 0.838\ mol\ H_2Ox \frac{18\ g\ H_2O}{mol\ H_2O} = 15.1\ g\ H_2O$$

$$0.625\ mol\ O_2 x \frac{1\ mol\ CO_2}{2\ mol\ O_2} = 0.313\ mol\ CO_2 x \frac{44\ g\ CO_2}{mol\ CO_2} = 13.8\ g\ CO_2$$

$$0.625\ mol\ O_2 x \frac{2\ mol\ H_2O}{2\ mol\ O_2} = 0.625\ mol\ H_2Ox \frac{18\ g\ H_2O}{mol\ H_2O} = 11.3\ g\ H_2O$$

The lower theoretical yields come from the 20 g of oxygen, so this is the limiting reagent, and the theoretical yields are 13.8 g of carbon dioxide and 11.3 g of water.

To determine how many grams of methane will be left over, set up another stoichiometry problem to determine how many moles are consumed.

$$0.625 \text{ mol } O_2 \text{ x } \frac{1 \text{ mol } CH_4}{2 \text{ mol } O_2} = 0.313 \text{ mol } CH_4 \text{ consumed}$$

0.419 moles of methane were available, 0.313 were consumed, leaving 0.106 moles.

$$0.106 \text{ moles } CH_4 \text{ x } \frac{16 \text{ g } CH_4}{\text{mol } CH_4} = 1.7 \text{ g } CH_4 \text{ left over}$$

6. Using the definition of molarity, M = mol/V, rearrange to mol = M x V

$$\text{mol KI} = 0.35 \frac{\text{mol}}{\text{L}} \text{ x } 0.250 \text{ L} = 0.0875 \text{ mol KI x } \frac{166 \text{ g KI}}{\text{mol KI}} = 14.5 \text{ g}$$

Determine how many moles of potassium iodide are contained in 2.60g.

$$2.60 \text{ g KI x} \frac{1 \text{ mol KI}}{166 \text{ g KI}} = 0.0157 \text{ mol KI}$$

Rearrange the definition of molarity again to solve for volume: V = mol/M

$$V = \frac{0.0157 \text{ mol}}{0.35 \text{ M}} = 0.0447 \text{ L } = 44.7 \text{ mL}$$

To dilute the 0.35 M solution to provide 100 mL of 0.10 M solution, the formula $M_1 \text{ x } V_1 = M_2 \text{ x } V_2$ should be used. It can be rearranged to solve for V_1.

$$V_1 = \frac{M_2 \text{ x } V_2}{M_1} = \frac{0.10 \text{ M x} 100 \text{ mL}}{0.35 \text{ M}} = 28.6 \text{ mL stock}$$

7. Write and balance the equation.

$$Fe + CuSO_4 \rightarrow FeSO_4 + Cu$$

Determine how many moles of each reactant are present.

$$1.37 \text{ g Fe x} \frac{1 \text{ mol Fe}}{55.8 \text{ g}} = 0.0246 \text{ mol Fe}$$

$$0.100 \text{ L } CuSO_4 \text{ x } \frac{0.50 \text{ mol } CuSO_4}{\text{L}} = 0.050 \text{ mol } CuSO_4$$

The reactants interact in a 1:1 ratio, so in this case, the smaller number of moles will be the limiting reactant: all the Fe will be converted to Fe^{2+}, displacing an equal quantity of copper.

$$0.0246 \text{ mol Fe x } \frac{1 \text{ mol Cu}}{1 \text{ mol Fe}} = 0.0246 \text{ mol Cu x } \frac{63.5 \text{ g Cu}}{\text{mol Cu}} = 1.56 \text{ g Cu}$$

8. Balance the equation:

$$Zn + 2\,HCl \rightarrow ZnCl_2 + H_2$$

Determine how many moles of zinc are present.

$$1.63 \text{ g Zn x} \frac{1 \text{ mol Zn}}{65.4 \text{ g}} = 0.0249 \text{ mol Zn}$$

Set up a stoichiometric ratio for the reaction to determine how many moles of HCl are needed to completely consume this amount of zinc.

$$0.0249 \text{ mol Zn x} \frac{2 \text{ mol HCl}}{\text{mol Zn}} = 0.0498 \text{ mol HCl}$$

Rearrange the definition of molarity to solve for volume.

$$V = \frac{\text{mol}}{\text{M}} = \frac{0.0514 \text{ mol HCl}}{0.50 \text{ M HCl}} = 0.0996 \text{ L HCl} = 99.6 \text{ mL}$$

Any quantity over 100 mL of acid should consume the zinc. If you were doing this in the lab, you would know that this is the minimum quantity of acid required, and probably add a little more to make things go faster.

9. $Ba(NO_3)_2 + 2NaOH \rightarrow Ba(OH)_2 + 2NaNO_3$

Find the number of moles of each reactant from their concentrations and volumes.

$$0.050 \text{ L Ba(NO}_3)_2 \text{ x } 0.250 \text{ M Ba(NO}_3)_2 = 0.0125 \text{ mol Ba(NO}_3)_2$$

$$0.050 \text{ L NaOH x } 0.100 \text{ M NaOH} = 0.0050 \text{ mol NaOH}$$

Determine the limiting reactant: 2 moles of NaOH are needed for each mole of $Ba(NO_3)_2$, so NaOH is clearly the limiting reagent here. Set up a stoichiometry problem to solve for the moles of barium hydroxide that can be produced.

$$0.0050 \text{ mol NaOHx } \frac{1 \text{ mol Ba(OH)}_2}{\text{mol NaOH}} = 0.0050 \text{ mol Ba(OH)}_2 \text{x} \frac{171.3 \text{ g Ba(OH)}_2}{\text{mol Ba(OH)}_2} = 0.857 \text{ g Ba(OH)}_2$$

10. No equation is given, but none is needed: The information required is the fact that the sodium hydroxide and acetic acids react in a 1:1 ratio. The moles of sodium hydroxide will be equivalent to the number of moles of acetic acid, and the moles can be divided by volume to give the molarity of the acetic acid solution.

$$0.100 \text{ M NaOH x } 0.0173 \text{ L NaOH} = 0.00173 \text{ mol NaOH x} \frac{1 \text{mol acetic acid}}{\text{mol NaOH}} = 0.00173 \text{ mol acetic acid}$$

$$M = \frac{0.00173 \text{ mol acetic acid}}{0.050 \text{ L solution}} = 0.0346 \text{ M acetic acid}$$

11. Carbon dioxide

CO_2 molecular weight = 44 amu

$$\%C = \frac{12}{44} x100 = 27.3\%$$

$$\%O = \frac{32}{44} x100 = 72.7\%$$

Sodium phosphate

Na_3PO_4 molecular weight = 164 amu

$$\%Na = \frac{69}{164} x100 = 42.1\%$$

$$\%P = \frac{31}{164} x100 = 18.9\%$$

$$\%O = \frac{64}{164} x100 = 39.0\%$$

CH_3COOH

Acetic acid molecular weight = 60 amu

$$\%C = \frac{24}{60} x100 = 40\%$$

$$\%H = \frac{4}{60} x100 = 6.7\%$$

$$\%O = \frac{32}{60} x100 = 53.3\%$$

Adenosine triphosphate: $C_{10}H_{16}N_5O_{13}P_3$

Molecular weight = 507 amu

$$\%C = \frac{120}{507} x100 = 23.7\%$$

$$\%H = \frac{16}{507} x100 = 3.16\%$$

$$\%N = \frac{70}{507} x100 = 13.8\%$$

$$\%O = \frac{208}{507} x100 = 41.0\%$$

$$\%P = \frac{93}{507} x100 = 18.3\%$$

12. Convert percentages to mole ratios assuming a 100-gram sample of the unknown compound.

$$40\%\ C = 40\ g\ C\ x \frac{1\ mol\ C}{12\ g\ C} = 3.33\ mol\ \ C$$

$$6.7\%\ H = 6.7\ g\ H\ x \frac{1\ mol\ H}{1.008\ g\ H} = 6.65\ mol\ H$$

$$53.3\%\ O = 53.3\ g\ O\ x \frac{1\ mol\ O}{16\ g\ O} = 3.33\ mol\ \ O$$

Carbon and oxygen have a 1:1 ratio, and carbon and hydrogen are 1:2, making the empirical formula CH_2O (this empirical formula is the origin of the term "carbohydrate" – you know from this that you are looking at a sugar molecule). The empirical formula weight is 30, so there are six empirical formula units in the molecular formula, which must be $C_6H_{12}O_6$. For the record, this is glucose.

13. Convert grams of water and carbon dioxide to moles, then grams of carbon and hydrogen.

$$3.71\ g\ CO_2 x \frac{1\ mol\ CO_2}{44\ g\ CO_2} x \frac{1\ mol\ C}{mol\ CO_2} = 0.084\ mol\ Cx \frac{12\ g\ C}{mol\ C} = 1.01\ g\ C$$

$$1.51\ g\ H_2O\ x \frac{1\ mol\ H_2O}{18\ g\ H_2O} x \frac{2\ mol\ H}{mol\ H_2O} = 0.1667\ mol\ H\ x \frac{1.008\ g\ H}{mol\ H} = 0.169\ g\ H$$

From this, it is possible to solve for the grams of oxygen in the sample.

$$1.57g - 1.01g - 0.169g = 0.391\ g\ O$$

$$0.391\ g\ O\ x \frac{1\ mol\ O}{16\ g\ O} = 0.024\ mol\ O$$

It is now possible to develop mole ratios for the formula.

$$\frac{0.084\ mol\ C}{0.024\ mol\ O} = 3.5 \frac{C}{O}$$

$$\frac{0.167\ mol\ H}{0.024\ mol\ O} = 7 \frac{H}{O}$$

From this, an initial formula can be developed: $C_{3.5}H_7O$. This can now be refined to the proper formula: $C_7H_{14}O_2$.

Challenge Problem

To attack this problem, first solve for the mole quantity of hydroxide generated by the reaction.

$$0.0597\ L\ HCl\ x\ 7.5 \frac{mol}{L} x \frac{1mol\ OH}{mol\ HCl} = 0.448\ mol\ OH$$

This is also equivalent to the number of moles of the alkali metal. There is now enough information to set up a system of equations.

$$Li + Na = 4\ g$$

$$\frac{Li}{6.9} + \frac{Na}{23} = 0.448mol$$

This second equation is equivalent to $0.145Li + 0.0435Na = 0.448$
To solve by substitution,

$Li = 4 - Na$, so

$0.145(4-Na) + 0.0435Na = 0.448$

$0.58 - 0.145Na + 0.0435Na = 0.448$

$-0.145Na + 0.0435Na = 0.448 - 0.58$

$-0.102Na = -0.132$

$Na = 1.3$ g

$Na + Li = 4$

$Li = 2.7$ g

Chapter 4

True–False

1. F. An acid and a base react to form water and a salt.
2. F. Both sodium and nitrate indicate solubility.
3. T
4. T
5. T
6. F. Acids release hydrogen ions or hydronium ions into solution.
7. T
8. F. Oxidation numbers can be equal to ionic charge, but are not always.
9. F. An activity series makes it possible to predict the results of redox reactions.
10. T

Matching

Reduction	h
Net ionic equation	f
Oxidation number	l
Weak electrolyte	a
Redox reaction	b
Dissociate	o
Spectator	j
Hydroxide ion	n
Activity series	d
Electrolyte	m
Reducing agent	e
Precipitation reaction	k
Oxidation	i
Hydronium ion	c
Oxidizing agent	g
Strong electrolyte	r
Base	p
Acid	q

Fill–in–the–Blank
1. soluble, insoluble, precipitation
2. acid, base, water, salt
3. redox or oxidation/reduction
4. strong electrolytes, weak electrolytes
5. molecular, ionic, net ionic
6. hydrogen or hydronium, hydroxide
7. loses, gains

Problems

1. a. HNO_3 (*aq*) + KOH (*aq*) → KNO_3 (*aq*) + H_2O (*l*).

 Neutralization: net ionic equation is $H^+ + OH^- \rightarrow H_2O$

 b. $Cu(NO_3)_2$ (*aq*) + 2NaOH (*aq*) → $Cu(OH)_2$ (*aq*) + $2NaNO_3$ (*aq*)

 Precipitation: two soluble components generate and insoluble product.

 c. $Cu(NO_3)_2$ (*aq*) + Zn (*s*) → $Zn(NO_3)_2$ (*aq*) + Cu (*s*)

 Redox: Cu and Zn are transferring electrons – Zn is oxidized, Cu is reduced.

2. Convert grams to moles.

$$15 \text{ g } Na_3PO_4 \text{x} \frac{1 \text{ mol}}{164 \text{ g}} = 0.0915 \text{ mo}$$

 Convert to molar concentration of ions.

$$\frac{0.0915 \text{ mol } Na_3PO_4}{.250 \text{ L}} \text{ x } \frac{4 \text{ ions}}{Na_3PO_4} = 1.46 \text{ M}$$

3. a. $Pb(NO_3)_2$ (*aq*) + 2NaCl (*aq*) → $PbCl_2$ (*s*) + $2NaNO_3$ (*aq*)

 Pb^{2+} (*aq*) + $2NO_3^-$ (*aq*) + $2Na^+$ (*aq*) + $2Cl^-$ (*aq*) → $PbCl_2$ (*s*) + $2Na^+$ (*aq*) + $2NO_3^-$ (*aq*)

 Pb^{2+} (*aq*) + 2Cl (*aq*) → $PbCl_2$ (*s*)

 b. H_2SO_4 (*aq*) + 2LiOH (*aq*) → Li_2SO_4 (*aq*) + $2H_2O$ (*l*)

 $2H^+$ (*aq*) + SO_4^{2-} (*aq*) + $2Li^+$ (*aq*) + $2OH^-$ (*aq*) → $2Li^+$ (*aq*) + SO_4^{2-} (*aq*) + $2H_2O$ (*l*)

 H^+ (*aq*) + OH^- (*aq*) → H_2O (*l*)

 c. KCl (*aq*) + $AgCH_3CO_2$ (*aq*) → KCH_3CO_2 (*aq*) + AgCl (*s*)

 K^+ (*aq*) + Cl^- (*aq*) + Ag^+ (*aq*) + $CH_3CO_2^-$ (*aq*) → K^+ (*aq*) + $CH_3CO_2^-$ (*aq*) + AgCl (*s*)

 Ag^+ (*aq*) + Cl^- (*aq*) → AgCl (*s*)

4. Write a balanced net ionic equation.
 Ca^{2+} (*aq*) + $2OH^-$ (*aq*) → $Ca(OH)_2$ (*s*)

Determine the number of moles of calcium hydroxide required:

$$1.8\text{g Ca(OH)}_2 \text{ x} \frac{1 \text{ mol}}{74.1 \text{ g}} = 0.243 \text{ mol}$$

From the balanced equation above, 0.243 mol of Ca ions are required, and 0.0486 mol of hydroxide ions are required.

From the definition of molarity, solve for the volumes needed.

$$M = \frac{\text{mol}}{L} \therefore V = \frac{\text{mol}}{M}$$

$$V_{OH} = \frac{0.0486 \text{ mol}}{1 \text{ M}} = 0.0486 \text{ L or } 48.6 \text{ mL}$$

$$V_{Ca} = \frac{0.0243 \text{ mol}}{1 \text{ M}} = 0.0243 \text{ L or } 24.3 \text{ mL}$$

48.6 mL 1 M $Ca(NO_3)_2$ solution would need to be combined with 24.3 mL 1 M KOH to produce 1.8 g $Ca(OH)_2$.

5. a. No reaction. Possible products are lead acetate (soluble) and copper (II) nitrate (soluble).

   b. No reaction. Possible products are sodium sulfate (soluble) and ammonium nitrate (soluble).

   c. Ag^+ (*aq*) + Cl^- (*aq*) → AgCl (*s*)

6. a. Cl_2 Elemental forms, even in molecules, always have oxidation numbers of 0

   b. NaCl Na = +1, Cl = -1

   c. CO_3^{2-} O = -2, C = +4 4 + (3 x -2) = -2, which is the charge on the ion.

   d. Cr_2O_3 O = -2, Cr = +3 (3 x -2) + (2 x +3) = 0

   e. ClO_3^- O = -2, Cl = +5 (3 x -2) + 5 = -1

7. a. $C_2H_6 + O_2 \rightarrow CO_2 + H_2O$ (Combustion is a form of redox reaction).

   Oxidation number of C goes from -3 to +4 oxidized, reducing agent

   Oxidation number of O goes from 0 to -2 reduced, oxidizing agent

   b. $Na + Cl_2 \rightarrow NaCl$

   Oxidation number of Na goes from 0 to +1 oxidized, reducing agent

   Oxidation number of Cl goes from 0 to -1 reduced, oxidizing agent

   c. $Mg + Fe^{2+} \rightarrow Mg^{2+} + Fe$

   Oxidation number of Mg goes from 0 to +2 oxidized, reducing agent

Oxidation number of Fe goes from +2 to 0   reduced, oxidizing agent

d. $Ca + H^{+} \rightarrow Ca^{2+} + H_2$

Oxidation number of Ca goes from 0 to +2   oxidized, reducing agent

Oxidation number of H goes from +1 to 0   reduced, oxidizing agent

8. a. chromium and copper (II) nitrate

$2Cr\ (s) + 3Cu(NO_3)_2\ (aq) \rightarrow 2Cr(NO_3)_3\ (aq) + 3Cu\ (s)$

b. tin and silver (I) nitrate

$Sn\ (s) + 2AgNO_3 \rightarrow Sn(NO_3)_2 + 2Ag$

c. copper (II) nitrate and cobalt

$2Co\ (s) + 3Cu(NO_3)_2\ (aq) \rightarrow 2Co(NO_3)_3\ (aq) + 3Cu\ (s)$

d. silver and HCl – no reaction.

9. Not all of these require breaking into half-reactions.

a. $3Mg\ (s) + 2Cr(NO_3)_3\ (aq) \rightarrow 3Mg(NO_3)_2\ (aq) + 2Cr\ (s)$

b. $2Al\ (s) + 3FeCl_2\ (aq) \rightarrow 2AlCl_3\ (aq) + 3Fe\ (s)$

c. $BrO_3^-\ (aq) + N_2H_4 \rightarrow Br^-\ (aq) + N_2\ (g)$

$BrO_3^-\ (aq) \rightarrow Br^-\ (aq) + 3H_2O$ Balance O

$6\ H^+\ (aq) + BrO_3^-\ (aq) \rightarrow Br^-\ (aq) + 3H_2O$ Balance H

$6\ H^+\ (aq) + 6\ e^- + BrO_3^-\ (aq) \rightarrow Br^-\ (aq) + 3H_2O$ Balance e^-

$N_2H_4 \rightarrow N_2\ (g) + 4H^+\ (aq) + 4e^-$ Balance H and e^-

$12\ H^+\ (aq) + 12\ e^- + 2\ BrO_3^-\ (aq) \rightarrow 2Br^-\ (aq) + 6H_2O$

$3N_2H_4 \rightarrow 3N_2\ (g) + 12\ H^+\ (aq) + 12e^-$

$2BrO_3^-\ (aq) + 3N_2H_4 \rightarrow 2Br^-\ (aq) + 6H_2O + 3N_2\ (g)$

d. $MnO_4^-\ (aq) + Al\ (s) \rightarrow Mn^{2+}\ (aq) + Al^{3+}\ (aq)$

Balance half-reactions for O, H, and e^-.

$5e^- + 8H^+\ (aq) + MnO_4^-\ (aq) \rightarrow Mn^{2+}\ (aq) + 4H_2O\ (l)$

$Al\ (s) \rightarrow Al^{3+}\ (aq) + 3e^-$

Equalize numbers of electrons.

$15e^- + 24H^+ (aq) + 3MnO_4^- (aq) \rightarrow 3Mn^{2+} (aq) + 12H_2O (l)$

$5Al (s) \rightarrow 5Al^{3+} (aq) + 15e^-$

Add equations together.

$15e^- + 24H^+ (aq) + 3MnO_4^- (aq) + 5Al (s) \rightarrow 3Mn^{2+} (aq) + 12H_2O (l) + 5Al^{3+} (aq) + 15e^-$

Cancel:

$24H^+ (aq) + 3MnO_4^- (aq) + 5Al (s) \rightarrow 3Mn^{2+} (aq) + 12H_2O (l) + 5Al^{3+} (aq)$

10.

a. $NO_2^- (aq) + Al (s) \rightarrow NH_3 (aq) + AlO_2^- (aq)$

Balance half-reactions for O, H, and e^-.

$6e^- + 7H^+ (aq) + NO_2^- (aq) \rightarrow NH_3 (aq) + 2H_2O (l)$

$2H_2O (l) + Al (s) \rightarrow AlO_2^- (aq) + 4H^+ (aq) + 3e^-$

Equalize numbers of electrons.

$6e^- + 7H^+ (aq) + NO_2^- (aq) \rightarrow NH_3 (aq) + 2H_2O (l)$

$4H_2O (l) + 2Al (s) \rightarrow 2AlO_2^- (aq) + 8H^+ (aq) + 6e^-$

Add equations together.

$6e^- + 7H^+ (aq) + NO_2^- (aq) + 4H_2O (l) + 2Al (s) \rightarrow NH_3 (aq) + 2H_2O (l) + 2AlO_2^- (aq) + 8H^+ (aq) + 6e^-$

Cancel.

$NO_2^- (aq) + 2H_2O (l) + 2Al (s) \rightarrow NH_3 (aq) + 2AlO_2^- (aq) + H^+ (aq)$

Neutralize.

$OH^- (aq) + NO_2^- (aq) + 2H_2O (l) + 2Al (s) \rightarrow NH_3 (aq) + 2AlO_2^- (aq) + H^+ (aq) + OH^- (aq)$

Cancel.

$OH^- (aq) + NO_2^- (aq) + H_2O (l) + 2Al (s) \rightarrow NH_3 (aq) + 2AlO_2^- (aq)$

b. $Cl_2 (g) \rightarrow Cl^- (aq) + ClO^- (aq)$

$Cl_2 (g) + H_2O (l) \rightarrow Cl^- (aq) + ClO^- (aq) + 2H^+ (aq)$

$2OH^- (aq) + Cl_2 (g) + H_2O (l) \rightarrow Cl^- (aq) + ClO^- (aq) + 2H^+ (aq) + 2OH^- (aq)$

$2OH^- (aq) + Cl_2 (g) \rightarrow Cl^- (aq) + ClO^- (aq) + H_2O (l)$

c. $MnO_4^- (aq) + Br^- (aq) \rightarrow MnO_2 (s) + BrO_3^- (aq)$

Balance half-reactions for O, H, and e^-.

$3e^- + 4H^+ (aq) + MnO_4^- (aq) \rightarrow MnO_2 (s) + 2H_2O (l)$

$Br^- (aq) + 3H_2O (l) \rightarrow BrO_3^- (aq) + 6H^+ (aq) + 6e^-$

Equalize numbers of electrons.

$6e^- + 8H^+ (aq) + 2\ MnO_4^- (aq) \rightarrow 2\ MnO_2 (s) + 4\ H_2O (l)$

$Br^- (aq) + 3H_2O (l) \rightarrow BrO_3^- (aq) + 6H^+ (aq) + 6e^-$
Add reactions.

$6\ e^- + 8\ H^+ (aq) + 2\ MnO_4 (aq) + Br^- (aq) + 3\ H_2O (l) \rightarrow$
$2\ MnO_2 (s) + 4\ H_2O (l) + BrO_3^- (aq) + 6\ H^+ (aq) + 6e^-$

Cancel.

$2\ MnO_4 (aq) + 2\ H^+ (aq) + Br^- (aq) \rightarrow 2MnO_2 (s) + 4\ H_2O (l) + BrO_3^- (aq)$

d. $MnO_4^- (aq) + I^- (aq) \rightarrow I_2 + Mn^{2+} (aq)$

Balance half-reactions for O, H, and e^-.

$5e^- + 8H^+ (aq) + MnO_4^- (aq) \rightarrow Mn^{2+} (aq) + 4H_2O (l)$

$2I^- (aq) \rightarrow I_2 + 2e^-$

Equalize numbers of electrons.

$10e^- + 16H^+ (aq) + 2MnO_4^- (aq) \rightarrow 2Mn^{2+} (aq) + 8H_2O (l)$

$10I^- (aq) \rightarrow 5I_2 + 10e^-$

Add equations.

$10e^- + 16H^+ (aq) + 2MnO_4^- (aq) + 10I^- (aq) \rightarrow 2Mn^{2+} (aq) + 8H_2O (l) + 5I_2 + 10e^-$

Cancel.

$16H^+ (aq) + 2MnO_4^- (aq) + 10I^- (aq) \rightarrow 2Mn^{2+} (aq) + 8H_2O (l) + 5I_2$

Neutralize.

$16OH^- (aq) + 16H^+ (aq) + 2MnO_4^- (aq) + 10I^- (aq) \rightarrow 2Mn^{2+} (aq) + 8H_2O (l) + 5I_2 + 16OH^- (aq)$

Cancel.

$8H_2O (l) + 2MnO_4^- (aq) + 10I^- (aq) \rightarrow 2Mn^{2+} (aq) + 5I_2 + 16OH^- (aq)$

11. Balance the equation.

$SO_3^{2-} (aq) + Cr_2O_7^{2-} (aq) \rightarrow SO_4^{2-} (aq) + 2Cr^{3+} (aq)$

Balance half-reactions in H, O, and e^-.

SO_3^{2-} (*aq*) + H_2O (*l*) → SO_4^{2-} (*aq*) + $2H^+$ (*aq*) + $2e^-$

$Cr_2O_7^{2-}$(*aq*) + $14H^+$ (*aq*) + $6e^-$ → $2Cr^{3+}$ (*aq*) + $7H_2O$ (*l*)

Equalize numbers of electrons.

$3SO_3^{2-}$ (*aq*) + $3H_2O$ (*l*) → $3SO_4^{2-}$ (*aq*) + $6H^+$ (*aq*) + $6e^-$

$Cr_2O_7^{2-}$(*aq*) + $14H^+$ (*aq*) + $6e^-$ → $2Cr^{3+}$ (*aq*) + $7H_2O$ (*l*)

Add equations.

$3SO_3^{2-}$ (*aq*) + $3H_2O$ (*l*) + $Cr_2O_7^{2-}$(*aq*) + $14H^+$ (*aq*) + $6e^-$ →
$3SO_4^{2-}$ (*aq*) + $6H^+$ (*aq*) + $6e^-$ + $2Cr^{3+}$ (*aq*) + $7H_2O$ (*l*)

Cancel.

$3SO_3^{2-}$ (*aq*) + $Cr_2O_7^{2-}$(*aq*) + $8H^+$ (*aq*) → $3SO_4^{2-}$ (*aq*) + $2Cr^{3+}$ (*aq*) + $4H_2O$ (*l*)

Determine the moles of potassium chromate.

0.0237 L x 0.124 M = 0.00294 mole $K_2Cr_2O_7$

Determine moles of Na_2SO_3 using the balanced equation.

$$0.00294 \text{ mol } K_2Cr_2O_7 \text{ x } \frac{3 \text{ mol } Na_2SO_3}{\text{mol } K_2Cr_2O_7} = 0.00882 \text{ mol } Na_2SO_3$$

$$M = \frac{0.00882 \text{ mol } NaSO_3}{0.015 \text{ L}} = 0.588 \text{ M}$$

12. a. K_2SO_4 (*aq*) + $Pb(NO_3)_2$ → $2KNO_3$ (*aq*) + $PbSO_4$ (*s*)

$2K^+$ (*aq*) + SO_4^{2-} (*aq*) + Pb^{2+} (*aq*) + $2NO_3^-$ (*aq*) → $2K^+$ (*aq*) + $2NO_3^-$ (*aq*) + $PbSO_4$ (*s*)

Pb^{2+} (*aq*) + SO_4^{2-} (*aq*) → $PbSO_4$ (*s*)

b. Determine the number of moles of each reactant.

$$5.3 \text{ g } K_2SO_4 \text{ x } \frac{1 \text{mol}}{174 \text{ g } K_2SO_4} = 0.0304 \text{ mol } K_2SO_4$$

$$4.9 \text{ g } Pb(NO_3)_2 \text{ x } \frac{1 \text{mol}}{331 \text{ g } Pb(NO_3)_2} = 0.0148 \text{ mol } Pb(NO_3)_2$$

Reactants react in a 1:1 ratio: $Pb(NO_3)_2$ is the limiting reagent.

$$0.0148 \text{ mol } Pb(NO_3)_2 \text{ x } \frac{1 \text{ mol } PbSO_4}{\text{mol } Pb(NO_3)_2} \text{ x } 303 \frac{\text{g}}{\text{mol } PbSO_4} = 4.48 \text{ g } PbSO_4$$

c. Subtract the moles of limiting reagent used from the moles of excess reagent:

0.0304 mol – 0.0148 mol = 0.0156 mol K_2SO_4

$$0.0156 \text{ mol } K_2SO_4 \text{ x } \frac{174 \text{ g } K_2SO_4}{1 \text{ mol}} = 2.71 \text{ g } K_2SO_4$$

d. $\text{percent yield} = \frac{4.37}{4.48} \text{ x } 100 = 98\%$

e. For this, the number of moles of K_2SO_4 in excess and the moles of KNO_3 need to be converted to ions and diluted.

$$0.0156 \text{ mol } K_2SO_4 \text{ x } \frac{3 \text{ mol ions}}{\text{mol } K_2SO_4} = 0.0468 \text{ mol ions}$$

$$0.0148 \text{ mol } Pb(NO_3)_2 \text{ x } \frac{2 \text{ mol } KNO_3}{\text{mol } Pb(NO_3)_2} = 0.0296 \text{ mol } KNO_3$$

$$0.0296 \text{ mol } KNO_3 \text{ x } \frac{2 \text{ mol ions}}{\text{mol } KNO_3} = 0.0592 \text{ mol ions}$$

$$M_{ions} = \frac{0.0592 \text{ mol } + 0.0468 \text{ mol}}{.200 \text{ L}} = 0.53 \text{ M}$$

Challenge Problem

As always, the first step is to write and balance an equation that will describe what is happening and allow for stoichiometric determinations. The trick to this problem is realizing that tin that is oxidized to +2 can be further oxidized to +4.

$$Sn^{2+}(aq) + NO_3^-(aq) \rightarrow Sn^{4+}(aq) + NO(g)$$

Separate and balance the half-reactions.

$$3e^- + 4H^+(aq) + NO_3^-(aq) \rightarrow NO(g) + 2H_2O(l)$$

$$Sn^{2+}(aq) \rightarrow Sn^{4+}(aq) + 2e^-$$

Equalize the numbers of electrons, add, and cancel.

$$8H^+(aq) + 3Sn^{2+}(aq) + 2NO_3^-(aq) \rightarrow 2NO(g) + 4H_2O(l) + 3Sn^{4+}(aq)$$

Determine the number of moles of NO_3^- required in titration.

$$0.0276 \text{ L x } 0.0563 \text{ M} = 0.00155 \text{ mol } NO_3^-$$

Determine the grams of tin titrated.

$$0.00155 \text{ mol } NO_3^- \text{ x } \frac{3 \text{ mol Sn}}{2 \text{ mol } NO_3^-} \text{ x } \frac{118.7 \text{ g Sn}}{\text{mol Sn}} = 0.276 \text{ g Sn}$$

Determine the percentage of tin in the sample.

$$\% \text{ Sn} = \frac{0.276 \text{ g Sn}}{5.36 \text{ g sample}} \text{ x } 100 = 5.14\% \text{ Sn}$$

Chapter 5

True–False

1. F. Amplitude is the height of a wave from the midline.
2. T
3. T
4. F. Above the atomic level, deBroglie wavelengths are irrelevant.
5. F. The Bohr model only holds for hydrogen.
6. T
7. F. The angular momentum quantum number must be less than the principal quantum number.
8. F. *f*–orbitals are only found in the lanthanide and actinide metals.
9. T
10. F. Electrons that share an orbital must have opposite spins.

Multiple Choice

1. d
2. b
3. c
4. a
5. b
6. c
7. b
8. a
9. d
10. c

Matching

Degenerate	b
Wavelength	g
Frequency	e
Amplitude	d
Hertz	c
Photon	i
Orbital	a
Aufbau principle	f
Photoelectric effect	h

Fill–in– the–Blank

1. wavelength, frequency
2. line spectra
3. photons
4. emission
5. quantum numbers
6. *s*-orbitals
7. *p*-orbitals, two, six
8. degenerate, parallel spins
9. valence electrons
10. atomic radius

Problems

1. First, convert from frequency to wavelength.

$$c = \lambda\nu, \text{ so } \lambda = \frac{c}{\nu} = \frac{3.00 \text{ x } 10^8 \text{ m/s}}{835.6 \text{ x } 10^6\text{/s}} = 0.359 \text{ m or } 359 \text{ mm}$$

Then calculate the energy of one mol of these photons.

$$E = h\nu = 6.626 \text{ x } 10^{-34} \text{ J} \bullet \text{s x } 835.6 \text{ x } 10^{6}/\text{s} = 5.54 \text{ x } 10^{-25} \text{ J/photon}$$

$$5.54 \text{ x } 10^{-25} \text{ J/photon x } 6.022 \text{ x } 10^{23} \text{ photons/mole} = 0.333 \text{ J/mol}$$

2. First, convert from wavelength to frequency, taking care with the units of wavelength.

$$c = \lambda\nu, \text{ so } \nu = \frac{c}{\lambda} = \frac{3.00 \text{ x } 10^{8} \text{ m/s}}{2.15 \text{ x } 10^{-14} \text{ m}} = 1.40 \text{ x } 10^{22}/\text{s}$$

Then calculate the energy of one mol of these photons.

$$E = h\nu = 6.626 \text{ x } 10^{-34} \text{ J} \bullet \text{s x } 1.40 \text{ x } 10^{22}/\text{s} = 9.28 \text{ x } 10^{-12} \text{ J/photon}$$

$$9.28 \text{ x } 10^{-12} \text{ J/photon x } 6.02 \text{ x } 10^{23} \text{ photons/mole} = 5.59 \text{ x } 10^{12} \text{ J/mol}$$

3. First, determine the energy of one photon at this wavelength.

$$E = \frac{hc}{\lambda} = \frac{6.626 \text{ x } 10^{-34} \text{ J} \bullet \text{s x } 3.00 \text{ x } 10^{8} \text{ m/s}}{432 \text{ x } 10^{-9} \text{ m}} = 4.60 \text{ x } 10^{-19} \text{ J}$$

From this, determine the number of photons in 2.5 mJ of energy.

$$\frac{2.5 \text{ x } 10^{-3} \text{ J}}{4.60 \text{ x } 10^{-19} \text{ J/photon}} = 5.43 \text{ x } 10^{15} \text{ photons}$$

4. First, calculate the energy needed to dislodge one electron. This will be the same energy found in a photon capable of dislodging an electron.

$$495.8 \frac{\text{kJ}}{\text{mol}} \text{ x } \frac{1 \text{mol}}{6.022 \text{ x } 10^{23} \text{ photons}} \text{ x } \frac{1000 \text{ J}}{\text{kJ}} = 8.233 \text{ x } 10^{-19} \frac{\text{J}}{\text{photon}}$$

A photon with this much or more energy can dislodge an electron from sodium. From this, the threshold frequency and wavelength can be calculated.

$$E = h\nu, \upsilon = \frac{E}{h} = \frac{8.233 \text{ x } 10^{-19} \text{ J}}{6.626 \text{ x } 10^{-34} \text{ J} \cdot \text{s}} = 1.243 \text{ x } 10^{15}/\text{s}$$

$$\lambda = \frac{c}{\nu} = \frac{3.00 \text{ x } 10^{8} \text{ m/s}}{1.243 \text{ x } 10^{15}/\text{s}} = 2.414 \text{ x } 10^{-7} \text{ m} = 241.4 \text{ nm}$$

If the threshold frequency corresponds to a wavelength of 241 nm, any wavelength that gives a higher energy photon, that is, any wavelength lower than 241 nm will eject electrons. 540 nm is below the threshold frequency and will not eject electrons. 195 nm is above the threshold frequency and will eject photons.

5.

$$(\Delta x)(\Delta mv) \geq \frac{h}{4\pi}$$

$$\Delta v \geq \frac{h}{4\pi m \Delta x} \geq \frac{6.626 \text{ x } 10^{-34} \frac{kg \bullet m^2}{s}}{4 \text{ x } 3.14 \text{ x } 9.11 \text{ x } 10^{-34} \text{ kg x } 327 \text{ x } 10^{-12} \text{ m}} \geq 1.77 \text{x} 10^8 \text{ m/s}$$

6.

$$\frac{1}{\lambda} = R\left[\frac{1}{1} - \frac{1}{6^2}\right] = 1.097 \text{x} 10^{-2} \text{ /nm}\left[1 - \frac{1}{36}\right] = 0.0107 \text{ /nm}$$

$$\lambda = 93.8 \text{ nm}$$

$$E = \frac{hc}{\lambda} = \frac{6.626 \text{x} 10^{-34} \text{ J} \bullet \text{s x } 3.00 \text{x} 10^{8} \text{ m/s}}{93.8 \text{ x } 10^{-9} \text{ m}} = 2.12 \text{ x } 10^{-18} \text{ J}$$

7. Y = $1s^2 2s^2 2p^6 3s^2 3p^6 4s^2 3d^{10} 4p^6 5s^2 4d^1$

8. Neon has electrons in 2 energy levels, *n* = 1 and *n* = 2. It has both s and p electrons, and a total of 10 electrons. They will have the following quantum numbers:

1	0	0	+½	$1s^1$
1	0	0	-½	$1s^2$
2	0	0	+½	$2s^1$
2	0	0	-½	$2s^2$
2	1	-1	+½	$2p^1$
2	1	-1	-½	$2p^2$
2	1	0	+½	$2p^3$
2	1	0	-½	$2p^4$
2	1	+1	+½	$2p^5$
2	1	+1	-½	$2p^6$

9. The highest energy electron in bromine will be a 4*p* electron. The quantum numbers associated with a 4p electron are *n* = 4, and *l* = 1.

10. Si = ↑↓ (1*s*) ↑↓ (2*s*) ↑↓ ↑↓ ↑↓ (2*p*) ↑↓ (3*s*) ↑ ↑ _ (3*p*)

P = ↑↓ (1*s*) ↑↓ (2*s*) ↑↓ ↑↓ ↑↓ (2*p*) ↑↓ (3*s*) ↑ ↑ ↑ (3*p*)

Cl = ↑↓ (1*s*) ↑↓ (2*s*) ↑↓ ↑↓ ↑↓ (2*p*) ↑↓ (3*s*) ↑↓ ↑↓ ↑ (3*p*)

Phosphorus has three unpaired electrons, making it the most paramagnetic. Chlorine has one unpaired electron, and is therefore the least paramagnetic element.

11. Solve for the wavelength of the photon, then solve the Balmer-Rydberg equation for *n*.

$$E = \frac{hc}{\lambda},\ \lambda = \frac{hc}{E} = \frac{6.626 \text{ x } 10^{-34} \text{ J} \bullet \text{s x } 3.00 \text{ x } 10^{8} \text{ m/s}}{4.58 \text{ x } 10^{-19} \text{ J}} = 434 \text{ nm}$$

$$\frac{1}{434\text{nm}} = 1.097\text{x}10^{-2}/\text{nm}\left[\frac{1}{2^2} - \frac{1}{n^2}\right]$$

$$\frac{1}{434\text{nm x } 1.097 \text{ x } 10^{-2}/\text{nm}} = \left[\frac{1}{4} - \frac{1}{n^2}\right]$$

$$0.210 = .25 - \frac{1}{n^2}$$

$$.04 = \frac{1}{n^2}$$

$$n^2 = 25$$

$$n = 5$$

12. $$\lambda = \frac{h}{mv} = \frac{6.626 \text{ x } 10^{-34} \frac{\text{kg} \bullet \text{m}^2}{\text{s}}}{9.109 \text{ x } 10^{-31} \text{kg x } 2.57 \text{ x } 10^{6} \text{ m/s}} = 2.83 \text{ x } 10^{-10} \text{m} = 0.283 \text{ nm}$$

13. $[Ar]4s^1 3d^{10}$ = 29 electrons - Cu
$[Xe]6s^2 4f^{14} 5d^{10} 6p^5$ = 85 electrons - At
$[Kr]\ 4d^{10}$ = 46 electrons - Pd
$[Ar]4s^1 3d^5$ = 24 electrons - Cr

14. These elements are in different periods, so need to be ranked from lowest to highest in the periodic table.

Rn > Os > Ru > Fe

15. These elements are in different groups, so need to be ranked from right to left in the periodic table.

Mg > S > Cl > Ne

Challenge Problem

1. First, solve for the energy of one photon at 550 nm.

$$E = \frac{hc}{\lambda} = \frac{6.626\text{x}10^{-34} \text{ J} \bullet \text{s x } 3.00 \text{ x } 10^{8} \text{ m/s}}{550 \text{ x } 10^{-9} \text{ m}} = 3.61 \text{ x } 10^{-19} \text{ J/photon}$$

The light bulb is emitting 40 J/s, and is allowed to do so for four hours. The energy emitted would be

$$40 \frac{\text{J}}{\text{s}} \text{x 4 hours x } \frac{3600 \text{ s}}{\text{h}} = 576{,}000 \text{ J}$$

Solve for photons emitted.

$$\frac{576,000 \text{ J}}{3.61 \text{ x } 10^{-19} \text{ J/photon}} = 1.60 \text{ x } 10^{24} \text{ photons}$$

For the second part of the problem, solve for the energy needed to heat the water.

$$\text{energy} = 200 \text{ g x } 4.184 \frac{\text{J}}{\text{g} \bullet {}^{\circ}\text{C}} \text{ x } (95\ {}^{\circ}\text{C} - 20\ {}^{\circ}\text{C}) = 62,760 \text{ J}$$

If the lightbulb is emitting 576,000J of energy in 4 hours, and 95% of this is heat, the bulb is emitting 547,000 J of heat in 4 hours, or 137,000 J/hr.

If all of this heat went into the 200 g of water, solve for the time.

$$\text{time} = \frac{62,760\text{J}}{137,000 \text{ J/hr}} = 0.459 \text{ hours x} \frac{60 \text{ min}}{\text{hr}} = 27.5 \text{ min}$$

2. 15 W = 15 J/s.

If the laser were 100% efficient, $2.73 \text{ x } 10^{27}$ photons would equal 15 J. The energy of each photon would therefore be

$$\frac{15 \text{ J}}{2.73 \text{ x } 10^{19} \text{ photons}} = 5.49\text{x}10^{-19} \text{ J / photon}$$

$$E = \frac{hc}{\lambda}, \lambda = \frac{hc}{E} = \frac{6.626 \text{ x } 10^{-34} \text{ J} \bullet \text{s x } 3.00 \text{ x } 10^{8} \text{ m/s}}{5.49 \text{ x } 10^{-19} \text{ J}} = 3.62 \text{ x } 10^{-7} \text{ m or } 362 \text{ nm}$$

If the laser were 80% efficient, the wavelength would be

$$\lambda = \frac{hc}{E} = \frac{6.626 \text{ x } 10^{-34} \text{ J} \bullet \text{s x } 3.00 \text{ x } 10^{8} \text{ m/s}}{5.49 \text{ x } 10^{-19} \text{ J x } 0.8} = 4.52 \text{ x } 10^{-7} \text{ m or } 452 \text{ nm}$$

If 20%

$$\lambda = \frac{hc}{E} = \frac{6.626 \text{ x } 10^{-34} \text{ J} \bullet \text{s x } 3.00 \text{ x } 10^{8} \text{ m/s}}{5.49 \text{ x } 10^{-19} \text{ J x } 0.2} = 1.81 \text{ x } 10^{-6} \text{ m or } 1810 \text{ nm}$$

To solve for the efficiency if its wavelength is 1542, rearrange the equation to solve for the percentage.

$$x\% = \frac{hc}{E\lambda} = \frac{6.626 \text{ x } 10^{-34} \text{ J} \bullet \text{s x } 3.00 \text{ x } 10^{8} \text{ m/s}}{5.49 \text{ x } 10^{-19} \text{ J x } 1.542 \text{ x } 10^{-6} \text{ m}} = 0.235$$

The laser is 23.5% efficient.

Chapter 6

True–False

1. T
2. F. Nonmetals gain electrons to obtain noble-gas electron structures.
3. F. Transition metals will sometimes form more than one ion.
4. T
5. T

6. F. Ionization energies decrease down the periodic table.
7. F. Electron affinity applies primarily to nonmetals and is difficult to determine for nonmetals.
8. T
9. F. The Born—Haber cycle provides information about ionic bonding.
10. F. Lattice energies are related to atomic size and charge.
11. F. Alkali metals are powerful reducing agents, as they are happy to give up electrons.
12. T
13. T
14. T

Multiple Choice
1. d
2. b
3. c
4. b
5. a
6. c
7. d
8. a
9. b
10. d
11. a
12. c

Matching

Ionization energy	e
Coulomb's law	d
Electron affinity	a
Lattice energy	b
Born–Haber cycle	h
Core electron	g
Ionic bond	f
Octet rule	c

Fill–in–the–Blank
1. argon
2. decreases
3. core, easy
4. increase, valence
5. anions
6. charge, Coulomb's
7. eight, octet rule
8. oxidized, reduced
9. noble gases, 8A or 18
10. more

Problems

1. P:

1s	2s	2p	3s	3p
↑↓	↑↓	↑↓ ↑↓ ↑↓	↑↓	↑ ↑ ↑

P^{3-}:

1s	2s	2p	3s	3p
↑↓	↑↓	↑↓ ↑↓ ↑↓	↑↓	↑↓ ↑↓ ↑↓

Electron structure is that of argon

Ti:

1s	2s	2p	3s	3p	4s	3d
↑↓	↑↓	↑↓ ↑↓ ↑↓	↑↓	↑↓ ↑↓ ↑↓	↑↓	↑ ↑ _ _ _

	1s	2s	2p	3s	3p	4s	3d	
Ti^{2+}:	↑↓	↑↓	↑↓ ↑↓ ↑↓	↑↓	↑↓ ↑↓ ↑↓	__	↑ ↑ __ __ __	most common
Ti^{4+}:	↑↓	↑↓	↑↓ ↑↓ ↑↓	↑↓	↑↓ ↑↓ ↑↓	__	__ __ __ __ __	also possible
Mn:	↑↓	↑↓	↑↓ ↑↓ ↑↓	↑↓	↑↓ ↑↓ ↑↓	↑↓	↑ ↑ ↑ ↑ ↑	
Mn^{2+}:	↑↓	↑↓	↑↓ ↑↓ ↑↓	↑↓	↑↓ ↑↓ ↑↓	__	↑ ↑ ↑ ↑ ↑	most common many others

	1s	2s	2p	3s	3p	4s	
K:	↑↓	↑↓	↑↓ ↑↓ ↑↓	↑↓	↑↓ ↑↓ ↑↓	↑	
K^{+}:	↑↓	↑↓	↑↓ ↑↓ ↑↓	↑↓	↑↓ ↑↓ ↑↓		Electron structure is that of argon

		4s	3d	4p	
Se:	[Ar]	↑↓	↑↓ ↑↓ ↑↓ ↑↓ ↑↓	↑↓ ↑ ↑	
Se^{2-}:	[Ar]	↑↓	↑↓ ↑↓ ↑↓ ↑↓ ↑↓	↑↓ ↑↓ ↑↓	Electron structure is that of krypton

2. Remembering that cations are smaller than neutral atoms, anions are larger than neutral atoms, and that atoms decrease in size from left to right across the periodic table and increase in size down the periodic table:

$$Li < Li^{+} < Al < Al^{3+} < Cl < Cl^{-}$$

3. There is a very large jump (~4x) between the third and fourth ionization energies. The fourth electron removed is a core electron, meaning that there are three valence electrons. In the third period of the periodic table, this is aluminum.

4.

		4s	3d	4p
Ga:	[Ar]	↑↓	↑↓ ↑↓ ↑↓ ↑↓ ↑↓	↑ __ __
Ge:	[Ar]	↑↓	↑↓ ↑↓ ↑↓ ↑↓ ↑↓	↑ ↑ __
Br:	[Ar]	↑↓	↑↓ ↑↓ ↑↓ ↑↓ ↑↓	↑↓ ↑↓ ↑

Based on its electron structure, gallium is far more likely to lose an electron to form a +1 ion, than to gain an electron. Looking at the periodic table, this is also confirmed by the position of gallium, which is a main-group metal. Germanium can adopt a half-filled shell, which is also a fairly stable arrangement, but it does not have the stability of a noble gas configuration. Bromine is the closest to having a noble gas configuration, so it will have the highest electron affinity.

5. Use a Born–Haber cycle. However, this time the net energy change is known, and we need to solve for the heat of sublimation of aluminum.

$Al\ (s) \rightarrow Al\ (g)$? kJ/mol

$Al\ (g) \rightarrow Al^{+}\ (g) + e^{-}$	578 kJ/mol	578 kJ/mol
$Al^{+}\ (g) \rightarrow Al^{2+}\ (g) + e^{-}$	1,817 kJ/mol	1,817 kJ/mol
$Al^{3+}\ (g) \rightarrow Al^{3+}\ (g) + e^{-}$	2,745 kJ/mol	2,745 kJ/mol
$3/2\ Cl_2\ (g) \rightarrow 3\ Cl\ (g)$	3/2(244 kJ/mol)	366 kJ/mol
$3\ Cl\ (g) \rightarrow 3\ Cl^{-}\ (g)$	3(-348.6 kJ/mol)	-1046 kJ/mol
$Al^{3+}\ (g)\ +\ 3\ Cl^{-}\ (g) \rightarrow AlCl_3\ (s)$	–5,492 kJ/mol	–5,492 kJ/mol

Overall net energy change –704.2 kJ/mol

–704.2 = ? + 578 kJ/mol + 1,817 kJ/mol + 2,745 kJ/mol + 366 kJ/mol - 1046 kJ/mol – 5,492 kJ/mol

? = 328 kJ/mol

6. $Mg\ (s) + Cl_2\ (g) \rightarrow MgCl_2\ (s)$
   $2\ Mg\ (s) + O_2\ (g) \rightarrow 2\ MgO\ (s)$
   $3\ Mg\ (s) + 2\ P\ (s) \rightarrow Mg_3P_2\ (s)$

   $2\ Li\ (s) + Cl_2\ (g) \rightarrow 2\ LiCl\ (s)$
   $4\ Li\ (s) + O_2\ (g) \rightarrow 2\ Li_2O\ (s)$
   $3\ Li\ (s) + P\ (s) \rightarrow LiP_3\ (s)$

   $2Al\ (s) + 3Cl_2\ (g) \rightarrow 2\ AlCl_3\ (s)$
   $4Al\ (s) + 3O_2\ (g) \rightarrow 2\ Al_2O_3\ (s)$
   $Al\ (s) + P\ (s) \rightarrow AlP\ (s)$

7.

$K\ (s) \rightarrow K\ (g)$	+89 kJ/mol
$K\ (g) \rightarrow K^{+}\ (g) + e^{-}$	+418.8 kJ/mol
$\frac{1}{2}\ F_2\ (g) \rightarrow F\ (g)$	½(+158 kJ/mol)
$F\ (g) \rightarrow F\ (g)$	- 328 kJ/mol
$K^{+}\ (g)\ +\ F^{-}\ (g) \rightarrow KF\ (s)$	- 821 kJ/mol
Total:	-562 kJ/mol

8. $AlCl_3 > MgCl_2 > LiCl > KCl > CsCl$

   Using Coulomb's law, an increase in ionic charge on the cation will dominate other factors, so the cation with the highest charge will have the highest lattice energy. With ions that have charges of 1, size becomes a factor: smaller ions can get closer together and will thus have a larger lattice energy than larger ions, so these can be ranked in order of cation size, smallest to largest.

Challenge Problem

The chemical equation for the reaction of the alkaline earth chloride with silver nitrate is

$$MCl_2\ +\ 2\ AgNO_3\ \rightarrow\ M(NO_3)_2\ +\ 2\ AgCl$$

a. The mass of chlorine can be determined from the mass of silver chloride produced.

$$18.8g\ AgCl \times \frac{1\ mol\ AgCl}{143.3g} \times \frac{1\ mol\ Cl}{1\ mol\ AgCl} \times \frac{35.5g\ Cl}{mol\ Cl} = 4.66g\ Cl$$

$$\text{mass\% Cl in metal chloride} = \frac{4.66\text{g Cl}}{5.25\text{g MCl}_2} = 88.7\%$$

b. The identity of the metal can be determined by calculating the molar mass of the metal chloride, then subtracting out the mass of the chloride.

$$\text{mol MCl}_2 = 18.8\text{g AgCl x} \frac{1\text{mol AgCl}}{143.3\text{g AgCl}} \text{x} \frac{1\text{mol MCl}_2}{2\text{mol AgCl}} = 0.0656\text{mol MCl}_2$$

$$\text{molar mass MCl}_2 = \frac{5.25\text{ g}}{0.0656\text{ mol}} = 80\frac{\text{g}}{\text{mol}}$$

Molar mass of M = 80 g/mol – 71 g/mol = 9 g/mol

From the periodic table, the metal must be beryllium.

c. For the reaction of beryllium with chlorine,

$Be\ (s) + Cl_2\ (g) \rightarrow BeCl_2\ (s)$

For the reaction of $BeCl_2$ with $AgNO_3$,

$BeCl_2 + 2\ AgNO_3 \rightarrow Be(NO_3)_2 + 2\ AgCl$

d. To determine which was the limiting reactant, the moles of beryllium and the moles of chlorine gas must be calculated.

$$1.53\text{g Be x } \frac{1\text{ mol}}{9.01\text{g}} = 0.170\text{ mol}$$

$$4.75\text{L Cl}_2 \text{ x } 3.17\frac{\text{g}}{\text{L}} \text{x} \frac{1\text{mol}}{71\text{g}} = 0.212\text{ mol}$$

Chlorine was in excess: the liters left unreacted can now be calculated.

0.212 mol Cl_2 present – 0.170 mol Cl_2 reacted = 0.042 mol Cl_2 unreacted.

$$0.042\text{ mol Cl}_2\text{x}\frac{71\text{g}}{\text{mol}}\text{x}\frac{1\text{ L}}{3.17\text{g}} = 0.94\text{ L}$$

Chapter 7

True–False

1. F. Covalent bonds involve sharing of electrons.
2. T
3. T
4. T
5. T
6. F. Valences of main group elements can be easily predicted using the periodic table.
7. F. Multi-atom structures can be predicted using electron-dot models.
8. F. The structures of charged species can be predicted by removing or adding electrons as needed.
9. T
10. T

11. F. Methane molecules are tetrahedral.
12. T
13. F. When orbitals overlap, π-bonds are formed.
14. T
15. T

Fill–in–the–Blank

1. bond lengths, bond, nuclei
2. bond dissociation energy, positive, released
3. polar covalent, ionic
4. nonbonding, bonding
5. octet rule, expanded octets
6. formal charges
7. tetrahedral, bent, trigonal pyramidal
8. sp^2
9. σ, π
10. electronegativity

Matching

Bond length	j
Bond dissociation energy	c
Coordinate covalent bond	o
Polar covalent bond	e
Lone pair	k
Electronegativity	n
VSEPR model	p
Hybrid orbitals	a
σ bond	d
π bond	f
Molecular orbital	g
Paramagnetic	b
Diamagnetic	h
Bonding pair	m
Formal charge	l
Resonance hybrid	i

Problems

1. LiBr – ionic (metal/nonmetal)
   HCl – covalent (nonmetal/nonmetal)
   MgO – ionic (metal/nonmetal)
   NF_3 – covalent (nonmetal/nonmetal)

2. H—O—O—H

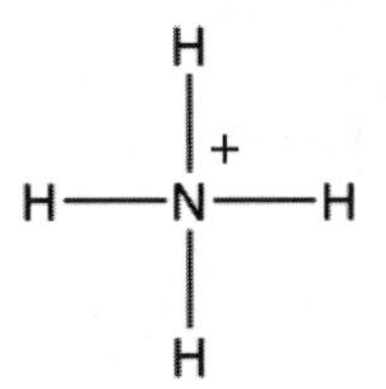

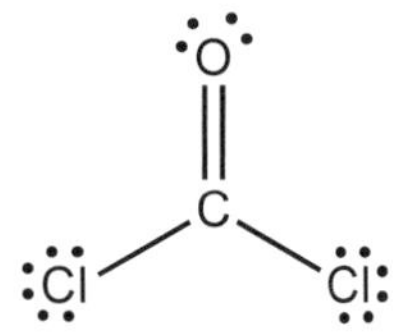

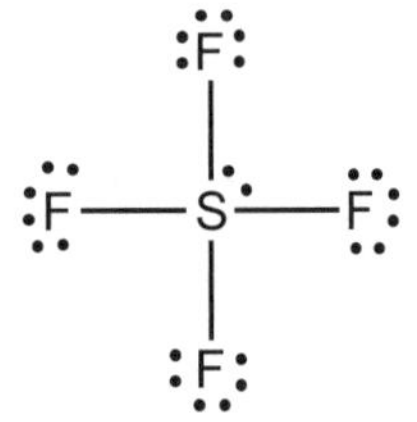

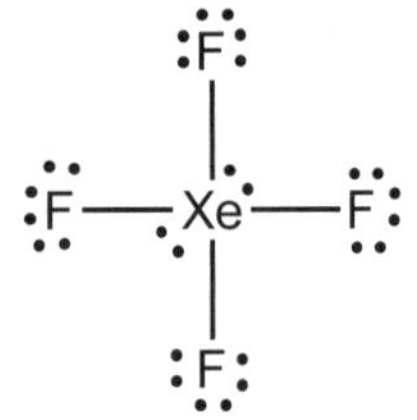

3. The high melting point and conductivity in solution indicate that the first compound is ionic. The second compound, with its low melting point and non-conducting solution, is covalent. The solubilities of the compounds provide no information about bonding.

4.

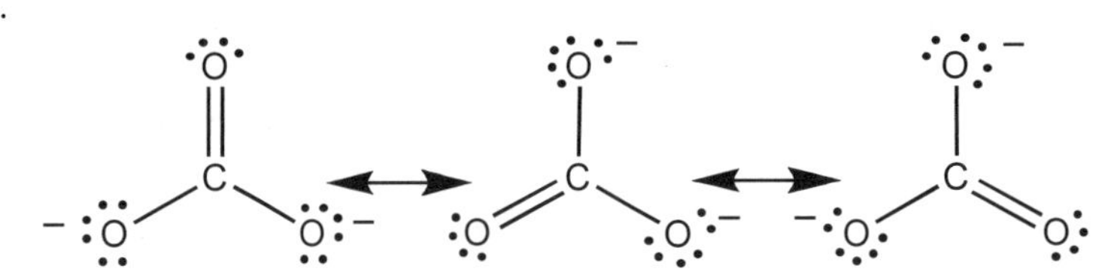

The three resonance structures are all equivalent.

Formal charges:	Central C	= 4 – 4 – 0 = 0
	Double-bonded O	= 6 – 2 – 4 = 0
	Single-bonded O	= 6 – 1 – 6 = -1

The formal charges match the ionic charge.

5. Li-Cl > Al-F > H-Cl > C-H > Cl-Cl. The atoms that are farther apart on the periodic table will have more polar bonds.

6. Cs < Fe < O < F. Fluorine is the most electronegative element: proximity to fluorine gives a good approximation of electronegativity of an element.

7. The most symmetrical structure is usually the most stable, and formal charges bear this out.

Leftmost structure:

Leftmost nitrogen = 5 – 1 – 6 = -2
Central nitrogen = 5 – 4 – 0 = -1
Oxygen = 6 – 3 – 2 = +1

Center Structure:

Leftmost nitrogen = 5 – 2 – 4 = -1

Central nitrogen = 5 – 4 – 0 = -1
Oxygen = 6 – 2 – 4 = 0

Rightmost structure:
Leftmost nitrogen = 5 – 3 – 2 = 0
Central nitrogen = 5 – 4 – 0 = -1
Oxygen = 6 – 1 – 6 = -1

None of these structures is perfectly satisfactory, but the rightmost structure is the only one in which there is a negative formal charge on oxygen, the most electronegative element in the structure.

8. NF_3 – four electron clouds provides tetrahedral electron geometry. Three bonds lead to trigonal pyramidal molecular geometry.

   H_2S – four electron clouds again provides tetrahedral electron geometry, with two bonds leading to bent molecular geometry.

   IF_4^+ – here, there is an expanded octet to consider with 34 electrons spread over 5 atoms. There are six electron clouds in this molecule, with two non-bonding pairs on the iodine. This molecule will have octahedral electron geometry, and square planar molecular geometry.

   PF_5 – here, there is another expanded octet. There are five electron clouds in this molecule, with no non-bonding pairs. Both the electronic and molecular geometries are trigonal bipyramidal.

9. The carbon with four bonds has tetrahedral geometry, which indicates sp^3 hybridization and σ binding. The carbon bonded to the oxygen has three electron groups, which indicates trigonal planar geometry and sp^2 hybridization. This atom has σ bonds to carbon and hydrogen, and one σ and one π bond to oxygen.

10. constructive interference

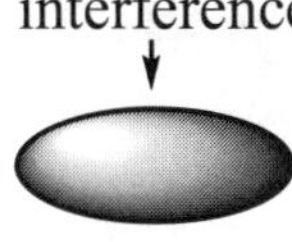

Bonding

destructive interference

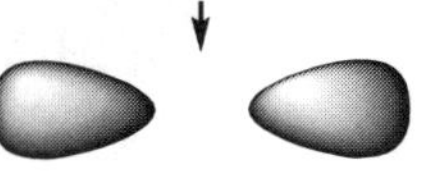

Anti-bonding

Challenge Problem

Determine the empirical formula of the acid.

26.1 g C = 2.18 mol C
69.7 g O = 4.36 mol O
4.35 g H = 4.35 mol H

Empirical formula = CO_2H_2

Determine the molecular formula.

The moles of NaOH needed to neutralize the 0.36 grams of acid are

0.07824 L x 0.10 mol/L = 0.007824 mol

The molecular weight of the compound is

0.360 g/0.007824 mol = 46.01

The empirical formula weight is 46.01, so the empirical formula is the same as the molecular formula.

Lewis structure:

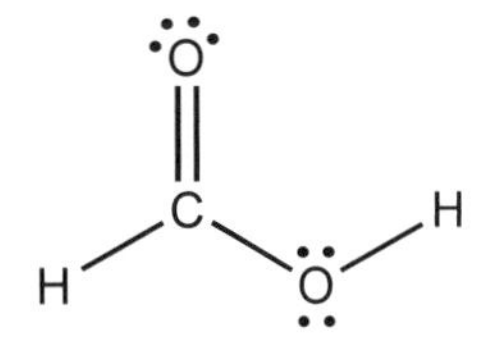

The central carbon atom is sp^2 hybridized, with trigonal planar geometry

The oxygen bonded to the C and H atoms is sp^3 hybridized, with tetrahedral geometry.

C-H = σ

C-O = σ

O-H = σ

C=O = π

Chapter 8

True–False

1. T
2. F. Changes in energies are always calculated *final – initial.*
3. F. State functions are path independent.
4. T
5. F. Standard state is measured at 25 °C – room (or, lab) temperature
6. F. It is possible for a reaction to both transfer heat and do work.
7. F. Melting is an endothermic process – heat must be added.
8. F. Specific heat is an intensive property: it can be used to identify materials.
9. T
10. T
11. T
12. F. A positive change in entropy – an increase in disorder - is favorable.
13. F. Gases have higher entropy than liquids, as they are more disordered.
14. T
15. T

Matching

Conservation of energy	e
Energy	h
Temperature	d
System	m
Entropy	g
State function	l

Heat	q
Work	n
Enthalpy	p
Heat of fusion	i
Sublimation	c
Specific heat	k
Exothermic	j
Hess's law	o
Heat of combustion	a
Spontaneous	b
Endothermic	f

Fill–in–the–Blank

1. kinetic energy, potential energy
2. temperature, heat
3. negative, exothermic
4. reversible
5. volume, work
6. thermodynamic standard state
7. specific heat
8. surroundings, system, positive
9. bond dissociation energy, positive
10. entropy, increase
11. Gibbs free energy, equilibrium, negative, positive

Problems

1. To easily solve this problem, set the two energies equal to one another to solve for the needed velocity of the marble. Watch your units: they must match for the answer to come out correctly.

$$\frac{1}{2} \text{ x } 1.3 \text{ kg x } \left(5\frac{\text{m}}{\text{s}}\right)^2 = \frac{1}{2} \text{ x } 0.002 \text{ kg x } \left(\text{v}\frac{\text{m}}{\text{s}}\right)^2$$

$$16.25 \ \frac{\text{kg} \cdot \text{m}^2}{\text{s}^2} = 0.001 \ \text{v}^2 \text{ kg}$$

$$16250 \ \frac{\text{m}^2}{\text{s}^2} = \text{v}$$

$$\text{v} = 127 \ \frac{\text{m}}{\text{s}}$$

2. Work is being done on the system since the change in volume is negative. Knowing q and w makes it possible to calculate ΔE.

   $w = - P\Delta V = $ -8atm x (0.550L – 0.950L) = 3.2 L•atm x 101 J/(L•atm) = 323 J = 0.323 kJ

   q = 12 kJ (positive, as it is being absorbed by the system)

   For the system, $\Delta E = q + w$; ΔE = 12 kJ + 0.323 kJ = 12.3 kJ.

   For the surroundings, ΔE = –12.3 kJ

3. Using $q = \text{m x c x } \Delta T$:

$$150 \text{ J} = 15 \text{ g x } 0.140 \text{ J/g•°C x } \Delta T$$

$$\Delta T = 150 \text{ J}/(15 \text{ g x } 0.140 \text{ J/g•°C})$$

$$\Delta T = 71.4 \text{ °C}$$

4. For this reaction, both w and q need to be calculated to be able to determine the overall energy change of the system.

   $w = - P\Delta V = -1\text{atm x } (1\text{L} - 3\text{L}) = 2 \text{ L•atm x } 101 \text{ J/(L•atm)} = 202 \text{ J} = 0.202 \text{ kJ}$

   To calculate q it is necessary to know how many moles of chlorine gas were involved in the reaction:

$$3.17 \frac{\text{g}}{\text{L}} \text{x } 1.00 \text{ L } = 3.17 \text{ g}$$

$$3.17 \text{ gx } \frac{1\text{mol}}{71 \text{ g}} = 0.0446 \text{ mol}$$

   Multiply by the heat involved per mole of chlorine.

$$0.00446 \text{ mol x } 80.3 \text{ kJ/mol} = 3.58 \text{ kJ}$$

   $\Delta E = q + w$; $\Delta E = 3.58 \text{ kJ} + 0.202 \text{ kJ} = 3.79 \text{ kJ}$.

5. From these data, it is possible to calculate ΔS for the reaction, and thus the free energy of the reaction.

   ΔS = products – reactants = 266.1 J/mol•K – [223.0 J/mol•K + 0.5(205 J/mol•K)] = -59.4 J/mol•K

   This is the entropy change for the overall reaction. For our reaction, which involved 0.0446 moles of chlorine, the entropy change is

$$-59.4 \text{ J/mol•K x } 0.0446 \text{ mol} = -2.65 \text{ J/K}$$

   Free energy is calculated $\Delta G = \Delta H - T\Delta S$.

$$\Delta G = 3.58 \text{ kJ} - 298\text{K } •(-0.00265 \text{ kJ/K}) = 4.37 \text{ kJ}$$

   This is not a spontaneous reaction at 25° C.

   As the enthalpy terms and the entropy terms are both unfavorable, this will not be a spontaneous reaction at any temperature.

6. Heat is being absorbed from the water by the salt as it dissolves (indicating an entropy-driven process). The ΔH for the solution is per mole, not per gram, so it is first necessary to convert the grams of Epsom salts to moles.

$$15\text{g x} \frac{1 \text{ mol}}{246.46 \text{ g}} = 0.0609 \text{ mol x } 16.11 \frac{\text{kJ}}{\text{mol}} = 0.980 \text{ kJ} = 980 \text{ J}$$

   $q_{\text{rxn}} = -q_{\text{soln}} = -[(\text{sp. heat}) \times (\text{mass of soln}) \times (\Delta T)]$

$$980 \text{ J} = -4.184 \frac{\text{J}}{\text{g}\cdot{}^{\circ}\text{C}} \times 100 \text{ g} \times \Delta\text{T};\ \Delta T = -2.34\ {}^{\circ}\text{C}$$

The final temperature will be 24.36 °C + (–2.34 °C) = 22.0 °C

7. A number of calculations are required for this problem: the ice and water must be warmed, and two phase changes completed.

   50 g H_2O = 2.8 moles H_2O

   -10 °C to 0 °C: $q = \text{mc}\Delta T$ = 50 g x 2.108 J/g°C x 10 °C = 1054 J = 1.054 kJ

   ice to water: q = 2.8 moles x 6.01 kJ/mol = 16.8 kJ

   0 °C to 100 °C: q = 50 g x 4.184 J/g°C x 100 °C = 20900 J = 20.9 kJ

   water to steam: q = 2.8 moles x 40.7 kJ/mol = 114 kJ

   Total energy = 153 kJ. Of this energy, 75% of it is used in the phase change from liquid to gas. This makes sense, because this phase transition requires a breaking of all intermolecular attractions. Warming ice or water is simply adding kinetic energy, and the phase change from solid to liquid loosens, but does not break, the interactions between the molecules.

8. This problem requires a bit of algebra.

   Let the amount of copper = "Cu," and the amount of gold = "Au."

   The specific heat of the metal chunk can be determined by seeing how much heat it released to the water.

   $$q = 25.0 \text{ g x } 4.184 \text{ J/g}\bullet{}^{\circ}\text{C x } (26.65\ {}^{\circ}\text{C} - 25.00\ {}^{\circ}\text{C}) = 173 \text{ J}$$

   Switching to the metal

   $$-173 \text{ J} = 6.73 \text{ g x c x } (26.65\ {}^{\circ}\text{C} - 100\ {}^{\circ}\text{C})$$

   $$\text{c} = 0.350 \text{ J/g}\bullet{}^{\circ}\text{C}$$

   To solve for the amount of each, it is necessary to set up a system of equations.

   Cu + Au = 6.73g; Cu = 6.73g - Au

   $$(6.73 - \text{Au})\left(0.385 \frac{\text{J}}{\text{g}\bullet{}^{\circ}\text{C}}\right) + \frac{\text{Au}}{6.73\text{g}}\left(0.129 \frac{\text{J}}{\text{g}\bullet{}^{\circ}\text{C}}\right) = 0.350 \frac{\text{J}}{\text{g}\bullet{}^{\circ}\text{C}}$$

Substituting:

$$\frac{6.73\text{g}-\text{Au}}{6.73\text{g}}\left(0.385\frac{\text{J}}{\text{g}\bullet{}^{\circ}\text{C}}\right)+\frac{\text{Au}}{6.73\text{g}}\left(0.129\frac{\text{J}}{\text{g}\bullet{}^{\circ}\text{C}}\right)=0.350\frac{\text{J}}{\text{g}\bullet{}^{\circ}\text{C}}$$

$$6.73\text{g}-\text{Au}\left(0.385\frac{\text{J}}{\text{g}\bullet{}^{\circ}\text{C}}\right)+\frac{\text{Au}}{6.73\text{g}}\left(0.129\frac{\text{J}}{\text{g}\bullet{}^{\circ}\text{C}}\right)=0.350\frac{\text{J}}{\text{g}\bullet{}^{\circ}\text{C}}\text{x}6.73\text{g}$$

$$2.59\frac{\text{J}}{{}^{\circ}\text{C}}-0.385\text{Au}\frac{\text{J}}{\text{g}\bullet{}^{\circ}\text{C}}+0.129\text{Au}\frac{\text{J}}{\text{g}\bullet{}^{\circ}\text{C}}=2.36\frac{\text{J}}{{}^{\circ}\text{C}}$$

$$2.59\frac{\text{J}}{{}^{\circ}\text{C}}-2.36\frac{\text{J}}{{}^{\circ}\text{C}}=0.385\text{Au}\frac{\text{J}}{\text{g}\bullet{}^{\circ}\text{C}}-0.129\text{Au}\frac{\text{J}}{\text{g}\bullet{}^{\circ}\text{C}}$$

$$0.23\frac{\text{J}}{{}^{\circ}\text{C}}=0.256\text{Au}\frac{\text{J}}{\text{g}\bullet{}^{\circ}\text{C}}$$

$$\text{Au}=0.90\text{g}$$

$$\text{Cu}=5.83\text{g}$$

9. $2\,Fe\,(s) + 3/2\,O_2\,(g) \rightarrow Fe_2O_3\,(s)$

Given the following:

$$Fe_2O_3\,(s) + 3CO\,(g) \rightarrow 2\,Fe\,(s) + 3CO_2\,(g) \qquad \Delta H^\circ = -26.7\text{ kJ}$$

$$CO + \tfrac{1}{2}\,O_2 \rightarrow CO_2\,(g) \qquad \Delta H^\circ = -283.0\text{ kJ}$$

The first reaction needs to be turned so as to give the proper product. Once this is done, the second reaction is fine in the direction it is written, but needs to be multiplied through by 3 to give the proper number of CO_2 molecules.

$$2\,Fe\,(s) + 3CO_2\,(g) \rightarrow Fe_2O_3\,(s) + 3CO\,(g) \qquad \Delta H^\circ = +26.7\text{ kJ}$$

$$\underline{3CO + 3/2\,O_2 \rightarrow 3CO_2\,(g) \qquad \Delta H^\circ = -849.0\text{ kJ}}$$

$$2\,Fe\,(s) + 3CO_2\,(g) + 3CO + 3/2\,O_2 \rightarrow Fe_2O_3\,(s) + 3CO\,(g) + 3CO_2\,(g) \qquad \Delta H^\circ = -822.3\text{ kJ}$$

Cancel terms that appear on both sides.

$$2\,Fe\,(s) + 3/2\,O_2 \rightarrow Fe_2O_3\,(s) \quad \Delta H^\circ = -822.3\text{ kJ}$$

10. $2\,NaHCO_3\,(s) \rightarrow Na_2CO_3\,(s) + H_2O\,(l) + CO_2\,(g)$

Given the following heats of formation, what is the approximate ΔH of this decomposition reaction?

$\Delta H_f^\circ\ NaHCO_3 = -947.7$ kJ/mol

$\Delta H_f^\circ\ Na_2CO_3 = -1131$ kJ/mol

$\Delta H_f^\circ\ H_2O = -285.9$ kJ/mol

$\Delta H_f^\circ\ CO_2 = -393.5$ kJ/mol

The approximate ΔH is calculated by subtracting the sum of the heats of formation of the reactants from the sum of the heats of formation of the products.

[(-1131 kJ/mol) + (-285.9 kJ/mol) + (-393.5 kJ/mol)] – (2 x -947.7 kJ/mol) = 85.00 kJ/mol

11. Using the bond dissociation energies given, calculate the enthalpy of the following reaction:

$2\,Cl_2\,(g) + CH_4\,(g) \rightarrow 2\,H_2\,(g) + CCl_4\,(g)$

Cl-Cl = 243 kJ/mol
H-H = 436 kJ/mol
C-H = 410 kJ/mol
C-Cl = 330 kJ/mol

Products:

2 H-H bonds:	872 kJ/mol
4 C-Cl bonds:	1320 kJ/mol
	2192 kJ/mol bonds formed

Reactants:

2 Cl-Cl bonds:	486 kJ/mol
4 C-H bonds:	1640 kJ/mol
	2126 kJ/mol bonds broken

reactants – products = -66 kJ/mol

12. $\Delta G° = \Delta H° - T\Delta S°$, when $\Delta G° = 0$, $T = \Delta H/\Delta S$

ΔH = +66 kJ/mol for the reverse reaction, solve for ΔS as follows:

$S°\ Cl_2$ = 223 J/mol•K
$S°\ H_2$ = 130.6 J/mol•K
$S°\ CH_4$ = 188 J/mol•K
$S°\ CCl_4$ = 214 J/mol•K

Products – reactants = [2(223 J/mol•K) + 188 J/mol•K)] - [2(130.6 J/mol•K) + 214 J/mol•K] = 158.8 J/mol•K = 0.1588 kJ/mol•K

T = 66,000/158.8 = 415.6 K = 143 °C

13. First, write and balance the equation for the combustion of pentane.

$$C_5H_{12}\,(l) + 8\,O_2\,(g) \rightarrow 5\,CO_2\,(g) + 6\,H_2O\,(l)$$

ΔH_f^o pentane = −146.3 kJ/mol
$\Delta H_f^o\ H_2O$ = −285.9 kJ/mol
$\Delta H_f^o\ CO_2$ = −393.5 kJ/mol

[5(-393.5 kJ/mol) + 6(-285.9 kJ/mol)] – [-146.3 kJ/mol] = -3537 kJ/mol

$$-3537\,\frac{\text{kJ}}{\text{mol}}\ \text{x}\ \frac{1\,\text{mol}}{72\,\text{g}} = -49\,\frac{\text{kJ}}{\text{g}}$$

$$-49\,\frac{\text{kJ}}{\text{g}}\ \text{x}\,0.626\,\frac{\text{g}}{\text{mL}} = -30.8\,\frac{\text{kJ}}{\text{mL}}$$

Chapter 9

True–False

1. F. The atmosphere is approximately 70% nitrogen.
2. T
3. T
4. T
5. T
6. F. The smaller a gas particle is, the faster it travels.
7. F. The ratio is inverse.
8. F. The ideal gas law begins to fail at high pressures.
9. F. We breathe troposphere.
10. T

Matching

Barometer	j
Pressure	a
Atmospheric pressure	l
Boyle's law	i
Charles's law	h
Avogadro's law	o
Ideal gas law	m
Dalton's law	b
Partial pressures	e
Kinetic-molecular theory	n
Graham's law	c
Diffusion	f
Effusion	d
Ideal gas	p
STP	r
Standard molar volume	q
Mole fraction	g
Manometer	k

Fill–in–the–Blank

1. pressure, collisions
2. pressure, volume, temperature, moles
3. mole fraction, partial pressure
4. volume, negligible, pressure
5. low, faster

Problems

1. To determine the mm of silicone oil that the pressure will support, first determine the mm of Hg that it will support by converting from atmospheres to mmHg.

$$0.983 \text{ atm x} \frac{760 \text{ mm Hg}}{1 \text{ atm}} = 747 \text{ mm Hg}$$

Then convert from mm Hg to mm silicone oil using the relative densities as a conversion factor.

$$747 \text{ mm Hg x} \frac{13.6 \text{ mm oil}}{0.970 \text{ mm Hg}} = 10,500 \text{ mm oil}$$

2. a) 340 mm ethyl alcohol greater than atmospheric pressure.

$$\text{Pressure in bulb} = 340\text{ mm EtOH x}\frac{0.789\text{ mm Hg}}{13.6\text{ mm EtOH}} = 19.7\text{ mm Hg above atmospheric pressure.}$$

Total pressure in the bulb = 760 mm Hg + 19.7 mm Hg = 780 mmHg

b) 257 mm ethyl alcohol less than atmospheric pressure.

$$\text{Pressure in bulb} = 257\text{ mm EtOH x}\frac{0.789\text{ mm Hg}}{13.6\text{ mm EtOH}} = 14.9\text{ mm Hg below atmospheric pressure.}$$

Total pressure in the bulb = 760 mm Hg – 14.9 mm Hg = 745 mmHg

3. $$n = \frac{5.78\text{atm} \times 0.050\text{ L}}{0.08206\ \frac{\text{L} \cdot \text{atm}}{\text{mol} \cdot \text{K}} \times (\text{-}20\ ^\circ\text{C}\ +273.15)\text{K}} = 0.0139\text{ mol}$$

If the mass of 0.0139 mol is 0.281 g, the molecular weight is 0.281 g/0.0139 mol = 20.2 g/mol. The gas is neon.

4. Assume that you have 1 L of the gas and determine the number of moles in that liter. Alternatively, you can do the reverse: assume that you have 1 mole of gas, and determine the number of liters. Both will get you the correct answer.

$$n = \frac{4.00\text{atm} \times 1.00\text{ L}}{0.08206\ \frac{\text{L} \cdot \text{atm}}{\text{mol} \cdot \text{K}} \times (100\ ^\circ\text{C}\ +273.15)\text{K}} = 0.131\text{ mol}$$

$$\text{density} = \frac{0.131\text{ mol}}{\text{L}}\text{ x }\frac{16.0\text{ g}}{\text{mol}} = 2.10\frac{\text{g}}{\text{L}}$$

5. a. Assume a 100 g sample, and convert 92.3 g C and 7.69 g H to moles.

$$92.3\text{ g C} \times \frac{1\text{ mol C}}{12.0\text{ g C}} = 7.69\text{ mol C; } 7.69\text{ g H} \times \frac{1\text{ mol H}}{1.0\text{ g H}} = 7.69\text{ mol H}$$

This gives rise to a 7.69:7.69 C to H mole ratio.

There is one carbon atom for every hydrogen atom in the molecule, so the empirical formula is CH.

b. To solve for molar mass, we need to determine the moles of gas.

$$n = \frac{\frac{763}{760}\text{ atm} \times 0.250\text{ L}}{0.08206\ \frac{\text{L} \cdot \text{atm}}{\text{mol} \cdot \text{K}} \times (35 + 273)\text{ K}} = 0.00993\text{ mol}\ ;\ \frac{0.258\text{ g}}{0.00993\text{ mol}} = 26.0\text{ g/mol}$$

c. The empirical formula weight is 13 g/mol, so the molecular formula is 2 empirical formula units: C_2H_2.

6. Determine the number of moles in 1 L under these conditions.

$$n = \frac{15\text{ atm} \times 1.00\text{ L}}{0.08206\frac{\text{L} \cdot \text{atm}}{\text{mol} \cdot \text{K}} \times 573\text{ K}} = 0.319\text{ mol}$$

14.04 g/0.319 mol = 44.0 g/mol

7. Write and balance the equation.

$$2\ C_4H_{10} + 13\ O_2\,(g) \rightarrow 8\ CO_2\,(g)\ + 10\ H_2O\,(g)$$

Determine the number of moles of butane burned.

$$25\text{mL x } 0.6014\frac{\text{g}}{\text{mL}}\text{ x}\frac{1\text{ mol}}{58.12\text{ g}} = 0.26\text{ mol}$$

Determine the moles of oxygen consumed, then convert to L.

$$0.26\text{ mol } C_4H_{12}\text{ x}\frac{13\text{ mol } O_2}{2\text{ mol } C_4H_{12}} = 1.7\text{ mol } O_2;$$

$$V = \frac{nRT}{P} = \frac{1.7\text{ mol x } 0.08206\ \frac{\text{L}\cdot\text{atm}}{\text{mol}\cdot\text{K}} \times (30+273)\text{ K}}{1\text{ atm}} = 42.2\text{ L}$$

Determine the number of moles of carbon dioxide produced, then convert to L.

$$0.26\text{ mol } C_4H_{12}\text{ x}\frac{8\text{ mol } CO_2}{2\text{ mol } C_4H_{12}} = 1.04\text{ mol } CO_2;$$

$$V = \frac{nRT}{P} = \frac{1.04\text{ mol x } 0.08206\ \frac{\text{L}\cdot\text{atm}}{\text{mol}\cdot\text{K}} \times (30+273)\text{ K}}{1\text{ atm}} = 25.9\text{ L}$$

Determine the number of moles of water vapor produced, then convert to L.

$$0.26\text{ mol } C_4H_{12}\text{ x}\frac{10\text{ mol } H_2O}{2\text{ mol } C_4H_{12}} = 1.30\text{ mol } H_2O;$$

$$V = \frac{nRT}{P} = \frac{1.30\text{ mol x } 0.08206\ \frac{\text{L}\cdot\text{atm}}{\text{mol}\cdot\text{K}} \times (30+273)\text{ K}}{1\text{ atm}} = 32.3\text{ L}$$

Determine the overall energy change of the reaction.

Volume of products – volume of reactants = (32.3 L + 25.9 L) – (42.2 L + 0.26 L) = +15.97 L

Work is done by this reaction.

8. Find the number of moles of each gas.

$$7.03\text{ g Ne x}\frac{1\text{ mol}}{20.2\text{ g}} = 0.348\text{ mol };11.7\text{ g } N_2\text{ x}\frac{1\text{ mol}}{28.0\text{ g}} = 0.418\text{ mol};\ 15.7\text{ g He x}\frac{1\text{ mol}}{4.00\text{ g}} = 3.93\text{ mol}$$

The total moles of gas = 0.348 mol + 0.418 mol + 3.93 mol = 4.70 moles

Find the total pressure.

$$P = \frac{nRT}{V} = \frac{(0.348 \text{ mol} + 0.418 \text{ mol} + 3.93 \text{ mol}) \text{ x } 0.08206 \frac{\text{L} \cdot \text{atm}}{\text{mol} \cdot \text{K}} \text{ x } 298 \text{ K}}{5.000 \text{ L}} = 22.97 \text{ atm}$$

Find the mole fraction and partial pressure of each component.

$$\text{Ne}: \frac{0.348 \text{ mol}}{4.70 \text{ mol}} \text{ x } 23.0 \text{ atm} = 1.70 \text{ atm}$$

$$\text{N}_2: \frac{0.418 \text{ mol}}{4.70 \text{ mol}} \text{ x } 23.0 \text{ atm} = 2.05 \text{ atm}$$

$$\text{He}: \frac{3.93 \text{ mol}}{4.70 \text{ mol}} \text{ x } 23.0 \text{ atm} = 19.2 \text{ atm}$$

9. $\frac{\text{rate He}}{\text{rate H}_2} = \frac{\sqrt{1.00794\text{x}2}}{\sqrt{4.002602}} = \frac{1.4198}{2.00065} = 0.7096$ or

$$\frac{\text{rate H}_2}{\text{rate He}} = \frac{\sqrt{4.002602}}{\sqrt{1.00794\text{x}2}} = \frac{2.00065}{1.4198} = 1.409$$

10. $\frac{P_1V_1}{T_1} = \frac{P_2V_2}{T_2}$; $T_2 = \frac{P_2V_2T_1}{P_1V_1} = \frac{1 \text{ atm x } 0.375 \text{ L x } 298 \text{ K}}{7.8 \text{ atm x } 0.050 \text{ L}} = 287 \text{ K} = 13.4 \text{ °C}$

Chapter 10

True–False

1. F. The polarity of individual bonds sometimes cancels out, leaving a molecule with polar bonds without a dipole.
2. F. Intermolecular forces range from very weak dispersion forces to fairly strong.
3. T
4. T
5. F. Vapor pressure is temperature dependent. The closer the liquid is to its boiling point, the higher the vapor pressure will be.
6. T
7. F. X-ray crystallography is used to examine the structure of crystalline solids.
8. F. The coordination number is the number of other atoms each atom is touching.
9. F. Fullerene and graphite are both allotropes of carbon.
10. T

Matching

Allotrope	g
Critical point	d
Dipole	k
Heat of fusion	c
Molecular solid	i
London dispersion force	b
Normal boiling point	h
Viscosity	e
Hydrogen bonding	j
Vapor pressure	m
Triple point	a
Amorphous solid	f
Crystalline solid	l

Multiple Choice

1. a
2. b
3. d
4. d
5. c
6. c
7. a
8. b
9. b
10. a

Fill–in–the–Blank

1. dipole-dipole, London dispersion forces, hydrogen bonds
2. ion-dipole
3. enthalpy
4. phase changes
5. sublimation
6. decreases, increase
7. ionic solid, molecular solid, covalent network solid

Problems

1. $CH_3CH_2CH_2OH$ – this molecule is capable of hydrogen bonding. As this is the strongest intermolecular force, this will predominate.
   $CH_3(CH_2)_4CH_3$ – this molecule is nonpolar, so will be unable to interact except with London dispersion forces.

   PCl_3 – this is a polar molecule with a dipole. The most important interactions will be dipole-dipole.

2. a. **Br_2** or Cl_2 – Bromine is larger, so will have more dispersion forces and a higher boiling point.
   b. CH_3OH or **CH_3CH_2OH** – both will hydrogen bond, but the second will also have higher dispersion forces.
   c. CH_4 or **CH_3OH** – the second molecule can hydrogen bond, while the first cannot: it will have the highest boiling point.

3. $T = \frac{\Delta H_{vap}}{\Delta S_{vap}}$; $\quad T = \frac{565,300 \text{ J/mol}}{150.8 \text{ J/K} \cdot \text{mol}} = 3749 \text{ K} = 3476°\text{C}$

4. $0.237 \text{ g x} \frac{1 \text{ mol}}{18.0 \text{ g}} \text{x} - 40.7 \frac{\text{kJ}}{\text{mol}} = -0.536 \text{ kJ}$ released by the condensing water.

   $$536 \text{ J} = 50.0 \text{ g x } 0.449 \frac{\text{J}}{\text{g} \bullet °\text{C}} \text{ x } \Delta T$$

   $\Delta T = 23.9$ °C

   The final temperature is 25.0 °C + 23.9 °C = 48.9 °C

5. To heat the ice cube from -20 °C to 0.0 °C:

   q = 45g x 1mol/18g x 36.57 J/g•°C x 20 °C = 1830 J = 1.83 kJ

   To melt the ice cube at 0.0 °C:

$$45 \text{ g x} \frac{1 \text{ mol}}{18 \text{ g}} \text{x } 6.01 \frac{\text{kJ}}{\text{mol}} = 15 \text{ kJ}$$

To warm the water to body temperature:

$q = 45$ g x 4.184 J/g•°C x 37 °C = 7.0 kJ

Total heat absorbed by water/lost by victim = 23.8 kJ.

6. $$\ln P_2 = \ln P_1 + \frac{\Delta H_{vap}}{R}\left(\frac{1}{T_1} - \frac{1}{T_2}\right)$$

$$\ln P_2 = \ln(760 \text{ mm Hg}) + \frac{30{,}800 \frac{\text{J}}{\text{mol}}}{\left(8.3145 \frac{\text{J}}{\text{mol} \cdot \text{K}}\right)}\left(\frac{1}{(80.1 + 273.15)\text{ K}} - \frac{1}{(75.0 + 273.15)\text{ K}}\right) = 6.48;$$

$$P_2 = e^{6.48} = 653 \text{ mm Hg}$$

7. Face-centered cubic unit cells contain four total atoms.

Mass of unit cell: $4 \text{ atoms x } \frac{1 \text{ mol}}{6.022\text{x}10^{23} \text{ atoms}} \text{ x } \frac{196.97 \text{ g}}{\text{mol}} = 1.308\text{x}10^{-21} \text{ g}$

Volume of unit cell: $\frac{1.308\text{x}10^{-21} \text{ g}}{19.3 \frac{\text{g}}{\text{cm}^3}} = 6.779\text{x}10^{-23} \text{ cm}^3$

Edge length of unit cell = $\sqrt[3]{6.779 \text{ x } 10^{-23} \text{ cm}^3} = 4.147 \text{ x } 10^{-8} \text{ cm} = 415 \text{ pm}$

A face-centered cubic cell has an edge diagonal equal to 4r. Using the Pythagorean theorem:

$(4r)^2 = 2(415 \text{ pm})^2$

$r = 146$ pm

8. A face-centered cubic unit cell has 4 total atoms, and the diagonal of a side is equal to 2x the atomic diameter.

Mass of unit cell = $4 \text{ atoms x } \frac{1 \text{ mol}}{6.022\text{x}10^{23} \text{ atoms}} \text{ x } \frac{106.42 \text{ g}}{\text{mol}} = 7.069 \text{ x } 10^{-22} \text{ g}$

Volume of unit cell:

Diagonal = 137 pm x 4 = 548 pm = $(548 \text{ x } 10^{-10} \text{ cm})$
Edge length:
$2l^2 = (548 \text{ x } 10^{-10} \text{ cm})^2$
$l = 390 \text{ x } 10^{-10}$ cm
Volume = $(390 \text{ x } 10^{-10} \text{ cm})^3 = 5.81 \text{ x } 10^{-23} \text{ cm}^3$

Density of metal = $\frac{7.069\text{x}10^{-22} \text{ g}}{5.81\text{x}10^{-23} \text{ cm}^3} = 12.15 \frac{\text{g}}{\text{cm}^3}$

9. normal freezing point = 197.5 K
   triple point = 216.6 K
   critical point = 304.25 K
   boiling = line C
   sublimation = line A
   melting = line B

   The forward slope of the central line (the freezing/melting line) shows that the solid form of the substance is more dense than the liquid. As pressure increases, it is more likely to be a solid.

Challenge Problem

To solve this problem, first determine the number of joules available.

$$320 \text{ g x} \frac{1 \text{ mol}}{44.0 \text{ g}} \text{ x } 2220 \frac{\text{kJ}}{\text{mol}} = 16{,}145 \text{ kJ}$$

Now set up an algebraic equation to solve for the mass of water through the phase changes.

x = mass of water

$x/18$ = moles of water

16145 kJ =

$$\left(0.00203 \frac{\text{kJ}}{\text{g} \cdot {}^\circ\text{C}}\right) 20\,{}^\circ\text{Cx} + \left(6.01 \frac{\text{kJ}}{\text{mol}}\right) \frac{\text{x}}{18 \frac{\text{g}}{\text{mol}}} + \left(0.004184 \frac{\text{kJ}}{\text{g} \cdot {}^\circ\text{C}}\right) 100\,{}^\circ\text{Cx} + \left(40.67 \frac{\text{kJ}}{\text{mol}}\right) \frac{\text{x}}{18 \frac{\text{g}}{\text{mol}}} + \left(0.00208 \frac{\text{kJ}}{\text{g} \cdot {}^\circ\text{C}}\right.$$

$$16145 \text{kJ} = 0.406 \frac{\text{kJ}}{\text{g}} \text{x} + 0.334 \frac{\text{kJ}}{\text{g}} \text{x} + 0.414 \frac{\text{kJ}}{\text{g}} \text{x} + 2.26 \frac{\text{kJ}}{\text{g}} \text{x} + 0.0312 \frac{\text{kJ}}{\text{g}} \text{x}$$

$$16145 \text{kJ} = 3.445 \frac{\text{kJ}}{\text{g}} \text{x}$$

$$\text{x} = 4690 \text{g}$$

Chapter 11

True–False
1. T
2. T
3. F. ppm or ppb are the most convenient units for environmental work.
4. F. Potassium nitrate will dissolve readily in water (polar solvent).
5. T
6. F. Supersaturated solutions are always unstable.
7. F. Increasing pressure causes gases to become more soluble.
8. T
9. F. Concentrated solutions have a lower freezing point.
10. T

Multiple Choice
1. a
2. d
3. c

4. b
5. c

Matching

Colligative property	k
Colloid	f
Fractional distillation	a
Heat of solution	m
Miscible	d
Molality	l
Osmosis	b
Semipermeable membrane	g
Saturated	c
Solubility	n
Solute	h
Solution	i
Solvent	e
Supersaturated	j

Fill–in–the–Blank

1. solvent-solvent, solute-solute
2. positive, increased disorder of both solvent and solute
3. density
4. more
5. drop, rise, drop, osmotic pressure
6. molecular mass
7. solvated, hydrated
8. exothermic, endothermic

Problems

1. The rankings will be from most polar (least soluble) to most nonpolar (most soluble). Iodine is completely nonpolar, so will be readily soluble in benzene. Butanol $CH_3(CH_2)_2CH_2OH$ is primarily nonpolar, but has a polar oxygen group that will reduce its solubility. NaCl is ionic, and so will be essentially insoluble in benzene.

$$NaCl < CH_3(CH_2)_2CH_2OH < I_2$$

2. a. $\text{mass \%} = \dfrac{28.4\text{ g}}{(28.4\text{ g} + 350\text{ g})} \text{x } 100 = 7.50\%$

b. mole fraction:

moles sucrose = 28.4 g/342.3 g/mol = 0.0830 mol
moles water = 350 g/ 18.0 g/mol = 19.4 mol

$$\text{mol fraction} = \frac{0.0830\text{ mol}}{(19.4\text{ mol} + 0.0830\text{ mol})} = 0.00425$$

c. $\text{molality} = \dfrac{0.0830\text{ mol}}{0.350\text{ kg}} = 0.237\text{ m}$

d. $\text{molarity} = \dfrac{0.0830\text{ mol}}{0.374\text{ L}} = 0.222\text{ M}$

3. 100 g of this solution will contain 3 g of hydrogen peroxide and have a volume of 100 g/1.01 g/mL = 99 mL

$$\text{molarity} = \frac{3\text{ g x }\frac{1\text{ mol}}{34\text{ g}}}{0.099\text{ L}} = 0.891\text{ M}$$

4. $$15\text{ppb} = \frac{0.015\text{ mg}}{\text{L}}\text{ x }\frac{1\text{ g}}{1000\text{ mg}}\text{ x }\frac{1\text{ mol As}}{74.9\text{ g}} = 2.00\text{ x }10^{-7}\text{ M}$$

$$15\text{ppb} = \frac{0.015\text{ mg}}{\text{L}}\text{ x }\frac{1\text{ g}}{1000\text{ mg}}\text{ x }100{,}000\text{ L} = 1.5\text{ g}$$

5. $$\text{solubility} = 1.32\text{ x }10^{-3}\ \frac{\text{mol}}{\text{L}\cdot\text{atm}}\text{ x }0.21\text{ atm} = 2.77\text{ x }10^{-4}\text{ M}$$

6. $$\text{solubility} = 6.25\text{ x }10^{-4}\ \frac{\text{mol}}{\text{L}\cdot\text{atm}}\text{ x }976\text{ mm Hg x }\frac{1\text{ atm}}{760\text{ mm Hg}} = 8.03\text{ x }10^{-4}\text{ M}$$

7. moles of glucose = 4.5 g/180 g/mol = 0.025 mol
   moles of water = 55 g/18 g/mol = 3.06 mol

$$P_{soln} = 42.175\text{ mm Hg x }\frac{3.06\text{ mol}}{3.06\text{ mol} + 0.025\text{ mol}} = 41.9\text{ mm Hg}$$

8. moles of $FeCl_3$ = 5.8 g /162.2 g/mol = 0.36 mol $FeCl_3$
   moles of ions = 0.36 mol x 3.4 = 0.122 mol ions
   moles of water = 72.1 mol /18 g/mol = 4.01 mol

$$P_{soln} = 23.8\text{ mm Hg x }\frac{4.01\text{ mol}}{4.01\text{ mol} + 0.122\text{ mol}} = 23.1\text{ mm Hg}$$

9. moles of benzene: 10g/78.0 g/mol = 0.128 mol
   moles of tolulene: 10g/92.0 g/mol = 0.109 mol

$$\text{mole fraction benzene} = \frac{0.128\text{ mol}}{(0.128\text{ mol} + 0.109\text{ mol})} = 0.540$$

$$\text{mole fraction tolulene} = \frac{0.109\text{ mol}}{(0.128\text{ mol} + 0.109\text{ mol})} = 0.460$$

$$P_{total} = (0.540\text{ x }93.4\text{ mm Hg}) + (0.460\text{ x }26.9\text{ mm Hg}) = 62.8\text{ mm Hg}$$

10.

$$\Delta T_b = 80.0\ °\text{C} - 76.7\ °\text{C} = 3.3\ °\text{C} = 5.03\ °\text{C}/m\text{ x X}$$

$$\text{X} = 0.656\text{ m}$$

$$0.656\text{ m} = \frac{\text{x mol napthalene}}{1.0\text{ kg}} = 0.656\text{ mol naphthalene}\text{x}\frac{128\text{ g}}{\text{mol}} = 84.0\text{g}$$

11.

$$\text{m} = \frac{10.0\text{ g NaCl x }\frac{1\text{ mol}}{58.5\text{ g}}}{0.090\text{ kg}} = 1.90\text{ m x }\frac{2\text{ ions}}{\text{mol}} = 3.80\text{ m ions}$$

$$\Delta T_b = 0.51 \frac{°C}{m} \times 3.80\ m = 1.94\ °C$$

$$\Delta T_f = 1.86 \frac{°C}{m} \times 3.80\ m = 7.07\ °C$$

Boiling point = 101.94 °C
Freezing point = -7.07 °C

12. $\Delta T_f = |3.5 - 5.5| = 2.0\ °C$

$$2.0\ °C = 5.12 \frac{°C}{m} \times X$$

X = 0.390 m

0.390 m = x mol/0.075 kg

x = 0.0292 mol

M.W. = 2.50 g/0.0292 mol = 85.33 g/mol

13. moles glucose = 18.0g/180 g/mol = 0.10 mol

molarity = 0.10 mol/1.2 L= 0.0833 M

$$\Pi = 0.0833\ M \times 0.08206 \frac{L \cdot atm}{K \cdot mol} \times (273.15+15)K = 1.97 atm$$

14. $16.4\ mm\ Hg \times \frac{1\ atm}{760\ mm\ Hg} = X\ M \times 0.08206 \frac{L \cdot atm}{K \cdot mol} \times (273.15+15)K$

$X = 9.13 \times 10^{-4}\ M \times 1.00\ L = 9.13 \times 10^{-4}\ mol$

$$\frac{0.250\ g}{9.13 \times 10^{-4}\ mol} = 274 \frac{g}{mol}$$

Challenge Problem

First, determine the mol fraction of pentane in the vapor.

100 g vapor will be 43.2 g pentane, 56.8 g hexane

43.2 g pentane /72 g/mol = 0.60 mol
56.8 g hexane /86 g/mol = 0.66 mol

$$X_{pentane} = \frac{0.60\ mol}{0.60\ mol + 0.66\ mol} = 0.476$$

$$X_{hexane} = \frac{0.66\ mol}{0.60\ mol + 0.66\ mol} = 0.524$$

Now set up a two-variable system to solve for the mole fractions in the solution.

$0.476P_{total} = 425\ x$, where x = mole fraction of pentane in solution

$0.524P_{total} = 151\ (1 - x)$, where $1-x$ = mole fraction of hexane in solution

$P = 425\ x/0.476 = 893\ x$

$-\quad P = [151\ (1 - x)]/0.524 = 288 - 288x$

$0 = -288 + 1181\ x$

$x = 0.243$ = mole fraction of pentane

$1 - x = 0.757$ = mole fraction of hexane

P_{total} = (151 mmHg x 0.757) + (425 mmHg x 0.243) = 218 mmHg

Chapter 12

True–False
1. T
2. F. The exponents must be experimentally determined.
3. F. The order of a reaction is determined by adding the exponents in the rate law.
4. T
5. T
6. T
7. F. The half-life is the time required for the concentration to drop to ½ its initial level.
8. T
9. F. Zeroth-order reactions are quite uncommon.
10. F. Reaction steps can occur at all different rates.
11. T
12. F. Activation energy is always positive, although reactions can be net exothermic.
13. T
14. F. Biological reactions almost invariably involve catalysts.
15. T

Multiple Choice
1. b
2. c
3. a
4. a
5. c
6. d
7. d
8. c
9. b
10. d

Matching

Activation energy	k
Bimolecular reaction	d
Catalyst	n
Kinetics	b
Decay constant	f
Enzyme	a
Half-life	m

Heterogeneous catalyst c
Homogeneous catalyst o
Initial rate i
Molecularity p
Rate-determining step g
Steric factor h
Thermolecular reaction j
Transition state l
Zeroth-order reaction e

Fill–in–the–Blank

1. reaction rate, decrease, increase, product
2. rate constant, *k*
3. 2, 4, 8, 1
4. reverse reaction
5. first–order
6. elementary reactions, rate-determining step, slowest
7. activation energy, transition state
8. lower energy
9. double
10. adsorb, release/desorb

Problems

1. $\frac{\Delta[C]}{\Delta t}$ can be calculated arithmetically.

$$\text{initial rate} = \frac{\Delta[C]}{\Delta t} = \frac{0.005-0.000}{15-0} = 3.3\times10^{-4}\ \text{M/s}$$

$$\text{rate from 30 to 75s} = \frac{\Delta[C]}{\Delta t} = \frac{0.019-0.010}{75-30} = 2.0\times10^{-4}\ \text{M/s}$$

2. Rate of appearance of H_2 = $\left[1.67\times10^{-2}\ \text{mol PH}_3/(\text{L}\cdot\text{s})\right]\times\frac{6\ \text{mol H}_2}{4\ \text{mol PH}_3} = 2.50\times10^{-2}\ \text{mol H}_2/(\text{L}\cdot\text{s})$

3. When the value of [NO] is doubled, the reaction rate increases by a factor of four. The reaction is second order in NO. When the concentration of Cl_2 is doubled, the reaction rate doubles. The reaction is first order in Cl_2.

   The rate law is

$$\text{Rate} = k[\text{NO}]^2[\text{Cl}_2]$$

   Using the date in the second experiment:

$$63.6\ \text{mol/(L}\cdot\text{s)} = \text{k}(0.050\ \text{mol/L})^2(0.025\ \text{mol/L});$$

$$\text{k} = \frac{63.6\ \text{mol/(L}\cdot\text{s)}}{6.25\times10^{-5}\ \text{mol}^3/\text{L}^3} = 1.02\times10^6\ \text{L}^2/(\text{mol}^2\cdot\text{s})$$

4. When the concentration of A is doubled, the reaction rate increases by a factor of eight, so the reaction is third order in A. When the concentration is B is doubled, the rate is unaffected: the reaction is zero order in B.

   The reaction is third order overall.

The rate law is

$$\text{Rate} = k[\text{A}]^3$$

Using the data in second experiment:

$$0.134 \text{ mol/(L}\cdot\text{s)} = \text{k}(0.300 \text{ mol/L})^3$$

$$\text{k} = \frac{0.134 \text{ mol/(L}\cdot\text{s)}}{(0.027 \text{ mol}^3\text{/L}^3)} = 4.96 \text{ L}^2\text{/(mol}^2\cdot\text{s)}$$

$$\text{Rate} = 4.96 \text{ L}^2\text{/(mol}^2\cdot\text{s)} \times (0.542 \text{ mol/L})^3 = 0.790 \text{ mol/(L}\cdot\text{s)}$$

5. Use the equation for half-life to determine k:

$$k = \frac{0.693}{24{,}000 \text{ years}} = 2.89\text{x}10^{-5} \text{/year}$$

Substitute into the integrated rate equation for a first-order reaction.

$$\ln [200 \text{ atoms}] = -2.89 \text{ x } 10^{-5}\text{/year x t} + \ln [6.022 \text{ x } 10^{23} \text{ atoms}]$$

$$\text{t} = 1.71 \text{ x } 10^6 \text{ years}$$

6. $$\frac{1}{[O_3]} = [1.40 \times 10^{-2} \text{L/(mol}\cdot\text{s)}]\left(12 \text{ h} \times \frac{3600 \text{ s}}{1 \text{ h}}\right) + \frac{1}{2.37 \text{ M}} = 605 \text{ L/mol}$$

$$[O_3] = 1.65 \times 10^{-3} \text{ M}$$

7. The reaction order can be determined graphically by plotting the data as [HI] versus time, ln [HI] versus time, and1/[HI] versus time. The graph that provides a straight line will also provide the order of the reaction.

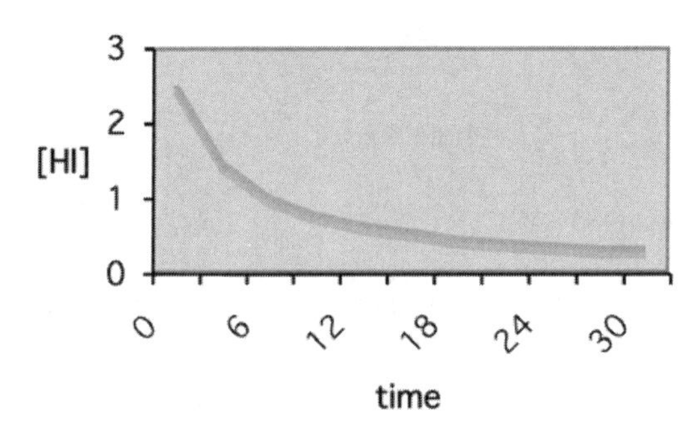

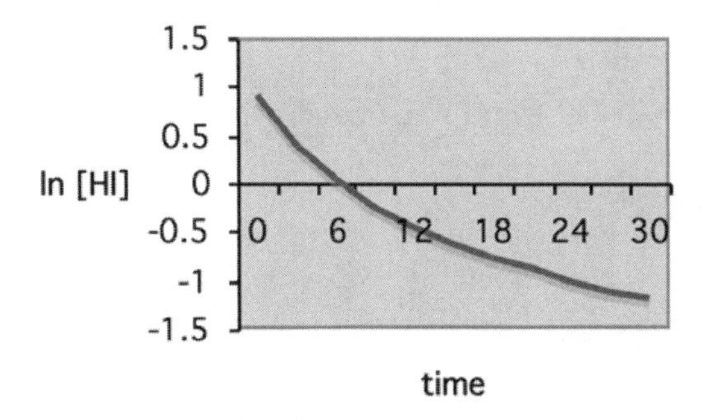

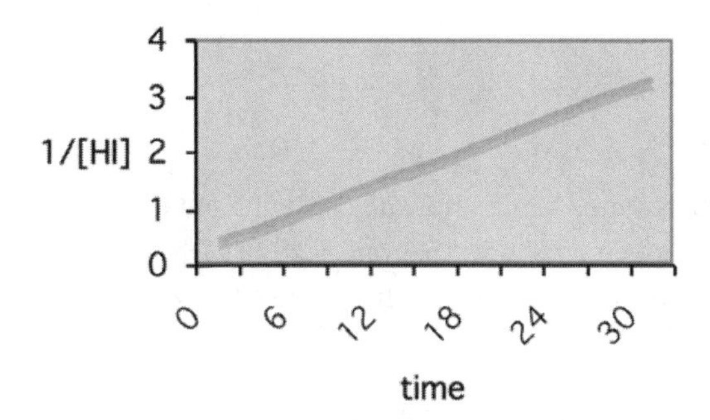

The plots show that the reaction is second order. Using the integrated rate law and the initial rate,

$$\frac{1}{1.45\ \text{M}} = \text{k}(3\,\text{min}) + \frac{1}{2.50\ \text{M}}$$

$$k = \frac{\dfrac{1}{1.45\ \text{M}} - \dfrac{1}{2.50\ \text{M}}}{3\,\text{min}} = 9.66 \times 10^{-2}\ \text{L}/(\text{mol}\cdot\text{min})$$

$$t_{1/2} = \frac{1}{\left[9.66 \times 10^{-2}\ \text{L/(mol}\cdot\text{min)}\right] \text{x } 2.50\ \text{mol/L}} = 4.14\ \text{min}$$

8. The overall reaction is

$$2\ NO_2\ (g) + Cl_2\ (g) \rightarrow 2\ NO_2Cl\ (g)$$

Both steps are bimolecular, and Cl (*g*) is a reaction intermediate: it does not appear in the final equation.

The expected rate will be: Rate = $k[NO_2][Cl_2]$.

This is a second-order reaction.

9. Overall reaction: $2\ H_2O_2 \rightarrow 2\ H_2O + O_2$; OH^- and HO_2 are reaction intermediates. The first step is unimolecular. The second and third steps are bimolecular. The first step is the slowest step.

10. $k = Ae^{\frac{-E_a}{RT}}$

$$A = \frac{k}{e^{\frac{-E_a}{RT}}} = \frac{3.79\text{x}10^{-5}\,\text{L/mol}\bullet\text{s}}{e^{\frac{-1.18\text{x}10^4\,\text{J/mol}}{8.314\frac{\text{J}}{\text{mol}\bullet\text{K}}\text{x}381\text{ K}}}} = \frac{3.79\text{x}10^{-5}\,\text{L/mol}\bullet\text{s}}{e^{-3.73}} = 1.57\text{x}10^{-3}$$

11. $\ln\left(\frac{3.2\times10^{-3}}{7.2\times10^{-5}}\right) = \left(\frac{-9.862\times10^4\text{ J/mol}}{8.314\text{ J/(mol}\cdot\text{K)}}\right)\left(\frac{1}{T_2} - \frac{1}{340\text{ K}}\right)$

$$3.79 = -1.186\times10^4\text{ K}\left(\frac{1}{T_2} - \frac{1}{340\text{ K}}\right)$$

$$-3.195\times10^{-4}\text{ K}^{-1} = \frac{1}{T_2} - \frac{1}{340\text{ K}}$$

$$\frac{1}{T_2} = 3.26\times10^{-3}\text{ K}^{-1} \qquad T_2 = 307\text{ K}$$

Challenge Problem

To calculate the initial rate requires a rate law, rate constant, and the initial concentrations of C_2H_5I and/or OH^-. From the information given, we know that rate = $k[C_2H_5I][OH^-]$ since the doubling of the concentration of each reactant while the other is held constant results in a doubling of the rate: it is first order in each reactant.

The rate constant, *k*, is calculated using the Arrhenius equation, $k = Ae^{-E_a/RT}$.

$$k = 1.09\times10^9\text{ M}^{-1}\text{s}^{-1}e^{-(7.97\times10^4\text{ J/mol})/(8.314\text{J/mol}\cdot\text{K})(318.15\text{K})} = 8.956\times10^{-5}\text{ M}^{-1}\text{s}^{-1}.$$

Remembering that the two 250 mL solutions are going to be combined, and assuming that they sum to 500 mL, the molar concentrations of the reactants are:

$$\frac{0.475\text{ g KOH}}{0.500\text{ L}}\text{ x }\frac{1\text{ mol}}{56.1\text{ g}} = 1.69\text{ x }10^{-2}\text{ M OH}$$

$$\frac{1.378\text{ g }C_2H_5I}{0.500\text{ L}}\text{ x }\frac{1\text{ mol}}{156\text{ g}} = 1.76\text{ x }10^{-2}\text{ M }C_2H_5I$$

$$\text{Rate} = (8.956\times10^{-5}\text{ M}^{-1}\text{s}^{-1})(1.692\times10^{-2}\text{ M})(1.768\times10^{-3}\text{ M}) = 2.68\times10^{-8}\text{ M}\cdot\text{s}^{-1}$$

Chapter 13

True–False

1. T
2. F. Reaction needs to be run at different concentrations, not temperatures.
3. F. Equilibrium constants are unitless.
4. T
5. F. There will be more products than reactants.
6. F. The reaction will go in the direction of fewer moles of gas.
7. T
8. T
9. T
10. T

Multiple Choice

1. d
2. b
3. c
4. a
5. d
6. a
7. d
8. c
9. a
10. b

Matching

Chemical equilibrium	g
Equilibrium constant, K_c	d
Equilibrium constant K_p	a
Equilibrium mixture	j
Homogeneous equilibria	b
Heterogeneous equilibria	h
Reaction quotient, Q_c	c
Reversible reaction	i
Dynamic state	f
Le Châtelier's principle	e

Fill–in–the–Blank

1. dynamic, concentration, constant
2. heterogeneous, liquids, solids
3. minimize, products/right, reactants/left, temperature, kinetic, catalyst, increases
4. direction, equilibrium, products, reactants, reactants, products
5. lowering

Problems

1. a. $K_p = \frac{[O_2][ClO]}{[O_3][Cl]}$ b. $K_p = \frac{[NOBr]^2}{[NO]^2[Br_2]}$ c. $K_p = \frac{1}{[O_2]}$

   d. $K_p = \frac{[N_2]^2\,[O_2]}{[N_2O]^2}$

2. $K_p = \frac{(p_{NO})^2}{(p_{N_2})(p_{O_2})}$ $K_p = \frac{(0.045)^2}{(0.27)(0.187)} = 4.0 \times 10^{-2}$

3. $K_c = \frac{K_p}{(RT)^{\Delta n}}$; $\Delta n = 2 - 2 = 0$; $K_c = K_p = 5.1 \times 10^{-2}$

4. a. mainly products b. slightly more products, reactants also present c. mainly reactants

5. To determine if a system is at equilibrium, you must first solve for the value of Q_c and then compare it to the value of K_c.

   $Q_c = \frac{[PCl_3][Cl_2]}{[PCl_5]}$ $Q_c = \frac{(0.83)(0.83)}{(2.43 \times 10^{-3})} = 2.83 \times 10^2$

Since $Q_c > K_c$, the system is not at equilibrium. The reaction will shift from right to left to achieve equilibrium.

6. To solve this problem, make a table of initial, change, and final concentrations.

   The equilibrium concentrations of N_2O and O_2 can be determined by allowing x to be the concentration of N_2O that reacts on going to the equilibrium state. Since the mole ratios are 2:3:4, if $2x$ mol/L of N_2O reacts, then $3x$ mol/L of O_2 reacts and $4x$ mol/L of NO_2 are produced.

   This can be summarized in the following table.

Principal Reaction	$2\ N_2O\ (g)$	+	$3\ O_2\ (g)$	⇆	$4\ NO_2\ (g)$
Initial Concentration (M)	0.0342		0.0415		0
Change (M)	$-2x$		$-3x$		$+4x$
Eq Concentration (M)	$0.0342 - 2x$		$0.0415 - 3x$		$+4x$

   We know that the equilibrium concentration of NO_2 is 0.0298 M and can, therefore, solve immediately for the value of x.
   $4x = 0.0298;\ x = 7.45 \times 10^{-3}$

   The value of x can be substituted back into the expressions for the equilibrium concentrations of N_2O and O_2.

   $[N_2O] = 0.0342\ M - (2 \times 7.45 \times 10^{-3}\ M) = 1.93 \times 10^{-2}\ M;$

   $[O_2] = 0.0415\ M - (3 \times 7.45 \times 10^{-3}\ M) = 1.91 \times 10^{-2}\ M$

   Now calculate K_c by substituting the equilibrium concentrations of all species into the equilibrium expression.

$$K_c = \frac{[NO_2]^4}{[N_2O]^2[O_2]^3} = \frac{\left(7.45 \times 10^{-3}\right)^4}{\left(1.93 \times 10^{-2}\right)^2\left(1.91 \times 10^{-2}\right)^3} = 1.19$$

7. The initial concentrations of SO_2 and NO_2 are 0.100 M. Let x be the concentration of SO_2 that reacts on going to the equilibrium state. Since the mole ratios are 1:1:1:1, if x mol/L of SO_2 reacts, then x mol/L of NO_2 also reacts and x mol/L of SO_3 and NO are produced. This can be summarized in the following table:

Principal Reaction	$SO_2\ (g)$	+	$NO_2\ (g)$	⇆	$SO_3\ (g)$	+	$NO\ (g)$
Initial Concentration (M)	0.100		0.100		0		0
Change	$-x$		$-x$		$+x$		$+x$
Eq Concentration (M)	$0.100 - x$		$0.100 - x$		$+x$		$+x$

   The algebraic expressions for the equilibrium concentrations from the table can be substituted into this expression, and we can solve for x.

$$K_c = \frac{[NO][SO_3]}{[SO_2][NO_2]}$$

$$85.0 = \frac{(x)(x)}{(0.100 - x)(0.100 - x)} = \left(\frac{x}{0.100 - x}\right)^2$$

Taking the square root of both sides gives

$$\pm 9.22 = \frac{x}{0.100 - x}$$

Solving for x, there are two possible solutions. The equation with the positive square root of 85 gives

$+9.22(0.100 - x) = x$ $+0.922 = x + 9.22x$ $x = \frac{0.922}{10.22} = 9.02 \times 10^{-2}$

The equation with the negative square root of 85 gives

$-9.22(0.100 - x) = x$; $-0.922 = x - 9.22x$ $x = \frac{-0.922}{-8.22} = 0.112$

Because the initial concentrations of SO_2 and NO_2 are 0.100, x cannot exceed 0.100. Therefore, x = 0.112 can be discarded as unreasonable and $x = 9.02 \text{ x } 10^{-2}$ is the solution. The equilibrium concentrations are:

$[SO_2] = [NO_2] = 0.100 - (9.02 \text{ x } 10^{-2}) = 0.0098$ M; $[SO_3] = [NO] = 0.0902$ M

8. a. reaction shifts to the left
   b. shifts left (reaction that absorbs heat, reverse reaction generates heat)
   c. shifts to the left (greater number of moles of gas)
   d. no effect
   e. shifts to the right

9. $K_c = \frac{k_f}{k_r}$; $K_c = \frac{6.8 \times 10^{-5}}{7.2 \times 10^{-7}} = 94.4$

10. The initial concentration of $HCONH_2$ is 0.125 M. Let x be the concentration of $HCONH_2$ that reacts on going to the equilibrium state. Since the mole ratios are 1:1:1, if x mol/L of $HCONH_2$ reacts, then x mol/L of NH_3 and CO are produced. This can be summarized in the following table:

Principal Reaction	$HCONH_2$ (g)	⇆	NH_3 (g)	+	CO (g)
Initial Concentration (M)	0.125		0		0
Change	$-x$		$+x$		$+x$
Eq Concentration (M)	$0.125 - x$		$+x$		$+x$

The equilibrium expression for the reaction is $K_c = \frac{[NH_3][CO]}{[HCONH_2]}$. The equilibrium concentrations from the table can be substituted into this expression.

$$4.84 = \frac{(x)(x)}{(0.125 - x)} = \frac{x^2}{0.125 - x}$$

To solve for x, we need to rearrange this expression and use the quadratic equation.

$4.84(0.125 - x) = x^2$ $x^2 + 4.84x - 0.605 = 0$;

$$x = \frac{-4.84 \pm \sqrt{(4.84)^2 - 4(-0.605)}}{2} \qquad x = -4.96 \text{ or } 0.120$$

The mathematical solution that makes chemical sense is 0.120.

This can be used to calculate the equilibrium concentrations.

$[HCONH_2] = 0.125 - 0.120 = 0.005$ M $\qquad [NH_3] = [CO] = 0.120$ M

11. The position of equilibrium is to the left, given the very small value of K_c. We can calculate the value of Q_c from the concentrations given.

$$Q_c = \frac{[H_2]^4[CS_2]}{[H_2S]^2[CH_4]} = \frac{(0.450)^4(0.085)}{(2.00)^2(1.75)} = 4.98 \times 10^{-4}$$

$Q_c > K_c$; therefore, the reaction mixture is not at equilibrium. The reaction will proceed from right to left to reach equilibrium.

Challenge Problem

The equilibrium expression is

$$K_c = \frac{[N_2O][O_2]}{[NO_2][NO]}$$

The initial concentrations (mol/L) for each species are

$[NO_2] = 0.100$ M; $[NO] = 0.050$ M; $[N_2O] = 0.0125$ M; $[O_2] = 0.00125$ M

The mole ratio of NO_2 to NO is 1:1. Therefore, the amount of NO_2 that reacts equals the amount of NO that reacts. Define this amount as x. We can summarize the initial and equilibrium concentrations in the following table:

Principal Reaction	$NO_2\ (g)$ +	$NO\ (g)$ ⇆	$N_2O\ (g)$ +	$O_2\ (g)$
Initial Concentration (M)	0.100	0.050	0.0125	0.00125
Change (M)	$-x$	$-x$	$+x$	$+x$
Eq. Concentration (M)	$0.100 - x$	$0.050 - x$	$0.0125 + x$	$0.00125 + x$

The algebraic equations for the equilibrium concentrations can be substituted back into the equation for the equilibrium constant:

$$0.914 = \frac{(0.0125 + x)(0.00125 + x)}{(0.100 - x)(0.050 - x)}$$

$$0.914 = \frac{\left(1.56\text{x}10^{-5} + 0.0138\text{x} + \text{x}^2\right)}{\left(0.0050 - 0.150\text{x} + \text{x}^2\right)}$$

$$0.00457 - 0.137\text{x} + 0.914\text{x}^2 = 1.56\text{x}10^{-5} + 0.0138\text{x} + \text{x}^2$$

$$0 = 0.086\text{x}^2 + 0.151\text{x} - 0.00455$$

$$\text{x}^2 + 1.76\text{x} - 0.0529 = 0$$

$$\text{x} = \frac{-1.76 \pm \sqrt{1.76^2 - 4(-0.0529)}}{2}$$

$$\text{x} = 0.0296, -1.60$$

Substituting the value of *x* into the equilibrium concentrations leads to

$[NO_2]_e = 0.100 - 0.0296 = 0.0704$ M
$[NO]_e = 0.050 - 0.0296 = 0.0204$ M
$[N_2O]_e = 0.0125 + 0.0296 = 0.0421$ M
$[O_2]_e = 0.00125 + 0.0296 = 0.0308$ M

Chapter 14

True–False

1. F. The Lewis theory deals with electron transfers, but an electron acceptor is an acid, and an electron donor is a base.
2. F. A strong acid fully dissociates. A concentrated solution of a weak acid can be far more harmful than a dilute solution of a strong acid.
3. T
4. T
5. F. The strength of an acid and the strength of its conjugate base are inversely related (the stronger the acid, the weaker the conjugate base).
6. T
7. F. The principal reaction for a diprotic acid is the dissociation of H_2A.
8. T
9. T
10. F. $Mg(OH)_2$ is a strong base.

Multiple Choice

1. c
2. d
3. a
4. b
5. c
6. d
7. d
8. a

Matching

Acid-base indicator	f
Acid dissociation constant	a
Arrhenius acid	j
Arrhenius base	d
Base dissociation constant	g
Brønsted-Lowry acid	b
Brønsted-Lowry base	k
Conjugate acid-base pair	c
Hydronium ion	m
Ion-product for water	o
Lewis acid	e
Lewis base	n
Percent dissociation	h
pH	i
Polyprotic acid	l
Strong acid	q
Weak acid	p

Fill–in–the Blank

1. weak, proton
2. hydrogen or hydronium, hydroxide, neutral, equals
3. increases, decreases
4. stepwise, smaller
5. electrons, hydrogen

Problems

1. We first need to solve for the concentration of the H_3O^+ ion, then solve for $[OH^-]$.

$$[H_3O^+] = \text{antilog}\ (-11.78) = 1.66 \times 10^{-12}\ M$$
$$[OH^-] = \frac{1.0 \times 10^{-14}}{1.66 \times 10^{-12}} = 6.0 \times 10^{-3}\ M$$

2. $Ba(OH)_2$ is a strong base and completely dissociates in water according to the reaction

$$Ba(OH)_2\ (aq) \rightarrow Ba^{2+}\ (aq) + 2\ OH^-\ (aq)$$

Determine the molar concentration of OH^-.

$$1.83\ g\ Ba(OH)_2 \times \frac{1\ mol\ Ba(OH)_2}{171.3\ g\ Ba(OH)_2} \times \frac{2\ mol\ OH^-}{mol\ Ba(OH)_2} = 2.14 \times 10^{-2}\ mol\ OH^-;$$

$$\frac{2.14 \times 10^{-2}\ mol\ OH^-}{0.150\ L} = 0.142\ M$$

$$[H_3O^+] = \frac{1.0 \times 10^{-14}}{0.142} = 7.02 \times 10^{-14}; \qquad pH = -\log(7.02 \times 10^{-14}) = 13.15$$

3. $$0.51\% = \frac{[\text{acid}]\ \text{dissociated}}{7.35 \times 10^{-3}} \times 100 \qquad [\text{acid}]\ \text{dissociated} = 3.74 \times 10^{-5}$$

The dissociated acid represents the $[H_3O^+]$ and $[A^-]$ at equilibrium.

$$pH = -\log(3.74 \times 10^{-5}) = 4.427$$

The equilibrium concentration of acid is $7.35 \times 10^{-3} - 3.74 \times 10^{-5} = 7.31 \times 10^{-3}$ M.

$$K_a = \frac{(3.74 \times 10^{-5})(3.74 \times 10^{-5})}{(7.31 \times 10^{-3})} = 1.91 \times 10^{-7}$$

4. Calculate $[H_3O^+]$ at equilibrium from the pH.

   $[H_3O^+]$ = antilog (–6.03) = 9.33×10^{-7}.

   This concentration also represents the amount of histidine that dissociated and can be used to calculate the equilibrium concentration of histidine.

   $(1.30 \times 10^{-3}) - (9.33 \times 10^{-7}) = 1.30 \times 10^{-3}$

   $[H_3O^+] = [A^-]$ at equilibrium. Calculate K_a.

$$K_a = \frac{(9.33 \times 10^{-7})(9.33 \times 10^{-7})}{(1.30 \times 10^{-3})} = 6.70 \times 10^{-10}$$

5. This is an equilibrium problem.

Principal Reaction	$C_6H_5N\ (aq) + H_2O\ (l)$	⇆	$C_6H_5NH^+\ (aq)$	+	$OH^-\ (aq)$
Initial Concentration	1.24		0		0
Change	$-x$		$+x$		$+x$
Equilibrium Concentration	$1.24 - x$		$+x$		$+x$

Substituting these values into the equilibrium expression gives

$$K_b = 1.8 \times 10^{-9} = \frac{[C_6H_5NH^+][OH^-]}{[C_6H_5N]} = \frac{(x)(x)}{(1.24 - x)}$$

Assume that $1.24 - x \approx 1.24$. The above expression then simplifies to

$$x^2 = (1.8 \times 10^{-9})(1.24); \quad x = 4.72 \times 10^{-5}\ M$$

$[OH^-] = x = 4.72 \times 10^{-5}$

Solve for $[H_3O^+]$ from $[OH^-]$

$$[H_3O^+] = \frac{1.0 \times 10^{-14}}{4.72 \times 10^{-5}} = 2.12 \times 10^{-10}\ M$$

Calculate pH.

$pH = -\log(2.12 \times 10^{-10}) = 9.67$

6. Oxalic acid is a diprotic acid, and has two possible dissociations:

Principal Reaction	$H_2C_2O_4\ (aq) + H_2O\ (l)$	⇆	$H_3O^+\ (aq)$	+	$HC_2O_4^-\ (aq)$
Initial Concentration	0.125		0		0

Change	$-x$	$+x$	$+x$
Equilibrium Concentration	$0.125 - x$	$+x$	$+x$

Substituting these values into the equilibrium expression gives

$$K_{a1} = 5.9 \times 10^{-2} = \frac{[H_3O^+][HC_2O_4^-]}{[H_2C_2O_4]} = \frac{(x)(x)}{(0.125 - x)}$$

Due to the value of K_{a1}, we cannot assume that the value of x is negligible. Therefore, we must use the quadratic equation to solve for x. Rearranging the above expression gives

$$(7.375 \times 10^{-3}) - (5.9 \times 10^{-2})x = x^2$$

$$x^2 + (5.9 \times 10^{-2})x - (7.375 \times 10^{-3}) = 0$$

$$x = \frac{-(5.9\times10^{-2}) \pm \sqrt{(5.9\times10^{-2})^2 - 4(-7.375\times10^{-3})}}{2}$$

$$x = 6.13\times10^{-2}$$

$[H_2C_2O_4] = 0.125 - 0.061 = 0.064$ M; $\quad [H_3O^+] = [HC_2O_4^-] = x = 0.0613$

To calculate the $[C_2O_4^{2-}]$, use the second dissociation step of $H_2C_2O_4$, remembering that the dissociation takes place in the presence of 0.0613 M H_3O^+ from the first dissociation.

$$HC_2O_4^- (aq) + H_2O (l) \leftrightarrows H_3O^+ (aq) + C_2O_4^{2-} (aq)$$

$$K_{a2} = 6.4 \times 10^{-5} = \frac{[H_3O^+][C_2O_4^{2-}]}{[HC_2O_4^-]}$$

Use the concentration of $HC_2O_4^-$ calculated above in the calculation of $[C_2O_4^-]$. The quantities of $C_2O_4^-$ generated will be negligible compared to the ions already present.

$$6.5\times10^{-5} = \frac{(0.061+x)x}{0.061-x} = \frac{0.061x}{0.061}$$

$$x = 6.5 \text{ x } 10^{-5}$$

$[C_2O_4^{2-}] = x = 6.5 \text{ x } 10^{-5}$ M

pH = –log (0.0613) = 1.21

7. Let the formula B represent the formula for benzylamine.

Principal Reaction	$B (aq) + H_2O (l) \leftrightarrows$	$BH^+ (aq)$ +	$OH^- (aq)$
Initial Concentration	0.350	0	0
Change	$-x$	$+x$	$+x$
Equilibrium Concentration	$0.350 - x$	$+x$	$+x$

Substituting these values into the equilibrium expression gives

$$K_b = 2.14 \times 10^{-5} = \frac{[BH^+][OH^-]}{[B]} = \frac{(x)(x)}{(0.350 - x)}$$

Assume that $0.350 - x \approx 0.350$. The above expression then simplifies to:

$$x^2 = (2.14 \times 10^{-5})(0.350) \qquad x = 2.74 \times 10^{-3}$$

$[OH^-] = 2.74 \times 10^{-3}$

$$[H_3O^+] = \frac{1.0 \times 10^{-14}}{2.74 \times 10^{-3}} = 3.65 \times 10^{-12}$$

$pH = -\log(3.65 \times 10^{-12}) = 11.44$

8. a. Ascorbic acid: $K_b = \dfrac{1.0 \times 10^{-14}}{8.0 \times 10^{-5}} = 1.2 \times 10^{-10}$

   b. Hydrazine: $K_a = \dfrac{1.0 \times 10^{-14}}{8.9 \times 10^{-7}} = 1.1 \times 10^{-8}$

9. KCN is a salt derived from a weak acid (HCN) and a strong base (KOH). This solution will be basic. The hydrolysis reaction is

$$CN^- (aq) + H_2O (l) \leftrightarrows HCN (aq) + OH^- (aq)$$

The K_b for this reaction can be calculated from K_w and K_a for HCN.

$$K_b = \frac{1.0 \times 10^{-14}}{4.9 \times 10^{-10}} = 2.04 \times 10^{-5}$$

To determine the HCN equilibrium concentration and the pH of the solution, determine the initial concentration of CN^-.

$$1.75 \text{ g KCN} \times \frac{1 \text{ mol KCN}}{65.1 \text{ g KCN}} \times \frac{1 \text{ mol } CN^-}{1 \text{ mol KCN}} = 0.0269 \text{ mol } CN^-$$

$$\frac{0.0269 \text{ mol } CN^-}{0.150 \text{ L}} = 0.179 \text{ M}$$

Principal Reaction	$CN^- (aq) + H_2O (l)$	$\leftrightarrows$	$HCN (aq)$ +	$OH^- (aq)$
Initial Concentration	0.179		0	0
Change	$-x$		$+x$	$+x$
Equilibrium Concentration	$0.179 - x$		$+x$	$+x$

Substituting these values into the equilibrium expression gives

$$K_b = 2.04 \times 10^{-5} = \frac{[HCN][OH^-]}{[CN^-]} = \frac{(x)(x)}{(0.179 - x)}$$

Assume that $0.179 - x \approx 0.179$. The above expression then simplifies to:

$$x^2 = (2.04 \times 10^{-5})(0.179) \qquad x = 1.91 \times 10^{-3}$$

$x = [HCN] = [OH^-] = 1.91 \times 10^{-3}$

$$[H_3O^+] = \frac{1.0 \times 10^{-14}}{1.91 \times 10^{-3}} = 5.23 \times 10^{-12}$$

$pH = -\log(5.23 \times 10^{-12}) = 11.281$

10. a. stronger acid: HI – For binary acids, acid strength increases down a group due to the increase in the size of the atom.

    b. Stronger acid: H_3PO_4—For oxoacids, acid strength increases with increasing number of oxygen atoms.

Challenge Problem

Assume 100 g of solution, which is 25 g of H_3PO_4. The volume of the solution can be calculated from the mass and density of the solution.

$$\text{mL of solution} = \frac{100\text{ g}}{1.1667\text{ g/mL}} = 85.7\text{ mL}$$

$$\text{mol } H_3PO_4 = 25\text{ g x}\frac{1\text{ mol } H_3PO_4}{98\text{ g}} = 0.255\text{ mol}$$

$$M = \frac{0.255\text{ mol } H_3PO_4}{0.0857\text{ L}} = 2.98\text{ M}$$

1. Write the stepwise dissociation for phosphoric acid.

$H_3PO_4\,(aq) + H_2O\,(l) \leftrightarrows H_3O^+\,(aq) + H_2PO_4^-\,(aq)$ $\qquad K_{a1} = 7.5 \times 10^{-3}$

$H_2PO_4^-\,(aq) + H_2O\,(l) \leftrightarrows H_3O^+\,(aq) + HPO_4^{2-}\,(aq)$ $\qquad K_{a2} = 6.3 \times 10^{-8}$

$HPO_4^{2-}\,(aq) + H_2O\,(l) \leftrightarrows H_3O^+\,(aq) + PO_4^{3-}\,(aq)$ $\qquad K_{a3} = 4.8 \times 10^{-13}$

Since $K_{a1} > K_{a2}, K_{a3}$, we know that the principal reaction is the first dissociation step and that all of the H_3O^+ present is produced from this step. Therefore, we need only to consider the first dissociation step when calculating the pH of the solution.

Make a table showing the principal reaction, initial concentration, change in concentration, and equilibrium concentration of the reactant and products.

Principal Reaction	$H_3PO_4\,(aq) \leftrightarrows$	H_3O^+	+ $H_2PO_4^-$
Initial Concentration	2.98	0	0
Change	$-x$	$+x$	$+x$
Eq Concentration	$2.98 - x$	$+x$	$+x$

Substitute the equilibrium concentrations into the equilibrium expression. Since $K_{a1} > 1.0 \times 10^{-3}$, you cannot assume that x is negligible.

$$7.5\times10^{-3}=\frac{x^2}{2.98-x}$$

$$x^2+\left(7.5\times10^{-3}\right)x-\left(2.2\times10^{-2}\right)=0$$

$$x=\frac{-\left(7.5\times10^{-3}\right)\pm\sqrt{\left(7.5\times10^{-3}\right)^2+4\left(2.2\times10^{-2}\right)}}{2}$$

$x = 0.145$ M. This value represents the equilibrium concentration of H_3O^+ and $H_2PO_4^-$.

Calculate the pH from this value.

pH = –log (0.145) = 0.839

Equilibrium concentrations are

$[H_3PO_4]_e = 2.98$ M – 0.145 M = 2.83 M
$[H_3O^+]_e = [H_2PO_4^-]_e = 0.145$ M

We now repeat the process, using the chemical equation for $H_2PO_4^-$ and the equilibrium concentration calculated above.

Principal Reaction	$H_2PO_4^-$ (*aq*) ⇆	H_3O^+ (*aq*) +	HPO_4^{2-} (*aq*)
Initial Concentration	0.145	0.145	0
Change	$-x$	$+x$	$+x$
Eq Concentration	$0.145 - x$	$0.145 + x$	$+x$

The value of K_{a2} is small enough so that x is negligible; therefore $0.145 - x \approx 0.145$ and $0.145 + x \approx 0.145$.

$$6.2\times10^{-8}=\frac{(0.145)x}{0.145}$$

$$x=6.2\times10^{-8}$$

This value represents the equilibrium concentration of $[H_2PO_4^-]$.

$[HPO_4^{2-}]_e = 6.2 \times 10^{-8}$ M

Repeat the process to calculate the equilibrium concentration of PO_4^{2-}.

Principal Reaction	HPO_4^{2-} (*aq*)	⇆	H_3O^+ (*aq*) +	PO_4^{3-} (*aq*)
Initial concentration	6.2×10^{-8}		0.145	0
Change	$-x$		$+x$	$+x$
Eq. concentration	$(6.2 \times 10^{-8}) - x$		$0.145 + x$	$+x$

The value of K_{a3} is small enough so that x is negligible; therefore $(6.2 \times 10^{-8}) - x \approx 6.2 \times 10^{-8}$.

$$4.8 \times 10^{-13} = \frac{(0.145)x}{6.2 \times 10^{-8}}$$

$$x = 2.05 \times 10^{-19}$$

This value of x represents the equilibrium concentration of PO_4^{3-}.

$[PO_4^{3-}]_e = 2.05 \times 10^{-19}$

Chapter 15

True—False

1. T
2. F. When a weak acid reacts with a strong base the pH of the solution is greater than 7.00.
3. T
4. F. The equilibrium shifts to more dissociated acid to neutralize the base.
5. F. Half of the acid is dissociated.
6. T
7. F. Equivalence point will be lower than 7, as there is conjugate acid left in solution.
8. F. Molar solubility and K_{sp} can be mathematically interconverted.
9. T
10. F. Na and K ions cannot be precipitated out, but can be identified by flame tests.

Matching

Amphoteric	d
Buffer capacity	i
Buffer solution	b
Common-ion effect	g
Complex ion	a
Equivalence point	c
Formation constant	h
Henderson-Hasselbalch equation	e
Ion product	j
Qualitative analysis	f

Fill–in–the–Blank

1. strong, completion, protons
2. weak, conjugate, pH, equilibrium, buffers
3. metal, coordinate, molecules

Problems

1. a. HNO_3 and NaOH: This is a reaction between a strong acid and a strong base. The net ionic equation is $H_3O^+ (aq) + OH^- (aq) \rightarrow 2\ H_2O\ (l)$; pH = 7.

   b. HCl and NH_2NH_2: NH_2NH_2 is a weak base. The net ionic equation is $H_3O^+ (aq) + NH_2NH_2 \rightarrow H_2O\ (l) + NH_2NH_3^+ (aq)$; pH < 7.00.

   c. Acetic acid (CH_3COOH) and NaOH: Acetic acid is a weak acid. The net ionic equation is $CH_3COOH + OH^- (aq) \rightarrow CH_3COO^- (aq) + H_2O\ (l)$; pH > 7.00.

2. The salt NH_4Cl is 100% dissociated, so the species present initially are NH_3, NH_4^+, Cl^-, and H_2O. NH_3 is the strongest base present ($K_b >> K_w$). This gives the following table:

Principal Reaction	$NH_3\ (aq) + H_2O\ (l) \leftrightarrows$	$NH_4^+ (aq) +$	$OH^- (aq)$

Initial Concentration (M)	0.10	0.25	0
Change (M)	$-x$	$+x$	$+x$
Eq Concentration (M)	$0.10-x$	$0.25+x$	$+x$

The common ion in this problem is NH_4^+. The equilibrium equation for the principal reaction is

$$K_b = 1.8\times10^{-5} = \frac{[NH_4^+][OH^-]}{[NH_3]} = \frac{(0.25+x)(x)}{(0.10-x)} \approx \frac{(0.25)(x)}{(0.10)}$$

x is assumed to be negligible.

$$x = [OH^-] = \frac{(1.8\times10^{-5})(0.10)}{(0.25)} = 7.2\times10^{-6}$$

The assumption concerning the size of x is justified. Since we now know the $[OH^-]$, we can calculate the $[H_3O^+]$.

$$[H_3O^+] = \frac{1.0\times10^{-14}}{7.2\times10^{-6}} = 1.4\times10^{-9}$$

pH = –log (1.4×10^{-9}) = 8.85

3. The principal reaction and equilibrium concentrations for this solution are

Principal Reaction	$HClO\ (aq)\ +\ H_2O\ (l) \leftrightarrows$	$H_3O^+\ (aq)\ +$	$ClO^-\ (aq)$
Initial Concentration (M)	0.45	0	0.25
Change (M)	$-x$	$+x$	$+x$
Equilibrium Concentration (M)	$0.45-x$	$+x$	$0.25+x$

If we solve the equilibrium equation for H_3O^+, we obtain

$$K_a = \frac{[H_3O^+][ClO^-]}{[HClO]}$$

$$[H_3O^+] = K_a\frac{[HClO]}{[ClO^-]}$$

K_a for HClO is 3.5×10^{-8}. Substituting gives

$$[H_3O^+] = (3.5\times10^{-8})\frac{(0.45)}{(0.25)} = 6.3\times10^{-8}$$

pH = –log (6.3×10^{-8}) = 7.20

After addition of NaOH:

Neutralization Reaction	$HClO\ (aq)\ +$	$OH^-\ (aq) \leftrightarrows$	$H_2O\ (l)\ +$	$ClO^-\ (aq)$
Before Reaction (mol)	0.45	0.10		0.25
Change (mol)	– 0.10	–0.10		+ 0.10
After Reaction (mol)	0.35	0		0.35

Substituting these values into the expression for $[H_3O^+]$, we can then calculate the pH.

$$[H_3O^+] = \left(3.5 \times 10^{-8}\right)\frac{(0.35)}{(0.35)} = 3.5 \times 10^{-8}$$

pH = $-\log(3.5 \times 10^{-8}) = 7.46$

To determine the pH of the solution after addition of HCl, we must take into account the neutralization reaction that takes place.

Neutralization Reaction	ClO^- (*aq*) +	H_3O^+ (*aq*) ⇆	H_2O (*l*) +	HClO (*aq*)
Before reaction (mol)	0.30	0.10		0.30
Change (mol)	–0.10	–0.10		+0.10
After Reaction (mol)	0.20	0		0.40

Substituting these values into the expression for $[H_3O^+]$, we can then calculate the pH.

$$[H_3O^+] = \left(3.5 \times 10^{-8}\right)\frac{(0.40)}{(0.20)} = 7.0 \times 10^{-8}$$

pH = $-\log(7.0 \times 10^{-8}) = 7.15$

4. Rearrange the Henderson-Hasselbalch equation:

$$\log \frac{[\text{base}]}{[\text{acid}]} = \text{pH - p}K_a$$

The pK_a for lactic acid is $-\log(1.4 \times 10^{-4}) = 3.85$.

Substituting the pH and pK_a values gives

$$\log \frac{[\text{lactate}]}{[\text{lactic acid}]} = 4.75 - 3.85 = 0.90$$

Taking the antilog of both sides gives

$$\frac{[\text{lactate}]}{[\text{lactic acid}]} = 7.94$$

This gives the ratio of lactate ion to lactic acid. To determine the ratio of lactic acid to the lactate ion, take the reciprocal.

lactic acid = 0.126 x lactate

$$\frac{[\text{lactic acid}]}{[\text{lactate}]} = 0.126$$

If the solution is to be 0.750 M overall, this can be rearranged to solve for the amount of each.

lactic acid + lactate = 0.750 mol

$x + 0.126x = 0.750$

$x = 0.666$ mol

0.666 mol lactic acid and 0.084 mol lactate are required. To check, 0.084/0.666 = 0.126.

5. Calculate the pH of the buffer solution using the Henderson-Hasselbalch equation.

$$\mathrm{pH} = \mathrm{p}K_a + \log \frac{[\mathrm{HCO_2^-}]}{[\mathrm{HCOOH}]} \qquad \mathrm{pH} = 3.74 + \log \frac{0.25}{0.50} = 3.44$$

Account for the reaction on addition of HCl.

Neutralization Reaction	HCO_2^- (*aq*) +	H_3O^+ (*aq*) →	HCOOH (*aq*) + H_2O (*l*)
Before Reaction (mol)	0.250	0.300	0.50
Change (mol)	–0.250	–0.250	0.250
After Reaction (mol)	0	0.050	0.75

The limiting reactant in this problem is the HCO_2^- ion, which means that this amount of HCl exceeds the capacity of the buffer. The pH will be determined by the excess HCl present.

pH = –log(0.050) = 1.30

6. a. 0.0 mL of NaOH: $[H_3O^+]$ = 0.0400 M; pH = 1.398

b. 10.0 mL of NaOH = 0.010 L x 0.100 M = 1 mmol

75 mL of HCl = 0.075 L x 0.040 M = 3 mmol

2 mmol HCl will remain after the reaction.

$$[\mathrm{H_3O^+}] \text{ after neutralization} = \frac{(2.00 \text{ mmol})}{(85.0 \text{ mL})} = 0.0235 \qquad \mathrm{pH} = -\log(0.0235) = 1.629$$

c. 30.0 mL of NaOH = 3 mmol. This is the equivalence point: the HCl will be completely neutralized, and there is no NaOH in excess. The pH will be 7.0

d. 50.0 mL of NaOH = 5 mmol of NaOH.

2 mmol of NaOH will remain after the reaction.

$$[\mathrm{OH^-}] \text{ after neutralization} = \frac{2.00 \text{ mmol}}{125 \text{ mL}} = 0.016 \text{ M}$$

$$[\mathrm{H_3O^+}] = \frac{1.0 \times 10^{-14}}{0.016} = 6.25 \times 10^{-13} \qquad \mathrm{pH} = -\log(6.25 \text{ x } 10^{-13}) = 12.20$$

7. a. 0.0 mL NaOH: pH is calculated in the same manner as a weak acid.

Principal Reaction	CH_3COOH (*aq*) + H_2O (*l*) ⇆	H_3O^+ (*aq*) +	CH_3COO^- (*aq*)
Initial concentration (M)	0.0500	0	0
Change (M)	$-x$	$+x$	$+x$
Equilibrium Concentration (M)	$0.0500 - x$	$+x$	$+x$

$$K_a = 1.8 \times 10^{-5} = \frac{[H_3O^+][CH_3COO^-]}{[CH_3COOH]} = \frac{(x)(x)}{(0.0500 - x)} \approx \frac{(x)^2}{(0.0500)}$$

$x = 9.48 \times 10^{-4}$

$\text{pH} = -\log(9.48 \times 10^{-4}) = 3.02$

b. 10.0 mL of NaOH = 0.010 L x 0.100 M = 1 mmol

50 mL of acetic acid = 0.050 L x 0.050 M = 2.5 mmol

1.5 mmol acetic acid will remain after the reaction, and 1 mmol of sodium acetate will have been formed.

Using the Henderson-Hasselbalch equation, using the pK_a for acetic acid (4.74):

$$\text{pH} = 4.74 + \log\frac{1.5 \times 10^{-3}}{1.0 \times 10^{-3}} = 4.92$$

c. 25.0 mL NaOH = 2.5 mmol. This is the equivalence point of the titration. All the acetic acid is neutralized, and what remains is a solution containing 2.5 mmol sodium acetate.

2.5 mmol/75 mL = 0.033 M acetate

$$K_b = \frac{K_w}{K_a} = \frac{1 \times 10^{-14}}{1.8 \times 10^{-5}} = 5.56 \times 10^{-10}$$

The pH is determined from the equilibrium expression for sodium acetate.

$$K_b = 5.56 \times 10^{-10} = \frac{(x)^2}{3.33 \times 10^{-2}} \qquad x = [OH^-] = 4.35 \times 10^{-6}$$

$[H_3O^+] = K_w/4.35 \times 10^{-6} = 2.32 \times 10^{-9}$.

$\text{pH} = -\log(2.32 \times 10^{-9}) = 8.63$

d. 30.0 mL of NaOH = 0.030 L x 0.100 M = 3 mmol

50 mL of acetic acid = 0.050 L x 0.050 M = 2.5 mmol

0.5 mmol NaOH will remain after the reaction. This will determine the pH of the solution.

$$[OH^-] = \frac{0.50 \text{ mmol}}{80.0 \text{ mL}} = 6.3 \times 10^{-3}$$

$$[H_3O^+] = \frac{1 \times 10^{-14}}{6.3 \times 10^{-3}} = 1.59 \times 10^{-12} \qquad \text{pH} = -\log(1.59 \times 10^{-12}) = 11.80$$

8. The solubility equilibrium for $CaCO_3$ is

$CaCO_3\,(s) \leftrightarrows Ca^{2+}\,(aq) + CO_3^{2-}\,(aq)$

If 7.07×10^{-5} mol is the amount of $CaCO_3$ that dissolves in 1.0 L of solution then the $[Ca^{2+}] = 7.07 \times 10^{-5}$ M and the $[CO_3^{2-}] = 7.07 \times 10^{-5}$ M. Substituting these values into the equilibrium expression gives

$$K_{sp} = [Ca^{2+}][CO_3^{2-}] = (7.07 \times 10^{-5})^2 = 5.0 \times 10^{-9}$$

9.

Solubility Equilibrium	$Ca_3(PO_4)_2\ (s) \leftrightarrows$	$3\ Ca^{2+}\ (aq)$	$2\ PO_4^{3-}\ (aq)$
Equilibrium Concentration (M)		$+3x$	$+2x$

$$K_{sp} = 2.1 \times 10^{-33} = [Ca^{2+}]^3[PO_4^{3-}]^2 = (3x)^3(2x)^2 = 108x^5$$

$x = 1.14 \times 10^{-7}$

10.

Solubility Equilibrium	$CaSO_4\ (s) \leftrightarrows$	$Ca^{2+}\ (aq)$	+	$SO_4^{2-}\ (aq)$
Equilibrium Concentration (M)		$+x$		$0.50 + x$

$$K_{sp} = 7.1 \times 10^{-5} = [Ca^{2+}][SO_4^{2-}] = (x)(0.50 + x) \approx (x)(0.50)$$

$x = 1.42 \text{ x } 10^{-4}$

11. a. AgBr: Not more soluble. Br^- is the conjugate base of a very strong acid and is very unreactive.
    b. Na_2S: More soluble. S^{2-} is the conjugate base of a weak acid.
    c. LiCN: More soluble. CN^- is the conjugate base of a very weak acid.
    d. CaF_2: More soluble. F^- is the conjugate base of a weak acid.

12. Because K_f for $Zn(NH_3)_4^{2+}$ is large, nearly all the Zn^{2+} from $Zn(NO_3)_2$ will be converted to $Zn(NH_3)_4^{2+}$.

    $Zn^{2+}\ (aq) + 4\ NH_3\ (aq) \leftrightarrows Zn(NH_3)_4^{2+}$

    Conversion of 0.50 mol/L of Zn^{2+} to $Zn(NH_3)_4^{2+}$ consumes 2.00 mol/L of NH_3 (due to a 1:4 mole ratio of Zn^{2+} to NH_3). Assuming 100% conversion to $Zn(NH_3)_4^{2+}$ gives the following concentrations:

    $[Zn^{2+}] = 0$ M

    $[Zn(NH_3)_4^{2+}] = 0.50$ M

    $[NH_3] = 4.0 - 2.00 = 2.0$ M

    After conversion of Zn^{2+} to $Zn(NH_3)_4^{2+}$, assume that a small amount of the reverse reaction occurs, producing Zn^{2+}.

    $Zn(NH_3)_4^{2+} \leftrightarrows Zn^{2+}\ (aq) + 4\ NH_3\ (aq)$

    Dissociation of x mol/L of $Zn(NH_3)_4^{2+}$ produces x mol/L of Zn^{2+} and $4x$ mol/L of NH_3. The equilibrium concentrations are

    $[Zn(NH_3)_4^{2+}] = 0.50 - x$

    $[Zn^{2+}] = x$

$[NH_3] = 2.0 + 4x$

The following table summarizes this reasoning under the balanced equation.

Principal Reaction	Zn^{2+} (*aq*) +	$4\ NH_3$ (*aq*) $\leftrightarrows$	$Zn(NH_3)_4^{2+}$ (*aq*)
Initial Concentration (M)	0.50	4.0	0
After 100% Reaction (M)	0	2.0	0.50
Equilibrium Concentration (M)	x	$2.0 + 4x$	$0.50 - x$

Substitute the equilibrium concentrations into the expression for K_f and make the approximation that x is negligible compared with 0.50.

$$K_f = 7.8 \times 10^8 = \frac{[Zn(NH_3)_4^{2+}]}{[Zn^{2+}][NH_3]^4} = \frac{(0.50 - x)}{(x)(2.0 + 4x)^4} \approx \frac{(0.50)}{(x)(2.0)^4}$$

$$[Zn^{2+}] = x = \frac{(0.50)}{(7.8 \times 10^8)(2.0)^4} = 4.01 \times 10^{-11}\ M$$

$$[Zn(NH_3)_4] = 0.50 - (4.01 \times 10^{-11})M = 0.50\ M$$

13. The equation for this reaction is

$$3\ CaCl_2\ (aq) + 2\ Na_3PO_4\ (aq) \rightarrow Ca_3(PO_4)_2\ (s) + 6\ NaCl\ (aq)$$

After the two solutions are mixed, the total volume is 550 mL, making the concentrations of the relevant ions

$$[Ca^{2+}] = \frac{350\ \text{mL x } 0.75}{550\ \text{mL}} = 0.477M$$

$$[PO_4^{3-}] = \frac{200\ \text{mL x } 1.50}{550\ \text{mL}} = 0.545M$$

To determine if a precipitate will form, calculate the ion product and compare its value to K_{sp}.

$$IP = [Ca^{2+}]^3[PO_4^{3-}]^2$$

$$IP = (0.477)^3(0.545)^2 = 0.0322$$

K_{sp} for $Ca_3(PO_4)_2 = 2.1 \times 10^{-33}$; IP >> K_{sp}; therefore, a precipitate will form.

14. K_{spa} (CuS) = 6×10^{-16}; K_{spa} (FeS) = 6×10^{2}

$$Q_c = \frac{[Cu^{2+}][H_2S]^2}{[H_3O^+]^2} = \frac{(0.007)(0.10)^2}{(0.20)^2} = 1.75 \times 10^{-3}$$

Since K_{spa} (CuS) < Q_c < K_{spa} (FeS), Cu^{2+} will selectively precipitate out of solution.

Challenge Problem

For CuS

$$CuS\ (s) \leftrightharpoons Cu^{2+}\ (aq) + S^{2-}\ (aq) \qquad K_{spa} = 6.0 \times 10^{-16}$$

The acidic buffer will cause S^{2-} ions to be removed from the solution to form HS^{-} and H_2S (*aq*).

$$H^{+}\ (aq) + S^{2-}\ (aq) \leftrightharpoons HS^{-}\ (aq)$$

$$H^{+}\ (aq) + HS^{-}\ (aq) \leftrightharpoons H_2S\ (aq)$$

Adding the three chemical equations together, we have

$$CuS\ (s) + 2\ H^{+}\ (aq) \leftrightharpoons Zn^{2+}\ (aq) + H_2S\ (aq)$$

This equation is the definition of K_{spa}.

$$K_{spa} = \frac{[Cu^{2+}][H_2S]}{[H_3O^{+}]^2}$$

To determine the molar solubility of CuS, we need to determine the concentration of Cu^{2+} at equilibrium. The concentration of H_3O^{+} can be calculated from the information given about the buffer.

$[HCHO_2] = 0.25$ M and $[CHO_2^{-}] = 0.35$ M; $K_a = 1.8 \times 10^{-4}$

$$pH = pK_a + \log\frac{[CH_2O^{-}]}{[HCH_2O]} = 3.74 + \log\frac{0.25}{0.45} = 3.48$$

$[H^{+}] = 3.31 \times 10^{-4}$

Principal Reaction	CuS (*s*) +	$2\ H^{+}\ (aq) \leftrightharpoons$	$Cu^{2+}\ (aq)$ +	$H_2S\ (aq)$
Initial concentration		3.31×10^{-4}	0	0
Change		$-2x$	$+x$	$+x$
Equilibrium Conc.		$(3.31 \times 10^{-4}) - 2x$	$+x$	$+x$

Substituting the equilibrium concentrations into the equilibrium expression, we have

$$6.0 \times 10^{-16} = \frac{(x)(x)}{[(3.31 \times 10^{-4}) - 2x]^2}$$

The value of x can be considered negligible.

$x = 8.11 \text{ x } 10^{-12}$

$x = [Cu^{2+}] = 8.11 \times 10^{-12}$. $[Cu^{2+}]$ represents the molar solubility of CuS.

Chapter 16

True–False

1. F. A reaction can be spontaneous and very slow.
2. T
3. T
4. F. Standard entropies of reaction can be calculated from standard entropies of formation.
5. T
6. T. A lower temperature will reduce the contribution of the entropy term, allowing the enthalpy to dominate.
7. T
8. F. Standard enthalpy of formation for a pure element is always zero.
9. F. ΔG determines the spontaneity of a reaction.
10. T

Matching

Entropy	c
First law of thermodynamics	a
Free energy	f
Second law of thermodynamics	e
Spontaneous process	d
Standard state	h
Thermodynamics	b
Third law of thermodynamics	g

Fill–in–the–Blank

1. spontaneous, increases
2. increases, decreases
3. temperature, motion, entropy
4. larger, smaller, gases, solids
5. entropy/enthalpy, entropy/enthalpy, free energy, temperature
6. positive, unstable
7. reverse/leftward

Problems

1. a. The toy box has the higher entropy. A dishwasher can be arranged in a finite number of ways that could be considered orderly, but toys can be tumbled into a toybox in a multitude of ways.
   b. The sugar dissolved in tea has the higher entropy. The crystals in the bowl are relatively pure and very constrained in their arrangement.
   c. The 1 mol of O_2 gas in 35.5 has the higher entropy. The 32 g of oxygen gas will have a volume of 22.4 L, as they are at STP, and more volume means less order.

2. a. ΔS is positive; an increase in the volume of a gas at constant temperature leads to a decrease in the pressure, which leads to an increase in the entropy
   b. ΔS is negative; Solids have lower entropy than aqueous ions.
   c. ΔS is positive; production of a gas increases disorder.
   d. ΔS is negative; reactant side of the reaction has more moles of gas.
   e. ΔS is negative; separating the isotopes decreases the randomness of the system.

3. a. $2\ NH_4NO_3\ (s) \rightarrow 2\ N_2\ (g) + O_2\ (g) + 4\ H_2O\ (g)$

$$\Delta S^\circ_{rxn} = [2 \times S^\circ(N_2) + S^\circ(O_2) + 4 \times S^\circ(H_2O)] - [2 \times S^\circ(NH_4NO_3)]$$

$$\Delta S^\circ_{rxn} = [(2 \times 191.5) + 205.0 + (4 \times 188.7)] - (2 \times 151.1)$$

$$\Delta S^\circ_{rxn} = 1040.6\ J/K$$

b. $2\ S\ (s) + 3\ O_2\ (g) \rightarrow 2\ SO_3\ (g)$

$$\Delta S^{o}_{rxn} = [2 \times S^{o}(SO_3)] - [2 \times S^{o}(S) + 3 \times S^{o}(O_2)]$$
$$\Delta S^{o}_{rxn} = (2 \times 256.6) - [(2 \times 31.8) + (3 \times 205.0)]$$
$$\Delta S^{o}_{rxn} = -165.4\ J/K$$

4. Use the equation $\Delta S^{o}_{surr} = -\frac{\Delta H^{o}_{rxn}}{T}$.

Solve for ΔS^{o}_{surr}:

$$\Delta S^{o}_{surr} = \Delta S^{o}_{total} - \Delta S^{o}_{rxn} = 1957 - 515.7 = 1441\ J/K$$
$$\Delta H^{o}_{rxn} = -T\Delta S_{surr} = -298 \times 1441 \frac{J}{K} = -4.295 \times 10^{5} \frac{J}{mol} = -429.5 \frac{kJ}{mol}$$

5. Using the equation $\Delta S = R \ln \frac{V_{final}}{V_{initial}}$, we can calculate the value of ΔS.

$$\Delta S = 8.314 \frac{J}{mol \cdot K} \ln \frac{45.3\ L}{22.4\ L} = 5.85 \frac{J}{mol \cdot K} \times 1\ mol = 5.85 \frac{J}{K}$$

6. The temperature at which ethanol spontaneously goes from liquid to gas (and back again) can be calculated using the formula

$$T = \frac{\Delta H^{o}}{\Delta S^{o}} = \frac{-235.1\ kJ/mol - (-277.7\ kJ/mol)}{282.6\ J/mol \cdot K - 161\ J/mol \cdot K} = \frac{42.6\ kJ/mol}{121.6\ J/mol \cdot K}$$

Take care with the units!

$$T = \frac{42.6\ kJ/mol}{121.6\ J/mol \cdot K} = \frac{42,600\ J/mol}{121.6\ J/mol \cdot K} = 350\ K = 77.2\ °C$$

7. ΔS_{sys} is the same as ΔS_{rxn}, and can be calculated using the standard enthalpies given.

$$\Delta S^{o}_{rxn} = [2 \times S^{o}(HCl)] - [S^{o}(H_2) + S^{o}(Cl_2)]$$

$$\Delta S^{o}_{rxn} = [2 \times 186.8] - [130.6 + 223.0] = 20.0\ J/(mol \cdot K)$$

The entropy of the surroundings can be calculated using the formula

$$\Delta S^{o}_{surr} = -\frac{\Delta H^{o}_{rxn}}{T} = -\frac{-92.3\ kJ/mol}{298K} = 310\ J/(mol \cdot K)$$

Once again, when going from enthalpy values to entropy values, be careful with units.

$$\Delta S^{o}_{total} = \Delta S^{o}_{surr} + \Delta S^{o}_{rxn} = 330\ J/(mol \cdot K)$$

8. $$\Delta G^{o}_{rxn} = [2 \times \Delta G^{o}_{f}(CO_2) + 2 \times \Delta G^{o}_{f}(H_2O)] - [\Delta G^{o}_{f}(C_2H_4) + 3 \times \Delta G^{o}_{f}(O_2)]$$
$$\Delta G^{o}_{rxn} = [(2 \times -394.4) + (2 \times -237.2)] - [68.1 + 0] = -1331\ kJ/mol$$

9. $$\Delta G^{o} = [2 \times \Delta G^{o}_{f}(CO_2)] - [2 \times \Delta G^{o}_{f}(CO) + \Delta G^{o}_{f}(O_2)]$$

$\Delta G^{\circ} = (2 \text{ x } -394.4) - [(2 \text{ x } -137.2) + 0] = -514.4 \text{ kJ/mol}$

$$\Delta G = \Delta G^{\circ} + RT \ln Q = -514{,}400 \text{ J/mol} + 8.314 \text{ J/K} \ln \frac{(15)^2}{(0.5)^2(0.1)} = -514{,}300 \text{ J/mol} = -514.3 \text{ kJ/mol}$$

10. $\Delta G^{\circ} = -RT \ln K$

$$\Delta G^{\circ}_{rxn} = [\Delta G^{\circ}_f(CO_2) + \Delta G^{\circ}_f(H_2O)] - [\Delta G^{\circ}_f(H_2CO_3)]$$

$$\Delta G^{\circ}_{rxn} = [(-386.0) + (-237.2)] - (-623) = -0.2 \text{ kJ/mol}$$

$$\ln K = -\frac{\Delta G^{\circ}}{RT} = -\frac{-200 \text{ J/mol}}{8.314 \text{ J/K x } 310 \text{ K}} = 7.76\text{x}10^{-2}$$

$$K = e^{7.76\text{x}10^{-2}} = 1.08$$

11. To solve this problem, you need to use the data found in Appendix B in your textbook.

$$\Delta G^{\circ}_{rxn} = [0 + 0] - [(2 \times -384.2)] = 768.4 \text{ kJ/mol}$$

To calculate the temperature at which the reaction becomes spontaneous, we need to first calculate the ΔS°_{rxn} and ΔH°_{rxn}.

$$\Delta S^{\circ}_{rxn} = [223.0 + (2 \times 51.2)] - [(2 \times 72.1)] = 181.2 \text{ J/mol·K}$$

$$\Delta H^{\circ}_{rxn} = [0 + 0] - [(2 \times -411.2)] = 822.4 \text{ kJ/mol}$$

Solve for the temperature (**watch your units**).

$$T = \frac{\Delta H}{\Delta S} = \frac{822{,}400 \frac{\text{J}}{\text{mol}}}{181.2 \frac{\text{J}}{\text{mol} \cdot \text{K}}} = 4538 \text{ K}$$

$$\ln K = \frac{\Delta G^{\circ}_{rxn}}{-RT} = \frac{822{,}400 \frac{\text{J}}{\text{mol}}}{-\left[\left(8.314 \frac{\text{J}}{\text{mol} \cdot \text{K}}\right)(4538 \text{ K})\right]} = -21.8$$

$$K = e^{-21.8} = 3.41\text{x}10^{-10}$$

Challenge Problem

$$\Delta G^{\circ} = [(2 \times 86.6 \text{ kJ/mol})] = 173.2 \text{ kJ}$$

$$\Delta G^{\circ} = -RT\ln K$$

$$\ln K = \frac{1.732 \times 10^5 \text{ J}}{-\left(8.314 \frac{\text{J}}{\text{mol} \cdot \text{K}}\right)(298 \text{ K})} = -69.9$$

$$K = 4.39 \times 10^{-31}$$

$\Delta H° = [2 \text{ mol NO} \times 90.2 \text{ kJ/mol}] = 180.4 \text{ kJ}$

$\Delta S° = [2 \text{ mol NO} \times 210.7 \text{ J/mol·K}] – [(1 \text{ mol } N_2 \times 191.5 \text{ J/mol·K}) + (1 \text{ mol } O_2 \times 205.0 \text{ J/mol·K})]$

$\Delta S° = 24.9 \text{ J/mol}$

$$\Delta G° = \Delta H - T\Delta S = 1.804 \times 10^5 \text{ J} - (773 \text{ K})\left(24.9 \frac{\text{J}}{\text{K}}\right) = 1.61 \times 10^5 \text{ J}$$

$$\Delta G° = -RT \ln K$$

$$\ln K = \frac{-1.21 \times 10^4 \text{ J}}{-\left(8.314 \frac{\text{J}}{\text{mol} \cdot \text{K}}\right)(773 \text{ K})} = -25.05$$

$$K = e^{-25.05} = 1.32 \times 10^{-11}$$

To determine the partial pressure of N_2, O_2, and NO at equilibrium, we need to first calculate the initial pressures of N_2 and O_2 using the ideal gas law.

$$P_{N_2} = P_{O_2} = \frac{(3.00 \text{ mol})\left(\frac{0.08206 \text{ L} \cdot \text{atm}}{\text{mol} \cdot \text{K}}\right)(723 \text{ K})}{15.0 \text{ L}} = 11.9 \text{ atm}$$

Principal Reaction	N_2 (*g*)	+	O_2 (*g*) ⇆	2 NO (*g*)
Initial Pressure	11.9		11.9	0
Change	$-x$		$-x$	$+2x$
Equilibrium Pressure	$11.9 - x$		$11.9 - x$	$+2x$

Using the equilibrium constant calculated above, we can now solve for the equilibrium pressures.

$$1.32 \times 10^{-11} = \frac{P_{NO}^2}{P_{N_2} P_{O_2}} = \frac{2x^2}{(11.9 - x)(11.9 - x)} = \frac{2x^2}{(11.9 - x)^2}$$

Taking the square root of both sides gives

$$3.63 \times 10^{-6} = \frac{2x}{11.9 - x}$$

$$x = 2.16 \times 10^{-5}$$

The equilibrium partial pressures are $P_{N_2} = P_{O_2} = 11.9 \text{ atm}$ and $P_{NO} = 2.16 \times 10^{-5}$ atm.

Chapter 17

True–False

1. T
2. F. A wire allows the transfer of electrons, the salt bridge maintains electrical neutrality.
3. F. Positive *E* corresponds to negative free-energy change. Both indicate spontaneity.

4. T
5. F. The Nernst equation allows calculation of E under non-standard conditions.
6. T
7. T
8. T
9. F. Corrosion can be prevented by contact with a metal *below* the metal of interest in the table of standard reduction potentials.
10. F. Overvoltage can lead electrolysis reactions in aqueous solutions to be difficult to predict.

Multiple Choice
1. b
2. a
3. a
4. c
5. b
6. c
7. a
8. d
9. a
10. b

Fill–in–the–Blank
1. lead storage, alkaline dry cells
2. corrosion, galvanization, sacrificial electrode
3. sodium or aluminum, aluminum or sodium, electricity, power plants
4. reduction, decreasing, increasing, standard hydrogen electrode
5. equilibrium, cell potentials, small or large, small or large, completion

Matching

Anode	f
Cathode	c
Corrosion	h
Electrolysis	a
Electroplating	b
Electrorefining	i
Galvanizing	d
Overvoltage	e
Salt bridge	g

Problems
1. Half-reactions:

Anode: $Mg\ (s) \rightarrow Mg^{2+}\ (aq) + 2\ e^-$
Cathode: $Ni^{2+}\ (aq) + 2\ e^- \rightarrow Ni\ (s)$

Shorthand notation:

$Mg\ (s)\ |\ Mg^{2+}\ (aq)\ ||\ Ni^{2+}\ (aq)\ |\ Ni\ (s)$

2. a. $2\ Al\ (s) \rightarrow 2\ Al^{3+}\ (aq) + 6\ e^-$ $E° = 1.66$ V

$3\ Pb^{2+}\ (aq) + 6\ e^- \rightarrow 3\ Pb\ (s)$ $E° = 0.13$ V

overall $E° = 1.79$ V – spontaneous as written

b. $Br_2\ (aq) + 2\ e^- \rightarrow 2\ Br^-\ (aq)$ $E° = 1.09$ V

$2\ Cl^-\ (aq) \rightarrow Cl_2\ (g) + 2\ e^-$ $E° = -1.36$ V

overall $E°$ = -0.27 V – spontaneous in reverse

c. $Ni\ (s) \rightarrow Ni^{2+}\ (aq) + 2\ e^-$ $E° = 0.26$ V

$Pb^{2+}\ (aq) + 2\ e^- \rightarrow Pb\ (s)$ $E° = -0.13$ V

overall $E°$ = 0.13 V – spontaneous as written

d. $Fe\ (s) \rightarrow Fe^{2+}\ (aq) + 2\ e^-$ $E° = 0.45$ V

$2\ H^+ + 2\ e^- \rightarrow H_2\ (g)$ $E° = 0.00$ V

overall $E°$ = 0.45 V – spontaneous as written

3. The two half-reactions (balanced in e^-) and their potentials for this cell are

Anode: $3\ Mg\ (s) \rightarrow 3\ Mg^{2+}\ (aq) + 6\ e^-$ $E^o = 2.37$ V

Cathode: $2\ Fe^{3+}\ (aq) + 6\ e^- \rightarrow 2\ Fe\ (s)$ $E^o = -0.04$ V

a. At standard state, $E^o = 2.33$ V

b. $[Mg^{2+}] = 0.1$ M, $[Fe^{3+}] = 2.0$ M

$$E_{cell} = E^o_{cell} - \frac{0.0592}{6}\log\frac{[Mg^{2+}]^3}{[Fe^{3+}]^2} = 2.33 - \frac{0.0592}{6}\log\frac{[0.1]^3}{[2]^2} = 2.37\text{ V}$$

c. $[Mg^{2+}] = 2.0$ M, $[Fe^{3+}] = 0.1$ M

$$E_{cell} = E^o_{cell} - \frac{0.0592}{6}\log\frac{[Mg^{2+}]^3}{[Fe^{3+}]^2} = 2.33 - \frac{0.0592}{6}\log\frac{[2.0]^3}{[0.1]^2} = 2.30\text{ V}$$

4. $$\Delta G° = -nFE = -2 \text{ x } 96,500 \frac{C}{\text{mol } e^-} \text{ x } 0.27 \text{ V x } \frac{1\text{ J}}{C \bullet V} = -52.1\ \frac{\text{kJ}}{\text{mol}}$$

$$E° = \frac{0.0592V}{n}\log K$$

$$\log K = \frac{2 \text{ x } 0.27V}{0.0592\text{ V}} = 9.12$$

$$K = 10^{9.12} = 1.32 \text{ x } 10^9$$

5. The half-reactions for the cell are:

$Zn^{2+}\ (aq) + 2\ e^- \rightarrow Zn\ (s)$ $E° = -0.76$ V

$E^o_{cell} = E^o_{anode} + E^o_{cathode}$ $+1.61\text{ V} = E^o_{anode} + (-0.76\text{ V})$ $E^o_{anode} = 2.37\text{ V}$

From the table of standard reduction potentials we find that Mg (*s*) has a reduction potential of –2.37 V. The half–reaction at the anode is

$Mg\ (s) \rightarrow Mg^{2+}\ (aq) + 2\ e^-$

6. E_{ref} = -0.26 V for $Ni^{2+}\ (aq) + 2\ e^- \rightarrow Ni\ (s)$

$$pH = \frac{E_{cell} - E_{ref}}{0.0592\ V} = \frac{0.14 - (-0.26)}{0.0592\ V} = 6.75$$

7. For a metal to be able to protect iron from corrosion, it must have a reduction potential below that of Fe^{2+}. Any metal below that reaction will be able to protect iron, as any iron atoms that are oxidized will be reduced again by the metal.

$Fe^{2+}\ (aq) + 2\ e^- \rightarrow Fe\ (s)$ $\qquad E° = -0.45$

a. $Sn^{2+}\ (aq) + 2\ e^- \rightarrow Sn\ (s)$ $E° = -0.14$ V. This is *above* the reduction potential of iron, so would speed corrosion.

b. $Mn^{2+}\ (aq) + 2\ e^- \rightarrow Mn\ (s)$ $E° = -1.18$ V. This is *below* the reduction potential of iron, so would protect against corrosion.

c. $Cu^{2+}\ (aq) + 2\ e^- \rightarrow Cu\ (s)$ $E° = 0.34$ V. This is *above* the reduction potential of iron, so would speed corrosion.

8. $AgCl\ (s) \rightarrow Ag^+\ (aq) + Cl^-\ (aq)$

Using the table of standard reduction potentials, we can break this overall reaction into the following two half–reactions:

$AgCl\ (s) + e^- \rightarrow Ag\ (s) + Cl^-\ (aq)$ $\qquad E° = 0.22$ V

$Ag\ (s) \rightarrow Ag^+\ (aq) + e^-$ $\qquad E° = –0.80$ V

$$E^{o}_{cell} = 0.22\ V + (-0.80\ V) = -0.58\ V$$

$$-0.58\ V = \frac{0.0592}{1} \log K$$

$$\log K = -9.80$$

$$K = 10^{-9.80}$$

$K_{sp} = 1.58 \times 10^{-10}$

9. $Na^+\ (l) + e^- \rightarrow Na\ (s)$

$2\ Br^-\ (l) \rightarrow Br_2\ (g) + 2\ e^-$

To determine the current required to produce 3 g of sodium per hour, set up a unit analysis to convert 3 g Na/hr to C/s.

$$\frac{3\ g\ Na}{hr} \times \frac{1\ mol\ Na}{23.0\ g} \times \frac{96,500\ C}{mol\ e^-} \times \frac{1\ mol\ e^-}{mol\ Na} \times \frac{1\ hr}{3600\ s} = 3.5\frac{C}{s} = 3.5\ A$$

10. $15 \text{ g Ag} \times \frac{1 \text{ mol Ag}}{108 \text{ g}} \times \frac{96,500 \text{ C}}{\text{mol e}^-} \times \frac{1 \text{ mol e}^-}{\text{mol Ag}} \times \frac{1 \text{ s}}{3 \text{ C}} \times \frac{1 \text{ min}}{60 \text{ s}} = 74 \text{ min}$

11. To begin this problem, it is necessary to write and balance the equation. To make it easier, take the half-reaction for the reduction of MnO_4^- straight from a table to determine the number of electrons involved.

$$5 \text{ Zn } (s) \rightarrow 5 \text{ Zn}^{2+} (aq) + 10 \text{ e}^-$$

$$2 \text{ MnO}_4^- + 16 \text{ H}^+ (aq) + 10 \text{ e}^- \rightarrow 2 \text{ Mn}^{2+} (aq) + 8 \text{ H}_2\text{O } (l)$$

The overall reaction is $5 \text{ Zn } (s) + 2 \text{ MnO}_4^- + 16 \text{ H}^+ (aq) \rightarrow 5 \text{ Zn}^{2+} (aq) + 2 \text{ Mn}^{2+} (aq) + 8 \text{ H}_2\text{O } (l)$

The problem can now be set up as a unit-analysis problem.

$3.27 \text{ g Zn} \times \frac{1 \text{ mol Zn}}{65.4 \text{ g Zn}} \times \frac{2 \text{ mol MnO}_4^-}{5 \text{ mol Zn}} \times \frac{142 \text{ g NaMnO}_4}{\text{mol NaMnO}_4} = 2.84 \text{ g}$ are the minimum needed.

12. To calculate the volume of O_2, first calculate the moles of O_2 produced and then use the ideal gas law. The half-reaction for the electrolysis of water that is relevant here is

$$2 \text{ H}_2\text{O } (l) \rightarrow \text{O}_2 (g) + 4 \text{ H}^+ (aq) + 4 \text{ e}^-$$

$$2.97 \times 10^3 \text{ C} \times \frac{1 \text{ mol e}^-}{96,500 \text{ C}} \times \frac{1 \text{ mol O}_2}{4 \text{ mol e}^-} = 7.69 \times 10^{-3} \text{ mol O}_2$$

$$V = \frac{nRT}{P} = \frac{(7.69 \times 10^{-3})(0.0821)(298)}{\left(356 \text{ mm Hg} \times \frac{1 \text{ atm}}{760 \text{ mm Hg}}\right)} = 0.402 \text{ L}$$

Challenge Problem

Separate the overall reaction into half reactions and calculate K for the reduction of lead. $K_{red} \times K_{sp} = K_{overall}$. From this, $E°$ for the reaction can be calculated.

$PbSO_4 (s) \rightarrow Pb^{2+} (aq) + SO_4^{2-} (aq)$	K_{sp}
$Pb^{2+} (aq) + 2 e^- \rightarrow Pb (s)$	K_{red} (calculated from $E°$)
$PbSO_4 (s) + 2 e^- \rightarrow Pb (s) + SO_4^{2-} (aq)$	$K_{overall} = K_{sp} \times K_{red}$

Calculate $K_{overall}$, then $E_{overall}$.

$$E^\circ_{red} = \frac{0.0592}{n} \log K_{red}$$

$n = 2$ for this reaction.

$$\log K_{red} = \frac{2(-0.126)}{0.0592} = -4.26$$

$$K_{red} = 5.50 \times 10^{-5}$$

Calculate K_{net} for the overall reaction.

$$K_{net} = (1.8 \times 10^{-8}) \times (5.5 \times 10^{-5}) = 9.9 \times 10^{-13}$$

Now calculate $E^{o}_{overall}$.

$$E^{o}_{net} = \frac{0.0592}{n} \log K_{net}$$

$$E^{o}_{net} = \frac{0.0592}{2} \log(9.9x10^{-13}) = -0.355 \text{ V}$$

Chapter 18

True–False

1. T
2. F. D_2O is more difficult to hydrolyze than H_2O.
3. T
4. T
5. F. Oxygen levels in the atmosphere are very stable, despite being used and replenished constantly.
6. F. The Earth's crust is made up largely of oxides, especially silicon oxides.
7. F. Oxygen is an inexpensive and readily available oxidizing agent.
8. T
9. T
10. F. Ozone filters UV in the upper atmosphere, but is a pollutant at ground level.

Multiple Choice

1. c
2. d
3. b
4. a
5. d
6. c
7. b
8. c
9. a
10. b

Fill–in–the–Blank

1. halogen or alkali metal, halogen or alkali metal, electron
2. ionic, covalent, or metallic
3. sun, chemical, releases or produces, consumes or requires
4. ionic, metal, covalent
5. double, larger, overlap
6. -2, -1, -1/2
7. acidic or covalent, acidic or covalent, basic or ionic, basic or ionic
8. heavier group 1A and 2A metals, heating the metals in an excess of air
9. divalent, ion exchange
10. ionic, hydrates, drying agents or dessicants, water

Matching

Isotope effect	e
Hygroscopic	d
Binary hydride	b
Hydrate	c
Nonstoichiometric	f
Disproportionation	a

Problems

1. a. Fe (*s*) + 2 HCl (*aq*) → $FeCl_2$ (*aq*) + H_2 (*g*)

   b. Mg (*s*) + H_2SO_4 (*aq*) → $MgSO_4$ (*aq*) + H_2 (*g*)

   c. 2 Al (*s*) + 6 HNO_3 (*aq*) → 2 $Al(NO_3)_3$ (*aq*) + 3 H_2 (*g*)

2. Zn (*s*) + 2 HCl (*aq*) → $ZnCl_2$ (*aq*) + H_2 (*g*).

   Determine the number of moles of hydrogen gas needed.

$$n = \frac{(1\text{ atm})(0.100\text{ L})}{\left(0.08206\frac{\text{L}\cdot\text{atm}}{\text{mol}\cdot\text{K}}\right)(308\text{ K})} = 3.96\times10^{-3}\text{ mol}$$

   Calculate the mass of Zn.

$$(3.96\times10^{-3}\text{ mol H}_2)\times\frac{1\text{ mol Zn}}{1\text{ mol H}_2}\times\frac{65.4\text{ g Zn}}{1\text{ mol Zn}} = 0.259\text{ g Zn}$$

3. a. LiH - ionic
   b. PH_3 - covalent
   c. HCl - covalent
   d. CaH_2 - ionic

4. CaH_2 (*s*) + 2 H_2O (*l*) → 2 H_2 (*g*) + $Ca(OH)_2$ (*aq*)

$$1400\text{ g CaH}_2\text{ x }\frac{1\text{ mol CaH}_2}{42.0\text{ g CaH}_2}\text{ x }\frac{2\text{ mol H}_2}{\text{mol CaH}_2} = 66.7\text{ mol H}_2$$

$$V = \frac{66.7\text{ mol x }0.08206\ \frac{\text{L}\bullet\text{atm}}{\text{mol}\bullet\text{K}}\text{ x }275\text{ K}}{1\text{ atm}} = 1514\text{ L}$$

5. a. K_2O – basic oxide, will dissolve in water by the following equation:

   K_2O (*s*) + H_2O (*l*) → 2 K^+ (*aq*) + 2 OH^- (*aq*)

   b. Al_2O_3 – amphoteric oxide, will dissolve in acid and base by the following equations:

   Al_2O_3 (*s*) + 6 H^+ (*aq*) → 2 Al^{3+} (*aq*) + 3 H_2O (*l*)

   Al_2O_3 (*s*) + 2 OH^- (*aq*) + 3 H_2O (*l*) → 2 $Al(OH)_4^-$ (*aq*)

   c. Cl_2O_7 – acidic oxide, will react with water by the following equation:

$Cl_2O_7 + H_2O\ (l) \rightarrow 2HClO_4\ (aq)$

6. a. Rb has an oxidation number of +1, so O has an oxidation number of -1/2 – superoxide.
   b. Mg has an oxidation number of +2, so O has an oxidation number of -1 – peroxide.
   c. H has an oxidation number of 1, so O has an oxidation number of -2 – oxide.
   d. Ca has an oxidation number of +2, so O has an oxidation number of -2 – oxide.

7. a. Fe goes from an oxidation state of +2 to +3, so it is oxidized, and H_2O_2 is an oxidizing agent.

   b. Mn goes from an oxidation state of +7 to +4, so it is reduced, and H_2O_2 is a reducing agent.

8. $$1\ km^3 x \left(\frac{1000\ m}{km}\right)^3 x \left(\frac{100\ cm}{m}\right)^3 x\ 1.025\ \frac{g}{cm^3} x\ 0.035 = 3.6 x 10^{13}\ g \text{ or } 3.6 x 10^{10}\ kg \text{ salt}.$$

9. The final weight is anhydrous $CoCl_2$, which has a formula weight of 129.9 g/mol. This can be converted directly to moles.

$$2.72\ g\ x \frac{1\ mol\ CoCl_2}{129.8\ g} = 0.02096\ mol\ CoCl_2$$

The difference in the masses is the amount of water driven off: 5.00 g - 2.72 g = 2.28 g. This can now be converted to moles of water.

$$2.28\ g\ x \frac{1\ mol\ H_2O}{18.0\ g} = 0.127\ mol\ H_2O$$

The mole ratio can be converted to the formula of the hydrate.

$$\frac{0.127\ mol\ H_2O}{0.02096\ mol\ CoCl_2} = 6.0\ \frac{H_2O}{CoCl_2}$$

There are six waters per formula unit: $CoCl_2 \bullet 6\ H_2O$

10. Convert the volume to moles of hydrogen.

$$n = \frac{1\ atm\ x\ 1.30\ L}{0.08206\ \frac{L \bullet atm}{mol \bullet K} x 298\ K} = 0.0532\ mol\ H_2$$

Balance the theoretical equation to determine stoichiometry.

$XH_2 + 2\ HCl \rightarrow XCl_2 + 2\ H_2\ (g)$

Determine the number of moles of X.

$$0.0532\ mol\ H_2\ x\ \frac{1\ mol\ XH_2}{2\ mol\ H_2} = 0.0266\ mol\ XH_2$$

Determine the molecular weight of XH_2.

$$\frac{3.7\ g}{0.0266\ mol} = 139.2 \frac{g}{mol}$$

As the hydrogen has a molecular weight of 2, the molecular weight of the metal is 137.2 – the hydride is BaH_2.

Challenge Problem

First, write and balance the equation between the two reactants to determine the reaction stoichiometry.

$$NaOCl\ (aq) + H_2O_2\ (aq) \rightarrow NaCl\ (aq) + H_2O\ (l) + O_2\ (g)$$

Next, determine the number of moles of each reagent and which, if any, is the limiting reagent.

$$\text{Bleach: } 500\text{ mL x } \frac{1\text{ g}}{\text{mL}} \text{ x } 0.06 \text{ x } \frac{1\text{ mol}}{74.5\text{ g}} = 0.0403\text{ mol NaOCl}$$

$$\text{Peroxide: } 500\text{ mL x } \frac{1\text{ g}}{\text{mL}} \text{ x } 0.03 \text{ x } \frac{1\text{ mol}}{34.0\text{ g}} = 0.442\text{ mol } H_2O_2$$

In the equation, these are in a 1:1 ratio, so NaOCl is the limiting reagent. The volume of oxygen gas evolved will be

$$0.0403\text{ mol NaOCl x } \frac{1\text{ mol } O_2}{\text{mol NaOCl}} = 0.0403\text{ mol } O_2$$

$$V = \frac{0.0403\text{ mol x } 0.08206\ \frac{\text{L} \bullet \text{atm}}{\text{mol} \bullet \text{K}} \text{ x} 298\text{ K}}{1\text{ atm}} = 0.985\text{ L } O_2$$

The moles of NaCl in the solution will be

$$0.0403\text{ mol NaOCl x } \frac{1\text{ mol } O_2}{\text{mol NaOCl}} = 0.0403\text{ mol NaCl}$$

The molarity of the solution is 0.0403 mol/1 L = 0.0403 M

There is excess H_2O_2 in the solution: this will cause the solution to be weakly acidic.

Chapter 19

True–False

1. F. Semimetals are on a zigzag line from boron to astatine, not in the center of the periodic table.
2. F. Metallic character increases down a column of the periodic table.
3. T
4. F. Due to their size, it is uncommon to find double bonds in elements outside of the second row.
5. T
6. T
7. F. Silicon is too large to form double bonds: the chains consist of single bonds.
8. T
9. T
10. F. Liquid sulfur decreases in viscosity until the S_8 rings begin to break and reform into long chains. The viscosity increases at this point, and decreases again when the chains begin to break as well.

Multiple Choice

1. b
2. c
3. a

4. d
5. d
6. c
7. d
8. a
9. b
10. b

Fill–in–the–Blank

1. increases, increases, decreases, decreases
2. diamond, graphite, fullerene
3. white phosphorus, bent, red phosphorus
4. SiO_4
5. solid, liquid, S_8 rings, chains, viscosity

Matching

Aluminosilicate	c
Borane	d
Carbide	g
Contact process	f
Ostwald process	a
Silicate	e
Zone refining	b

General Questions

1. a) Rb b) In c) Cs

2. a) K b) In c) Sb

3. Ionization energy— decreases down a group; atomic radius—increases down a group; electronegativity—decreases down a group; basicity of oxides—increases down a group

4. Each boron in diborane uses an sp^3 hybrid orbital to overlap with a terminal hydrogen 1*s* orbital. Each bridging H atom is joined to both boron atoms through a three–center two–electron bond in which the two electrons in the B–H–B bridge are spread out over three atoms.

5. Diamond: tetrahedral bonding with sp^3 hybridization. Non-conducting, hardest known substance, and highest melting point of any element.

   Graphite: trigonal planar bonding with sp^2 hybridization, with delocalized π bonds to adjacent molecules. Conducts electricity and forms sheets that can slide past one another, making it useful as a dry lubricant.

   Graphene: prepared by the "Scotch tape method," graphene is a two-dimensional array of carbon molecules with a hexagonal arrangement. It is an excellent conductor of electricity.

   Fullerene: this covers a number of structures from carbon nanotubes to spheres, which share a common bonding. They are molecular structures and dissolve in nonpolar solvents, as well as conducting electricity.

6. -3 = NH_3; –2 = N_2H_4; –1 = NH_2OH; +1 = N_2O; +2 = NO; +3 = HNO_2; +4 = NO_2; +5 = HNO_3

7. $HClO_4$—perchloric acid; $HClO_3$—chloric acid; $HClO_2$—chlorous acid; HClO—hypochlorous acid
   $HBrO_4$—perbromic acid; $HBrO_3$—bromic acid; $HBrO_2$—bromous acid; HBrO—hypobromous acid
   HIO_4—periodic acid; HIO_3—iodic acid; HIO_2—iodous acid; HIO—hypoiodous acid

8. $Cu\ (s) + 4\ HNO_3\ (aq) \rightarrow 2\ NO_2\ (g) + 2\ H_2O\ (l) + Cu(NO_3)_2\ (aq)$

Challenge Problem

$CaC_2\ (s) + H_2O\ (l) \rightarrow C_2H_2\ (g) + CaO\ (s)$

Determine the moles present of each reactant.

$$15.0 \text{ g } CaC_2 \text{ x } \frac{1 \text{ mol } CaC_2}{64.10 \text{ g}} = 0.234 \text{ mol } CaC_2$$

$$7.45 \text{ g } H_2O \text{ x } \frac{1 \text{ mol } H_2O}{18.0 \text{ g}} = 0.413 \text{ mol } H_2O$$

Calcium carbide is the limiting reactant, as the reactants react in a 1:1 ratio.

$$0.234 \text{ mol } C_2H_2 \text{ x } \frac{26.04 \text{ g } C_2H_2}{\text{mol}} = 6.09 \text{ g } C_2H_2$$

Write the balanced reaction for the burning of acetylene.

$$C_2H_2\ (g) + 5/2\ O_2\ (g) \rightarrow 2\ CO_2\ (g) + H_2O\ (g)$$

Determine overall ΔH_{rxn} for acetylene.

$$[(-241.8 \text{ kJ/mol}) + 2\ (-393.5 \text{ kJ/mol})\ ] - (227.4 \text{ kJ/mol}) = -1256 \text{ kJ/mol}$$

Determine the heat generated from this reaction.

$$0.234 \text{ mol } C_2H_2 \text{ x } -1256 \text{ kJ/mol} = -294 \text{ kJ}$$

Chapter 20

True–False

1. T
2. F. Mercury (Hg) is a liquid.
3. F. Densities have a maximum in the center of the row.
4. F. Transition metals are commonly found in multiple oxidation states.
5. T
6. T
7. F. The oxidation number is always given as part of the complex name.
8. T
9. T
10. F. Crystal field theory assumed that all interactions are electrostatic.

Multiple Choice

1. b
2. c
3. a
4. d
5. c
6. c
7. a
8. d

9. b
10. b

Fill–in–the–Blank
1. oxidation, oxidizing, reducing
2. Coordination, ligands coordinate covalent
3. isomers, geometric, stereoisomers
4. wavelength or frequency, energy, color
5. valence bond theory, hybrid, crystal field theory, electrostatic

Matching

Absorption spectrum	h
Achiral	m
Bidentate ligand	a
Chelate	f
Chiral	b
Cis isomer	g
Coordination number	c
Crystal-field splitting	k
Enantiomer	n
High-spin complex	e
Ligand	q
Ligand donor atom	i
Low-spin complex	p
Racemic mixture	j
Strong-field ligand	l
Trans isomer	d
Weak-field ligand	o

General Questions

1. Mn^{5+}: [Ar] $4s^2$ Mo: [Kr] $5s^2\ 4d^4$ Cr^{3+}: [Ar] $4s^2 3d^1$ Fe^{2+}: [Ar] $4s^1 3d^5$

2. Mn, Rh, La

3. Zn > Cu > Co > Mn. These metals are all better reducing agents than oxidizing agents.

4. MnO_4^- > CrO_3^- > Fe_2O_3 > CoO

5. 3, 8, 5, 2

6. a. Ligands: H_2O (donor atom—O); NH_3 (donor atom—N).
   Coordination number is 6; oxidation state of metal is +3; has the general formula MA_2B_4; therefore can be geometrical isomers.
   b. Ligands: NH_3 (donor atom—N); Br (donor atom—Br); H_2O (donor atom—O).
   Coordination number is 6; oxidation state of metal is +3; has the general formula MA_4BC; therefore can be geometrical isomers. This compound can also have ionization isomers were the Br and NO_3 to switch places.
   c. Ligands: NH_3 (donor atom—N), SCN^- (donor atom either N or S).
   Coordination number is 6; oxidation state of metal is +2. Because SCN^- can bond through either the N or S, this compound can display linkage isomerism.
   d. Ligands: ox (donor atom—O)
   Coordination number is 6 (ox is a bidentate ligand); oxidation state of metal is +3;
   Charge on complex ion is –3. Because the ligands can spiral either to the right or to the left, this compound can exist as enantiomers.

e. Ligands: en (donor atoms—N)
   Coordination number is 6 (en is a bidentate ligand); oxidation state of metal is +3;
   The en ligand can spiral either to the right or to the left, indicating the compound can exist as enantiomers.

7. a. diammineatetraaquacobalt(III) chloride
   b. tetraammineaquabromocobalt(II) nitrate
   c. tetraamminedithiocyanatochromate(II) or tetraamminediisothiocyanatochromate(II) depending on the linkage isomer.
   d. ammonium trioxaloferrate(III)
   e. tris(ethylenediamine)cobalt(III) iodide

8. a. $[Cr(NH_3)_4(H_2O)_4]Cl_3$
   b. $[Cu(en)(SCN)_2]$
   c. $[Mn(CO)_5Cl]$
   d. $K_4[Fe(CN)_4(Cl)_2] \cdot 2\ H_2O$
   e. $[Cu(H_2O)_4]SO_4$

9. a. Since $[Zn(NH_3)_4]^{2+}$ is tetrahedral, the hybridization on the metal ion is sp^3. The Zn^{2+} ion has the electron configuration [Ar] $4s^2 3d^8$. The orbital diagram for this ion is

[Ar] ↑↓ ↑↓ ↑↓ ↑ ↑ (3d) ↑↓ (4s) _ _ _ (4p)

To share the 4 electron pairs provided by the ammonia molecules, Zn^{2+} must use the empty 4*s* and 4*p* orbitals to form hybrid orbitals. The orbital diagram for the complex is

[Ar] ↑↓ ↑↓ ↑↓ ↑↓ ↑↓ (3d) ↑↓ (4s) ↑↓ ↑↓ ↑↓ (4p)

This complex is low spin since it has no unpaired electrons.

b. $[Ni(CN)_4]^{2-}$ is square planar; therefore, the hybrid orbitals being used by the Ni^{2+} ion are dsp^2. The Ni^{2+} ion has the electron configuration [Ar] $3d^8$. The orbital diagram for this ion is

[Ar] ↑↓ ↑↓ ↑↓ ↑ ↑ (3d) _ (4s) _ _ _ (4p)

For Ni^{2+} to use dsp^2 hybrid orbitals, the two unpaired electrons must first pair up. Ni^{2+} can then accept a share in the 4 electron pairs provided by the CN^- ion. The orbital diagram for the complex is

[Ar] ↑↓ ↑↓ ↑↓ ↑↓ _ (3d) ↑↓ ↑↓ ↑↓ ↑↓ (dsp^2) _ (4p)

The complex is low spin since there are no unpaired electrons.

10. The crystal field splitting in tetrahedral complexes is half of the crystal field splitting in octahedral complexes because none of the *d* orbitals point directly at the ligands, and there are only four ligands, as opposed to six in an octahedral complex.

Challenge Problem

Chromium forms octahedral complexes, and there are 9 possible ligands. Therefore, the compound can be a hydrate or a salt: all 9 ligands will not be complexed to the chromium. The experimental data must be used to determine what is bonded, and what is hydrate water or balancing ions.

In each case, the moles of compound dehydrated are the same. The molecular weight of all three is 266.45 g/mol.

$$0.500 \text{ g x } \frac{1 \text{ mol}}{273.5 \text{ g}} = 1.88 \text{ x } 10^{-3} \text{mol}$$

The moles of compound in each solution is also the same.

$$0.125 \text{ mol/L x } 0.050 \text{ L} = 6.25 \text{ x } 10^{-3} \text{ mol}$$

If the mass of the compound does not change when it is dehydrated, all six waters are complexed to the chromium. This is seen with compound C. Therefore, it is likely that this is $[Co(H_2O)_6]Cl_3$. This can be confirmed by determining whether 3 moles of AgCl are produced per mol of compound.

$$0.809 \text{ g AgCl x } \frac{1 \text{ mol}}{143.4 \text{ g}} = 5.64 \text{ x } 10^{-3} \text{ mol AgCl}$$

$$\frac{5.64 \text{ x } 10^{-3} \text{mol AgCl}}{1.88 \text{ x } 10^{-3} \text{mol C}} = 3.00$$

This confirms that the chlorides are balancing ions, and separate in solution. This is $[Co(H_2O)_6]Cl_3$.

If there is a mass change on dehydration, the number of hydrate waters can be determined.

The mass change for compound A is 0.500 g – 0.432 g = 0.068 g water. The number of hydrate waters can be determined by comparing the moles of water to the moles of compound.

$$0.068 \text{ g } H_2O \text{ x } \frac{1 \text{ mol}}{18.01 \text{ g}} = 3.76 \text{ x } 10^{-3} \text{mol } H_2O$$

$$\frac{3.76 \text{ x } 10^{-3} \text{mol } H_2O}{1.88 \text{ x } 10^{-3} \text{mol A}} = 2.00$$

If there are 2 hydrate waters, the compound will likely be $[Co(H_2O)_4Cl_2]Cl\cdot 2H_2O$. This can be confirmed by the number of moles of AgCl produced per mole of compound.

$$0.269 \text{ g AgCl x } \frac{1 \text{ mol}}{143.4 \text{ g}} = 1.88 \text{ x } 10^{-3} \text{ mol AgCl}$$

$$\frac{1.88 \text{ x } 10^{-3} \text{mol AgCl}}{1.88 \text{ x } 10^{-3} \text{mol A}} = 1.00$$

This confirms one balancing chloride ion. This is $[Co(H_2O)_4Cl_2]Cl\cdot 2H_2O$.

The mass change for compound B is 0.500 g – 0.466 g = 0.034 g water. The number of hydrate waters can be determined by comparing the moles of water to the moles of compound.

$$0.034 \text{ g } H_2O \text{ x } \frac{1 \text{ mol}}{18.01 \text{ g}} = 1.88 \text{ x } 10^{-3} \text{ mol } H_2O$$

$$\frac{1.88 \text{ x } 10^{-3} \text{ mol } H_2O}{1.88 \text{ x } 10^{-3} \text{ mol B}} = 1.00$$

If there is 1 hydrate water, the compound is likely to be $[Co(H_2O)_5Cl]Cl_2 \cdot H_2O$. This can be confirmed by the number of moles of AgCl produced per mole of compound.

$$0.539 \text{ g AgCl x } \frac{1 \text{ mol}}{143.4 \text{ g}} = 3.76 \text{ x } 10^{-3} \text{ mol AgCl}$$

$$\frac{3.76 \text{ x } 10^{-3} \text{ mol AgCl}}{1.88 \text{ x } 10^{-3} \text{ mol B}} = 2.00$$

This confirms two balancing chloride ions. This is $[Co(H_2O)_5Cl]Cl_2 \cdot H_2O$.

Chapter 21

True–False

1. F. Most silicates are poor metal sources.
2. T
3. T
4. F. Reduction is by carbon monoxide.
5. F. There is no good theory for superconductivity at the present time.
6. T
7. T
8. F. Heating semiconductors increases conductivity. Heating metals decreases conductivity.
9. T
10. F. The primary drawback of ceramics is that they are brittle.

Multiple Choice

1. c
2. d
3. b
4. a
5. d
6. b
7. a
8. c
9. c
10. b

Fill–in–the–Blank

1. malleability and ductility, localized, brittle
2. current, conductance, antibonding, empty
3. below, resistance, nitrogen
4. metals, density, insulators, corrosion, malleable
5. (In any order): ceramics, metals, fibers, epoxy

Matching

Alloy	m
Band gap	g
Band theory	a

Ceramic	l
Conduction band	c
Doping	j
Electron-sea model	p
Flotation	t
Gangue	e
Metallurgy	k
Mineral	f
Ore	b
p-type semiconductor	r
Semiconductor	d
Roasting	h
Sintering	i
Slag	o
Sol-gel method	n
Superconducting transition temperature	q
Superconductor	u
Valence band	s

Problems:

1. Write and balance the equations.

$$2\ \text{MnCO}_3\ (s) + \text{O}_2\ (g) \rightarrow 2\ \text{MnO}_2\ (s) + \text{CO}_2\ (g)$$

$$3\ \text{MnO}_2\ (s) + 4\ \text{Al}\ (s) \rightarrow 3\ \text{Mn}\ (s) + 2\ \text{Al}_2\text{O}_3\ (s)$$

Determine the number of moles and kg of Mn in 1500 kg of $MnCO_3$.

$$1500\ \text{kg MnCO}_3\ \text{x}\ \frac{1000\ \text{g}}{1\ \text{kg}}\ \text{x}\ \frac{1\ \text{mol MnCO}_3}{115\ \text{g}} = 13{,}050\ \text{mol}$$

$$13{,}050\ \text{mol MnCO}_3\ \text{x}\ \frac{1\ \text{mol Mn}}{\text{mol MnCO}_3}\ \text{x}\ \frac{54.94\ \text{g}}{\text{mol Mn}}\ \text{x}\ \frac{1\ \text{kg}}{1000\ \text{g}} = 717\ \text{kg}$$

Determine the moles, then kg of aluminum needed.

$$13{,}050\ \text{mol MnCO}_3\ \text{x}\ \frac{1\ \text{mol MnO}_2}{\text{mol MnCO}_3}\ \text{x}\ \frac{4\ \text{mol Al}}{3\ \text{mol MnO}_2}\ \text{x}\ \frac{26.98\ \text{g Al}}{\text{mol Al}}\ \text{x}\ \frac{1\ \text{kg}}{1000\ \text{g}} = 469\ \text{kg Al}$$

2. As fourth-period elements, these all have access to 4 *s* orbitals and 3 *d* orbitals. 12 electrons can be accommodated, 6 in bonding orbitals, and 6 in nonbonding orbitals. The more electrons there are in bonding orbitals, and the fewer electrons in the nonbonding orbitals, the harder the element and the higher the melting point.

Cr: [Ar] $4s^2 3d^4$ – 6 electrons in bonding orbitals, no electrons in antibonding orbitals

Co: [Ar] $4s^2 3d^7$ – 6 electrons in bonding orbitals, 3 electrons in antibonding orbitals

Cu: [Ar] $4s^2 3d^9$ – 6 electrons in bonding orbitals, 5 electrons in antibonding orbitals

Highest to lowest melting point: Cr > Co > Cu

The observed melting points are
Cr: 1857 °C

Co: 1495 °C
Cu: 1083 °C

This confirms the determination made using molecular orbital theory.

3. Determine the number of valence electrons in each of the materials.

Si – Group 4A – 4 valence electrons
Ge – Group 4A – 4 valence electrons
P – Group 5A – 5 valence electrons
As – Group 5A – 5 valence electrons
B – Group 3A – 3 valence electrons

Identify the combinations and their types.

SiP - 4A5A – *n*-type
SiAs - 4A5A – *n*-type
SiB - 4A3A – *p*-type

GeP - 4A5A – *n*-type
GeAs - 4A5A – *n*-type
GeB - 4A3A – *p*-type

Challenge Problem

The feasibility of using cyanidation to remove silver from argentite can be determined by calculating ΔG for the reaction. If ΔG is negative, the process may be a means of obtaining silver from the ore. Determine ΔG for the process from the K_{net} for the reaction. The two equations involved are

$$Ag_2S\,(s) \rightarrow 2\,Ag^+\,(aq) + S^{2-}\,(aq) \qquad K_{sp} = 6 \times 10^{-51}$$

$$2\,Ag^+\,(aq) + 4\,CN^-\,(aq) \rightarrow 2\,[Ag(CN)_2]^- \qquad K_f^2 = 1 \times 10^{42}$$

(Square the value of K_f because the stoichiometry of the chemical equation is two times that of the chemical equation for the formation of $[Ag(CN)_2]^-$.)

The overall reaction and K_{net} are

$$Ag_2S\,(s) + 4\,CN^-\,(aq) \rightarrow 2\,[Ag(CN)_2]^-\,(aq) + S^{2-}\,(aq) \qquad K_{net} = 6 \times 10^{-9}$$

Calculate ΔG (assume 25°C).

$$\Delta G = -RT(\ln K)$$

$$\Delta G = -(8.314\ \text{J/mol}\cdot\text{K})(298\text{K})\ln 6\times10^{-9} = 4.69\times10^{4}\ \text{J/mol} = +46.9\ \text{kJ/mol}$$

Both the small value of K_{net} and the positive value of ΔG indicate that the cyanidation process for the removal of silver from argentite is not practical.

Use the same approach to determine if it is possible to use cyanidation to remove silver from horn silver.

$$AgCl\,(s) \rightarrow Ag^{+}\,(aq) + Cl^{-}\,(aq) \qquad K_{sp} = 1.8 \times 10^{-10}$$

$$Ag^{+}\,(aq) + 2\,CN^{-}\,(aq) \rightarrow [Ag(CN)_2]^{-}\,(aq) + Cl^{-}\,(aq) \qquad K_f = 1 \times 10^{21}$$

The overall chemical equation is

$$AgCl\,(s) + 2\,CN^{-}\,(aq) \rightarrow [Ag(CN)_2]^{-}\,(aq) + Cl^{-}\,(aq) \qquad K_{net} = 1.8 \times 10^{11}$$

$$\Delta G = -(8.314\ J/mol \cdot K)(298\ K)\ln 1.8 \times 10^{11} = -6.42 \times 10^{4}\ J/mol = -64.2\ kJ/mol$$

Both the large value of K_{net} and the negative value of ΔG indicate that cyanidation is practical for extracting silver from horn silver.

Chapter 22

True–False

1. F. Radioactive decay is a first-order process.
2. F. Light elements have a 1:1 ratio of protons:neutrons, but heavier elements require more "glue" and thus more neutrons.
3. F. A small amount of mass is lost by being converted to energy.
4. T
5. T
6. T
7. F. The oxidation number is always given as part of the complex name.
8. T
9. T
10. T

Multiple Choice

1. b
2. c
3. a
4. d
5. c
6. c
7. a
8. d
9. c
10. b

Fill–in–the–Blank

1. glue, proton, proton, nucleus, mass defect, $\Delta E = \Delta mc^2$
2. maximum, fusion, fission, decay
3. transmutation, particles, nucleus, element
4. alpha, beta, internal, gamma, X, external, stopped, lead
5. in vivo, diseased, internal, external, imaging

Matching

Binding energy	e
Chain reaction	g
Critical mass	d
Fission	a
Fusion	c
Mass defect	b

Nuclear transmutation f

Problems

1. Europium-130 Mass of 63 protons = (63)(1.007 28 amu) = 63.458 64 amu
Mass of 67 neutrons = (67)(1.008 66 amu) = 67.580 22 amu
Mass of 63 protons and 67 neutrons = 131.038 86 amu

Mass of europium-130 atom = 129.963 57 amu
Mass of electrons = (63) (5.486 x 10^{-4}) = 0.034 56 amu
Mass of europium-130 nucleus = 129.929 01 amu

Mass defect: 131.038 86 amu – 129.929 01 amu= 1.109 85 amu

Europium-153 Mass of 63 protons = (63)(1.007 28 amu) = 63.458 64 amu
Mass of 90 neutrons = (90)(1.008 66 amu) = 90.779 40 amu
Mass of 63 protons and 90 neutrons = 154.238 04 amu

Mass of europium-153 atom = 152.921 23 amu
Mass of electrons = (63) (5.486 x 10^{-4}) = 0.034 56 amu
Mass of europium-153 nucleus = 152.886 67 amu

Mass defect: 154.238 04 amu – 152.886 67 amu= 1.351 37 amu

Europium-167 Mass of 63 protons = (63)(1.007 28 amu) = 63.458 64 amu
Mass of 104 neutrons = (104)(1.008 66 amu)= 104.900 64 amu
Mass of 63 protons and 104 neutrons = 168.359 28 amu

Mass of europium-167 atom = 166.953 21 amu
Mass of electrons = (63) (5.486 x 10^{-4}) = 0.034 56 amu
Mass of europium-167 nucleus = 166.918 65 amu

Mass defect: 168.359 28 amu – 166.918 65 amu = 1.440 64 amu

Convert this mass to kilograms:

Europium-130: $1.109\ 85 \text{ amu} \times \frac{1.6605 \times 10^{-27} \text{ kg}}{\text{amu}} = 1.842\ 91 \times 10^{-27} \text{kg}$

Europium-153: $1.351\ 37 \text{ amu} \times \frac{1.6605 \times 10^{27} \text{ kg}}{\text{amu}} = 2.243\,95 \times 10^{27} \text{kg}$

Europium-167: $1.440\ 64 \text{ amu} \times \frac{1.6605 \times 10^{-27} \text{ kg}}{\text{amu}} = 2.392\ 18 \times 10^{-27} \text{kg}$

Use the Einstein equation to convert the mass defect to binding energy:

Europium-130: $\Delta E = 1.842\ 91 \times 10^{-27} \text{kg} \times (3.00 \times 10^{8} \text{m/s})^2 = 1.6586 \times 10^{-10}$ J

Europium-153: $\Delta E = 2.243\ 95 \times 10^{-27} \text{kg} \times (3.00 \times 10^{8} \text{m/s})^2 = 2.0196 \times 10^{-10}$ J

Europium-167: $\Delta E = 2.392\ 18 \times 10^{-27} \text{kg} \times (3.00 \times 10^{8} \text{m/s})^2 = 2.152\ 96 \times 10^{-10}$ J

Convert to MeV/nucleon.

E-130: $1.6586 \times 10^{-10} \dfrac{\text{J}}{\text{nucleus}} \times \dfrac{1\ \text{MeV}}{1.60 \times 10^{-13}\ \text{J}} \times \dfrac{1\ \text{nucleus}}{130\ \text{nucleons}} = 7.97 \dfrac{\text{MeV}}{\text{nucleon}}$

E-153: $2.0195 \times 10^{-10} \dfrac{\text{J}}{\text{nucleus}} \times \dfrac{1\ \text{MeV}}{1.60 \times 10^{-13}\ \text{J}} \times \dfrac{1\ \text{nucleus}}{153\ \text{nucleons}} = 8.25 \dfrac{\text{MeV}}{\text{nucleon}}$

E-167: $2.152\,96 \times 10^{-10} \dfrac{\text{J}}{\text{nucleus}} \times \dfrac{1\ \text{MeV}}{1.60 \times 10^{-13}\ \text{J}} \times \dfrac{1\ \text{nucleus}}{167\ \text{nucleons}} = 8.06 \dfrac{\text{MeV}}{\text{nucleon}}$

According to this data, from most to least stable these are

E-153 > E-167 > E-130

The observed half-lives of these isotopes are

E-130: 1.1 ms
E-153: stable
E-167: 200 ms

This bears out the calculated results.

2. Rearrange the equation $E = mc^2$ and solve.

$$\Delta m = \frac{\Delta E}{c^2} = \frac{-498{,}400\ \text{J/mol}}{(3.00 \times 10^8\ \text{m/s})^2} = -5.54 \times 10^{-12}\ \text{kg/mol} = -5.54 \times 10^{-9}\ \text{g/mol}$$

3. mass of products = 91.9262 amu + 140.9144 amu +3(1.008 66 amu) = 235.867 amu

mass of reactants = 235.0439 amu + 1.008 66 amu = 236.053 amu

Δm = 236.053 amu - 235.867 amu = 0.186 amu

Convert mass to kg: $0.186\ \text{amu} \times \dfrac{1.6605 \times 10^{-27}\ \text{kg}}{\text{amu}} = 3.08 \times 10^{-28}\ \text{kg}$

Convert mass to energy: $E = 3.08 \times 10^{-28}\ \text{kg} \times (3.00 \times 10^8\ \text{m/s})^2 = 2.77 \times 10^{-11}\ \text{J/reaction}$

Convert to J/mol: $2.77 \times 10^{-11}\ \text{J/reaction} \times \dfrac{6.022 \times 10^{23}\ \text{reactions}}{\text{mol}} = 1.67 \text{x} 10^{13}\ \text{J/mol}$

4. mass of products = 4.002 60 amu + 1.008 66 amu = 5.011 26 amu
mass of reactants = 2 (3.016 049 amu) = 6.032 10 amu

Δm = 6.032 10 amu - 5.011 26 amu = 1.020 08 amu

Convert mass to kg: $1.020\,08\ \text{amu} \times \dfrac{1.6605 \times 10^{-27}\ \text{kg}}{\text{amu}} = 1.6949 \times 10^{-27}\ \text{kg}$

Convert mass to energy: $E = 1.6949 \times 10^{-27}\ \text{kg} \times (3.00 \times 10^8\ \text{m/s})^2 = 1.525 \times 10^{-10}\ \text{J/reaction}$

Convert to J/mol: $1.525 \times 10^{-10}\ \text{J/reaction} \times \dfrac{6.022 \times 10^{23}\ \text{reactions}}{\text{mol}} = 9.184 \times 10^{13}\ \text{J/mol}$

5. a. ${}^{9}_{4}\text{Be} + {}^{1}_{1}\text{H} \rightarrow {}^{6}_{3}\text{Li} + {}^{4}_{2}\text{He}$

   b. ${}^{239}_{94}\text{Pu} + {}^{4}_{2}\text{He} \rightarrow {}^{242}_{96}\text{X} + {}^{1}_{0}\text{n}$. With an atomic number of 96, this is Curium, Cm-96.

   c. ${}^{208}_{82}\text{Pb} + {}^{58}_{26}\text{X} \rightarrow {}^{265}_{108}\text{Hs} + {}^{1}_{0}\text{n}$. With an atomic number of 26, this is Fe-58.

6. The decay constant, *k* can be calculated from the half life.

$$t_{1/2} = \frac{0.693}{k};\ k = \frac{0.693}{t_{1/2}} = \frac{0.693}{6.3\text{x}10^{26}\ \text{s}} = 1.1 \times 10^{-27}\ /\text{s}$$

7. Use the integrated first-order rate law:

$$\ln\frac{N_t}{N_0} = -0.623\left(\frac{t}{t_{1/2}}\right)$$

   Substituting:

$$\ln\frac{3.5}{15.3} = -0.693\left(\frac{t}{5715\ \text{years}}\right)$$

$$\frac{\ln\ 0.229 \times 5715\ \text{years}}{-0.693} = 12{,}165\ \text{years}$$

   The bone fragment is approximately 12,000 years old.

8. The mol ration of lead to uranium needs to be found.

   56 g Pb-207 = 0.271 mol

   27 g U-235 = 0.115 mol

   The total number of moles of uranium initially will be the moles of uranium remaining, plus the moles of lead found in the sample.

   Substitute into the rate law.

$$\ln\frac{0.115}{(0.271 + 0.115)} = -0.693\left(\frac{t}{7.04 \times 10^{8}\ \text{years}}\right)$$

$$\frac{\ln\ 0.298 \times 7.04 \times 10^{8}\ \text{years}}{-0.693} = 1.23 \times 10^{9}\ \text{years}$$

Challenge Problem

The information given in the problem can be summarized by the partial equation.

$${}^{10}_{5}\text{B} + {}^{1}_{0}\text{n} \rightarrow {}^{?}_{3}\text{Li} + ?$$

No protons are being added to the system, but two are being lost by the boron in this reaction, so the radiation produced must contain 2 protons. Only α particles meet this requirement.

$$^{10}_{5}B + ^{1}_{0}n \rightarrow ^{?}_{3}Li + ^{4}_{2}He$$

Balancing the masses leaves Li-7.

Radiation therapies are constantly being refined so that damage to surrounding tissues can be minimized. This therapy allows a nonradioactive material to be injected, removing the need for surgery. The neutron bombardment can be tightly focused to avoid as much as possible areas around the tumor, and the α-particles that are produced will be destructive in the local area only, allowing for destruction of the tumor with minimum collateral damage.

It may not be surprising that this therapy is being investigated primarily for inoperable brain cancers. Unfortunately, as elegant as it is, it does not appear to improve outcomes.

Chapter 23

True–False

1. T
2. F. Double and triple bonds are considered functional groups.
3. F. *Cis-trans* isomerization has tremendous practical consequences.
4. T.
5. F. Carboxylic acids are weak acids.
6. T
7. T.
8. T
9. F. Cellulose is an indigestible long-chain polysaccharide.
10. F. Nucleic acids carry genetic information and provide instructions for protein production.

Multiple Choice

1. b
2. c
3. a
4. c
5. d
6. b
7. d
8. c
9. b
10. d

Fill–in–the–Blank

1. bonding, functional groups
2. hydrocarbons, carbon, alkanes, alkenes, alkynes, triple, double, aromatic or benzene
3. carbohydrates, energy
4. amino acids, enzymes
5. replication, mRNA, translated, proteins

Matching

Alcohol	k
Aldehyde	d
Alkane	s
Alkene	i
Alkyne	w
Amino acid	a

Amide	m
Amine	h
Anabolism	b
Aromatic	r
Carbonyl group	f
Catabolism	c
Cycloalkane	e
Enzyme	t
Ester	p
Ether	g
Fatty acid	j
Hydrocarbon	u
Ketone	l
Lipid	y
Metabolism	n
Monosaccharide	o
Peptide	cc
Replication	bb
Saturated	x
Transcription	v
Translation	z
Unsaturated	aa

General Questions

1.

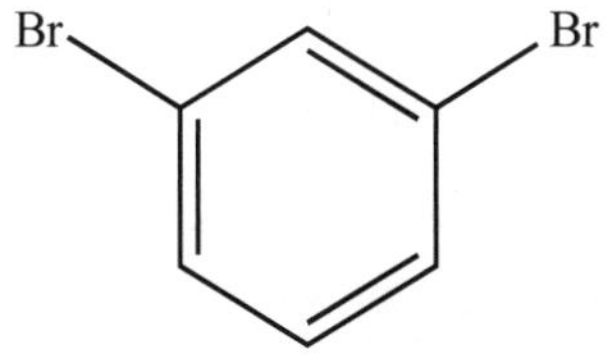

Br Br

meta-dibromobenzene

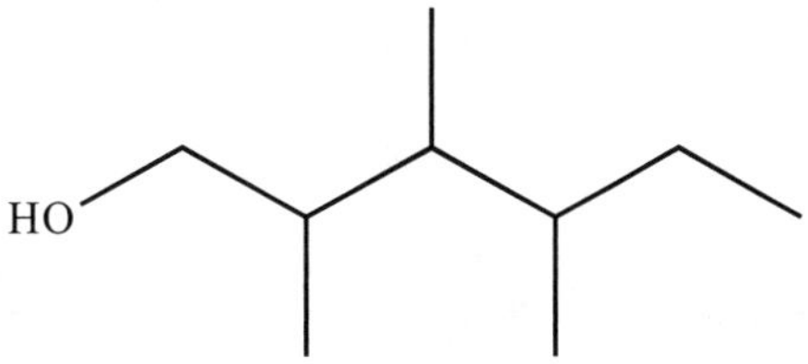

HO

2,3,4-trimethyl-1-hexanol

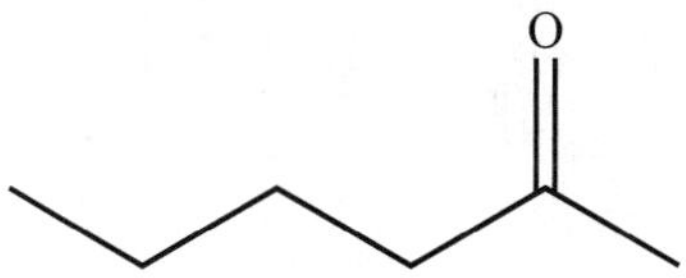

O

2-hexanone

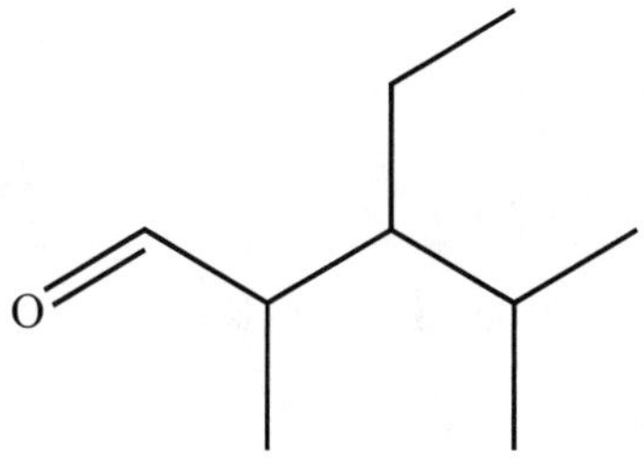

O

3-ethyl-2,4-dimethylpentanal

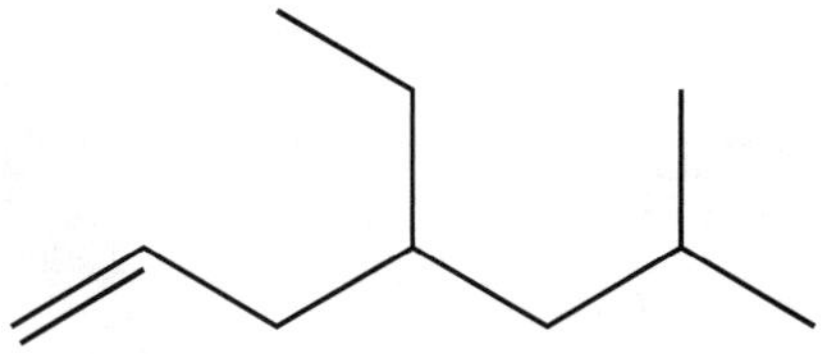

4-ethyl-6-methyl-1-heptene

2. a. 2-methyl hexanamide
   b. methyl butanoate
   c. cyclohexylamine
   d. cyclopentanone
   e. *cis*-3-hexene

3.

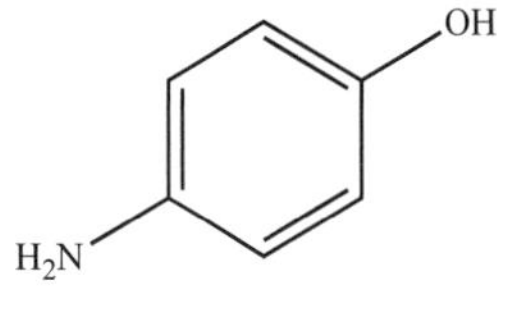

amine, alcohol

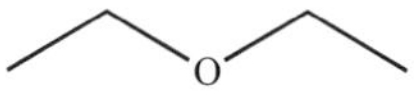

ether

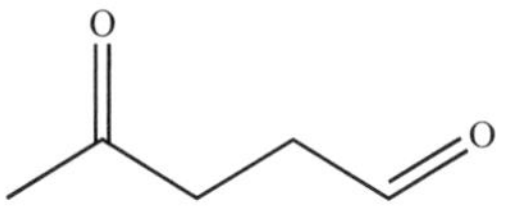

ketone, aldehyde

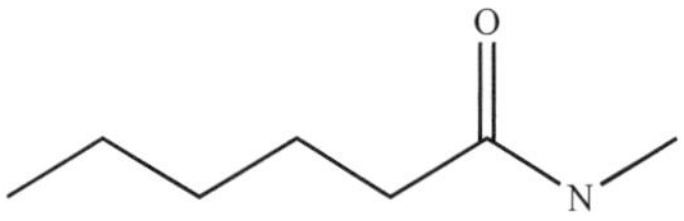

amide

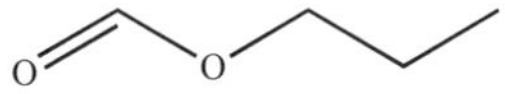

ester

4. a. nitrobenzene

   b. *o*-chloroethylbenzene, *p*-chloroethylbenzene, *m*-chloroethylbenzene

   c. 3-hexanol

   d. butane

   e. dichloroethane

   f. ethyl acetate or ethyl ethanoate

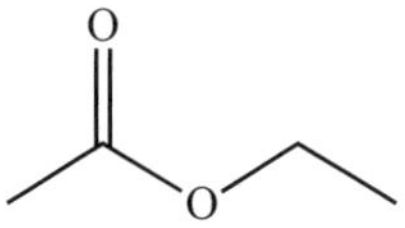

   ethyl acetate

   g. acetic acid or ethanoic acid and methylamine

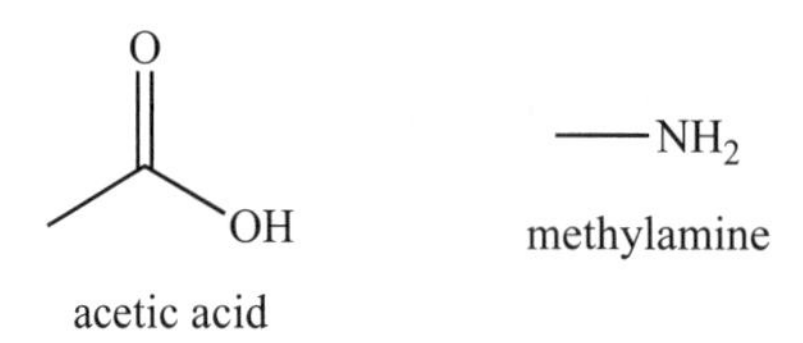

acetic acid methylamine

5. 1. Food is digested in the stomach and small intestine to yield small molecules.

   2. Small molecules are broken down to form acetyl CoA.

   3. Acetyl CoA is oxidized in the citric acid cycle to yield CO_2 and reduced coenzymes.

4. Reduced coenzymes are oxidized by the electron transport chain and energy released as ATP.

6.

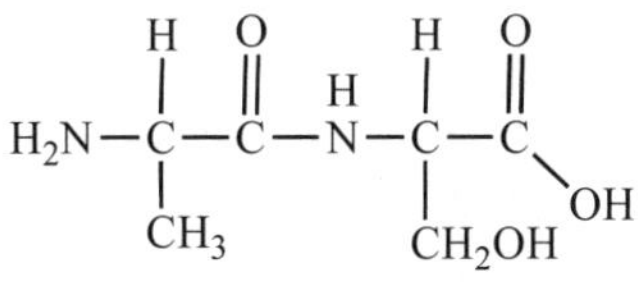

alanine serine

ala-ser

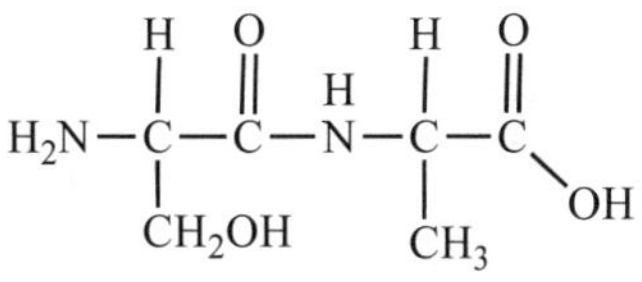

serine alanine

ser-ala

7. Ala-Leu-His, Ala-His-Leu, His-Leu-Ala, His-Ala-Leu, Leu-Ala-His, Leu-His-Ala

8.

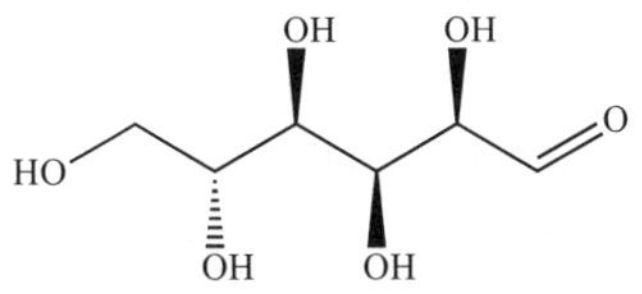

aldohexose

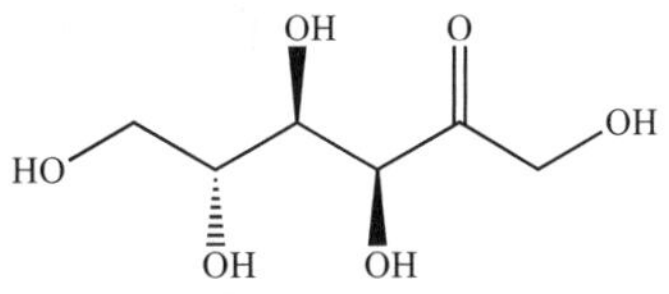

ketohexose

9.

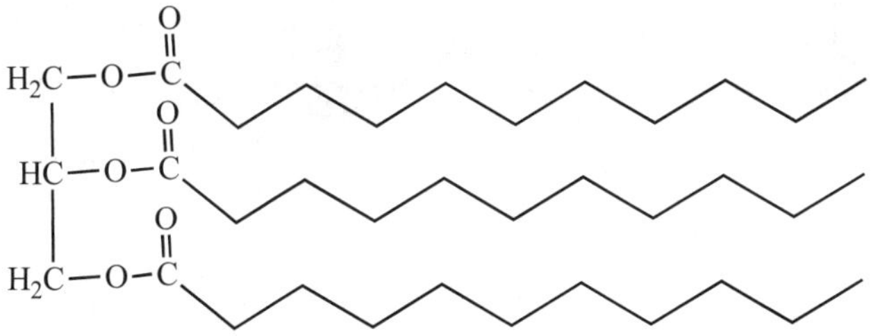

10. DNA sequence = ATCGTGACTAGTCATTCG

Challenge Problem

6.87 grams of CO_2 x 1 mol/12.01g = 0.156 mol of CO_2 = 0.156 mol C x 12 g/mol = 1.87 g C

1.87 grams of H_2O x 1 mol/18.01 g = 0.104 mol H_2O x 2 H/mol H_2O x 1.00 g/mol = 0.208 g H

5.00 grams of original compound – 0.208g – 1.87 g = 2.92 g left for oxygen.

2.92 g oxygen x 1 mol/16.0 g = 0.182 mol O

$$\text{moles C} = \frac{0.156}{0.156} = 1 = \frac{3}{3} = \frac{6}{6}$$

$$\text{moles H} = \frac{0.208}{0.156} = 1.3 = \frac{4}{3} = \frac{8}{6}$$

$$\text{moles O} = \frac{0.182}{0.156} = 1.17 = \frac{7}{6}$$

Empirical formula = $C_6H_8O_7$

The molecular molar mass is determined from the moles of NaOH used in the titration and the fact that citric acid is a triprotic acid. Three moles of NaOH will react for each mol of citric acid.

$$\text{mol citric acid} = 0.0488\ \text{L} \times \frac{0.100\ \text{mol NaOH}}{1\ \text{L}} \times \frac{1\ \text{mol citric acid}}{3\ \text{mol NaOH}} = 1.63 \times 10^{-3}\ \text{mol citric acid}$$

$$\text{molar mass} = \frac{0.312\ \text{g citric acid}}{1.63 \times 10^{-3}\ \text{mol maleic acid}} = 192\ \text{g/mol}$$

The molecular weight corresponds to the empirical formula. Therefore, the empirical and molecular formulas are the same.

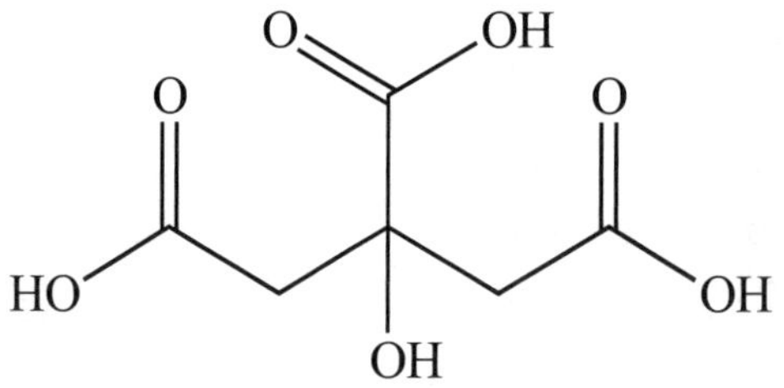

INQUIRY-BASED PROBLEMS SOLUTIONS

Chapter 1

Knowing the definition of density, what two measurements do you need to make?

Mass and volume

Are there any mathematical formulas which may be helpful in this experiment?

Volume of a cylinder = $\pi r^2 \times h$, where r = radius and h = height. Remember that 2 x r = diameter.

Is there more than one way to determine these measurements given the equipment provided?

Volume can be determined with the volume displacement method or by calculations using caliper measurements.

How many measurements should you make?

It is important to take an odd number of measurements for easier calculation of error if needed. One is too few, so three is the least number you should take, with five being even better.

Which method is more accurate?

The micron calipers are going to give you much better volume measurements, even given the fact that multiplying together three numbers will magnify any small errors they may contain. Volume displacement in a graduated cylinder that only has mL markings will require a great deal of estimating, and will be good only to two significant figures.

Once you have written the two procedures, you collected the following data. From this data and the two procedures, determine the identity of the metal cylinder.

Mass	**Volume – H_2O displaced**	**Diameter**	**Length**	***CRC Handbook Data***	
				Element/Alloy	**Density**
47.83 g	5.5 mL	1.0633 cm	6.3945 cm		
47.52 g	5.3 mL	1.0052 cm	6.3855 cm	Stainless steel Type 304	7.9 g/cm^3
47.99 g	5.4 mL	1.0608 cm	6.3891 cm	Yellow brass (high brass)	8.47 g/cm^3
				Aluminum bronze	7.8 g/cm^3
				Beryllium copper 25	8.23 g/cm^3
				Red brass, 85%	8.75 g/cm^3

Average mass: 47.78 g

Average volume displacement: 5.4 mL

Displacement density: 47.78 g ÷ 5.4 mL = 8.85 g/mL

Average diameter: 1.0431 cm

Average radius (diameter ÷ 2): 0.52155 cm

Average length: 6.3897 cm

Average volume (π x r^2 x h): 5.46054 cm^3

Measured density: 8.750 g

Chapter 3

$HNO_3 + NaOH \rightarrow NaNO_3 + H_2O$ - nitric acid

$HCl + NaOH \rightarrow NaCl + H_2O$ - hydrochloric acid

$H_2SO_4 + 2NaOH \rightarrow Na_2SO_4 + 2H_2O$ - sulfuric acid

$H_3PO_4 + 3NaOH \rightarrow Na_3PO_4 + 3H_2O$ - phosphoric acid

$$0.005\text{ L HNO}_3 \text{ x } 5\frac{\text{mol}}{\text{L}} = 0.025\text{ mol HNO}_3\text{x}\frac{1\text{mol NaOH}}{\text{mol HNO}_3} = 0.025\text{ mol NaOH} = 0.025\text{ L} = 25\text{ mL NaOH}$$

$$0.005\text{ L HCl x } 5\frac{\text{mol}}{\text{L}} = 0.025\text{ mol HCl x}\frac{1\text{mol NaOH}}{\text{mol HCl}} = 0.025\text{ mol NaOH} = 0.025\text{ L} = 25\text{ mL NaOH}$$

$$0.005\text{ L H}_2\text{SO}_4 \text{ x } 5\frac{\text{mol}}{\text{L}} = 0.025\text{ mol H}_2\text{SO}_4\text{x}\frac{2\text{mol NaOH}}{\text{mol H}_2\text{SO}_4} = 0.050\text{ mol NaOH} = 0.050\text{ L} = 50\text{ mL NaOH}$$

$$0.005\text{ L H}_3\text{PO}_4 \text{ x } 5\frac{\text{mol}}{\text{L}} = 0.025\text{ mol H}_3\text{PO}_4\text{x}\frac{3\text{mol NaOH}}{\text{mol H}_3\text{PO}_4} = 0.075\text{ mol NaOH} = 0.075\text{ L} = 75\text{ mL NaOH}$$

If it is true that no other acids have ever been ordered, and that this isn't a stray bottle of something borrowed, a titration that took about 50mL of 1M NaOH could be tentatively identified as sulfuric acid, pending further tests. A nice next step would be a selective precipitation, which will be discussed in Chapter Four.

Chapter 4

a. First, what are the steps?

1. reaction of penny with nitric acid to form copper nitrate and nitrogen dioxide
2. reaction of copper nitrate with sodium hydroxide to form copper hydroxide
3. dehydration of copper hydroxide to form water and copper oxide
4. copper oxide reacted with sulfuric acid to form copper sulfate
5. copper ions reacting with aluminum

Equations:

$Cu\ (s) + 4HNO_3\ (aq) \rightarrow Cu(NO_3)_2\ (aq) + 2NO_2\ (g) + 2H_2O\ (l)$

$$Cu(NO_3)_2\ (aq) + 2NaOH\ (aq) \rightarrow Cu(OH)_2\ (s) + 2NaNO_3\ (aq)$$

$$Cu(OH)_2\ (s) \xrightarrow{\text{heat}} CuO\ (s) + H_2O\ (l)$$

$$CuO\ (s) + H_2SO_4\ (aq) \rightarrow CuSO_4\ (aq) + H_2O\ (l)$$

$$3Cu^{2+}\ (aq) + 2Al\ (s) \rightarrow 2Al^{3+}\ (aq) + 3Cu\ (s)$$

b. Determine the moles of nitric acid in 18 mL of 5.0 M nitric acid

$$\text{mol } HNO_3 = 5.0 \text{ M x } 0.036 \text{ L} = 0.18 \text{ mol } HNO_3$$

$$0.18\text{mol } HNO_3 \text{ x } \frac{1 \text{ mol Cu}}{4 \text{ mol } HNO_3} = 0.045\text{mol Cu x } \frac{63.55 \text{ g}}{\text{mol}} = 2.86 \text{ g Cu}$$

$$\% \text{ Cu} = \frac{2.86 \text{ g Cu}}{3.01\text{g penny}} \text{ x } 100 = 95\% \text{ Cu}$$

c. $\% \text{ yield} = \frac{1.94 \text{ g}}{2.86 \text{ g}} \text{ x } 100 = 68\%$

Although this may look shockingly low, for a procedure that has five steps, it's mighty good!

Chapter 8

To calculate the heat of dissociation of the unknown, the 5 grams will need to be dissolved in a known mass of water, and the maximum temperature change measured precisely. From this, and the mass of the solution, the heat of dissolution can be calculated in J/g by using the heat absorbed or released by the water to calculate the heat given off or absorbed by the salt, the two being equal in magnitude,but opposite in sign. It is not possible to calculate a molar heat of dissolution for an unknown, as there is not a molecular weight available.

Any salt with a negative enthalpy of solution will release heat into the water as it dissolves. So calcium chloride and sodium acetate will increase the temperature of the solution. The other two will lower the temperature of the solution.

$CaCl_2$ $$\Delta H_{sol} = -81.3 \frac{kJ}{mol} \text{ x } \frac{1 \text{ mol}}{110.98 \text{ g}} \text{ x } \frac{1000 \text{ J}}{kJ} = -733 \frac{J}{g}$$

NH_4NO_3 $$\Delta H_{sol} = +25.8 \frac{kJ}{mol} \text{ x } \frac{1 \text{ mol}}{80.4 \text{ g}} \text{ x } \frac{1000 \text{ J}}{kJ} = +321 \frac{J}{g}$$

$NaCH_3CO_2$ $$\Delta H_{sol} = -17.4 \frac{kJ}{mol} \text{ x } \frac{1 \text{ mol}}{82.03 \text{ g}} \text{ x } \frac{1000 \text{ J}}{kJ} = -212 \frac{J}{g}$$

$NaCl$ $$\Delta H_{sol} = +3.88 \frac{kJ}{mol} \text{ x } \frac{1 \text{ mol}}{58.5 \text{ g}} \text{ x } \frac{1000 \text{ J}}{kJ} = +66.3 \frac{J}{g}$$

q_{water} = 50 g x 4.184 J/g•°C x -1.6 °C = -335 J

$q_{water} = - q_{salt}$.

$$q_{salt.} = \frac{335\text{ J}}{5.0\text{ g}} = 67\frac{\text{J}}{\text{g}}$$

This most closely matches sodium chloride.

The heat changes involved in the dissolution of the other salts are calculated thus:

$$5.0\text{ g } CaCl_2 \text{ x } -733\ \frac{\text{J}}{\text{g}} = -3655\text{ J}$$

$$5.0\text{ g } NH_4NO_3 \text{ x } 321\ \frac{\text{J}}{\text{g}} = 1605\text{ J}$$

$$5.0\text{ g } NaCH_3COO \text{ x } -212\ \frac{\text{J}}{\text{g}} = -1020\text{ J}$$

When each of these is dissolved in 50 g of water, the following temperature changes are expected:

$CaCl_2$ +3665 J = 50 g x 4.184 J/g•°C x ΔT; ΔT = 17.5 °C

NH_4NO_3 -1605 J = 50 g x 4.184 J/g•°C x ΔT; ΔT = -7.67 °C

$NaCH_3COO$ +1020 J = 50 g x 4.184 J/g•°C x ΔT; ΔT = 4.88 °C

Chapter 9

To figure out how much zinc in grams the students can safely be allowed to use, while still getting good results from the experiments, it is necessary to write and balance an equation for the reaction that is occurring when the zinc and hydrochloric acid are mixed.

$$Zn\ (s) + 2\ HCl\ (aq) \rightarrow ZnCl_2\ (aq) + H_2\ (g)$$

This shows that the zinc and hydrogen are in a 1:1 ratio.

Now to figure out how many moles of gas at 32 °C need to be generated: The lower limit is 200 mL, the upper limit 800 mL.

Lower limit:

$$n = \frac{1\text{ atm x } 0.2\text{ L}}{0.08206\dfrac{\text{L} \bullet \text{atm}}{\text{mol} \bullet \text{K}}\text{ x }(273+32)\text{K}} = 0.00799\text{ mol}$$

Upper limit:

$$n = \frac{1\text{ atm x } 0.80\text{ L}}{0.08206\dfrac{\text{L} \bullet \text{atm}}{\text{mol} \bullet \text{K}}\text{ x }(273+32)\text{K}} = 0.0320\text{ mol}$$

Now convert to grams of zinc.

Lower limit:

$$0.00799 \text{ mol } H_2 \text{x} \frac{1 \text{ mol Zn}}{\text{mol } H_2} \text{x} \frac{65.39 \text{ g Zn}}{\text{mol Zn}} = 0.522 \text{ g Zn}$$

Upper limit:

$$0.0320 \text{ mol } H_2 \text{x} \frac{1 \text{ mol Zn}}{\text{mol } H_2} \text{x} \frac{65.39 \text{ g Zn}}{\text{mol Zn}} = 2.09 \text{ g Zn}$$

The lab should be written to state that the students will need quantities of zinc between 0.6 grams and 2 grams.

Chapter 11

Given that the experimental design involves the use of a gentle flame, it seems clear that the boiling-point elevation of the unknown will be calculated. The equation for boiling-point elevation is

$$\Delta T_b = K_b \cdot m \text{ where } m = \frac{\text{mol solute}}{\text{kg solvent}}$$

Given that $\text{molar mass} = \frac{\text{g solute}}{\text{mol solute}}$ or $\text{mol solute} = \frac{\text{g solute}}{\text{molar mass}}$; we can re-write the equation for freezing point depression as

$$\Delta T_f = K_f \cdot \left(\frac{\text{g solute} / \text{molar mass}}{\text{kg solvent}} \right)$$

$$\frac{\Delta T_f \cdot \text{kg solvent}}{K_f} = \frac{\text{g solute}}{\text{molar mass}}$$

$$\text{molar mass} = \frac{\text{g solute} \cdot K_f}{\Delta T_f \cdot \text{kg solvent}}$$

From this re-arranged equation, we now know that we need to have the mass of the unknown solid and water as well as ΔT_b (Remember ΔT_b = solution boiling point – water boiling point).

To carry out this experiment, we need to measure the boiling point of water and the boiling point of the solution: during a phase change, the temperature remains constant: taking measurements of temperature at timed intervals will make it possible to locate the plateau that indicates a phase change.

A brief outline of the experimental procedure:

1. Weigh the clean, dry empty test tube in a beaker.
2. Add water and reweigh the test tube and the beaker. (The amount of water will depend upon the size of the test tube. You want to add enough so that you can easily submerge the thermometer without touching the bottom of the test tube.)
3. Clamp the test tube to a ring stand and place over a gentle flame.
4. Place the stopper with the thermometer and stirrer on the test tube.
5. With constant stirring, record the temperature at 60 *s* intervals.
7. Plot temperature vs. time to determine the boiling point of water.
8. Repeat the experiment for the solution keeping in mind that the amount of unknown should be much less than the water – perhaps 10% by mass, but no more.

Chapter 12

The overall rate law for this reaction is

Rate = $k[IO_3^-]^m[SO_3^{2-}]^n$

To determine the value of the exponents, it is necessary to hold one of the reactants at a constant concentration, and vary the concentration of the other to see how this affects the initial rate of the reaction.

The rate can be monitored by watching for the color of the iodine-stach complex: based on the reaction mechanism provided, this will not appear until the SO_3^{2-} is consumed, and the reaction complete. Therefore, the time until the color appears can be used to estimate the initial rate of the reaction at a variety of concentrations. The unit of the rate will be s^{-1}.

A table can be created to allow the data at various concentrations to be compared (see the example below):

Test Reaction	0.1 M HIO_3	Starch	0.05 M H_2SO_3	Time
1	40 mL	5 mL	25 mL	
2	20 mL	5 mL	25 mL	
3	10 mL	5 mL	25 mL	
4	5 mL	5 mL	25 mL	
5	2.5 mL	5 mL	25 mL	

After the rates are recorded, the value of the exponents in the rate law can be determined and the value of k for the overall rate law solved for.

There are still two variables in the Arrhenius equation – A and E_a – so the two-point form needs to be used. To do this, the experiment must be repeated at a higher temperature and the following equation used:

$$\ln\left(\frac{k_2}{k_1}\right) = \left(\frac{-E_a}{R}\right)\left(\frac{1}{T_2} - \frac{1}{T_1}\right)$$